Peptide-membrane interactions

Online virtual event

8–10 September 2021

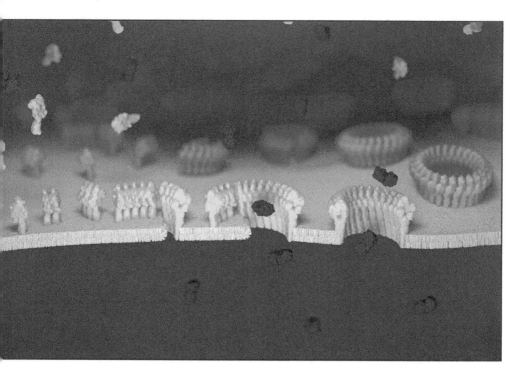

FARADAY DISCUSSIONS
Volume 232, 2021

ROYAL SOCIETY
OF **CHEMISTRY**

The **Faraday Division** of the Royal Society of Chemistry, previously the Faraday Society, was founded in 1903 to promote the study of sciences lying between chemistry, physics and biology.

Faraday Discussions

Faraday Discussions are unique international discussion meetings that focus on rapidly developing areas of chemistry and its interfaces with other scientific disciplines.

Faraday Discussions (Print ISSN 1359-6640, Electronic ISSN 1364-5498) is published 8 times a year by the Royal Society of Chemistry, Thomas Graham House, Science Park, Milton Road, Cambridge, UK CB4 0WF.

Volume 232 ISBN 978-1-78801-914-9

2021 annual subscription price: print+electronic £1220 US $2148; electronic only £1124, US $1978.
Customers in Canada will be subject to a surcharge to cover GST. Customers in the EU subscribing to the electronic version only will be charged VAT.

All orders, with cheques made payable to the Royal Society of Chemistry, should be sent to the Royal Society of Chemistry Order Department, Royal Society of Chemistry, Thomas Graham House, Science Park, Milton Road, Cambridge, CB4 0WF, UK
Tel +44 (0)1223 432398; E-mail orders@rsc.org

If you take an institutional subscription to any Royal Society of Chemistry journal you are entitled to free, site-wide web access to that journal. You can arrange access via Internet Protocol (IP) address at www.rsc.org/ip

Customers should make payments by cheque in sterling payable on a UK clearing bank or in US dollars payable on a US clearing bank.

Information for Authors

ROYAL SOCIETY OF CHEMISTRY

MIX
Paper from responsible sources
FSC
www.fsc.org FSC® C013604

Peptide-membrane interactions

Faraday Discussions

www.rsc.org/faraday_d

A General Discussion on Peptide-membrane interactions was held virtually on the 8th, 9th and 10th of September 2021.

The Royal Society of Chemistry is the world's leading chemistry community. Through our high impact journals and publications we connect the world with the chemical sciences and invest the profits back into the chemistry community.

CONTENTS

ISSN 1359-6640; ISBN 978-1-78801-914-9

Faraday
Discussions
Volume: 232

Peptide-membrane
interactions

ROYAL SOCIETY
OF CHEMISTRY

Cover
See Voskoboinik, Hoogenboom *et al.*, *Faraday Discuss.*, 2021, **232**, 236–255.

"In the synapse between a killer lymphocyte and its target, perforin monomers (light blue) bind to the target cell membrane, where they assemble into oligomeric transmembrane pores that facilitate the entry of toxic granzymes (red) into the cell."

Image reproduced by permission of Adrian W. Hodel, Jesse A. Rudd-Schmidt, Joseph A. Trapani, Ilia Voskoboinik and Bart Hoogenboom from *Faraday Discuss.*, 2021, **232**, 236.

INTRODUCTORY LECTURE

PAPERS AND DISCUSSIONS

CONCLUDING REMARKS

ADDITIONAL INFORMATION

Faraday Discussions

PAPER

Spiers Memorial Lecture: Analysis and *de novo* design of membrane-interactive peptides

Huong T. Kratochvil, [iD] *[a] Robert W. Newberry, [iD] †*[a] Bruk Mensa, [iD] [a] Marco Mravic [iD] [b] and William F. DeGrado [iD] *[a]

Received 1st September 2021, Accepted 3rd September 2021

DOI: 10.1039/d1fd00061f

Membrane–peptide interactions play critical roles in many cellular and organismic functions, including protection from infection, remodeling of membranes, signaling, and ion transport. Peptides interact with membranes in a variety of ways: some associate with membrane surfaces in either intrinsically disordered conformations or well-defined secondary structures. Peptides with sufficient hydrophobicity can also insert vertically as transmembrane monomers, and many associate further into membrane-spanning helical bundles. Indeed, some peptides progress through each of these stages in the process of forming oligomeric bundles. In each case, the structure of the peptide and the membrane represent a delicate balance between peptide–membrane and peptide–peptide interactions. We will review this literature from the perspective of several biologically important systems, including antimicrobial peptides and their mimics, α-synuclein, receptor tyrosine kinases, and ion channels. We also discuss the use of *de novo* design to construct models to test our understanding of the underlying principles and to provide useful leads for pharmaceutical intervention of diseases.

Introduction

Peptide–membrane interactions play important roles in all organisms and viruses, performing a diverse array of functions. To name just a few, antimicrobial peptides (AMPs) and certain classes of antibiotics disrupt the membranes of their targets, apolipoproteins stabilize lipoprotein particles and bar-code them for internalization and processing, fusion peptides promote vesicle budding and fusion, and ion channel-forming peptides are found in both antibiotics and viruses. Indeed, membrane-interactive peptides comprise such a fundamental facet of the functional fabric of living systems that it is difficult to overstate the

[a]Department of Pharmaceutical Chemistry, University of California – San Francisco, San Francisco, CA 94158, USA. E-mail: william.degrado@ucsf.edu; huong.kratochvil@ucsf.edu; robert.newberry@ucsf.edu

[b]Department of Integrative Structural and Computational Biology, Scripps Research Institute, La Jolla, CA 92037, USA

† Current address: Department of Chemistry, The University of Texas at Austin, Austin, TX 78712, USA.

importance of improving our understanding of how their structures and dynamics lead to their remarkable properties. Here, we categorize distinct conformational themes and functional mechanisms of peptide–membrane interactions, drawing from peptides with important biological functions. We also discuss the use of *de novo* design to construct peptides that test hypotheses concerning the structures and functions of peptides. We will also discuss one example of the translation of an AMP mimetic to the clinic. The literature of peptide–membrane interactions is so immense that any review will necessarily be incomplete. To provide a modicum of focus, we will primarily limit our scope to ribosomally expressed peptides and proteins with native peptide backbones, rather than those synthesized *via* synthetases or extensive post-translational modification. We apologize in advance for drawing disproportionately from examples of work from our lab; they have been taken to illustrate principles rather than to claim any priority over many other contributions to this vast literature.

We first must define what a "peptide" is in this context. We will adopt the definition that membrane peptides are relatively short sequences (mostly <50 residues) that form autonomous units which are capable of interacting with membranes. In the case of larger proteins, we will consider a membrane-interactive domain as a "peptide" if it preserves the membrane-interacting function of the larger protein when studied as an isolated peptide. For example, fusion peptides of viral fusion proteins would be considered as a class of membrane-interacting peptides. Also, the protein α-synuclein, which has a large number of tandemly repeated membrane-binding units, will be included in our discussion.

Intrinsically disordered membrane peptides

Membrane-interacting peptides can be subdivided based on their conformations and localization in the membrane (Fig. 1). One prominent class of membrane peptides are "natively disordered" in solution and on membranes. They interact with the membrane surfaces through highly dynamic electrostatic and hydrophobic interactions. Frequently, the peptides are rich in cationic residues, Lys and Arg, which are attracted to anionic groups in the headgroups of phospholipids, as well as aromatic residues that dip deeper to the acyl glycerol region of

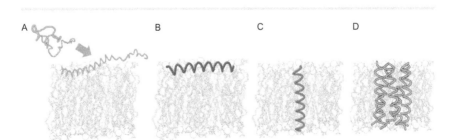

Fig. 1 Illustrations of the different ways peptides can associate with membranes. Peptides can interact with membranes in their (A) intrinsically disordered conformations or through the formation of secondary structures. (B) Amphiphilic helices, which have hydrophobic and hydrophilic faces, bind to membrane surfaces such that the helix lies parallel to the interface. Peptide–membrane interactions can influence the insertion of transmembrane peptides (C) and their assembly into helical bundles (D).

phospholipids.[1–3] Such conformational states have been identified as discrete kinetic intermediates in the process of the insertion of helical proteins into membranes.[4] In other cases, this dynamic disordered conformation represents the native mode of interaction observed at equilibrium. For example, a highly positively charged 25-residue peptide domain tethers the MARCKS protein to negatively charged lipids,[5] and similar motifs are found in GTPases, such as RAS, which helps localize the protein to phases rich in phosphatidylinositol lipids. It is likely that some non-helical AMPs similarly interact with membranes.[6] Intrinsically disordered proteins can also form networks that interact with lipid bilayers through liquid–liquid phase separation to remodel membranes.[7,8]

The amphiphilic helix

The amphiphilic helix is a ubiquitous motif that reprises frequently in the structures of membrane-interactive peptides. In this motif, hydrophobic and polar residues segregate onto opposing faces of the helix, imparting an amphiphilic character ideally suited for binding to apolar/water interfaces, such as those found at membrane surfaces. Such helices are also often termed "amphipathic helices". However, following Tanford,[9] we prefer the more euphonious and friendly term amphiphilic, etymologically stemming from "loving both", as opposed to "suffering both" when traced to the Greek roots.

As the apolipoproteins were first sequenced, their potential to form amphiphilic helices became clear from physical models or when their sequences were drawn on helical wheels (a popular way to appreciate the orientation of sidechains when viewed axially down a helix).[10,11] The distribution of residues was striking, with apolar residues lining one face of the helix, flanked by Asp, Glu, Lys and Arg, arranged so that the positive charges flank the hydrophobic sector, and the acidic residues are furthest from the apolar region. This configuration was expected to foster a close orientation of the zwitterionic and acidic head groups in ion pairs, and the interaction of the apolar residues with the acyl-glycerols and acyl chains of phospholipids. Segrest and coworkers demonstrated the validity of this hypothesis through the design of peptides representing single repeats, as well as consensus sequences.[12–16] Depending on the apolipoprotein, these peptides bound to membrane surfaces and also were able to stabilize disk-like assemblies (now designated as nanodiscs) by binding to the exposed apolar regions of the phospholipids.[17,18]

Kaiser and Kézdy took this strategy one step further by designing peptides that idealize the structural and physicochemical properties proposed to be responsible for their activity, rather than focusing on any natural sequence.[19–23] The resulting peptides were found to bind to and stabilize small unilamellar vesicles of a size and shape similar to that of high-density lipoproteins, and the concept was soon expanded to other peptides and proteins.[20,24,25]

Cytotoxic peptides and antimicrobial peptides (AMPs)

As a graduate student working in the labs of Kaiser and Kézdy, one of us (WFD) expanded this concept to test the structural basis for the cytotoxic function of the bee venom peptide, melittin (although spelled with a single l, melittin is named for the honey bee, *Apis mellifera*, often causing confusion about its spelling in the literature). At the time, it was unknown whether melittin elicited its toxic effects by a purely physical or receptor-based biochemical mechanism. Examining its structure on a helical wheel revealed an amphiphilic structure, but one that was distinct from apolipoproteins; in melittin 2/3 of the arc was composed of hydrophobic residues and the remaining 1/3 was primarily small neutral polar residues. A highly basic hexapeptide resides in the C-terminal of the helical region, and melittin is devoid of any acid residues. A peptide designed to idealize melittin's amphiphilic structure was several-fold more potent than melittin in its ability to lyse red blood cells and disrupt vesicles.[26,27] Also, as was the case for melittin, the peptide was in a random coil configuration in dilute solution, but became helical when allowed to self-associate or bind to apolar–water interfaces. Subsequently, sequences of a family of 14-residue peptides from wasp venom, the mastoparans,[28] were discovered to have an even simpler helical motif;[28,29] again about 2/3 of the residues were hydrophobic, but the positively-charged residues located to the polar side of the helix, rather in a distinct segment C-terminal of the helix, as in melittin. Thus, a minimal structural feature for the disruption of mammalian cells appeared to be a short, positively charged amphiphilic helix.

At about this time, Boman discovered the peptide cecropin[30] in the course of studying innate immunity in insects.[31] Cecropin was the first ribosomally synthesized AMP that was not toxic to eukaryotic cells. Cecropin had two basic amphiphilic helices; both were less hydrophobic than those found in melittin or the mastoparans. The labs of Merrifield[32,33] and DeGrado[34] noticed the amphiphilic character and showed the first helix alone to have antimicrobial activity. DeGrado also designed the first minimal synthetic model for AMPs with the sequence (LKKLLKL)$_2$.[29,34] Soon thereafter, Michael Zasloff made his seminal discovery of the first AMP, magainin.[35] Propelled by the importance of innate immunity to human health and Michael's charismatic leadership, the field rapidly grew. Soon, AMPs were found in many organisms, including humans. Many were found to form amphiphilic helices, but others were small disulfide-rich mini-proteins (e.g., defensins), Pro/Arg-rich peptides, Trp-rich peptides, anionic AMPs, and β-hairpins.[36–44] Because many AMPs work by a biophysical mechanism, rather than by targeting a specific protein or ribosomal target, they have less tendency to develop resistance. The field of both designed and natural antimicrobial peptides has exploded, and there are now books and entire conferences devoted to the subject.[45,46] AMPs have diverse targets, but most target membranes as at least a part of their mechanisms of action.[39,47–49]

Much biophysical work has been devoted to understanding the mechanism by which AMPs disrupt bacterial membranes, and what differentiates AMPs from more toxic peptides, such as melittin.[36–40] A reading of the literature highlights the diversity of the mechanisms, ranging from channel formation to more generalized membrane disruption. While many papers in this volume discuss very detailed measurements of AMP–membrane interactions, here we present a less detailed understanding of the principles of AMP–membrane interactions, which guided the design of both peptide and non-peptidic mimics of AMPs, described in the subsequent section.

AMPs disrupt membranes at relatively high peptide : lipid ratios, when the surface concentration exceeds a certain threshold[50] at which the peptides essentially carpet the membrane surface, leading to both generalized membrane disruption[51] and heterogeneous pore formation.[6,50,52–63] The asymmetric accumulation of such a large amount of a peptide on only one leaflet of the bilayer results in a large imbalance in the chemical potential. For example, at a peptide/lipid ratio of 1 : 100 (corresponding to 1 : 50 if the AMPs localize to only the outer leaflet), there is a strong thermodynamic driving force to equalize the concentration on both sides of the bilayer. The gradient is roughly equivalent to the osmotic gradient created by introducing a solute at 1 M on just one side of a water-permeable membrane (in which case the mole fraction of the solute to water would be approximately 1 : 50 for a small molecule or ion). In addition to this strong thermodynamic driving force, the physical properties of the peptide contribute to lowering the kinetic barrier for membrane translocation. Insertion leads to the expansion, disordering and thinning of the bilayer.[61,64] Moreover, asymmetric insertion leads to surface distortions, such as buckling. In each case, the bilayer is stressed, helping to reduce the energy of activation for AMP translocation and equilibration across the bilayer. Simulations suggest that the translocation step can also occur with a transient increase in ion permeability.[65,66]

AMPs also decrease the stability of bilayers and enhance the stability of non-lamellar phases, including lipidic cubic phases and cylindrical micelles, with head groups projecting into a water channel and the apolar groups radiating outward; a similar structure might be formed as a defect within a bilayer, creating a pore that is lined by both headgroups and surface-absorbed peptides.[52,53,55–60,67–70] The resulting toroidal pore has a shape similar to the inside of a donut with a topology designated as negative Gaussian curvature (a shape in which the curvature is negative in one direction and positive in the orthogonal dimension). Gerard Wong has shown that antimicrobial peptides of many classes stabilize negative Gaussian curvature.[71] His group has shown that Lys, Arg, and aromatic amino acids bias towards phases with negative Gaussian curvature and used this principle, together with machine learning, to design highly effective *de novo* AMPs.[72]

Fully peptide-lined pores can also be formed by peptides, such as alamethicin, which are sufficiently hydrophobic and long enough to span the bilayer (at least about 18–20 residues in α-helical peptides). For example, melittin binds initially as a random coil to acidic membranes, then it forms a laterally inserted helix, and ultimately it forms a membrane-spanning helix at sufficiently high peptide : lipid ratios.[4,73,74] The vertically inserted helix can next associate in a variety of stoichiometries to form heterogeneous transient pores, which then disrupt ionic gradients required for cell viability.

In summary, AMPs have a common feature of being able to bind to the surface of bilayers using a combination of electrostatic and hydrophobic interactions. They insert aromatic or hydrophobic groups sufficiently deep into the bilayer to lead to a surface pressure imbalance and modulation of the properties of the bilayer. The destabilization of the bilayer also decreases the activation energy for subsequent transitions of peptides into and/or across the bilayer. The accumulation of peptides on one side of the bilayer also leads to an imbalance in the chemical potential, providing a thermodynamic driving force for AMP translocation, which can occur with the concomitant diffusion of ions and disruption of the membrane potential. Full-fledged pore formation can occur through the formation of channels that are either lined solely by peptides or a combination of peptides and lipid head groups. Different AMPs can access different mechanisms.

The selectivity of AMPs for bacterial membranes contrasts with toxins like melittin, which disrupt membranes of both mammalian and bacterial cells –

a phenomenon that has been widely attributed to differences in the lipid compositions of different cell types. Bacterial cells are rich in acidic lipids compared to mammalian cells, and bacteria also lack cholesterol. In many cases,[36] subtle effects relating to phospholipid composition are involved; primary phospholipids in mammalian membranes include zwitterionic phosphatidyl-choline (PC) and negatively charged phosphatidylserine (PS) headgroups, while bacterial membranes are rich in zwitterionic phosphatidylethanolamine (PE) and negatively charged phosphatidylglycerol (PG) lipids, including cardiolipin. The differences in the physical properties of these various lipids can be as important as the overall charge of the membrane.[75-78]

A number of features of a peptide define its selectivity. We have already seen that cytotoxic peptides tend to have greater hydrophobicity, leading to more non-discriminate binding than selective AMPs. For helical AMPs, the insertion of the peptides is thermodynamically linked with helix formation, so peptides that strongly favor the helical state are intrinsically predisposed to binding more tightly than closely related peptides with a similar charge and hydrophobicity. Thus, for each scaffold there exists a delicate balance between charge, hydro-phobicity and rigidity that must be fine-tuned to optimize both the potency and selectivity. Finally, the response of a bilayer to bound AMPs likely varies with respect to differences in the lipid composition, so changes to not only the binding affinity, but also biological activity, likely contribute to the selectivity.[76]

Foldamers and polymers that mimic AMPs

Given the above mechanistic understanding of the structural features and mechanisms of action of AMPs, it became apparent that it should be possible to mimic their activities using molecules other than peptides which were composed of α-amino acids. Because the activities of many AMPs depend primarily on their overall physicochemical properties—rather than the fine details of their precise amino acid sequences—several groups designed "coarse–grained" molecules idealizing the amphiphilicity of natural AMPs.[29,37,79,80] Modifications of natural peptides have included the introduction of D-amino acids,[81,82] long acyl chains,[83,84] and cyclization.[81,85] The availability of β-peptides provided another avenue to probe the requirements for activity, as they can adopt distinct secondary struc-tures such as "12-helices" and "14-helices".[86-90] Our group,[91,92] as well as those of Gellman[93] and Seebach[94] independently showed that β-peptides capable of forming amphiphilic 14- or 12-helices had potent antimicrobial activity. Oligo-mers with lengths of 10–15 residues that achieved an appropriate hydrophilic/lipophilic balance were selective for killing bacteria vs. mammalian cells. Oligo-mers that were too short were inactive, and these short sequences also failed to adopt the desired conformation. Moreover, foldamers that were too long or hydrophobic were toxic towards mammalian cells.[95] These studies were extended to a variety of different helical types.[76-78,96,97] In a similar manner, Patch and Barron designed amphiphilic, helical, antibacterial N-substituted glycine oligo-mers (peptoids),[98] and Guichard and coworkers synthesized antimicrobial fol-damers based on a urea backbone.[99] Thus, by the early 2000s it had been well established that medium-sized molecules that adopt amphiphilic secondary structures were capable of acting as highly potent and selective antimicrobial agents.[100-102]

Going one step further, the groups of DeGrado, Tew and Klein collaborated on the design of antimicrobial oligomers that were more akin to small molecules than earlier foldamers.[103-108] Using a rigid arylamide as a framework, they systematically introduced cationic groups (cat. 1 and cat. 2 in Fig. 2) and hydrophobic groups to maximize activity against *Staphylococcus aureus*, while minimizing toxicity towards mammalian cells *in vitro* and in mouse models.[107,109]

Brilacidin has undergone two phase II clinical trials for *S. aureus* infections and was shown to have an efficacy on par with that of daptomycin.[109] There is currently great interest in AMPs as potential antiviral agents against enveloped viruses,[110-113] and brilacidin has similarly been shown to be active against SARS-CoV-2.[114] A phase II clinical trial for COVID is currently in progress (http://www.ipharminc.com/brilacidin-1).

The antibacterial mechanisms of action of brilacidin and closely related compounds have been studied by probing the transcriptional response and leakage of bacterial content in *E. coli*, as a Gram negative organism,[115] and *S. aureus*, as a Gram positive one.[116] These compounds cause significant changes in the permeability of the outer membrane of *E. coli*, similar to those observed for polymyxin B and nisin.[117,118] They cause comparatively little permeabilization of the inner membrane, although they lead to a loss of membrane integrity, reaching critical levels corresponding with the time required to bring about bacterial cell death. Transcriptional profiling of *E. coli* treated with sub-inhibitory concentrations of the arylamides also showed the induction of genes related to membrane and oxidative stresses. The induction of membrane-stress response regulons, coupled with morphological changes at the membrane observed by electron

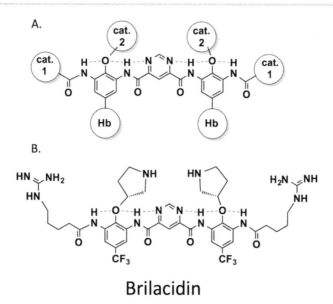

Brilacidin

Fig. 2   Generic structure of a rigid platform used to design antimicrobial peptide mimics. By systematically varying the cationic groups (cat. 1 and 2) and the hydrophobic groups (Hb), it was possible to maximize the potency and efficacy : safety ratio. Brilacidin is undergoing clinical trials for bacterial infections and SARS-CoV-2.

microscopy, indicate that the activity of the arylamides at the membrane is the primary mechanism of action.[115]

In *S. aureus*, brilacidin also causes membrane depolarization, to an extent comparable to that caused by the lipopeptidic drug daptomycin.[116] However, there was little leakage of cellular contents, ruling out mechanisms that involve large pores. Transcriptional profiling showed that the global response to brilacidin treatment is well-correlated with the corresponding response to daptomycin and a derivative of the cationic antimicrobial peptide, LL37, and induced regulons that respond to perturbation of the cell wall and membrane function. These stress responses were mainly orchestrated by three two-component systems: GraSR, VraSR and NsaSR, which have been implicated in virulence and drug resistance against other clinically available antibiotics.

Coarse–grained MD allowed simulations with a relevant number of arylamides per bilayer[65] for coarse–grained oligomer/phospholipid (CGO/PL) ratios from 1 : 256 to 1 : 14, spanning surface concentrations that experimentally give rise to results from no lysis to very rapid vesicle lysis. At low CGO/PL ratios, the anti-microbial inserts into the bilayer with its long axis parallel to the membrane surface and its apolar groups penetrating into the membrane. By contrast, at the high CGO/phospholipid ratios required for the lysis of bilayers in experimental systems, the drugs initially insert with an ensemble of different angles. The very large imbalance in the concentration of the oligomers between the proximal *versus* the distal leaflets of the bilayer leads to a metastable state with pronounced buckling of the proximal leaflet of the bilayer. At longer times, the molecules diffuse to the opposite side of the bilayer. The CGOs frequently diffuse in pairs across the bilayer, often with accompanying ions and water molecules. Following the translocation step, the oligomers are oriented predominantly parallel to the bilayer. However, they are less well ordered than in the simulations at low anti-microbial : phospholipid ratios. In the final configuration, the polarity gradient of the membrane is altered, and there is an increased permeability to water in the simulations. These simulations provide pictorial detail to the prior models described above.

We hope the above section is helpful in defining the principles required for the design of AMP mimetics. There are currently a number of AMPs and AMP mimetics undergoing clinical trials for various indications.[119,120] In addition to understanding the principles for design, which has been the primary focus here, this is also essential to understanding the biophysical bases for their action. The reader is pointed to other papers in this volume for a wealth of information on AMP–membrane interactions.

Dynamics, structure, and mechanism of membrane-bound α-synuclein

One of the most prominent examples of membrane-binding peptides, α-synuclein (Fig. 3A), is also one of the most complex.[121] Although α-synuclein is sufficiently long to be considered as a protein, it has a number of repeating peptide units, and different portions of its sequence interact differently with the membrane, showing the diversity of ways in which peptides can interact with membranes.

Originally discovered by screening cDNA expression libraries for proteins localized to synapses using antibodies raised against synaptic preparations from electric rays, α-synuclein was shown by electron microscopy to localize to the

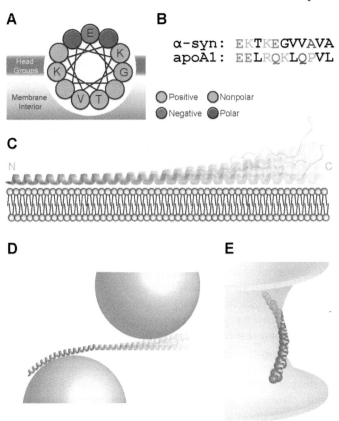

Fig. 3   Sequence and structures of α-synuclein. (A) Helical wheel depicting the amphiphilic nature of α-synuclein's 11-residue consensus sequence, which mediates membrane binding. (B) Alignment of the 11-residue consensus sequences of α-synuclein (α-syn) and apolipoprotein A1 (apoA1).[129] (C) Structural model of α-synuclein showing the C-terminal dynamics in an extended, membrane-bound helix, which is supported by data from FRET,[130] EPR,[131,132] NMR,[133,134] and DMS.[135] (D) Cartoon model of α-synuclein tethering nearby vesicles.[136] (E) Hypothetical model of α-synuclein interacting with membranes of negative Gaussian curvature, such as those in fusion pores.[137]

surface of synaptic vesicles, suggesting a direct peptide–membrane interaction.[122] This hypothesis was further supported by the similarity between the sequences of α-synuclein and apolipoproteins (Fig. 3B),[123] which use amphiphilic helices featuring unique 11-residue repeating segments to sequester and transport lipids.[124] α-synuclein localizes specifically to synaptic vesicles, rather than the plasma membrane, and disperses from nerve terminals after stimulation,[125] suggesting that it might participate in synaptic vesicle trafficking. Indeed, early experiments on knockout animals revealed changes in stimulated dopamine release,[126] suggesting that α-synuclein plays a key role in neurotransmission. These early studies, as well as genetic[127] and pathological[128] links to Parkinson's disease, motivated detailed study of the interactions of α-synuclein with lipid membranes, which have revealed a series of remarkable biophysical features that contribute directly to α-synuclein's physiological and pathological activities.

Plotting the distribution of residues on a helical wheel revealed a facial separation of hydrophilic and hydrophobic residues, with cationic seams in between to potentially support membrane binding (Fig. 3A).[123] Early experiments on purified α-synuclein confirmed this hypothesis, showing by circular dichroism that α-synuclein, which lacks defined structure in isolation,[138] adopts helical secondary structure upon association with lipid membranes.[139] Although these behaviors are shared with other membrane-binding peptides, α-synuclein's length and hydrophobicity make it unique, and these features likely contribute to both its physiological and pathological activities.

Compared to other amphiphilic helices, α-synuclein is only modestly hydrophobic.[140] Its hydrophobic face consists primarily of smaller residues like Gly, Ala and Val, which are interrupted by hydrophilic Thr residues. In contrast, the hydrophobic face of APLS (amphipathic lipid packing sensor) motifs, for example, often feature uninterrupted stretches of larger, greasier residues like Leu, Ile, Met, and Phe.[141] α-Synuclein's modest hydrophobicity tempers its affinity for lipid membranes. Unsurprisingly, for example, substituting the Thr residues with bulkier hydrophobic residues increases the affinity, which in turn increases their toxicity in cellular and animal models,[142] suggesting that the affinity of α-synuclein for lipid membranes is tightly optimized. Moreover, the prevalence of Gly residues in this helix further attenuates the membrane-binding affinity by increasing the entropic cost of helix formation, though the alignment of these Gly residues along one face of the helix suggests additional or more nuanced roles, perhaps mediating interactions with specific proteins or lipids. This tuning of α-synuclein's membrane affinity not only increases the rate of exchange between the membrane surface and solution, which could be involved in its reported roles in synaptic vesicle trafficking,[143] but it also enables α-synuclein to discriminate between membranes with different physical and chemical properties.

For example, α-synuclein has a higher affinity for membranes of increasingly negative charge,[139] likely owing to the relative importance of electrostatic interactions, rather than strictly hydrophobic associations, to drive membrane binding.[144] α-Synuclein's modest membrane affinity has also been suggested to contribute to its ability to discriminate between membranes with different packing or curvature.[139,145] The modest affinity likely provides insufficient energy to disrupt the packing of planar bilayers but could be sufficient to bind to the gaps between lipid molecules in highly curved membranes, such as those of synaptic vesicles. Moreover, theoretical models[146] and course–grained MD simulations[147] demonstrate that the partitioning of α-synuclein to a specific depth within the membrane, as tuned by its physicochemical properties, is sufficient to drive curvature formation.

α-Synuclein is also distinctive because of its length.[148] Preliminary sequence analysis suggested that the first 90–100 residues could form a continuous amphiphilic helix, far longer than most examples, the most notable exception being the perilipins that bind to lipid droplets.[149] Preliminary NMR experiments confirmed the membrane-binding region as the first 90–100 residues,[150] which include the 11-residue segments predicted to mediate lipid-binding; the remainder of the 140-residue protein remains disordered and unbound from the membrane. Subsequent studies disagreed about whether this region forms a long, single helix,[131] or whether the membrane-binding domain is broken into two segments.[151] This discrepancy highlights the effect that the membrane or membrane mimetic can have on the peptide and protein structure. The first high-resolution NMR structure, solved in SDS micelles, suggested such a break.[152] In contrast, studies using EPR,[132] FRET,[130] and solution-[133] and solid-state NMR,[134] conducted with the protein reconstituted in bilayers, instead favored a single, extended helix bound to the membrane. Consensus has emerged that this discrepancy likely arises from the curvature of the membrane mimetic; highly-curved species like SDS micelles induce breaking of the helix, while less curved membranes, such as LUVs, support helix formation across the entire membrane-binding domain.

The diversity of structures formed by α-synuclein has raised important questions about the functional and pathological roles of each individual species. To begin dissecting these roles, we adapted deep mutational scanning (DMS) to reveal the structure associated with a specific activity. In DMS, a pooled library of protein missense variants is screened for relative activity, which is inferred from changes in the frequency of each variant before and after selection.[153] By subjecting comprehensive libraries of missense variants to relatively weak selective pressure, such that most variants are retained in the population after selection, the relative effects of thousands of mutations can be determined and used to infer the structural features of the protein that are responsible for the activity.

We sought to determine the structural basis for a relatively simple activity of α-synuclein: its toxicity when ectopically expressed in yeast cells. By identifying mutations that interfere with formation of the toxic species and rescue the growth rate of yeast cells, we discovered that α-synuclein toxicity in yeast is driven by a dynamic, membrane-bound helix with increasing dynamics toward its C terminus.[135] For example, mutations that introduce polar groups onto the hydrophobic face reduce both membrane binding and toxicity, but this effect becomes less pronounced toward the dynamic C terminus. Similarly, helix-breaking proline residues reduce toxicity, again most potently at the N terminus. This result is in excellent agreement with models proposed to mediate vesicle clustering,[136] implicating this activity in toxicity, at least in yeast.

We then developed a simple structural model for this state (Fig. 3C),[135] which we used to make thermodynamic predictions of the effect of each mutation on membrane binding, based on the known energy associated with positioning each amino acid to varying depths in the lipid bilayer.[154] These predictions were remarkably successful at predicting the effects in our cellular assay, providing strong support for the role of this species in toxicity. Our model also demonstrated that the entire helix contacts the membrane to a similar depth, while the C terminus engages in increasingly frequent dynamic dissociation from the membrane surface. Again, these results from functional measurements were in remarkable agreement with inferences made from complementary spectroscopic methods, specifically NMR[133,134] and EPR.[131,132] The dynamics at the C-terminal end of the helix might result in part from the sequence of α-synuclein's uniquely long amphiphilic helix. Although each of the repeated 11-residue binding segments encodes nearly three turns of an α-helix, the slight mismatch between the sequence and its encoded structure causes the hydrophobic face of the helix to gradually wrap around the helix.[155] The energy associated with over-winding the helix to align the hydrophobic groups could prevent both ends from binding tightly.

The length of this helix creates interesting dynamical properties. In order to align the hydrophobic groups on one face of the helix, α-synuclein would have to overwind, and although the energy required to do so is small for individual residues or helical turns, that energy becomes significant for a helix of α-synuclein's length. This energetic penalty to adopting the structure that would optimize the hydrophobic groups for membrane binding prevents the protein from fully associating with the membrane surface across its entire length.

Dynamics in the membrane-bound helix are hypothesized to have a variety of physiological roles. One putative role is the tethering of membrane surfaces to one another; the release of the C-terminal region from the surface of one membrane allows it to associate with nearby surfaces,[136] which might mediate vesicle clustering (Fig. 3D), one known role of α-synuclein. Intriguingly, aberrant protein-rich inclusions in the brains of Parkinson's patients called Lewy bodies are enriched in both α-synuclein and membrane-bound organelles and vesicles,[156] suggesting that this clustering activity might contribute to the pathology, although this hypothesis is controversial.[157] Alternatively, dynamics in the membrane-bound state could support protein–protein interactions with other membrane-remodeling machinery, such as SNARE proteins.[158]

Dynamics in the membrane-bound state can also promote membrane curvature;[159,160] as the surface concentration of α-synuclein increases, clashes between monomers involving the dynamic end of the helix and the disordered tail would be relieved upon increased curvature. Electrostatic repulsion between the acid tails of each monomer could further drive repulsion and increased curvature. These biophysical mechanisms might underlie α-synuclein's ability to tubulate membranes.[131,161] Alternatively, the inability of the protein to associate with the membrane surface across its entire length might create an energetic driving force for the association with or formation of membranes with negative Gaussian curvature (Fig. 3E), which has been observed in course-grained simulations,[162] such as those that form during vesicle endocytosis or exocytosis. The energy release from the binding of α-synuclein to such a surface could contribute to its observed ability to accelerate fusion pore dilation.[137]

Membrane-binding has a strong influence on the aggregation of α-synuclein,[163] which is implicated in Parkinson's and related diseases. At low protein : lipid ratios, membrane binding suppresses α-synuclein aggregation by distracting individual monomers from self-association.[164] At higher concentrations, however, membrane-binding induces aggregation by concentrating monomers on the membrane surface. Moreover, the dynamic release of the C terminus of the amphiphilic helix from the membrane surface exposes the aggregation-prone NAC (non-amyloid component) region, which is then poised for self-association at high protein : lipid ratios.[165] Dysregulation of either the concentration of α-synuclein or its membrane affinity could therefore drive pathological aggregation.[166] In turn, α-synuclein aggregation can perturb its interaction with lipid membranes by either concealing[167] or revealing membrane-binding regions.[168]

In summary, α-synuclein exemplifies many of the features of peptide–lipid interactions considered in this volume. In one single protein, we see surface helices that specifically recognize and bind to membranes with high curvature, using a combination of peptide–lipid interactions as well as generalized crowding effects. We also see multiple types of interactions exemplified in a single protein:

α-synuclein adopts (1) a highly stable helix that is inserted into the bilayer surface near the N terminus, (2) a dynamic helix, which adopts a dynamic equilibrium between inserted helical conformations and unfolded structures, and (3) a fully unstructured C-terminal tail tethered to the membrane surface.

Insertion, assembly, and design of single-span TM helices

Insertion of transmembrane helices

Membrane proteins are generally inserted into membranes *via* the translocon, and they complete the folding process in the membrane environment.[169,170] The helical conformation is stabilized in the membrane because the backbone amide–carbonyl hydrogen bonds are satisfied within the helix and sequestered away from the nonpolar environment of the bilayer.[171,172] In the two-stage model of membrane protein folding,[173–175] once inserted in the bilayer, the protein is then able to fold *via* the coalescence of the helices to form the native tertiary structure. The study of designed and natural transmembrane peptides has greatly informed our understanding of the forces that stabilize both the insertion and coalescence of helices in membrane proteins. Thus, TM peptides are excellent models for multi-span membrane protein folding. For the biophysicist, self-associating TM helices allow the investigation of unconstrained inter-helical interactions with a clear unfolded state—a monomeric α-helix—where conformational specificity and thermodynamics can be simply evaluated by the oligomeric distribution. From a more biological perspective, half of all human membrane proteins have a single transmembrane helix.[176,177] Moreover, TM helices often do more than anchor proteins in membranes; they self-associate or interact with other membrane proteins to play vital roles from signaling to ion conduction, and often the aberrant assembly of TM helices is central to devastating diseases from cancer to Alzheimer's disease.[178,179] Thus, the study of TM peptides is a fruitful endeavor for biophysicists and biologists alike.

The features influencing the insertion of TM helices into membranes have been extensively researched. Sufficient hydrophobicity is required to enter the translocon and stabilize the helix in an apolar environment,[180] dictating the placement of apolar residues at most positions in the helix.[154,170,181–183] Additionally, aromatic and basic residues are well accommodated in the headgroup region of a bilayer, where they can contribute to stability.[184–187] The residues in the headgroup region also define the transverse shift of a helix in a membrane, as does the presence of polar residues within the hydrophobic region.[188] Charged residues in the hydrophobic region of a TM peptide also have a large effect on the ability of peptides to insert into bilayers. Early work from Caputo and London[189,190] focused on three model TM Leu-rich peptides, each containing Asp at different positions in their hydrophobic core. When the Asp residues were protonated at low pH, the peptides inserted vertically into vesicles composed of dioleoylphosphatidylcholine (DOPC). When the Asp residues were ionized at neutral or high pH, the topography was altered in a manner that would allow the charged Asp residues to reside near the bilayer surface. More recently, Engelman and coworkers have developed a "pHLIP" peptide, which is a 36-aa peptide derived from the bacteriorhodopsin C helix with an acidic residue near the center of the TM helix. This peptide exists in three states: soluble in aqueous solution, bound to the surface of a membrane, and inserted across the membrane as an α-helix; decreasing pH stabilizes the inserted state.[191,192] An "ATRAM" peptide from the Barrera lab has similar physical characteristics.[193,194] Both insert unidirectionally into bilayers, and are capable of bringing cargos into cells. Given the acidic environment of tumors, they show promise for the targeted delivery to and diagnosis of tumors.[195–197]

There is generally a minimum length of about 20 residues (30 Å) to stably span the bilayer in a vertical orientation, and there are significant effects of a hydrophobic mismatch between the membrane and the peptide.[198–201] Shorter TM helices will necessarily experience a hydrophobic mismatch with the bilayer, leading to destabilization of the membrane lamellar phase.[202] On the other hand, TM helices with hydrophobic lengths significantly longer than 20 residues will tend to insert at an angle relative to the membrane normal, as this orientation allows the more efficient burial of the TM helix.[203,204] Finally, while the rotation of a helix about its own axis does not affect the depth of insertion of an amino acid residue, rotation which is tilted about its axis places a sidechain into different positions in the bilayer, which can differ significantly near the headgroup region.[205–207] Thus, the membrane environment has a very significant impact on "setting up" a helix for favorable helix–helix interactions.[208]

Structural aspects of the assembly of transmembrane helices

Peptide–membrane interactions also play an essential role in defining helix–helix interactions in membranes.[209,210] The hydrophobic effect provided by the burial of apolar side chains in a protein's interior represents the predominant driving force for helix–helix association and folding in water, yet it is negligible in lipid membranes. What then drives the assembly and folding of membrane proteins? Lipid-specific effects, such as "solvophobic" exclusion, likely contribute.[211,212] Oligomerization also relies on matching the hydrophobic thickness of the TMD with that of the surrounding lipid. If the lipid bilayer is too thin or too thick, oligomerization can be reduced in model membranes.[213,214] Moreover, there can be very large effects when comparing native cell membranes with model membranes. These effects can even overwhelm stabilizing contributions from favorable helix–helix interactions, lateral packing pressure and excluded volume.[215] Thus, while the following section focuses on stabilizing helix–helix interactions, it is important to keep in mind that the association of helices represents a balance between membrane–protein and protein–protein interactions, which is sensitive to the membrane environment.

Structural informatics, peptide design, and the site-directed mutagenesis of natural proteins have greatly advanced our understanding of the mechanisms of association of TM helices. We identify three general categories of interactions that stabilize protein structures. In some packings, such as GX_3G motifs, the backbones of two helices approach closely, forming favorable weakly polar interactions and efficient van der Waals packing. In other cases, electrostatic interactions and hydrogen bonds involving polar or charged amino acid sidechains help drive association. Finally, tight packing of apolar sidechains can be a strong driving force for association. We will consider each separately.

GX$_3$G (and Small-X$_3$-Small) and Small-X$_6$-Small motifs. In the GX$_3$G motif, first identified in glycophorin, Gly residues spaced four residues apart mediate a very close association of the backbone of two helices stabilized in part by C–H hydrogen bonds between the alpha CH groups and carbonyl groups of adjacent helices.[216-218] This non-classical hydrogen bonded interaction is favorable, and it also allows very efficient van der Waals packing interactions between proximal groups in the interacting helices.[219-221] In the classical GX$_3$G motif, the two interacting helices are parallel, and they have a right-handed crossing angle of about +40°. Variations on the GX$_3$G motif are seen in which one or both of the Gly residues are instead small side chains, such as Ala. It is noteworthy that these residues lose no side-chain conformational entropy when the helices associate, as is the case with larger, more flexible side chains. By contrast, the release of lipid tails to the bulk will increase the entropy of the lipids. The GX$_3$G motif mediates critical TM interactions in many biologically important single-pass proteins that include receptor tyrosine kinases. An extensive literature attests to the importance of the subject, and the interested reader is directed to a number of classical reviews from the labs of Engelman, Fleming, MacKenzie, Langosch, Akin, Senes, Schneider, and others.[210,217,222-225] A more recent comprehensive review by Westerfield and Barrera is particularly insightful in its thoughtful treatment of the role of GX$_3$G and related motifs in receptor association.[226]

There are several variations on the GX$_3$G motif,[227] in both TM peptides and larger protein structures.[228] The classical GX$_3$G motif is C2 symmetric, and the interactions are identical on each helix, as in glycophorin A.[229] Additionally, small amino acids can occur repetitively every four residues to create extended "glycine zippers".[230] In other cases, an asymmetric interaction utilizing the GX$_3$G motif on just one of two helices is seen in heterodimers. One example occurs in the integrins. In these heterodimeric proteins, the TM helices of their alpha and beta subunits interact tightly in the resting state, but dissociate when they are activated for binding to extracellular proteins. Thus, the interaction strength between the two helices is finely balanced to allow the conformational change to occur in response to intracellular and extracellular signals. The interface of the αIIbβ3 TM domain heterodimer[231-233] consists of a tightly packed structure in which small and large residues interdigitate by efficient van der Waals packing along the heterodimer interface. Thus, a sequence motif in the αIIb TM domain, G–X$_3$–G–X$_3$–L, packs in a reciprocal manner with the β3 TM domain sequence V–X$_3$–I–X$_3$–G in such a way that bulky residues from one TM helix contact a hole formed by a small Gly residue on the neighboring helix.[227] A similar interaction occurs between the TM helices of a second β3 integrin, αvβ3, but using a different face of its TM helix to bind to the αv TM helix than the one used to interact with the αIIb subunit.[234]

The GX$_3$G motifs discussed above mediate interactions between parallel helices. An analogous type of interaction occurs between antiparallel helices with small residues spaced every 7 residues apart,[154,230,235] designated here as the Small-X$_6$-Small motif. The interaction is mediated by a mix of favorable electrostatic interactions between the peptide backbones, efficient van der Waals packing, and other interactions that depend on the sequence context. In this motif, the antiparallel helices have a left-handed crossing angle of about −10° to −30°, similar to classical coiled coils. Coiled coils have a characteristic 7-residue repeat, whose positions are designated "a" through "g" within a single heptad. By convention, the residues at the "a" and "d" positions of the heptad pack in the core of a coiled coil. The stability of water-soluble coiled coils scales with the size and hydrophobicity of the side chains at the "a" position, increasing over the series Gly < Ala < Val < Ile.[236] Just the opposite trend was seen in membrane-soluble coiled-coil peptides, Gly > Ala >> Val > Ile.[237] Furthermore, the peptides in which the "a" residues were either Gly or Ala had a strong preference to form antiparallel "Ala-coil" structures (Fig. 4) when in a membrane,[237,238] which appeared to be associated with the alignment of dipoles from the amides of adjacent helices. Similar results are seen with Ser at the "a" position. Interestingly, Ser can stabilize both antiparallel packing, as seen in protein structures[239] and model peptides (Fig. 4),[240] as well as a parallel helical dimerization motif found in the murine erythropoietin receptor.[241,242]

Polar interactions. Polar interactions, including salt bridges and hydrogen bonds, provide a significant driving force for the assembly of TM helices.[243–248] Numerous oncogenic mutations of receptor-mediated tyrosine kinases and other receptors involve the introduction of polar residues in the TM helices, which leads to enhanced association or alteration of the geometry of the TM helix–helix interaction (for reviews see ref. 224, 241, 249 and 250). Mutants with polar substitutions in their TM helices also figure largely in membrane protein misfolding diseases.[251]

Studies with model TM peptides have shown that Asn, Gln, Glu and Asp (and to a lesser extent His) enhance TM helix association.[2,244,245,247,248,252] Presumably, the Glu and Asp residues are protonated when inserted deep into a bilayer.[253] Thus, each of these residues that induce strong association have a common feature of having both hydrogen bond donors and acceptors, and so they are capable of mediating inter-helical interactions. Ser and Thr are exceptions, however. These

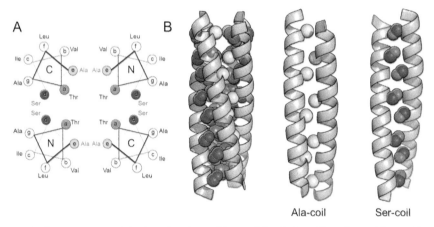

Fig. 4   Inner tetramer of the inactive channel form of the designed ion channel from pdb 6yb1. (A) Helical wheel diagram depicting key interactions, namely the Ser and Ala zippers, that stabilize the tetrameric assembly. (B) The inner tetramer is stabilized by a hydrogen-bonding network formed by Ser (purple), Thr (blue), and water molecules (not shown). Views of the Ala-coil and the Ser-coil interfaces reveal favorable interactions that mediate the formation of the tetramer.

residues generally form an intramolecular hydrogen bond with a carbonyl three or four residues above in the helix, so there is a reduced driving force for forming hydrogen bonds with adjacent TM helices.[171,254,255] The position of side-chain hydrogen bonds in a TM helix and their surrounding sequence have significant effects on stability.[256–259] As compared to an interaction between apolar residues, Asn–Asn interactions range from highly favorable in the apolar region of the bilayer to unfavorable when buried in the interior of a water-soluble structure.[260] Because Asn is capable of inducing or enhancing TM association, Asn residues have been scanned through the TM helices of several receptors, including the thrombopoietin receptor,[249] the erythropoietin receptor,[242] and the integrin αIIbβ3,[261,262] leading to both activation and inhibition of signaling.

Intramembrane salt bridges can also be a potent driving force for the association of TM helices. For example, signaling complexes of immune cells employ multi-component protein receptors composed of distinct binding and signaling subunits[263–265] that assemble *via* the helix–helix interactions of their transmembrane domains. The binding subunits associate with extracellular ligands, while the signaling subunits typically contain immunoreceptor tyrosine-based activation motifs (ITAMs), which couple to downstream phosphorylation and signaling pathways. The interaction between a pair of acidic Asp residues on the signaling dimer and a basic residue (Lys or Arg) on the ligand-binding coreceptor is required for complex assembly within the otherwise nonpolar membrane.[266,267] This 2-to-1 acidic to basic polar association motif is ubiquitous in multi-component receptor families, including Fc receptors[268] and T-cell receptor–CD3 complexes.[269] For example, in the T-cell receptor complex, the positively charged Arg and Lys residues on the TCRαβ TMs interact with negatively charged Asp and Glu residues on the CD3δε and CD3 δε heterodimers and the ζζ-homodimer.[263] These interactions are necessary for the proper assembly of the complex in the ER and subsequent transport to the cell surface.[263] Furthermore, the 2-to-1 acidic to basic association is observed in a family of more than 20 receptors that contain the signaling subunit DNAX-activation protein 12 (DAP12), a disulfide-linked homodimer with a minimal extracellular region.[265,267,270] Recently, MD simulations and density functional theory calculations of the transmembrane helices with differing protonation states of the Asp–Asp–Lys triad identified a structurally stable interaction in which a singly protonated Asp–Asp pair forms a hydrogen-bonded carboxyl–carboxylate clamp that clasps onto a charged Lys sidechain.[271] This polar motif was also frequently observed in a Protein Data Bank-wide search, indicating that it is a widely used motif.

van der Waals packing of apolar sidechains. In water, the burial of apolar sidechains in a protein interior gives rise to a large hydrophobic driving force that is not operative in a membrane. Thus, membrane proteins utilize many of the interactions and motifs described above. Nevertheless, the great majority of contacts in membrane proteins involve interactions between apolar sidechains.[211] Structural informatics suggests that these hydrophobic residues pack more efficiently in membrane proteins,[228,257,272–274] and the resulting van der Waals (vdW) interactions provide a primary driving force for folding. On the other hand, apolar side-chains in the folded state might provide little net stabilization, because lipids interact with these side-chains similarly in the unfolded state. In this view, vdW packing is secondary to stronger forces and structural restraints, including conventional and CH-mediated hydrogen bonds,[218,219,244,248,256] topology (*i.e.*, loops, water-soluble domains, α-helical insertion),[275] and weakly polar interactions.[210,276] Furthermore, experimental studies using site-directed mutagenesis suggest that apolar packing contributes similarly to stability in water-soluble and membrane proteins.[257] Finally, attempts to design proteins using membrane-solubilized versions of peptides that assembly favorably in water, such as leucine zippers, resulted in rather weak or structurally uncharacterized associations in the absence of polar interactions.[244,277]

Recent work has, however, shown that van der Waals interactions can play a dominant role in the assembly of a natural membrane homopentameric peptide, phospholamban. Mravic and coworkers[278] deciphered a steric code involving tight and regular packing that stabilizes natural membrane proteins, and used this insight to design self-associating pentameric peptides *de novo*. The structures of the resulting pentamers were solved to high resolution. Indeed, the assemblies were so stable that they remained partially associated after boiling in strong denaturant solutions. These packing motifs are frequently observed in numerous unrelated membrane protein families. These results reveal a substantial and widespread role for apolar packing in the folding, stabilization, and evolution of membrane proteins. Interestingly, a comparison of membrane-soluble and water-soluble sequences indicated that on average, the steric code was more stringent and the helices packed more tightly in the TM motifs *versus* water-soluble structures. Thus, achieving tight and specific packing would appear to be more essential in membrane proteins, which cannot rely on the hydrophobic effect.

A second recent example of the importance of van der Waals packing has been discovered through a random screening approach. DiMaio and coworkers have used clever genetic screens to discover "traptamers"[279] that interact in a sequence-specific manner with receptor tyrosine kinases, including the erythropoietin receptor and platelet-derived growth factor beta (PDGFβR).[280] Their work began with a small single-span TM protein, E5, a peptide from a tumor virus that targets and activates PDGFβR in a key step of oncogenesis.[281] E5 is a disulfide-bonded dimer and the dimeric conformation is required for the activation of PDGFβR. A single transmembrane Gln residue is required for activity in the context of the native sequence, but genetic selections[282,283] identified entirely hydrophobic sequences that are similarly able to activate PDGFβR in a highly specific manner.[284-287] Going one step further, DiMaio's group has designed traptamers that are composed of only Leu and Ile that target TM helices,[288] again in sequence and target-specific manners. Finally, they have extended this approach to an activator of the erythropoietin receptor[289] and other TM proteins.[287] It seems reasonable to assume that van der Waals forces provide the major driving force for assembly in these cases.

Combining a diversity of forces and motifs. In the above sections, we have focused on examples that illustrate individual forces and motifs, but in typical membrane peptides and proteins, a number of stabilizing interactions are combined to achieve stability and function.[211,290-292] We have already seen the intimate interconnection between non-conventional CH···O=C hydrogen bonds and van der Waals packing in GX$_3$G motifs.[221] Also, π-stacking[293] and hydrogen-bonded interactions can augment this motif. For example, strong hydrogen bonds between a Ser and His sidechain augments the affinity of a GX$_3$G motif in the transmembrane homodimer of BNIP.[291,292] Similarly, His37 forms multiple polar interactions in the homo-tetrameric four-helix bundle, M2 from influenza A virus.[294] While these interactions help stabilize the tetramer,[295] the tetrameric structure is retained in the H37A mutant, as is ion channel activity, although the ion selectivity is altered by this substitution.[296]

Multiple types of zippers of the Small-X_6-Small family can combine in protein assemblies larger than a dimer when they are placed at two, non-overlapping positions of a helix. A recent crystal structure of a designed channel peptide in a non-conducting state has illustrated this concept. Its structure displays two types of helix–helix interfaces, one featuring an Ala-coil and the other a Ser-coil. The designed sequence has a sequence repeat with small Ala or Ser residues of the heptad repeat (Fig. 4). As mentioned above, small residues spaced 7 residues apart are the signatures of Ala and Ser zippers,[228,239,297,298] in which two helices wrap around one another in an antiparallel coiled coil. In this structure, the packing of small residues provides a strong driving force for assembly in a phospholipid bilayer, while the Ser residues make polar interactions. Both Ala coils and Ser zippers are observed widely in membrane proteins and have been used in membrane protein design.[237,240,299]

Analysis and design of functional membrane-spanning TM helices and helical bundles

Modulation of protein–protein interactions in the membrane. The association of TM helices is often a critical feature in the assembly and function of single-span TM proteins. Peptides and small molecules that disrupt or modulate association provide very useful tools to probe the contribution of TM association to the function of membrane proteins—and they can translate to useful therapeutics. For example, Eltrombopag was the first small molecule drug approved that targets the TM domain of a single-pass membrane protein, and it is clinically approved for multiple indications.[300] This drug targets the TM domain of the thrombopoietin receptor near His99.[249,301] There are several approaches to designing peptides and small molecules that modulate receptor function, ranging from random screening/synthetic optimization to the computational design of TM peptides. Many excellent reviews of the field are available, so we will only introduce the topic briefly.[226,302]

The simplest way to design peptide modulators of TM proteins is to synthesize or express a peptide spanning the sequence of the TM helix of interest. This was first demonstrated for glycophorin A in 1976 by Furthmayr and Marchesi,[303,304] who showed that a TM peptide was able to shift the equilibrium of the full-length protein from a dimer to a monomer by hetero-association. Since then, this approach has been widely used to study the association of numerous proteins, as reviewed recently.[226,302] Going a step further, Barrera and coworkers have developed the introduction of a Glu residue into the TM helix of EphA2, rendering it water-soluble, but still capable of inserting into membranes in a controllable manner.[305–307] This peptide activates EphA2 by interacting with the TM helix and its juxtamembrane domain. Attesting to its specificity, this peptide only inserts into membranes at neutral pH in the presence of the TM region of EphA2.

A second approach involves the computational design of peptides that are designed to target the TM domains of proteins in a sequence- and structure-dependent manner. The GX_3G motif has been used as a basis to design

"Computed Helical Anti-Membrane Protein" (CHAMP) peptides, which bind to the TM helices of single-span membrane proteins that have this motif in their sequence.[308] A second designed CHAMP helix is docked to make good interactions with the peptide, then its sequence is optimized to recognize the specific sequence surrounding the GX$_3$G motif of the target. This method has been used to specifically recognize the αIIb, αv, and β1 subunits of integrins. They stabilize an activated form of the integrin by blocking the interaction of the alpha and beta TM helices[278,308–310] and have been helpful in determining the contribution of TM helix heterodimerization to the overall activation of the integrin αIIb.[311] Similar computational methods have now been expanded to a growing number of targets, as reviewed recently.[226,302]

The third approach involves genetic screening to identify modulators of a given activity. A classical method to determine associations is through the construction of *trans* dominant negative mutations. In early work, this approach was elegantly employed by Pinto and Lamb to determine the tetrameric association state of the M2 proton channel from the influenza A virus through quantitative analysis of the effects of a dominant negative mutant on the proton channel activity of the protein. The elegant work of DiMaio and coworkers is also covered above in the section on van der Waals interactions.

Irrespective of the method used to design TM peptides, there are a number of technical issues that need to be carefully considered and appropriate controls conducted. If TM peptides are synthesized and added exogenously, one must tackle the problem of their poor solubility and tendency to aggregate in solution. While adding short poly-Lys sequences has been employed, it is important to conduct appropriate controls with mutants or scrambled peptides as the addition of a poly-Lys tag can be perturbing. An alternative is to use a short polyethylene glycol sequence of defined length[308] in place of the Lys residues. Even after tagging, the peptides can have poor properties, which necessitates the addition of organic co-solvents or micelles. Again, rigorous controls for these variables need to be conducted. An alternative approach is to express the peptide genetically for biosynthetic incorporation into membranes. Again, however, care needs to be taken to ensure that the peptides are expressed appropriately in the desired target membrane, and to ensure that they are not triggering an unfolded protein response in the ER or other off-pathway effects.

Analysis and design of self-assembling peptide channels. Ion channels are essential for all electrical activity in the central nervous system, and for the translocation of ions, water, and small molecules across cellular membranes. Given our advanced understanding of the TM helices and helical bundles, we should now be able to design functional membrane proteins from first principles. Here, we focus primarily on designed peptide channels that transport protons and ions, highlighting natural model systems that have illuminated principles underlying channel activity and defining key considerations in peptide–membrane interactions that allow us to fine-tune ion conduction.

Our understanding of the detailed molecular mechanisms of ion conduction has progressed in a number of distinct stages.[312] Before the sequences and structures of ion channel proteins were available, detailed studies of the structural basis of ion conduction focused on simple peptides, primarily gramicidin A (GA) and alamethicin. Next, as the sequences, but not yet the atomic-level structures, of channels began to emerge, the *de novo* design of model peptides that mimic the pores of helical ion channels became an active area of research,[313,314] and studies of GA and alamethicin continued apace. As structures of natural ion channel proteins became increasingly available, they received increasing attention. Finally, we have reached a level of understanding of both membrane protein folding and ion channel function that it should now be possible to design channel proteins from first principles. The first forays into the field of *de novo* design have focused on testing and sharpening the principles of folding and conduction, but applications[315] for sensing, water purification, and ion separation could follow.

Several milestones in the field of the *de novo* design of ion channels have already been achieved. In early work, the design of peptides with well-defined rates of ion conduction and selectivity were reported, but high-resolution structures were not available.[313,314,316] Very recently, the design of peptides and proteins that assemble into channels has been reported by Baker and coworkers, but channel recordings were limited to macroscopic conductance measurements and the per protein or per channel rate of conductance was not reported.[317] The design of channels with well-defined channel properties and high-resolution structures is imminent, and we will probably see papers on this subject emerge in the next year. However, for now, the only published report focuses on a peptide that uses a dynamic rocking mechanism to transport $Zn(\text{II})$.[299,318] Below, we briefly summarize work on the natural peptide channels, gramicidin A and alamethicin, as well as a natural four-helical proton channel. We then highlight progress in the design of peptides that assemble into ion-conducting channels.

Gramicidin A and alamethicin. Gramicidin A from *Bracilus brevis*, a bacteriostatic peptide comprised of alternating D- and L-amino acids, forms head-to-head dimers in lipid membranes to selectively transport monovalent cations, including H^+, K^+, and Na^+.[319–323] Carbonyl groups of the peptide bonds line the pore and, although they are hydrogen-bonded to amide NH groups, they are able to tilt to stabilize transient cations. The conformational dynamics of the β-helices drive ion transport, and its conduction properties are affected by its interactions with the surrounding lipid.[324–326] Inspired by the β-helix structure of gramicidin, early peptide designs focused on synthetic polypeptides that formed similar helical architectures that open to allow for the permeation of ions.[327,328] Taking a different approach, Ghadiri and colleagues similarly designed cyclic tetra- and hexapeptides that assemble into cation-selective channels.[329]

Unlike gramicidin, alamethicin, an antibiotic peptide from *Trichoderma viride*, forms "barrel stave" pores lined by α-helices, which more closely resemble ion channel proteins.[300] This 20-residue peptide, consisting of interspersed α-aminoisobutyric acid residues, is postulated to associate into $n = 4\text{–}8$ α-helical bundles with its polar face, which contains the Gln and Glu residues, pointed to the center of the pore.[73,300,330–337] While a structure exists for the monomer, there is no structure available for the alamethicin channel.[338] Nevertheless, MD simulations of various oligomeric states ($n = 5\text{–}8$) reveals stability in all of the oligomeric states, which corroborates findings on the multiplicity of conductance levels observed for this channel.[339] Also, a lower-resolution model of the channel has been determined by atomic force microscopy and is consistent with the barrel stave model.[340]

M2, an example from nature that illustrates the requirements for selective proton transport. The proton channel from the influenza A virus is a small membrane protein, and the founding member of the viroporin family of proteins.[341-346] M2's proton channel is essential to the life cycle of the virus; it mediates the acidification of the interior of the virus,[347-349] which is necessary for the release of viral RNA. Following the discovery of M2, a large number of viruses have been found to also express small membrane proteins, many of which have ion channel activity.[341-346] Most topically, the structure of the E-protein, a viroporin from SARS-CoV-2, has been solved at moderate resolution by solid NMR,[350] and is an area of very active investigation.

The 24-residue TM helix of the M2 proton channel from the influenza A virus is the minimal construct of the M2 polypeptide that still retains its proton channel activity.[351] Like M2, it assembles into a parallel homotetramer with a water-filled channel lumen composed of apolar residues that leads to the proton-shuttling His37 tetrad (Fig. 5). Very high-resolution crystal structures of M2 (<1.1 Å),[294,352,353] together with solution[354] and solid NMR structures,[355,356] have revealed the interactions that stabilize the protein and the path taken by protons passing through the channel. Protons enter through a narrow hydrophobic sphincter and then travel along "water wires" to protonate a critical His37 residue. High-resolution crystallographic structures[352,353] reveal that the backbone carbonyls of the pore-facing hydrophobic residues, along with the interlumenal waters, constitute the extensive hydrogen-bonding network that both lowers the energy barrier for conductance and stabilizes excess positive charge during proton conduction.[357] Based on different solid NMR measurements, there is currently an unresolved question of whether or not the His37 residues are hydrogen bonded to

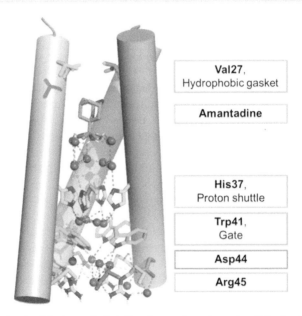

Fig. 5 Key structural features that define the proton channel activity for M2. The TM domain of M2 forms a proton-conductive channel. Here, we see the drug, amantadine, binding to the channel in the hydrophobic gasket with its ammonium group pointed towards the water-filled pore. The ammonium group acts as a proton and locks the interlumenal waters near the gating His in place, essentially blocking proton conduction.

one another.[358-360] In every independent crystallographic or solution NMR structure of the channel that we have solved, the His37 residues are either fully hydrated in one conducting state or indirectly involved in edge-on aromatic interactions with strong water-mediated hydrogen bonds in a second conformational state,[294,352,353,361-365] which agrees well with the model of Hong.[358] The conformation of the protein varies with the degree of protonation of His37 and, in the proton–conduction cycle, this region of the channel becomes more and less open as protons come on and off, respectively.[366] Voth's group has used classical MD, reactive MD, and QM calculations to evaluate the conduction mechanism, which were able to quantitatively predict the conductance of the channel, as well as the mechanism of binding to small molecule drugs.[357,367-369]

Just below the His37 residue, a ring of four Trp41 residues form tight aromatic–aromatic or aromatic–cation interactions with His37, which is further stabilized by a solvent-mediated hydrogen bond to Asp44.[370] This gate is interrupted when the His37 residues reach a critical protonation state of approximately +3 and a proton can move into the interior. Thus, M2 has a transporter-like mechanism.[351,371-374]

The contribution of each residue to the free energy of tetramerization of the channel has been evaluated in a series of mutants in micelles and bilayers. No single residue is absolutely required for tetramerization, although some substitutions result in a loss of thermodynamic stability,[295,375,376] particularly when Ala is substituted for His37. On the other hand, these mutants have substantially altered conductance and/or proton selectivity.[296] It is noteworthy that all the hydrogen bonds within the TM region are water-mediated rather than direct, with the only direct interaction being a salt bridge at the hydrated exit of the channel. Even the polar ammonium group of channel-blocking drugs form indirect, water-mediated hydrogen bonds with the channel (Fig. 5).[362-364,377] This finding underscores the importance of considering water in the stability and function of this channel.

Finally, no discussion of M2 would be complete without mentioning some early debates, which are now resolved. Early work questioned whether a second, cytoplasmic helix was needed for the structural integrity of the protein, but this was addressed thoroughly by deletion mutagenesis of M2 expressed in mammalian cells[351] and by solving the crystallographic[362] and NMR structures[378] with and without C-terminal extensions. Instead, the C-terminal cytoplasmic helix was found to be important for the budding of the virus.[71,379] Also, in early work, Chou et al. suggested that the channel-blocker amantadine (which was clearly observed in the crystal structures[361,364,377]) might instead bind on the outside of the channel.[354] This debate was also resolved when Chou was able to solve a structure with the drug bound in the interior pore.[380]

De novo designed ion channels. Designed peptide channels offer the opportunity to decipher the complexities of channel function in simple, minimalist systems. Beyond the considerations for peptide–peptide and peptide–membrane interactions in the design of membrane-soluble helical bundles, in designing functional ion channels, we need to account for the water- and ion-accessibility of the pore. In the late 1980s, DeGrado and coworkers designed the first TM α-helical peptide channel using a repeating sequence of Leu and Ser: H_2N-(Leu–Ser–Leu–Leu–Leu–Ser–Leu)$_3$-CONH$_2$ and H_2N-(Leu–Ser–Ser–Leu–Leu–Ser–Leu)$_3$-CONH$_2$, or LS2 and LS3, respectively.[313,316] The Leu residues were chosen to pack well at the helix–helix interfaces, while the Ser residues were important for both packing and the stabilization of a proton-conducting pore. The packing of small Ser residues near the core and larger Leu sidechains at the helical interfaces dictated a tetrameric arrangement for the proton channel LS2 (Fig. 6A). The Ser hydroxyls, together with water molecules, appeared to create a proton conduction pathway *via* a water-hopping mechanism. LS3 (Fig. 6B) formed hexameric channels with a pore large enough to accommodate a solvated ion in the hexameric bundle. While crystallographic structures were not available at the time, a large body of subsequent data, including the incorporation of LS2 peptides onto a rigid tetrameric scaffold,[381] has supported the hypothetical structures and conduction model.[313,314,316,382–384]

Recently, Woolfson and colleagues approached the design of α-helical channels by converting water-soluble coiled coils into membrane-spanning assemblies.[385] Unlike previous designs which borrowed from the D4 domain of the *E. coli* polysaccharide transporter Wza,[386] these designs were purely *de novo*. Their work began with a careful consideration of the features required to pack helices into various sized helical bundles of various sizes.[387] They showed that the packing of small residues at the positions designated "e" and "g" of a coiled coil dictate the formation of oligomeric bundles with sizes ranging from 5 to 8 helices, depending on the nature of the small residues and other features. In the initial work, the residues in the core at positions "a" and "d" were large Leu and Ile residues, respectively. They reasoned that at least some of these Leu and Ile residues could be changed to combinations of Ser and Thr to form water-filled pores, which was validated by crystal structures of the water-soluble peptides. Next, they converted the peptides to membrane peptides by introducing apolar residues at the exterior positions.[385] Single-channel recordings were consistent with the designs, suggesting that the designed structures had been realized. However, the peptides failed to crystallize in the active channel-forming states. Nevertheless, an unanticipated structure of a non-conducting antiparallel tetrameric state (Fig. 4) illustrated important design principles of *de novo* membrane protein design.

The most ambitious functional membrane protein designed to date is a TM four-helix bundle, Rocker (Fig. 6C), that transports the first row transition metal

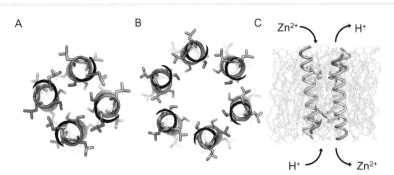

Fig. 6 Designed ion channels. Models of (A) the homotetrameric LS2, and (B) the homohexameric LS3 channels show pore-facing Ser residues with Leu at well-packed helix–helix interfaces. (C) Designed Zn^{2+}/H^+ antiporter Rocker (pdb 2MUZ) binds zinc through two 4-Glu, 4-His sites in the channel.

ion Zn^{2+} in exchange for protons.[299,318] Rocker has two 4-Glu, 4-His Zn^{2+} binding sites, which can alternately accommodate protons and metal ions. A computational design algorithm was used to stabilize two energetically degenerate asymmetric states of the protein while destabilizing a competing fully symmetrical state which might otherwise bind metal ions too tightly and impede motions required for ion transport. The computed TM bundle formed a dimer of dimers with two non-equivalent helix–helix interfaces (Fig. 6C). A "tight interface" had a small inter-helical distance (8.9 Å) stabilized by the efficient packing of small Ala residues in an Ala-coil motif, which were seen in the crystal structure of the protein. The "loose interface" had a larger interhelical distance of 12.0 Å and was less well packed. Solid NMR was thus used to define the structure of the dimer of dimers. The resulting membrane-spanning four-helical bundle transported the first–row transition metal ions Zn^{2+} and Co^{2+}, but not Ca^{2+}, across membranes. Vesicle flux experiments showed that, as Zn^{2+} ions diffused down their concentration gradients, protons were antiported. These experiments illustrate the feasibility of designing membrane proteins with predefined structural and dynamic properties. These studies illustrate how multiple forces, including structural motifs and diverse polar interactions can be combined to build functional assemblies.

Outlook

This review has given only the smallest glimpse of the vibrant and growing field of peptide–membrane interactions. Peptides have proven to be outstanding systems for understanding larger proteins, particularly for deciphering the rules of membrane peptide/protein association, insertion, and folding. This field has also spawned antimicrobial peptides and peptide mimetics currently in the clinic, while also inspiring new areas of antimicrobial polymer science.[103,106,388–395] Clearly, the future is very bright.

Funding

B. M. and W. F. D were supported by NIH grant R35-GM122603. H. T. K was supported by NIH grant K99-GM138753. R. W. N. was supported by NIH grant K99-NS116679.

Conflicts of interest

WFD is on the scientific advisory board of Innovation Pharma, who are pursuing clinical trials of Brilacidin.

References

1 S. H. White, *J. Gen. Physiol.*, 2007, **129**, 363–369.
2 N. M. Meindl-Beinker, C. Lundin, I. Nilsson, S. H. White and G. von Heijne, *EMBO Rep.*, 2006, 7, 1111–1116.
3 W. C. Wimley and S. H. White, *Biochemistry*, 2000, **39**, 4432–4442.
4 C. Wolfe, J. Cladera and P. O'Shea, *Mol. Membr. Biol.*, 1998, **15**, 221–227.
5 S. McLaughlin and A. Aderem, *Trends Biochem. Sci.*, 1995, **20**, 272–276.

6 L. T. Nguyen, E. F. Haney and H. J. Vogel, *Trends Biotechnol.*, 2011, **29**, 464–472.

7 K. J. Day, G. Kago, L. Wang, J. B. Richter, C. C. Hayden, E. M. Lafer and J. C. Stachowiak, *Nat. Cell Biol.*, 2021, **23**, 366–376.

8 F. Yuan, H. Alimohamadi, B. Bakka, A. N. Trementozzi, K. J. Day, N. L. Fawzi, P. Rangamani and J. C. Stachowiak, *Proc. Natl. Acad. Sci. U. S. A.*, 2021, **118**, e2017435118.

9 C. Tanford, *Science*, 1978, **200**, 1012–1018.

10 J. P. Segrest, R. L. Jackson, J. D. Morrisett and A. M. Gotto Jr, *FEBS Lett.*, 1974, **38**, 247–258.

11 R. L. Jackson, J. D. Morrisett, A. M. Gotto Jr and J. P. Segrest, *Mol. Cell. Biochem.*, 1975, **6**, 43–50.

12 J. P. Segrest, B. H. Chung, C. G. Brouillette, P. Kanellis and R. McGahan, *J. Biol. Chem.*, 1983, **258**, 2290–2295.

13 R. M. Epand, Y. C. Shai, J. P. Segrest and G. M. Anantharamaiah, *Biopolymers*, 1995, **37**, 319–338.

14 V. K. Mishra, M. N. Palgunachari, J. P. Segrest and G. M. Anantharamaiah, *J. Biol. Chem.*, 1994, **269**, 7185–7191.

15 J. P. Segrest, M. K. Jones, H. De Loof and N. Dashti, *J. Lipid Res.*, 2001, **42**, 1346–1367.

16 J. P. Segrest, M. K. Jones, A. E. Klon, C. J. Sheldahl, M. Hellinger, H. De Loof and S. C. Harvey, *J. Biol. Chem.*, 1999, **274**, 31755–31758.

17 B. H. Chung, G. M. Anatharamaiah, C. G. Brouillette, T. Nishida and J. P. Segrest, *J. Biol. Chem.*, 1985, **260**, 10256–10262.

18 G. M. Anantharamaiah, J. L. Jones, C. G. Brouillette, C. F. Schmidt, B. H. Chung, T. A. Hughes, A. S. Bhown and J. P. Segrest, *J. Biol. Chem.*, 1985, **260**, 10248–10255.

19 E. T. Kaiser and F. J. Kezdy, *Annu. Rev. Biophys. Biophys. Chem.*, 1987, **16**, 561–581.

20 E. T. Kaiser and F. J. Kezdy, *Science*, 1984, **223**, 249–255.

21 S. H. Lau, J. Rivier, W. Vale, E. T. Kaiser and F. J. Kezdy, *Proc. Natl. Acad. Sci. U. S. A.*, 1983, **80**, 7070–7074.

22 E. T. Kaiser and F. J. Kezdy, *Proc. Natl. Acad. Sci. U. S. A.*, 1983, **80**, 1137–1143.

23 D. Fukushima, S. Yokoyama, F. J. Kezdy and E. T. Kaiser, *Proc. Natl. Acad. Sci. U. S. A.*, 1981, **78**, 2732–2736.

24 K. Kornmueller, B. Lehofer, G. Leitinger, H. Amenitsch and R. Prassl, *Nano Res.*, 2018, **11**, 913–928.

25 E. J. Dufourc, *Biochim. Biophys. Acta, Biomembr.*, 2021, **1863**, 183478.

26 W. F. DeGrado, F. J. Kèzdy and E. T. Kaiser, *J. Am. Chem. Soc.*, 1981, **103**, 679–681.

27 W. F. DeGrado, G. F. Musso, M. Lieber, E. T. Kaiser and F. J. Kézdy, *Biophys. J.*, 1982, **37**, 329–338.

28 Y. Hirai, T. Yasuhara, H. Yoshida, T. Nakajima, M. Fujino and C. Kitada, *Chem. Pharm. Bull.*, 1979, **27**, 1942–1944.

29 W. F. DeGrado, *Adv. Prot. Chem.*, 1988, **39**, 51–124.

30 D. Hultmark, A. Engstrom, H. Bennich, R. Kapur and H. G. Boman, *Eur. J. Biochem.*, 1982, **127**, 207–217.

31 H. G. Boman, I. Faye, R. Gan, G. H. Gudmundsson, D. A. Lidholm, J. Y. Lee and K. G. Xanthopoulos, *Mem. Inst. Oswaldo Cruz*, 1987, **82**, 115–124.

32 D. Andreu, R. B. Merrifield, H. Steiner and H. G. Boman, *Proc. Natl. Acad. Sci. U. S. A.*, 1983, **80**, 6475–6479.

33 R. B. Merrifield, L. D. Vizioli and H. G. Boman, *Biochemistry*, 1982, **21**, 5020–5031.

34 W. F. DeGrado, Peptides: Structure and Function, *Proceedings of the Eighth American Peptide Symposium*, 1983, p. 1983.

35 M. Zasloff, *Proc. Natl. Acad. Sci. U. S. A.*, 1987, **84**, 5449–5453.

36 M. Zasloff, *Nature*, 2002, **415**, 389–395.

37 A. Tossi, L. Sandri and A. Giangaspero, *Biopolymers*, 2000, **55**, 4–30.

38 J. L. Bishop and B. B. Finlay, *Trends Mol. Med.*, 2006, **12**, 3–6.

39 K. A. Brogden, *Nat. Rev. Microbiol.*, 2005, **3**, 238–250.

40 Y. Shai, *Curr. Pharm. Des.*, 2002, **8**, 715–725.

41 R. D. Sarah, H. Frederick, M. Manuela and A. P. David, *Curr. Protein Pept. Sci.*, 2018, **19**, 823–838.

42 N. Shagaghi, E. A. Palombo, A. H. A. Clayton and M. Bhave, *World J. Microbiol. Biotechnol.*, 2016, **32**, 31.

43 D. I. Chan, E. J. Prenner and H. J. Vogel, *Biochim. Biophys. Acta, Biomembr.*, 2006, **1758**, 1184–1202.

44 M. Arias, K. B. Piga, M. E. Hyndman and H. J. Vogel, *Biomolecules*, 2018, **8**, 19.

45 G. Wang, *Antimicrobial Peptides: Discovery, Design and Novel Therapeutic Strategies*, CABI Publishing, 2010.

46 *Antimicrobial Peptides: Basics for Clinical Application*, ed. K. Matsuzaki, Springer, Singapore, 2019.

47 C. B. Park, H. S. Kim and S. C. Kim, *Biochem. Biophys. Res. Commun.*, 1998, **244**, 253–257.

48 G. Kragol, S. Lovas, G. Varadi, B. A. Condie, R. Hoffmann and L. Otvos Jr, *Biochemistry*, 2001, **40**, 3016–3026.

49 F. J. del Castillo, I. del Castillo and F. Moreno, *J. Bacteriol.*, 2001, **183**, 2137–2140.

50 H. W. Huang, *Biochim. Biophys. Acta, Biomembr.*, 2006, **1758**, 1292–1302.

51 J. E. Nielsen, V. A. Bjornestad, V. Pipich, H. Jenssen and R. Lund, *J. Colloid Interface Sci.*, 2021, **582**, 793–802.

52 Y. Shai, *Biopolymers*, 2002, **66**, 236–248.

53 Y. Shai, *Biochim. Biophys. Acta, Biomembr.*, 1999, **1462**, 55–70.

54 D. Roversi, V. Luca, S. Aureli, Y. Park, M. L. Mangoni and L. Stella, *ACS Chem. Biol.*, 2014, **9**, 2003–2007.

55 K. Matsuzaki, *Biochim. Biophys. Acta, Biomembr.*, 2009, **1788**, 1687–1692.

56 K. Matsuzaki, *Biochim. Biophys. Acta, Biomembr.*, 1999, **1462**, 1–10.

57 K. Matsuzaki, K. Sugishita, N. Ishibe, M. Ueha, S. Nakata, K. Miyajima and R. M. Epand, *Biochemistry*, 1998, **37**, 11856–11863.

58 K. Matsuzaki, *Biochim. Biophys. Acta, Rev. Biomembr.*, 1998, **1376**, 391–400.

59 K. Matsuzaki, O. Murase, N. Fujii and K. Miyajima, *Biochemistry*, 1996, **35**, 11361–11368.

60 K. Matsuzaki, K. Sugishita, N. Fujii and K. Miyajima, *Biochemistry*, 1995, **34**, 3423–3429.

61 H. Sato and J. B. Feix, *Biochim. Biophys. Acta, Biomembr.*, 2006, **1758**, 1245–1256.

62 W. C. Wimley and K. Hristova, *Aust. J. Chem.*, 2020, **73**, 96–103.

63 S. Guha, J. Ghimire, E. Wu and W. C. Wimley, *Chem. Rev.*, 2019, **119**, 6040–6085.

64 H. W. Huang, *Biochemistry*, 2000, **39**, 8347–8352.

65 C. F. Lopez, S. O. Nielsen, G. Srinivas, W. F. DeGrado and M. L. Klein, *J. Chem. Theory Comput.*, 2006, **2**, 649–655.

66 D. P. Tieleman, *Biophys. J.*, 2017, **113**, 1–3.

67 L. H. Yang, V. D. Gordon, A. Mishra, A. Sorn, K. R. Purdy, M. A. Davis, G. N. Tew and G. C. L. Wong, *J. Am. Chem. Soc.*, 2007, **129**, 12141–12147.

68 L. H. Yang, V. D. Gordon, D. R. Trinkle, N. W. Schmidt, M. A. Davis, C. DeVries, A. Som, J. E. Cronan, G. N. Tew and G. C. L. Wong, *Proc. Natl. Acad. Sci. U. S. A.*, 2008, **105**, 20595–20600.

69 N. W. Schmidt, A. Mishra, G. H. Lai, M. Davis, L. K. Sanders, D. Tran, A. Garcia, K. P. Tai, P. B. McCray, A. J. Ouellette, M. E. Selsted and G. C. L. Wong, *J. Am. Chem. Soc.*, 2011, **133**, 6720–6727.

70 N. W. Schmidt and G. C. L. Wong, *Curr. Opin. Solid State Mater. Sci.*, 2013, **17**, 151–163.

71 N. W. Schmidt, A. Mishra, J. Wang, W. F. DeGrado and G. C. Wong, *J. Am. Chem. Soc.*, 2013, **135**, 13710–13719.

72 E. Y. Lee, B. M. Fulan, G. C. L. Wong and A. L. Ferguson, *Proc. Natl. Acad. Sci. U. S. A.*, 2016, **113**, 13588.

73 B. Bechinger, *J. Membr. Biol.*, 1997, **156**, 197–211.

74 S. Pal, H. Chakraborty and A. Chattopadhyay, *J. Phys. Chem. B*, 2021, **125**, 8450–8459.

75 R. F. Epand, B. P. Mowery, S. E. Lee, S. S. Stahl, R. I. Lehrer, S. H. Gellman and R. M. Epand, *J. Mol. Biol.*, 2008, **379**, 38–50.

76 R. F. Epand, M. A. Schmitt, S. H. Gellman and R. M. Epand, *Biochim. Biophys. Acta*, 2006, **1758**, 1343–1350.

77 R. F. Epand, M. A. Schmitt, S. H. Gellman, A. Sen, M. Auger, D. W. Hughes and R. M. Epand, *Mol. Membr. Biol.*, 2005, **22**, 457–469.

78 R. F. Epand, T. L. Raguse, S. H. Gellman and R. M. Epand, *Biochemistry*, 2004, **43**, 9527–9535.

79 Y. Shai and Z. Oren, *Peptides*, 2001, **22**, 1629–1641.

80 L. Chen, H. Yin, B. Farooqi, S. Sebti, A. D. Hamilton and J. Chen, *Mol. Cancer Ther.*, 2005, **4**, 1019–1025.

81 E. J. Prenner, M. Kiricsi, M. Jelokhani-Niaraki, R. N. Lewis, R. S. Hodges and R. N. McElhaney, *J. Biol. Chem.*, 2005, **280**, 2002–2011.

82 A. Braunstein, N. Papo and Y. Shai, *Antimicrob. Agents Chemother.*, 2004, **48**, 3127–3129.

83 Y. Shai, A. Makovitzky and D. Avrahami, *Curr. Protein Pept. Sci.*, 2006, **7**, 479–486.

84 A. Makovitzki, J. Baram and Y. Shai, *Biochemistry*, 2008, **47**, 10630–10636.

85 S. Fernandez-Lopez, H. S. Kim, E. C. Choi, M. Delgado, J. R. Granja, A. Khasanov, K. Kraehenbuehl, G. Long, D. A. Weinber, K. M. wilcoxen and M. R. Ghadiri, *Nature*, 2001, **412**, 452–455.

86 W. S. Horne and S. H. Gellman, *Acc. Chem. Res.*, 2008, **41**, 1399–1408.

87 R. P. Cheng, S. H. Gellman and W. F. DeGrado, *Chem. Rev.*, 2001, **101**, 3219–3232.

88 S. H. Gellman, *Acc. Chem. Res.*, 1998, **31**, 173–180.

89 D. Seebach, D. F. Hook and A. Glattli, *Biopolymers*, 2006, **84**, 23–37.

90 D. Seebach, A. K. Beck and D. J. Bierbaum, *Chem. Biodiversity*, 2004, **1**, 1111–1239.

91 Y. Hamuro, J. P. Schneider and W. F. DeGrado, *J. Am. Chem. Soc.*, 1999, **121**, 12200–12201.

92 D. Liu and W. DeGrado, *J. Am. Chem. Soc.*, 2001, **123**, 7553–7559.

93 E. A. Porter, X. Wang, H. S. Lee, B. Weisblum and S. H. Gellman, *Nature*, 2000, **404**, 565.

94 P. I. Arvidsson, N. S. Ryder, H. M. Weiss, D. F. Hook, J. Escalante and D. Seebach, *Chem. Biodiversity*, 2005, **2**, 401–420.

95 M. A. Schmitt, B. Weisblum and S. H. Gellman, *J. Am. Chem. Soc.*, 2007, **129**, 417–428.

96 E. A. Porter, B. Weisblum and S. H. Gellman, *J. Am. Chem. Soc.*, 2005, **127**, 11516–11529.

97 M. A. Schmitt, B. Weisblum and S. H. Gellman, *J. Am. Chem. Soc.*, 2004, **126**, 6848–6849.

98 J. A. Patch and A. E. Barron, *J. Am. Chem. Soc.*, 2003, **125**, 12092–12093.

99 A. Violette, S. Fournel, K. Lamour, O. Chaloin, B. Frisch, J. P. Briand, H. Monteil and G. Guichard, *Chem. Biol.*, 2006, **13**, 531–538.

100 A. Kuroki, P. Sangwan, Y. Qu, R. Peltier, C. Sanchez-Cano, J. Moat, C. G. Dowson, E. G. L. Williams, K. E. S. Locock, M. Hartlieb and S. Perrier, *ACS Appl. Mater. Interfaces*, 2017, **9**, 40117–40126.

101 H. Takahashi, G. A. Caputo and K. Kuroda, *Biomater. Sci.*, 2021, **9**, 2758–2767.

102 R. Bhat, L. L. Foster, G. Rani, S. Vemparala and K. Kuroda, *RSC Adv.*, 2021, **11**, 22044–22056.

103 G. N. Tew, D. Liu, B. Chen, R. J. Doerksen, J. Kaplan, P. J. Carroll, M. L. Klein and W. F. DeGrado, *Proc. Natl. Acad. Sci. U. S. A.*, 2002, **99**, 5110–5114.

104 D. Liu, S. Choi, B. Chen, R. J. Doerksen, D. J. Clements, J. D. Winkler, M. L. Klein and W. F. DeGrado, *Angew. Chem., Int. Ed.*, 2004, **43**, 1158–1162.

105 I. Ivanov, S. Vemparala, V. Pophristic, K. Kuroda, W. F. DeGrado, J. A. McCammon and M. L. Klein, *J. Am. Chem. Soc.*, 2006, **128**, 1778–1779.

106 G. N. Tew, R. W. Scott, M. L. Klein and W. F. DeGrado, *Acc. Chem. Res.*, 2010, **43**, 30–39.

107 R. W. Scott, W. F. DeGrado and G. N. Tew, *Curr. Opin. Biotechnol.*, 2008, **19**, 620–627.

108 S. Choi, A. Isaacs, D. Clements, D. Liu, H. Kim, R. W. Scott, J. D. Winkler and W. F. DeGrado, *Proc. Natl. Acad. Sci. U. S. A.*, 2009, **106**, 6968–6973.

109 R. W. Scott and G. N. Tew, *Curr. Top. Med. Chem.*, 2017, **17**, 576–589.

110 H. Zhao, K. K. W. To, K. H. Sze, T. T. Yung, M. Bian, H. Lam, M. L. Yeung, C. Li, H. Chu and K. Y. Yuen, *Nat. Commun.*, 2020, **11**, 4252.

111 P. G. Barlow, P. Svoboda, A. Mackellar, A. A. Nash, I. A. York, J. Pohl, D. J. Davidson and R. O. Donis, *PLoS One*, 2011, **6**, e25333.

112 C. Salata, A. Calistri, C. Parolin, A. Baritussio and G. Palu, *Expert Rev. Anti-Infect. Ther.*, 2017, **15**, 483–492.

113 A. P. Gunesch, F. J. Zapatero-Belinchon, L. Pinkert, E. Steinmann, M. P. Manns, G. Schneider, T. Pietschmann, M. Bronstrup and T. von Hahn, *Antimicrob. Agents Chemother.*, 2020, 64.

114 A. Bakovic, K. Risner, N. Bhalla, F. Alem, T. L. Chang, W. K. Weston, J. A. Harness and A. Narayanan, *Viruses*, 2021, 13.

115 B. Mensa, Y. H. Kim, S. Choi, R. Scott, G. A. Caputo and W. F. DeGrado, *Antimicrob. Agents Chemother.*, 2011, **55**, 5043–5053.

116 B. Mensa, G. L. Howell, R. Scott and W. F. DeGrado, *Antimicrob. Agents Chemother.*, 2014, **58**, 5136–5145.

117 A. Goode, V. Yeh and B. B. Bonev, *Faraday Discuss.*, 2021, DOI: 10.1039/d1fd00036e.

118 B. B. Bonev, E. Breukink, E. Swiezewska, B. De Kruijff and A. Watts, *FASEB J.*, 2004, **18**, 1862–1869.

119 M. Mahlapuu, J. Håkansson, L. Ringstad and C. Björn, *Front. Cell. Infect. Microbiol.*, 2016, **6**, 194.

120 C. H. Chen and T. K. Lu, *Antibiotics*, 2020, **9**, 24.

121 C. M. Pfefferkorn, Z. Jiang and J. C. Lee, *Biochim. Biophys. Acta, Biomembr.*, 2012, **1818**, 162–171.

122 L. Maroteaux, J. T. Campanelli and R. H. Scheller, *J. Neurosci.*, 1988, **8**, 2804–2815.

123 J. M. George, H. Jin, W. S. Woods and D. F. Clayton, *Neuron*, 1995, **15**, 361–372.

124 J. P. Segrest, H. D. Loof, J. G. Dohlman, C. G. Brouillette and G. M. Anantharamaiah, *Proteins: Struct., Funct., Genet.*, 1990, **8**, 103–117.

125 D. L. Fortin, V. M. Nemani, S. M. Voglmaier, M. D. Anthony, T. A. Ryan and R. H. Edwards, *J. Neurosci.*, 2005, **25**, 10913–10921.

126 A. Abeliovich, Y. Schmitz, I. Farinas, D. Choi-Lundberg, W.-H. Ho, J. M. G. Verdugo, M. Armanini, A. Ryan, M. Hynes, H. Phillips, D. Sulzer and A. Rosenthal, *Neuron*, 2000, **25**, 239–252.

127 M. H. Polymeropoulos, C. Lavedan, E. Leroy, S. E. Ide, A. Dehejia, A. Dutra, B. Pike, H. Root, J. Rubenstein, R. Boyer, E. S. Stenroos, S. Chandrasekharappa, A. Athanassiadou, T. Papapetropoulos, W. G. Johnson, A. M. Lazzarini, R. C. Duvoisin, G. Di Iorio, L. I. Golbe and R. L. Nussbaum, *Science*, 1997, **276**, 2045–2047.

128 M. G. Spillantini, M. L. Schmidt, V. M. Lee, J. Q. Trojanowski, R. Jakes and M. Goedert, *Nature*, 1997, **388**, 839–840.

129 A. D. McLachlan, *Nature*, 1977, **267**, 465–466.

130 A. J. Trexler and E. Rhoades, *Biochemistry*, 2009, **48**, 2304–2306.

131 C. C. Jao, A. Der-Sarkissian, J. Chen and R. Langen, *Proc. Natl. Acad. Sci. U. S. A.*, 2004, **101**, 8331–8336.

132 C. C. Jao, B. G. Hegde, J. Chen, I. S. Haworth and R. Langen, *Proc. Natl. Acad. Sci. U. S. A.*, 2008, **105**, 19666–19671.

133 C. R. Bodner, C. M. Dobson and A. Bax, *J. Mol. Biol.*, 2009, **390**, 775–790.

134 G. Fusco, A. De Simone, T. Gopinath, V. Vostrikov, M. Vendruscolo, C. M. Dobson and G. Veglia, *Nat. Commun.*, 2014, **5**, 3827.

135 R. W. Newberry, T. Arhar, J. Costello, G. C. Hartoularos, A. M. Maxwell, Z. Z. C. Naing, M. Pittman, N. R. Reddy, D. M. C. Schwarz, D. R. Wassarman, T. S. Wu, D. Barrero, C. Caggiano, A. Catching, T. B. Cavazos, L. S. Estes, B. Faust, E. A. Fink, M. A. Goldman, Y. K. Gomez, M. G. Gordon, L. M. Gunsalus, N. Hoppe, M. Jaime-Garza, M. C. Johnson, M. G. Jones, A. F. Kung, K. E. Lopez, J. Lumpe, C. Martyn, E. E. McCarthy, L. E. Miller-Vedam, E. J. Navarro, A. Palar, J. Pellegrino, W. Saylor, C. A. Stephens, J. Strickland, H. Torosyan, S. A. Wankowicz,

D. R. Wong, G. Wong, S. Redding, E. D. Chow, W. F. DeGrado and M. Kampmann, *ACS Chem. Biol.*, 2020, **15**, 2137–2153.

136 G. Fusco, T. Pape, A. D. Stephens, P. Mahou, A. R. Costa, C. F. Kaminski, G. S. Kaminski Schierle, M. Vendruscolo, G. Veglia, C. M. Dobson and A. De Simone, *Nat. Commun.*, 2016, **7**, 12563.

137 T. Logan, J. Bendor, C. Toupin, K. Thorn and R. H. Edwards, *Nat. Neurosci.*, 2017, **20**, 681–689.

138 P. H. Weinreb, W. Zhen, A. W. Poon, K. A. Conway and P. T. Lansbury Jr, *Biochemistry*, 1996, **35**, 13709–13715.

139 W. S. Davidson, A. Jonas, D. F. Clayton and J. M. George, *J. Biol. Chem.*, 1998, **273**, 9443–9449.

140 B. Antonny, *Annu. Rev. Biochem.*, 2011, **80**, 101–123.

141 G. Drin and B. Antonny, *FEBS Lett.*, 2010, **584**, 1840–1847.

142 U. Dettmer, *Front. Neurosci.*, 2018, **12**, 623.

143 P. K. Auluck, G. Caraveo and S. Lindquist, *Annu. Rev. Cell Dev. Biol.*, 2010, **26**, 211–233.

144 I. M. Pranke, V. Morello, J. Bigay, K. Gibson, J. M. Verbavatz, B. Antonny and C. L. Jackson, *J. Cell Biol.*, 2011, **194**, 89–103.

145 E. R. Middleton and E. Rhoades, *Biophys. J.*, 2010, **99**, 2279–2288.

146 A. Zemel, A. Ben-Shaul and S. May, *J. Phys. Chem. B*, 2008, **112**, 6988–6996.

147 A. R. Braun, M. M. Lacy, V. C. Ducas, E. Rhoades and J. N. Sachs, *J. Am. Chem. Soc.*, 2014, **136**, 9962–9972.

148 A. R. Braun, M. M. Lacy, V. C. Ducas, E. Rhoades and J. N. Sachs, *J. Membr. Biol.*, 2017, **250**, 183–193.

149 A. Copic, S. Antoine-Bally, M. Gimenez-Andres, C. La Torre Garay, B. Antonny, M. M. Manni, S. Pagnotta, J. Guihot and C. L. Jackson, *Nat. Commun.*, 2018, **9**, 1332.

150 D. Eliezer, E. Kutluay, R. Bussell Jr and G. Browne, *J. Mol. Biol.*, 2001, **307**, 1061–1073.

151 S. Chandra, X. Chen, J. Rizo, R. Jahn and T. C. Südhof, *J. Biol. Chem.*, 2003, **278**, 15313–15318.

152 T. S. Ulmer, A. Bax, N. B. Cole and R. L. Nussbaum, *J. Biol. Chem.*, 2005, **280**, 9595–9603.

153 D. M. Fowler and S. Fields, *Nat. Methods*, 2014, **11**, 801–807.

154 C. A. Schramm, B. T. Hannigan, J. E. Donald, C. Keasar, J. G. Saven, W. F. Degrado and I. Samish, *Structure*, 2012, **20**, 924–935.

155 R. Bussell Jr and D. Eliezer, *J. Mol. Biol.*, 2003, **329**, 763–778.

156 S. H. Shahmoradian, A. J. Lewis, C. Genoud, J. Hench, T. E. Moors, P. P. Navarro, D. Castaño-Díez, G. Schweighauser, A. Graff-Meyer, K. N. Goldie, R. Sütterlin, E. Huisman, A. Ingrassia, Y. Gier, A. J. M. Rozemuller, J. Wang, A. Paepe, J. Erny, A. Staempfli, J. Hoernschemeyer, F. Grosserüschkamp, D. Niedieker, S. F. El-Mashtoly, M. Quadri, W. F. J. Van IJcken, V. Bonifati, K. Gerwert, B. Bohrmann, S. Frank, M. Britschgi, H. Stahlberg, W. D. J. Van de Berg and M. E. Lauer, *Nat. Neurosci.*, 2019, **22**, 1099–1109.

157 V. A. Trinkaus, I. Riera-Tur, A. Martinez-Sanchez, F. J. B. Bauerlein, Q. Guo, T. Arzberger, W. Baumeister, I. Dudanova, M. S. Hipp, F. U. Hartl and R. Fernandez-Busnadiego, *Nat. Commun.*, 2021, **12**, 2110.

158 J. Burré, M. Sharma, T. Tsetsenis, V. Buchman, M. R. Etherton and T. C. Südhof, *Science*, 2010, **329**, 1663–1667.

159 M. A. A. Fakhree, S. A. J. Engelbertink, K. A. van Leijenhorst-Groener, C. Blum and M. Claessens, *Biomacromolecules*, 2019, **20**, 1217–1223.

160 W. F. Zeno, A. S. Thatte, L. Wang, W. T. Snead, E. M. Lafer and J. C. Stachowiak, *J. Am. Chem. Soc.*, 2019, **141**, 10361–10371.

161 E. Jo, J. McLaurin, C. M. Yip, P. St George-Hyslop and P. E. Fraser, *J. Biol. Chem.*, 2000, **275**, 34328–34334.

162 A. R. Braun, E. Sevcsik, P. Chin, E. Rhoades, S. Tristram-Nagle and J. N. Sachs, *J. Am. Chem. Soc.*, 2012, **134**, 2613–2620.

163 C. Galvagnion, A. K. Buell, G. Meisl, T. C. T. Michaels, M. Vendruscolo, T. P. J. Knowles and C. M. Dobson, *Nat. Chem. Biol.*, 2015, **11**, 229–234.

164 M. Zhu and A. L. Fink, *J. Biol. Chem.*, 2003, **278**, 16873–16877.

165 D. Snead and D. Eliezer, *J. Biol. Chem.*, 2019, **294**, 3325–3342.

166 R. W. Newberry, J. T. Leong, E. D. Chow, M. Kampmann and W. F. DeGrado, *Nat. Chem. Biol.*, 2020, **16**, 653–659.

167 T. Bartels, J. G. Choi and D. J. Selkoe, *Nature*, 2011, **477**, 107–110.

168 G. Fusco, S. W. Chen, P. T. F. Williamson, R. Cascella, M. Perni, J. A. Jarvis, C. Cecchi, M. Vendruscolo, F. Chiti, N. Cremades, L. Ying, C. M. Dobson and A. D. Simone, *Science*, 2017, **358**, 1440–1443.

169 G. Von Heijne, *Adv. Protein Chem.*, 2003, **63**, 1–18.

170 S. H. White and G. von Heijne, *Curr. Opin. Struct. Biol.*, 2004, **14**, 397–404.

171 D. M. Engelman and T. A. Steitz, *Cell*, 1981, **23**, 411–422.

172 S. H. White and W. C. Wimley, *Annu. Rev. Biophys. Biomol. Struct.*, 1999, **28**, 319–365.

173 D. M. Engelman, Y. Chen, C. N. Chin, A. R. Curran, A. M. Dixon, A. D. Dupuy, A. S. Lee, U. Lehnert, E. E. Matthews, Y. K. Reshetnyak, A. Senes and J. L. Popot, *FEBS Lett.*, 2003, **555**, 122–125.

174 J. L. Popot and D. M. Engelman, *Annu. Rev. Biochem.*, 2000, **69**, 881–922.

175 J.-L. Popot and D. M. Engelman, *Biochemistry*, 1990, **29**, 4031–4037.

176 A. L. Lomize, M. A. Lomize, S. R. Krolicki and I. D. Pogozheva, *Nucleic Acids Res.*, 2017, **45**, D250–D255.

177 J. Kirrbach, M. Krugliak, C. L. Ried, P. Pagel, I. T. Arkin and D. Langosch, *Bioinformatics*, 2013, **29**, 1623–1630.

178 A. W. Partridge, S. Liu, S. Kim, J. U. Bowie and M. H. Ginsberg, *J. Biol. Chem.*, 2005, **280**, 7294–7300.

179 J. P. Schlebach and C. R. Sanders, *Q. Rev. Biophys.*, 2015, **48**, 1–34.

180 S. H. White and G. von Heijne, *Biochem. Soc. Trans.*, 2005, **33**, 1012–1015.

181 T. Hessa, N. M. Meindl-Beinker, A. Bernsel, H. Kim, Y. Sato, M. Lerch-Bader, I. Nilsson, S. H. White and G. von Heijne, *Nature*, 2007, **450**, 1026–1030.

182 A. Senes, D. C. Chadi, P. B. Law, R. F. Walters, V. Nanda and W. F. Degrado, *J. Mol. Biol.*, 2007, **366**, 436–448.

183 R. Worch, C. Bokel, S. Hofinger, P. Schwille and T. Weidemann, *Proteomics*, 2010, **10**, 4196–4208.

184 M. Lerch-Bader, C. Lundin, H. Kim, I. Nilsson and G. von Heijne, *Proc. Natl. Acad. Sci. U. S. A.*, 2008, **105**, 4127–4132.

185 C. Lundin, H. Kim, I. Nilsson, S. H. White and G. von Heijne, *Proc. Natl. Acad. Sci. U. S. A.*, 2008, **105**, 15702–15707.

186 A. Elazar, J. Weinstein, I. Biran, Y. Fridman, E. Bibi and S. J. Fleishman, *eLife*, 2016, **5**.

187 W. C. Wimley, T. P. Creamer and S. H. White, *Biochemistry*, 1996, **35**, 5109–5124.

188 S. S. Krishnakumar and E. London, *J. Mol. Biol.*, 2007, **374**, 1251–1269.

189 G. A. Caputo and E. London, *Biochemistry*, 2004, **43**, 8794–8806.

190 S. Lew, G. A. Caputo and E. London, *Biochemistry*, 2003, **42**, 10833–10842.

191 O. A. Andreev, A. D. Dupuy, M. Segala, S. Sandugu, D. A. Serra, C. O. Chichester, D. M. Engelman and Y. K. Reshetnyak, *Proc. Natl. Acad. Sci. U. S. A.*, 2007, **104**, 7893–7898.

192 F. N. Barrera, D. Weerakkody, M. Anderson, O. A. Andreev, Y. K. Reshetnyak and D. M. Engelman, *J. Mol. Biol.*, 2011, **413**, 359–371.

193 V. P. Nguyen, A. C. Dixson and F. N. Barrera, *Biophys. J.*, 2019, **117**, 659–667.

194 V. P. Nguyen, D. S. Alves, H. L. Scott, F. L. Davis and F. N. Barrera, *Biochemistry*, 2015, **54**, 6567–6575.

195 O. A. Andreev, D. M. Engelman and Y. K. Reshetnyak, *Front. Physiol.*, 2014, **5**, 97.

196 L. C. Wyatt, J. S. Lewis, O. A. Andreev, Y. K. Reshetnyak and D. M. Engelman, *Trends Biotechnol.*, 2018, **36**, 1300.

197 Y. K. Reshetnyak, A. Moshnikova, O. A. Andreev and D. M. Engelman, *Front. Bioeng. Biotechnol.*, 2020, **8**, 335.

198 E. Strandberg, S. Esteban-Martin, A. S. Ulrich and J. Salgado, *Biochim. Biophys. Acta, Biomembr.*, 2012, **1818**, 1242–1249.

199 J. Ren, S. Lew, J. Wang and E. London, *Biochemistry*, 1999, **38**, 5905–5912.

200 G. A. Caputo and E. London, *Biochemistry*, 2003, **42**, 3275–3285.

201 S. K. Kandasamy and R. G. Larson, *Biophys. J.*, 2006, **90**, 2326–2343.

202 S. Morein, I. R. Koeppe, G. Lindblom, B. de Kruijff and J. A. Killian, *Biophys. J.*, 2000, **78**, 2475–2485.

203 A. Stefansson, A. Armulik, I. Nilsson, G. von Heijne and S. Johansson, *J. Biol. Chem.*, 2004, **279**, 21200–21205.

204 J. A. Killian, *FEBS Lett.*, 2003, **555**, 134–138.

205 F. Afrose and R. E. Koeppe II, *Biomolecules*, 2020, 10.

206 M. J. McKay, A. N. Martfeld, A. A. De Angelis, S. J. Opella, D. V. Greathouse and R. E. Koeppe II, *Biophys. J.*, 2018, **114**, 2617–2629.

207 A. Holt, R. B. M. Koehorst, T. Rutters-Meijneke, M. H. Gelb, D. T. S. Rijkers, M. A. Hemminga and J. A. Killian, *Biophys. J.*, 2009, **97**, 2258–2266.

208 O. Soubias, W. E. Teague Jr, K. G. Hines and K. Gawrisch, *Biophys. J.*, 2015, **108**, 1125–1132.

209 B. Grasberger, A. P. Minton, C. DeLisi and H. Metzger, *Proc. Natl. Acad. Sci. U. S. A.*, 1986, **83**, 6258–6262.

210 D. Langosch and I. T. Arkin, *Protein Sci.*, 2009, **18**, 1343–1358.

211 H. Hong, *Arch. Biochem. Biophys.*, 2014, **564**, 297–313.

212 J. C. Nelson, J. G. Saven, J. S. Moore and P. G. Wolynes, *Science*, 1997, **277**, 1793–1796.

213 V. Anbazhagan and D. Schneider, *Biochim. Biophys. Acta, Biomembr.*, 2010, **1798**, 1899–1907.

214 L. Cristian, J. D. Lear and W. F. DeGrado, *Proc. Natl. Acad. Sci. U. S. A.*, 2003, **100**, 14772–14777.

215 H. Hong and J. U. Bowie, *J. Am. Chem. Soc.*, 2011, **133**, 11389–11398.

216 A. Senes, M. Gerstein and D. M. Engelman, *J. Mol. Biol.*, 2000, **296**, 921–936.

217 A. Senes, D. E. Engel and W. F. DeGrado, *Curr. Opin. Struct. Biol.*, 2004, **14**, 465–479.

218 M. G. Teese and D. Langosch, *Biochemistry*, 2015, **54**, 5125–5135.

219 A. Senes, I. Ubarretxena-Belandia and D. M. Engelman, *Proc. Natl. Acad. Sci. U. S. A.*, 2001, **98**, 9056–9061.

220 B. K. Mueller, S. Subramaniam and A. Senes, *Proc. Natl. Acad. Sci. U. S. A.*, 2014, **111**, E888–E895.

221 S. M. Anderson, B. K. Mueller, E. J. Lange and A. Senes, *J. Am. Chem. Soc.*, 2017, **139**, 15774–15783.

222 K. R. MacKenzie and K. G. Fleming, *Curr. Opin. Struct. Biol.*, 2008, **18**, 412–419.

223 F. Cymer and D. Schneider, *Cell Adhes. Migr.*, 2010, **4**, 299–312.

224 D. T. Moore, B. W. Berger and W. F. DeGrado, *Structure*, 2008, **16**, 991–1001.

225 M. A. Lemmon and D. M. Engelman, *Q. Rev. Biophys.*, 1994, **27**, 157–218.

226 J. M. Westerfield and F. N. Barrera, *J. Biol. Chem.*, 2020, **295**, 1792–1814.

227 B. W. Berger, D. W. Kulp, L. M. Span, J. L. DeGrado, P. C. Billings, A. Senes, J. S. Bennett and W. F. DeGrado, *Proc. Natl. Acad. Sci. U. S. A.*, 2010, **107**, 703–708.

228 S. Q. Zhang, D. W. Kulp, C. A. Schramm, M. Mravic, I. Samish and W. F. DeGrado, *Structure*, 2015, **23**, 527–541.

229 K. R. MacKenzie, J. H. Prestegard and D. M. Engelman, *Science*, 1997, **276**, 131–133.

230 S. Kim, T. J. Jeon, A. Oberai, D. Yang, J. J. Schmidt and J. U. Bowie, *Proc. Natl. Acad. Sci. U. S. A.*, 2005, **102**, 14278–14283.

231 J. Yang, Y. Q. Ma, R. C. Page, S. Misra, E. F. Plow and J. Qin, *Proc. Natl. Acad. Sci. U. S. A.*, 2009, **106**, 17729–17734.

232 J. Zhu, B. H. Luo, P. Barth, J. Schonbrun, D. Baker and T. A. Springer, *Mol. Cell*, 2009, **34**, 234–249.

233 D. G. Metcalf, D. W. Kulp, J. S. Bennett and W. F. DeGrado, *J. Mol. Biol.*, 2009, **392**, 1087–1101.

234 R. I. Litvinov, M. Mravic, H. Zhu, J. W. Weisel, W. F. DeGrado and J. S. Bennett, *Proc. Natl. Acad. Sci. U. S. A.*, 2019, **116**, 12295–12300.

235 R. F. Walters and W. F. DeGrado, *Proc. Natl. Acad. Sci. U. S. A.*, 2006, **103**, 13658–13663.

236 A. Acharya, V. Rishi and C. Vinson, *Biochemistry*, 2006, **45**, 11324–11332.

237 Y. Zhang, D. W. Kulp, J. D. Lear and W. F. DeGrado, *J. Am. Chem. Soc.*, 2009, **131**, 11341–11343.

238 J. D. Lear, A. L. Stouffer, H. Gratkowski, V. Nanda and W. F. Degrado, *Biophys. J.*, 2004, **87**, 3421–3429.

239 L. Adamian and J. Liang, *Proteins: Struct., Funct., Genet.*, 2002, **47**, 209–218.

240 B. North, L. Cristian, X. Fu Stowell, J. D. Lear, J. G. Saven and W. F. DeGrado, *J. Mol. Biol.*, 2006, **359**, 930–939.

241 K. F. Kubatzky, W. Ruan, R. Gurezka, J. Cohen, R. Ketteler, S. S. Watowich, D. Neumann, D. Langosch and U. Klingmuller, *Curr. Biol.*, 2001, **11**, 110–115.

242 W. Ruan, V. Becker, U. Klingmuller and D. Langosch, *J. Biol. Chem.*, 2004, **279**, 3273–3279.

243 J. U. Bowie, *Nat. Struct. Biol.*, 2000, **7**, 91–94.

244 C. Choma, H. Gratkowski, J. D. Lear and W. F. DeGrado, *Nat. Struct. Biol.*, 2000, **7**, 161–166.

245 H. Gratkowski, J. D. Lear and W. F. DeGrado, *Proc. Natl. Acad. Sci. U. S. A.*, 2001, **98**, 880–885.

246 A. Pasternak, J. Kaplan, J. D. Lear and W. F. Degrado, *Protein Sci.*, 2001, **10**, 958–969.

247 F. X. Zhou, M. J. Cocco, W. P. Russ, A. T. Brunger and D. M. Engelman, *Nat. Struct. Biol.*, 2000, **7**, 154–160.

248 F. X. Zhou, H. J. Merianos, A. T. Brunger and D. M. Engelman, *Proc. Natl. Acad. Sci. U. S. A.*, 2001, **98**, 2250–2255.

249 E. Leroy, J. P. Defour, T. Sato, S. Dass, V. Gryshkova, M. M. Shwe, J. Staerk, S. N. Constantinescu and S. O. Smith, *J. Biol. Chem.*, 2016, **291**, 2974–2987.

250 A. W. Partridge, A. G. Therien and C. M. Deber, *Proteins: Struct., Funct., Bioinf.*, 2004, **54**, 648–656.

251 D. P. Ng, B. E. Poulsen and C. M. Deber, *Biochim. Biophys. Acta, Biomembr.*, 2012, **1818**, 1115–1122.

252 M. Hermansson and G. von Heijne, *J. Mol. Biol.*, 2003, **334**, 803–809.

253 S. Choe, K. A. Hecht and M. Grabe, *J. Gen. Physiol.*, 2008, **131**, 563–573.

254 E. N. Baker and R. E. Hubbard, *Prog. Biophys. Mol. Biol.*, 1984, **44**, 97–179.

255 M. Mravic, J. L. Thomaston, M. Tucker, P. E. Solomon, L. Liu and W. F. DeGrado, *Science*, 2019, **363**, 1418–1423.

256 N. H. Joh, A. Min, S. Faham, J. P. Whitelegge, D. Yang, V. L. Woods and J. U. Bowie, *Nature*, 2008, **453**, 1266–1270.

257 N. H. Joh, A. Oberai, D. Yang, J. P. Whitelegge and J. U. Bowie, *J. Am. Chem. Soc.*, 2009, **131**, 10846–10847.

258 C. D. Tatko, V. Nanda, J. D. Lear and W. F. Degrado, *J. Am. Chem. Soc.*, 2006, **128**, 4170–4171.

259 J. Nordholm, D. V. da Silva, J. Damjanovic, D. Dou and R. Daniels, *J. Biol. Chem.*, 2013, **288**, 10652–10660.

260 J. D. Lear, H. Gratkowski, L. Adamian, J. Liang and W. F. DeGrado, *Biochemistry*, 2003, **42**, 6400–6407.

261 R. Li, N. Mitra, H. Gratkowski, G. Vilaire, R. Litvinov, C. Nagasami, J. W. Weisel, J. D. Lear, W. F. DeGrado and J. S. Bennett, *Science*, 2003, **300**, 795–798.

262 W. Li, D. G. Metcalf, R. Gorelik, R. Li, N. Mitra, V. Nanda, P. B. Law, J. D. Lear, W. F. Degrado and J. S. Bennett, *Proc. Natl. Acad. Sci. U. S. A.*, 2005, **102**, 1424–1429.

263 M. E. Call, J. Pyrdol, M. Wiedmann and K. W. Wucherpfennig, *Cell*, 2002, **111**, 967–979.

264 M. E. Call and K. W. Wucherpfennig, *Nat. Rev. Immunol.*, 2007, **7**, 841–850.

265 L. L. Lanier, *Immunol. Rev.*, 2009, **227**, 150–160.

266 J. Feng, D. Garrity, M. E. Call, H. Moffett and K. W. Wucherpfennig, *Immunity*, 2005, **22**, 427–438.

267 M. E. Call, K. W. Wucherpfennig and J. J. Chou, *Nat. Immunol.*, 2010, **11**, 1023–1029.

268 A. Blazquez-Moreno, S. Park, W. Im, M. J. Call, M. E. Call and H. T. Reyburn, *Proc. Natl. Acad. Sci. U. S. A.*, 2017, **114**, E5645–E5654.

269 M. W. Adams, F. E. Jenney Jr, M. D. Clay and M. K. Johnson, *JBIC, J. Biol. Inorg. Chem.*, 2002, **7**, 647–652.

270 M. E. Call, J. R. Schnell, C. Xu, R. A. Lutz, J. J. Chou and K. W. Wucherpfennig, *Cell*, 2006, **127**, 355–368.

271 L. K. Fong, M. J. Chalkley, S. K. Tan, M. Grabe and W. F. DeGrado, *Proc. Natl. Acad. Sci. U.S.A.*, 2021, 118.

272 M. Eilers, S. C. Shekar, T. Shieh, S. O. Smith and P. J. Fleming, *Proc. Natl. Acad. Sci. U. S. A.*, 2000, **97**, 5796–5801.

273 L. Adamian and J. Liang, *J. Mol. Biol.*, 2001, **311**, 891–907.

274 D. Langosch and J. Heringa, *Proteins: Struct., Funct., Genet.*, 1998, **31**, 150–159.

275 F. Cymer, G. von Heijne and S. H. White, *J. Mol. Biol.*, 2015, **427**, 999–1022.

276 R. M. Johnson, K. Hecht and C. M. Deber, *Biochemistry*, 2007, **46**, 9208–9214.

277 R. Gurezka and D. Langosch, *J. Biol. Chem.*, 2001, **276**, 45580–45587.

278 M. Mravic, H. Hu, Z. Lu, J. S. Bennett, C. R. Sanders, A. W. Orr and W. F. DeGrado, *Protein Eng., Des. Sel.*, 2018, **31**, 181–190.

279 J. Xie and D. DiMaio, *FEBS J.*, 2021, DOI: 10.1111/febs.15775.

280 K. Talbert-Slagle and D. DiMaio, *Virology*, 2009, **384**, 345–351.

281 D. DiMaio and L. M. Petti, *Virology*, 2013, **445**, 99–114.

282 L. L. Freeman-Cook, A. M. Dixon, J. B. Frank, Y. Xia, L. Ely, M. Gerstein, D. M. Engelman and D. DiMaio, *J. Mol. Biol.*, 2004, **338**, 907–920.

283 S. A. Marlatt, Y. Kong, T. J. Cammett, G. Korbel, J. P. Noonan and D. Dimaio, *Protein Eng., Des. Sel.*, 2011, **24**, 311–320.

284 K. Talbert-Slagle, S. Marlatt, F. N. Barrera, E. Khurana, J. Oates, M. Gerstein, D. M. Engelman, A. M. Dixon and D. Dimaio, *J. Virol.*, 2009, **83**, 9773–9785.

285 J. B. Ptacek, A. P. Edwards, L. L. Freeman-Cook and D. DiMaio, *Proc. Natl. Acad. Sci. U. S. A.*, 2007, **104**, 11945–11950.

286 T. J. Cammett, S. J. Jun, E. B. Cohen, F. N. Barrera, D. M. Engelman and D. Dimaio, *Proc. Natl. Acad. Sci. U. S. A.*, 2010, **107**, 3447–3452.

287 L. M. Petti, K. Talbert-Slagle, M. L. Hochstrasser and D. DiMaio, *J. Biol. Chem.*, 2013, **288**, 27273–27286.

288 E. N. Heim, J. L. Marston, R. S. Federman, A. P. Edwards, A. G. Karabadzhak, L. M. Petti, D. M. Engelman and D. DiMaio, *Proc. Natl. Acad. Sci. U. S. A.*, 2015, **112**, E4717–E4725.

289 L. He, E. B. Cohen, A. P. B. Edwards, J. Xavier-Ferrucio, K. Bugge, R. S. Federman, D. Absher, R. M. Myers, B. B. Kragelund, D. S. Krause and D. DiMaio, *iScience*, 2019, **17**, 167–181.

290 R. Guo, K. Gaffney, Z. Yang, M. Kim, S. Sungsuwan, X. Huang, W. L. Hubbell and H. Hong, *Nat. Chem. Biol.*, 2016, **12**, 353–360.

291 E. S. Sulistijo and K. R. Mackenzie, *Biochemistry*, 2009, **48**, 5106–5120.

292 C. M. Lawrie, E. S. Sulistijo and K. R. MacKenzie, *J. Mol. Biol.*, 2010, **396**, 924–936.

293 C. M. Deber, L. P. Liu and C. Wang, *J. Pept. Res.*, 1999, **54**, 200–205.

294 R. Acharya, V. Carnevale, G. Fiorin, B. G. Levine, A. L. Polishchuk, V. Balannik, I. Samish, R. A. Lamb, L. H. Pinto, W. F. DeGrado and M. L. Klein, *Proc. Natl. Acad. Sci. U. S. A.*, 2010, **107**, 15075–15080.

295 K. P. Howard, J. D. Lear and W. F. DeGrado, *Proc. Natl. Acad. Sci. U. S. A.*, 2002, **99**, 8568–8572.

296 J. A. Mould, H.-C. Li, C. S. Dudlak, J. D. Lear, A. Pekosz, R. A. Lamb and L. H. Pinto, *J. Biol. Chem.*, 2000, **275**, 8592–8599.

297 K. M. Gernert, M. C. Surles, T. H. Labean, J. S. Richardson and D. C. Richardson, *Protein Sci.*, 1995, **4**, 2252–2260.

298 B. North, C. M. Summa, G. Ghirlanda and W. F. DeGrado, *J. Mol. Biol.*, 2001, **311**, 1081–1090.

299 N. H. Joh, T. Wang, M. P. Bhate, R. Acharya, Y. Wu, M. Grabe, M. Hong, G. Grigoryan and W. F. DeGrado, *Science*, 2014, **346**, 1520–1524.

300 D. S. Cafiso, *Annu. Rev. Biophys. Biomol. Struct.*, 1994, **23**, 141–165.

301 M. J. Kim, S. H. Park, S. J. Opella, T. H. Marsilje, P. Y. Michellys, H. M. Seidel and S. S. Tian, *J. Biol. Chem.*, 2007, **282**, 14253–14261.

302 X. Zeng, P. Wu, C. Yao, J. Liang, S. Zhang and H. Yin, *Biochemistry*, 2017, **56**, 2076–2085.

303 H. Furthmayr and V. T. Marchesi, *Biochemistry*, 1976, **15**, 1137–1144.

304 B.-J. Bormann, W. J. Knowles and V. T. Marchesi, *J. Biol. Chem.*, 1989, **264**, 4033–4037.

305 J. M. Westerfield, A. R. Sahoo, D. S. Alves, B. Grau, A. Cameron, M. Maxwell, J. A. Schuster, P. C. T. Souza, I. Mingarro, M. Buck and F. N. Barrera, *J. Mol. Biol.*, 2021, **433**, 167144.

306 K. M. Stefanski, C. M. Russell, J. M. Westerfield, R. Lamichhane and F. N. Barrera, *J. Biol. Chem.*, 2021, **296**, 100149.

307 D. S. Alves, J. M. Westerfield, X. Shi, V. P. Nguyen, K. M. Stefanski, K. R. Booth, S. Kim, J. Morrell-Falvey, B. C. Wang, S. M. Abel, A. W. Smith and F. N. Barrera, *eLife*, 2018, **7**.

308 H. Yin, J. S. Slusky, B. W. Berger, R. S. Walters, G. Vilaire, R. I. Litvinov, J. D. Lear, G. A. Caputo, J. S. Bennett and W. F. DeGrado, *Science*, 2007, **315**, 1817–1822.

309 G. A. Caputo, R. I. Litvinov, W. Li, J. S. Bennett, W. F. DeGrado and H. Yin, *Biochemistry*, 2008, **47**, 8600–8606.

310 S. J. Shandler, I. V. Korendovych, D. T. Moore, K. B. Smith-Dupont, C. N. Streu, R. I. Litvinov, P. C. Billings, F. Gai, J. S. Bennett and W. F. DeGrado, *J. Am. Chem. Soc.*, 2011, **133**, 12378–12381.

311 K. P. Fong, H. Zhu, L. M. Span, D. T. Moore, K. Yoon, R. Tamura, H. Yin, W. F. DeGrado and J. S. Bennett, *J. Biol. Chem.*, 2016, **291**, 11706–11716.

312 B. Hille, *Ionic Channels of Excitable Membranes*, Sinauer Associates, Sunderland, MA, 3rd edn, 2001.

313 W. F. DeGrado, Z. R. Wasserman and J. D. Lear, *Science*, 1989, **243**, 622–628.

314 K. S. Åkerfeldt, J. D. Lear, Z. R. Waserman, L. A. Chung and W. F. DeGrado, *Acc. Chem. Res.*, 1993, **26**, 191–197.

315 T. K. Nguyen and T. Ueno, *Curr. Opin. Struct. Biol.*, 2018, **51**, 1–8.

316 J. D. Lear, Z. R. Wasserman and W. F. DeGrado, *Science*, 1988, **240**, 1177–1181.

317 C. Xu, P. Lu, T. M. Gamal El-Din, X. Y. Pei, M. C. Johnson, A. Uyeda, M. J. Bick, Q. Xu, D. Jiang, H. Bai, G. Reggiano, Y. Hsia, T. J. Brunette, J. Dou, D. Ma, E. M. Lynch, S. E. Boyken, P. S. Huang, L. Stewart, F. DiMaio, J. M. Kollman, B. F. Luisi, T. Matsuura, W. A. Catterall and D. Baker, *Nature*, 2020, **585**, 129–134.

318 N. H. Joh, G. Grigoryan, Y. Wu and W. F. DeGrado, *Philos. Trans. R. Soc. Lond. B Biol. Sci.*, 2017, 372.

319 W. Veatch and L. Stryer, *J. Mol. Biol.*, 1977, **113**, 89–102.

320 D. W. Urry, M. C. Goodall, J. D. Glickson and D. F. Mayers, *Proc. Natl. Acad. Sci. U. S. A.*, 1971, **68**, 1907–1911.

321 S. B. Hladky and D. A. Haydon, *Biochim. Biophys. Acta, Biomembr.*, 1972, **274**, 294.

322 R. R. Ketchem, W. Hu and T. A. Cross, *Science*, 1993, **261**, 1457–1460.

323 V. Berl, I. Huc, R. G. Khoury, M. J. Krische and J. M. Lehn, *Nature*, 2000, **407**, 720–723.

324 B. Roux and M. Karplus, *Annu. Rev. Biophys. Biomol. Struct.*, 1994, **23**, 731–761.

325 B. Roux and M. Karplus, *J. Phys. Chem.*, 1991, **95**, 4856–4868.

326 B. Roux and M. Karplus, *Biophys. J.*, 1988, **53**, 297–309.

327 M. C. Goodall and D. W. Urry, *Biochim. Biophys. Acta, Biomembr.*, 1973, **291**, 317–320.

328 S. J. Kennedy, R. W. Roeske, A. R. Freeman, A. M. Watanabe and H. R. Besche Jr, *Science*, 1977, **196**, 1341–1342.

329 M. R. Ghadiri, J. R. Granja, R. A. Milligan, D. E. McRee and N. Khazanovich, *Nature*, 1993, **366**, 324–327.

330 H. Vogel, *Biochemistry*, 1987, **26**, 4562–4572.

331 M. T. Lee, F. Y. Chen and H. W. Huang, *Biochemistry*, 2004, **43**, 3590–3599.

332 P. C. Biggin and M. S. Sansom, *Biophys. Chem.*, 1999, **76**, 161–183.

333 G. Boheim, *J. Membr. Biol.*, 1974, **19**, 277–303.

334 R. Nagaraj and P. Balaram, *Acc. Chem. Res.*, 1981, **14**, 356–362.

335 R. O. Fox and F. M. Richards, *Nature*, 1982, **300**, 325–330.

336 J. E. Hall, I. Vodyanoy, T. M. Balasubramanian and G. R. Marshall, *Biophys. J.*, 1984, **45**, 233–247.

337 G. A. Woolley and B. A. Wallace, *J. Membr. Biol.*, 1992, **129**, 109–136.

338 R. S. Cantor, *Biophys. J.*, 2002, **82**, 2520–2525.

339 D. P. Tieleman, J. Breed, H. J. Berendsen and M. S. Sansom, *Faraday Discuss.*, 1998, **111**, 209–223.

340 P. Pieta, J. Mirza and J. Lipkowski, *Proc. Natl. Acad. Sci. U. S. A.*, 2012, **109**, 21223–21237.

341 D. DiMaio, *Annu. Rev. Microbiol.*, 2014, **68**, 21–43.

342 C. Scott and S. Griffin, *J. Gen. Virol.*, 2015, **96**, 2000–2027.

343 J. To, W. Surya and J. Torres, *Adv. Protein Chem. Struct. Biol.*, 2016, **104**, 307–355.

344 D. Schoeman and B. C. Fielding, *Virol. J.*, 2019, **16**, 69.

345 N. S. Farag, U. Breitinger, H. G. Breitinger and M. A. El Azizi, *Int. J. Biochem. Cell Biol.*, 2020, **122**, 105738.

346 D. Schoeman and B. C. Fielding, *Front. Microbiol.*, 2020, **11**, 2086.

347 A. J. Hay, A. J. Wolstenholme, J. J. Skehel and M. H. Smith, *EMBO J.*, 1985, **4**, 3021–3024.

348 A. J. Hay, M. C. Zambon, A. J. Wolstenholme, J. J. Skehel and M. H. Smith, *J. Antimicrob. Chemother.*, 1986, **18**(Supplement_B), 19–29.

349 F. Ciampor, C. A. Thompson, S. Grambas and A. J. Hay, *Virus Res.*, 1992, **22**, 247–258.

350 V. S. Mandala, M. J. McKay, A. A. Shcherbakov, A. J. Dregni, A. Kolocouris and M. Hong, *Nat. Struct. Mol. Biol.*, 2020, **27**, 1202–1208.

351 C. Ma, A. L. Polishchuk, Y. Ohigashi, A. L. Stouffer, A. Schon, E. Magavern, X. Jing, J. D. Lear, E. Freire, R. A. Lamb, W. F. DeGrado and L. H. Pinto, *Proc. Natl. Acad. Sci. U. S. A.*, 2009, **106**, 12283–12288.

352 J. L. Thomaston, M. Alfonso-Prieto, R. A. Woldeyes, J. S. Fraser, M. L. Klein, G. Fiorin and W. F. DeGrado, *Proc. Natl. Acad. Sci. U. S. A.*, 2015, **112**, 14260–14265.

353 J. L. Thomaston, R. A. Woldeyes, T. Nakane, A. Yamashita, T. Tanaka, K. Koiwai, A. S. Brewster, B. A. Barad, Y. Chen, T. Lemmin, M. Uervirojnangkoorn, T. Arima, J. Kobayashi, T. Masuda, M. Suzuki, M. Sugahara, N. K. Sauter, R. Tanaka, O. Nureki, K. Tono, Y. Joti, E. Nango, S. Iwata, F. Yumoto, J. S. Fraser and W. F. DeGrado, *Proc. Natl. Acad. Sci. U. S. A.*, 2017, **114**, 13357–13362.

354 J. R. Schnell and J. J. Chou, *Nature*, 2008, **451**, 591–595.

355 S. D. Cady, K. Schmidt-Rohr, J. Wang, C. S. Soto, W. F. DeGrado and M. Hong, *Nature*, 2010, **463**, 689–692.

356 M. Sharma, M. Yi, H. Dong, H. Qin, E. Peterson, D. D. Busath, H. X. Zhou and T. A. Cross, *Science*, 2010, **330**, 509–512.

357 R. Liang, J. M. Swanson, J. J. Madsen, M. Hong, W. F. DeGrado and G. A. Voth, *Proc. Natl. Acad. Sci. U. S. A.*, 2016, **113**, 6955–6964.

358 J. K. Williams, A. A. Shcherbakov, J. Wang and M. Hong, *J. Biol. Chem.*, 2017, **292**, 17876–17884.

359 R. Fu, Y. Miao, H. Qin and T. A. Cross, *J. Am. Chem. Soc.*, 2020, **142**, 2115–2119.

360 R. Fu, Y. Miao, H. Qin and T. A. Cross, *J. Am. Chem. Soc.*, 2016, **138**, 15801–15804.

361 A. L. Stouffer, R. Acharya, D. Salom, A. S. Levine, L. Di Costanzo, C. S. Soto, V. Tereshko, V. Nanda, S. Stayrook and W. F. DeGrado, *Nature*, 2008, **451**, 596–599.

362 J. L. Thomaston, M. L. Samways, A. Konstantinidi, C. Ma, Y. Hu, H. E. Bruce Macdonald, J. Wang, J. W. Essex, W. F. DeGrado and A. Kolocouris, *Biochemistry*, 2021, **60**, 2471–2482.

363 J. L. Thomaston, A. Konstantinidi, L. Liu, G. Lambrinidis, J. Tan, M. Caffrey, J. Wang, W. F. Degrado and A. Kolocouris, *Biochemistry*, 2020, **59**, 627–634.

364 J. L. Thomaston, Y. Wu, N. Polizzi, L. Liu, J. Wang and W. F. DeGrado, *J. Am. Chem. Soc.*, 2019, **141**, 11481–11488.

365 J. L. Thomaston and W. F. DeGrado, *Protein Sci.*, 2016, **25**, 1551–1554.

366 M. Hong and W. F. DeGrado, *Protein Sci.*, 2012, **21**, 1620–1633.

367 L. C. Watkins, W. F. DeGrado and G. A. Voth, *J. Am. Chem. Soc.*, 2020, **142**, 17425–17433.

368 L. C. Watkins, R. Liang, J. M. J. Swanson, W. F. DeGrado and G. A. Voth, *J. Am. Chem. Soc.*, 2019, **141**, 11667–11676.

369 R. Liang, H. Li, J. M. Swanson and G. A. Voth, *Proc. Natl. Acad. Sci. U. S. A.*, 2014, **111**, 9396–9401.

370 C. Ma, G. Fiorin, V. Carnevale, J. Wang, R. A. Lamb, M. L. Klein, Y. Wu, L. H. Pinto and W. F. DeGrado, *Structure*, 2013, **21**, 2033–2041.

371 T. Ivanovic, R. Rozendaal, D. L. Floyd, M. Popovic, A. M. van Oijen and S. C. Harrison, *PLoS One*, 2012, **7**, e31566.

372 F. Hu, K. Schmidt-Rohr and M. Hong, *J. Am. Chem. Soc.*, 2012, **134**, 3703–3713.

373 E. Khurana, M. Dal Peraro, R. DeVane, S. Vemparala, W. F. DeGrado and M. L. Klein, *Proc. Natl. Acad. Sci. U. S. A.*, 2009, **106**, 1069–1074.

374 D. D. Busath, in *Advances in Planar Lipid Bilayers and Liposomes*, ed. A. Leitmannova Liu and A. Iglič, Academic Press, Burlington, 2009, vol. 10, pp. 161–201.

375 A. L. Stouffer, C. Ma, L. Cristian, Y. Ohigashi, R. A. Lamb, J. D. Lear, L. H. Pinto and W. F. DeGrado, *Structure*, 2008, **16**, 1067–1076.

376 A. L. Stouffer, V. Nanda, J. D. Lear and W. F. DeGrado, *J. Mol. Biol.*, 2005, **347**, 169–179.

377 J. L. Thomaston, N. F. Polizzi, A. Konstantinidi, J. Wang, A. Kolocouris and W. F. DeGrado, *J. Am. Chem. Soc.*, 2018, **140**, 15219–15226.

378 J. Wang, Y. Wu, C. Ma, G. Fiorin, J. Wang, L. H. Pinto, R. A. Lamb, M. L. Klein and W. F. DeGrado, *Proc. Natl. Acad. Sci. U. S. A.*, 2013, **110**, 1315–1320.

379 J. S. Rossman, X. Jing, G. P. Leser and R. A. Lamb, *Cell*, 2010, **142**, 902–913.

380 R. M. Pielak, K. Oxenoid and J. J. Chou, *Structure*, 2011, **19**, 1655–1663.

381 K. Åkerfeldt, R. M. Kim, D. Camac, J. T. Groves, J. D. Lear and W. F. DeGrado, *J. Am. Chem. Soc.*, 1992, **114**, 9656–9657.

382 G. R. Dieckmann, J. D. Lear, Q. Zhong, M. L. Klein, W. F. DeGrado and K. A. Sharp, *Biophys. J.*, 1999, **76**, 618–630.

383 T. H. Nguyen, Z. Liu and P. B. Moore, *Biophys. J.*, 2013, **105**, 1569–1580.

384 Q. Zhong, P. B. Moore, D. M. Newns and M. L. Klein, *FEBS Lett.*, 1998, **427**, 267–270.

385 A. J. Scott, A. Niitsu, H. T. Kratochvil, E. J. M. Lang, J. T. Sengel, W. M. Dawson, K. R. Mahendran, M. Mravic, A. R. Thomson, R. L. Brady, L. Liu, A. J. Mulholland, H. Bayley, W. F. DeGrado, M. I. Wallace and D. N. Woolfson, *Nat. Chem.*, 2021, **13**, 643–650.

386 K. R. Mahendran, A. Niitsu, L. Kong, A. R. Thomson, R. B. Sessions, D. N. Woolfson and H. Bayley, *Nat. Chem.*, 2017, **9**, 411–419.

387 J. L. Beesley and D. N. Woolfson, *Curr. Opin. Biotechnol.*, 2019, **58**, 175–182.

388 K. Kuroda, G. A. Caputo and W. F. DeGrado, *Chem. – Eur. J.*, 2009, **15**, 1123–1133.

389 B. P. Mowery, S. E. Lee, D. A. Kissounko, R. F. Epand, R. M. Epand, B. Weisblum, S. S. Stahl and S. H. Gellman, *J. Am. Chem. Soc.*, 2007, **129**, 15474–15476.

390 B. P. Mowery, A. H. Lindner, B. Weisblum, S. S. Stahl and S. H. Gellman, *J. Am. Chem. Soc.*, 2009, **131**, 9735–9745.

391 A. Munoz-Bonilla and M. Fernandez-Garcia, *Prog. Polym. Sci.*, 2012, **37**, 281–339.

392 D. Campoccia, L. Montanaro and C. R. Arciola, *Biomaterials*, 2013, **34**, 8533–8554.

393 C. H. B. Cruz, I. Marzuoli and F. Fraternali, *Faraday Discuss.*, 2021, DOI: 10.1039/d1fd00041a.

394 I. Marzuoli, C. H. B. Cruz, C. D. Lorenz and F. Fraternali, *Nanoscale*, 2021, **13**, 10342–10355.

395 V. Castelletto, E. de Santis, H. Alkassem, B. Lamarre, J. E. Noble, S. Ray, A. Bella, J. R. Burns, B. W. Hoogenboom and M. G. Ryadnov, *Chem. Sci.*, 2016, **7**, 1707–1711.

Faraday Discussions

PAPER

Molecular interactions of the M and E integral membrane proteins of SARS-CoV-2

Viviana Monje-Galvan ⓘ and Gregory A. Voth ⓘ *

Received 19th April 2021, Accepted 30th June 2021

DOI: 10.1039/d1fd00031d

Specific lipid–protein interactions are key for cellular processes, and even more so for the replication of pathogens. The COVID-19 pandemic has drastically changed our lives and caused the death of nearly four million people worldwide, as of this writing. SARS-CoV-2 is the virus that causes the disease and has been at the center of scientific research over the past year. Most of the research on the virus is focused on key players during its initial attack and entry into the cellular host; namely the S protein, its glycan shield, and its interactions with the ACE2 receptors of human cells. As cases continue to rise around the globe, and new mutants are identified, there is an urgent need to understand the mechanisms of this virus during different stages of its life cycle. Here, we consider two integral membrane proteins of SARS-CoV-2 known to be important for viral assembly and infectivity. We have used microsecond-long all-atom molecular dynamics to examine the lipid–protein and protein–protein interactions of the membrane (M) and envelope (E) structural proteins of SARS-CoV-2 in a complex membrane model. We contrast the two proposed protein complexes for each of these proteins, and quantify their effect on their local lipid environment. This ongoing work also aims to provide molecular-level understanding of the mechanisms of action of this virus to possibly aid in the design of novel treatments.

Introduction

Since the beginning of the COVID-19 pandemic, research progress has been unprecedented to elucidate and block the progression of SARS-CoV-2 infection, the virus that causes the disease. Scientists have been collaborating as never before in open platforms, sharing data as soon as available to aid in the fight against the virus. Though other coronaviruses have been around for several years,[1-3] the rate of transmission, infectivity, and unpredictability of SARS-CoV-2 make for an urgent need to understand its mechanisms at the molecular level.

Department of Chemistry, Chicago Center for Theoretical Chemistry, Institute for Biophysical Dynamics, and The James Franck Institute, The University of Chicago, Chicago, Illinois, 60637, USA. E-mail: gavoth@uchicago.edu

Several companies have provided viable vaccine candidates to the global market to combat the spread of the virus; yet, novel mutants and the rate of transmission clearly show the need for more and improved alternatives.

Viruses have developed mechanisms to use host cell proteins and lipids to propagate infection.[4,5] Understanding these interactions at the molecular level – and how they modulate the structure, aggregation, and function of viral proteins – will provide new insights about the mechanisms of viral replication. The design of successful novel therapeutic approaches against SARS-CoV-2 infection depends largely on the knowledge of such mechanisms and identifying the weak points of the viral life cycle.

Like other coronaviruses, SARS-CoV-2 has four structural envelope proteins: the spike (S), nucleocapsid (N), envelope (E), and membrane or matrix (M).[1] Of these, S, E, and M have transmembrane regions that are embedded in the lipid envelope that encloses the viral genome, a positive RNA strand. The N protein mainly interacts with the RNA itself, recruiting it to viral assembly sites and interacting with membrane lipids and other structural proteins to incorporate its cargo into new viral particles.[6,7] S is the key player during viral infection, inter-acting with ACE2 receptors on the surface of cells targeted by the virus and undergoing several conformational changes and proteolytic cleavages that enable viral fusion and release of the viral genome into the cellular host.[8] M has been identified as a main orchestrator of viral assembly, interacting with all the other structural proteins at some point to recruit them and stabilize the formation of new viral particles at the endoplasmic reticulum–Golgi intermediate compart-ment (ERGIC).[8–10] The E protein is classified as a viroporin,[11] working as an ion channel for the viral particle and blocking the immune response from the host during viral replication.[2,10,12,13] It is still unclear how the E protein is involved in the assembly process and/or scission of newly formed viral particles.[1,14]

Very few studies are currently available on the role and dynamics of the M and E integral membrane proteins of SARS-CoV-2.[11,15–17] These are both known to be key for viral assembly and infectivity,[9,16] and the M protein could potentially serve as an antigen.[18] Latest experimental studies examining the interactions among structural proteins of the virus identified that M and E are both required to recruit and retain S proteins at the viral assembly site.[17] Yet, their precise mechanisms and relevance in the viral life cycle remain largely unclear. By far, the main focus of current research is towards understanding and blocking the interaction of the S proteins on the surface of the virus. Most of the treatments consist of repurposing existing drugs to block different stages of the viral cycle to stop viral replication, and primarily to block the S-ACE2 initial interaction.[19] Current vaccine candidates also work by eliciting an immune response in the human body such that it produces antibodies that will recognize and bind to the S protein in case of an infection.[18,20] Despite great advances against the virus, we are still in urgent need to gain better understanding of the viral life cycle in its entirety to inspire and repurpose drug therapies more efficiently as the virus mutates.

The relevance and involvement of lipids in cellular and viral processes has been well established over the past two decades.[4,5,21] In this work, we take a closer look at the interplay between lipids and the M and E proteins of SARS-CoV-2 at early stages of viral assembly. We present the results of long all-atom molecular dynamics (MD) simulations to study molecular interactions in a complex membrane model, built to mimic the cell organelle where viral assembly takes place, the ERGIC.[3,22,23] We discuss protein–protein interactions in detail for a proposed M dimer as well as the E channel formed by five monomers. At the time of this study, only structures from homology modeling studies were publicly available at the MolSSi Covid-19 Hub.[24,25] To add to this emerging body of knowledge, we therefore simulated two proposed models for each protein, shown in Fig. 1, and characterized the differences in local membrane response based on protein conformation.

As this manuscript was being prepared, two experimental studies were published on the structure and function of the transmembrane region of the E channel,[11,15] which we comment on in the discussion section. To our knowledge, there are still no experimental structures for the M protein of SARS-CoV-2 that explicitly examine the macromolecular organization of either of these proteins during viral assembly or in formed viral particles. In light of the most recent

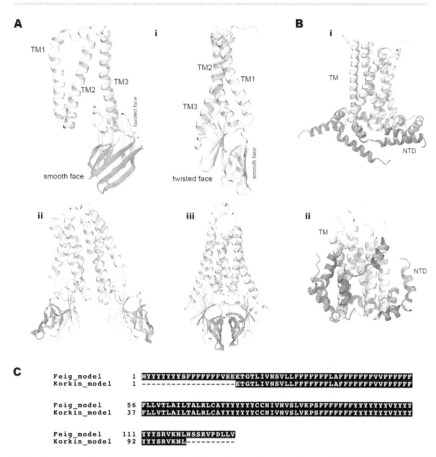

Fig. 1 Proposed homology models for integral membrane proteins of SARS-CoV-2. (A) M dimers (Feig lab[24]); (i) protein regions of an M monomer, three transmembrane helices (TM) and two distinct faces of the C-terminal domain (CTD), which we use to differentiate the (ii) open and (iii) closed conformations. (B) E protein channel from the (i) Feig lab[24] and (ii) Korkin lab;[25] the transmembrane (TM) helices are shown in grey and the C-terminus helices (CTD) in blue. (C) Sequence alignment between the Feig and Korkin proposed models for the E channel.

studies that clearly identify the high involvement of lipids in the pathogenesis and mechanisms of the virus,[26,27] we hope this work lays the groundwork to further examine the complex dynamics between viral proteins and lipids. Specifically, the roles of M and E – both integral membrane proteins of SARS-CoV-2 – during viral assembly, membrane deformation, virus maturation, and reshaping of the viral particle during binding and fusion to a new host. This knowledge is also critical for future advances in treatment and preventive care against Covid-19.

Methods

There are currently no lipidomic studies on the SARS-CoV-2 virion; we chose to model the endoplasmic reticulum (ER) as our starting point in the assembly pathway of the virus to capture the natural lipid sorting and local rearrangement caused by the proteins. Our membrane model has five representative lipid species for the ER of eukaryotic cells.[28-30] Incidentally, our lipid composition is nearly identical to that of a model bilayer used in an experimental study carried out to elucidate the structure and conformation of the transmembrane region of the E channel of SARS-CoV-2 in an ERGIC-like environment.[11] That study was published at the time when this manuscript was under preparation. Both the experimental and the simulation membrane models have a mixture of PC, PE, PI, PS, and cholesterol lipids. A very similar composition was also used to model the ER/ERGIC in another computational study published during our data curation. That study is a thorough examination and validation of homology models of the E-channel and its ion channeling function;[14] their model includes a small percentage of sphingomyelin lipids that our model does not include. Our bilayer model contains DYPC/DYPE/POPI/POPS/Chol in 45 : 10 : 12 : 23 : 10 molar ratio, with 610 lipids per leaflet over a 19 nm × 19 nm surface; DYPC: 3-palmitoleoyl-2-palmitoleoyl-D-*glycero*-1-phosphatidylcholine; DYPE: 3-palmitoleoyl-2-palmito-leoyl-D-*glycero*-1-phosphatidylethanolamine; POPI: 1-palmitoyl-2-oleoyl-inositol; POPS: 3-palmitoyl-2-oleoyl-D-*glycero*-1-phosphatidylserine; Chol: cholesterol. This composition was chosen to model the membrane thickness and unsaturation degree of the hydrophobic core in terms of number of double bonds per lipid tail. To our knowledge, this model is the most complex yet to be used in the study of SARS-CoV-2 transmembrane proteins and their dynamics with membrane lipids.

For the M and E proteins, we used structures proposed by the Feig lab from homology modeling and refined computational studies;[24] the second model for the E protein is based on a known structure for the E channel of SARS, and was proposed by the Korkin lab.[25] Given the degree of conservation between SARS and SARS-CoV-2,[14] the homology-derived structures used in this study provide a robust preliminary overview of the interplay between integral membrane proteins of the virus and their local lipid environment. The Feig lab proposed two possible conformations for an M dimer; we differentiate them as the open and closed conformations as shown in Fig. 1, depending on the relative orientation of the monomers. The C-terminal domain (CTD) of the proteins points towards the luminal side of the ER or ERGIC, and is located in the interior of formed viral particles or virions. The same group proposed a channel structure formed by five units of the E protein, with the N-terminus domains (NTD) located at the

Table 1 Simulation systems

System	Environment	System size (atoms)	Simulation length (ns)	Comp. resource
Open M dimer	Water	137 282	400	Frontera (TACC)
	Bilayer (610 lipids/leaflet)	616 117	4000	Frontera & Anton2 (PSC)
Closed M dimer	Water	190 643	400	Midway2 (RCC)
	Bilayer (610 lipids/leaflet)	667 837	4000	Frontera & Anton2
E channel – Feig	Bilayer (610 lipids/leaflet)	670 765	4000	Midway2 & Anton2
E channel – Korkin	Bilayer (610 lipids/leaflet)	551 741	4000	Midway2 & Anton2

membrane interface (refer to Fig. 1). The main difference between that model and the one proposed by the Korkin lab is the location of the NTD; the second structure has these helices inside the bilayer core. As we discuss later in this work, the position of the N-terminus helices results in marked differences in the membrane response. Table 1 summarizes our systems of study; the actual protein coordinates were obtained from the MolSSi Covid-19 Hub (https://covid.molssi.org).

All of our systems were built on CHARMM-GUI *Membrane Builder*, or *Quick-Solvator* for the M dimer in water.[31–35] Fully hydrated bilayers (40+ water molecules per lipid) were built around the M and E protein complexes, respectively, centering the transmembrane regions of the proteins as closely as possible to the bilayer center. We used KCl salt at 0.15 M concentration to neutralize the systems to run on the isobaric–isothermal ensemble on GROMACS simulation package,[36] with the CHARMM36m force field.[37] The protein–water systems were run on Frontera at the TACC, or Midway2 at the RCC of the University of Chicago, as listed in Table 1. Upon initial equilibration on these resources, the production runs of the protein–membrane systems were carried on the Anton 2 machine,[38] hosted at the Pittsburgh Supercomputing Center (PSC) (see Acknowledgments).

Initial minimizations of all systems were carried out following the 6-step protocol provided on CHARMM-GUI.[39] Followed by a 16-step restrained protocol over 123 ns to further equilibrate the E channels, to ensure proper relaxation of the lipids around it and prevent it from closing. This setup was adapted from previous studies of the influenza A M2 channel.[40,41] Initially, the alpha carbons of the transmembrane helices (TM) of the E channel were restrained and only the N-terminus helices, denoted as alpha-helices (AH) in this work, were allowed to move, followed by the reverse setup and final removal of all restraints prior to starting the production run of these systems. No water molecules were found in the bilayer region during the restrained protocol.

In all our trajectories we used a simulation timestep of 2 fs and periodic boundary conditions. The temperature was set to 310.15 K and controlled with the Nose–Hoover thermostat[42,43] with a coupling time constant of 1.0 ps in GROMACS. The pressure was set at 1 bar and controlled with the Berendsen barostat[44] during initial relaxation; the production runs were carried with the Parrinello–Rahman barostat semi-isotropically with a compressibility of 4.5×10^{-5} and a coupling time constant of 5.0 ps.[45,46] Non-bonded interactions were computed using a switching function between 1.0 and 1.2 nm, and Particle Mesh Ewald for long-range electrostatics.[47] The LINCS algorithm[48] was used to constrain hydrogen bonds in GROMACS.

Simulation parameters for the Anton2 machine were set by ark guesser files, which are automated internal scripts designed to optimize the parameters for the integration algorithms of this machine; as such, the cut-off values to compute interactions between neighboring atoms are set automatically during system preparation. Long-range electrostatics were computed using the Gaussian Split Ewald algorithm,[49] and hydrogen bonds constrained using the SHAKE algorithm.[50] Finally, the Nose–Hoover thermostat and MTK barostat were used to control the temperature and pressure during *NPT* dynamics on Anton2 using optimized parameters set for the *Multigrator* integrator of the machine.[51]

From the MD simulations, we characterized the interactions between protein units as well as with the lipids around them. For the M protein, we also simulated the dimers in water to examine the stability of the proposed structures and to quantify the effect of the bilayer on protein–protein interactions. We computed the root-mean-square-displacement (RMSD) per protein and root-mean-square-fluctuation (RMSF) per residue, and compared it across proposed structures for each protein complex, and *versus* the proteins in solution in the case of the M dimers. We examined the relative spatial conformations between the proposed models for the M and E complexes, and determined the lipid aggregation patterns around the proteins from micro-second-long simulation trajectories. We carried out one 4 μs long trajectory for each protein–membrane system, and 400 ns for the M dimers in water; our reported values are block-averages of the respective quantities. Finally, we compared the degree and extent of membrane deformation caused by each protein complex in terms of the relative position of the lipid headgroups in the cytosolic and luminal leaflets (top and bottom leaflets in our simulation setup). We further discuss below our data in the context of early viral assembly and outline potential directions of this work.

All the images included in this manuscript were rendered on Visual Molecular Dynamics (VMD) software package.[52] Internal GROMACS modules, and MDAnalysis[53,54] and MDTraj[55] python packages were used to carry out the trajectory analysis. Computational time for the trajectories in this work was generously provided by the COVID-19 HPC Consortium at the Frontera and Anton2 machines.

Results & discussion

From studies on other coronaviruses, it is known M and E are implicated in viral assembly, budding, infectivity, and evasion of the immune cell response.[12,13] However, little is known about their mechanisms of action, specially their effect on their surrounding lipid environment during viral assembly and in the viral particle itself. First, we report our insights from simulations of an open and a closed M dimer (as seen in Fig. 1). Then, we discuss the differences in molecular interactions and membrane response for two proposed models for the E channel. We present a detailed account on the molecular interactions of these integral membrane proteins and comment on plausible mechanisms of action for both proteins in the context of membrane reorganization during early stages of viral assembly of SARS-CoV-2.

The **M** protein is known as the orchestrator of SARS-CoV-2 viral assembly, preparing it for viral budding and scission. In this context, we examined the interactions between monomers in an M dimer in water and in a lipid environment. Both the open and closed conformations are stable and do not dissociate in water, as shown by the grey and blue structures in Fig. 2, and much less so in the bilayer. In fact, the closed conformation barely shifts position due to its tight configuration from the very beginning. The three TMs in the closed conformation form essentially a curved plane, thus have more surface to interact and stabilize the interaction between monomers in the dimer. Similarly, the CTDs of the monomers in this conformation interact very tightly through their smooth faces, as designated in Fig. 1. On the other hand, in the open conformation the TMs and CTDs have more flexibility to move. Only TM2 and TM3 interact between monomers, and TM1 is relatively free to move and rotate. The CTDs are separated with their twisted faces pointing towards the center of the dimer, and the smooth faces point outwards and can easily interact with surrounding lipid headgroups.

Fig. 3C and D show the number of contacts between heavy atoms of each monomer. The dimers in the closed conformation have more contacts than the one in the open conformation, both in water and in a bilayer. As expected from its tighter structure, the TMs in the closed conformation do not fluctuate much in a bilayer environment, and the CTDs come closer together. In fact, the presence of the bilayer restricts the motions of the TMs in the open conformation. Fig. 2 shows the TMs in this conformation collapse into each other when in water, bending near the mid-region. This is also shown in the accompanying RMSF in

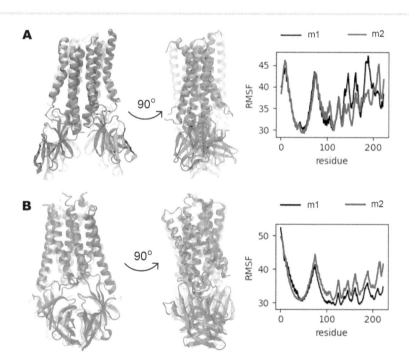

Fig. 2 Comparison of the initial (faded) and final (blue) structures of the (A) open and (B) closed conformations of an M dimer in water. Next to each structure is the RMSF computed for the 400 ns trajectory of each dimer in water.

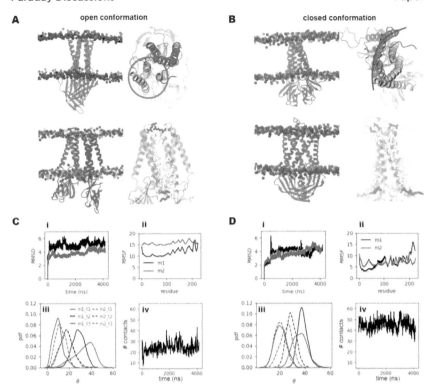

Fig. 3 Protein–protein interactions of M dimers in a bilayer. (A) Top and side view of the open dimer conformation highlighting the spatial arrangement of the transmembrane helices and the closest bulky non-polar residues between the protein units. Similarly, (B) the spatial conformation of the closed dimer in a bilayer and non-polar residues pointing towards the bilayer core. Quantitative differences between the (C) open and (D) closed dimers: (i) RMSD of each protein unit. (ii) RMSF per residue per protein unit. (iii) Probability distribution of the angle of each transmembrane helix with respect to the membrane normal. (iv) Number of contacts between M monomers in the respective conformation.

the figure, that shows the protein residues in TM1 and TM3 experience large shifts with respect to the initial structure. When the open dimer is embedded into a bilayer, the TMs retain a cylinder-like form throughout the simulation and the TMs of only one monomer experience a large shift from their initial position (see Fig. 3A and C). In the bilayer, the CTDs also fluctuate more in the open conformation *versus* the closed one, and the CTD from one monomer in the open conformation tilts toward the membrane surface as it interacts with the lipid headgroups.

A closer look at the TMs in both conformations inside the bilayer shows that bulky non-polar residues stabilize the interaction between TMs of each monomer in the open conformation, as shown in Fig. 3A, whereas these bulky residues point towards the bilayer core in the closed conformation. Looking at the protein motions throughout the simulation trajectory, most of the changes in the open conformation occurred within the first 100 ns of simulation (*cf.* Fig. 3C(i)), and the displacement of each monomer is not coupled to the movements of the other as shown by the RMSD over the course of the trajectory.

On the other hand, the displacement of the monomers in the closed conformation takes longer time and is more coordinated as shown in Fig. 3D(i). The conformational changes of the closed dimer occur within the first 1000 ns of trajectory, then again during the last microsecond. Finally, the orientation of the TMs in terms of their angle with respect to the bilayer normal shows TM1 and TM2 of each monomer align to each other in the closed conformation, and TM3 of one monomer aligns with TM3 from the opposite monomer. In the open conformation, TM1 and TM2 are not aligned in each individual monomer, but TM3 helices align between monomers as in the closed conformation. The angle of TM3 helices in the open conformation is smaller than in the closed conformation, *i.e.* these helices stand more vertical.

Examining a single M dimer already shows marked differences in the membrane response depending on the dimer conformation. Fig. 4 shows lipid density maps averaged over the last 50 ns of simulation along with a density map of the proteins in the respective conformation; darker regions indicate higher lipid content. There is more accumulation of cholesterol and inositol (PI) lipids around the open dimer, whereas phosphatidylserine (PS) accumulates near one of the monomers of the open conformation – notably, the monomer whose CTD shifts towards the bilayer. Interestingly, there is larger depletion of DYPE lipids around the open conformation. On the other hand, Fig. 4B shows PS lipids accumulate on both sides of the closed dimer, and DYPE localizes closely around the closed conformation. Charged amino acids at the extremes of the TMs of the closed conformation extend outward from the dimers interface and interact freely with charged lipid headgroups, shown in blue in Fig. 3B at the top and the base of each TM. The same amino acids in the open conformation remain mostly trapped between interacting TMs in the dimer (Fig. 3A), which can explain the more uneven distribution of charged lipids around the open conformation. Different patterns in lipid sorting also result in different degrees of membrane deformation around the dimers.

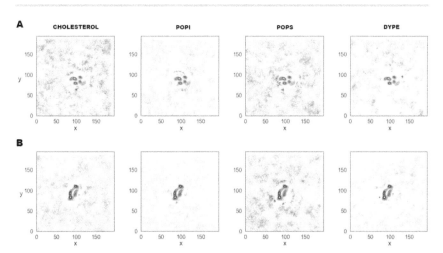

Fig. 4 Final lipid density maps relative to the location of the M dimers for the (A) open and (B) closed conformation, darker regions indicate higher density of the respective lipid species.

Fig. 5 shows colored maps of the relative height of each bilayer leaflet in terms of the position of the phosphate groups of each lipid. The red regions correspond to lower positions, and the blue sections are regions where the membrane surface is elevated. A color bar is included per leaflet to clearly distinguish between the lower and elevated regions of each membrane interface. Fig. 5A shows the open dimer causes greater deformation on the membrane plane, shifting upwards the lipids around it in both leaflets. From these maps, it is easier to observe the different volume the open dimer occupies across the bilayer. This dimer has a conical shape inside the hydrophobic core of the membrane, with the narrower region in the luminal leaflet and the wider region in the cytosolic leaflet, as observed in the white region at the center of each map. The CTDs of the open dimer are free and able to interact with the lipid headgroups of the luminal leaflet, and push them upwards upon binding. This provides insight into a possible mechanism to generate curvature that contributes to (i) stabilize the budding process of the virus towards the cytosol, (ii) further recruitment of additional viral proteins to the assembly site, and (iii) enhance lipid sorting to modify the mechanical and structural properties of the local membrane in favor of viral particle formation.

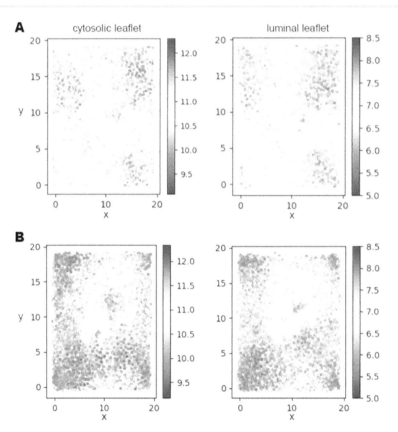

Fig. 5 Membrane deformation around the (A) open dimer, and (B) closed dimer. Each map shows the average location of the lipid phosphate groups during the last 50 ns of trajectory in the (i) cytosolic and (ii) luminal leaflets, the top and bottom leaflets in the simulation box.

The M protein is the structural protein present in highest concentration in the viral envelope.[9,12] Both dimer conformations contribute to membrane deformation towards the cytosol. Given the degree and extent of deformation induced by the open conformation, it is possible this conformation is prevalent at the very beginning to prime the site for viral assembly. As the process advances and the site is more crowded by additional M units and other viral proteins, it is plausible to suggest the open dimers are able to switch partners and adopt the closed conformation, especially since the interfaces of interaction are nearly opposite in both conformations (refer to Fig. 1). The closed conformation also pushes surrounding lipids upwards towards the cytosol, but the extent of deformation on the membrane plane is much more local. Fig. 5B shows the corresponding lipid elevation maps and an even distribution of the protein inside the bilayer core. In the context of viral assembly, a switch of conformation from open to closed dimers and a preference for the closed conformation as the assembly progresses would serve to stabilize a budding virus and further reshape the local lipid landscape to form an optimum assembly platform.

Along with the M protein, the E channel of coronaviruses is also known to actively modulate viral assembly and budding – among other roles.[1,14] From our simulations of two proposed conformations for this channel, we observe very different dynamics in the membrane response and the spatial orientation of the channel in the bilayer. We do not observe passage of ions or water in our trajectories for either model, but identified PHE26 in the TM region as potentially involved in the ion function or gating mechanism of the channel. This is the only residue inside the channel that flips its orientation repeatedly during the microsecond-long trajectories. As discussed earlier, PHE26 was also suggested as modulator of the open and closed conformations of the channel in another study[14] that characterized potential ion channeling mechanisms across homology models similar to the Korkin model used in this work.

Upon examination of Fig. 6A, it is clear the Korkin model for the channel causes greater membrane deformation. This occurs in part because the charged residues in the CTDs inside the bilayer core can interact with the lipid headgroups of both leaflets and pull on them like springs. Additionally, as discussed in the next paragraph, the local lipid distribution around the Korkin model is such that it results in a more flexible microenvironment. By contrast, the CTDs in the Feig model remain at the membrane interface and even detach from the surface momentarily to then bind again. The Feig CTDs that detach from the membrane surface can rotate in solvent and bind in a different conformation, which results in the sharp increases in the RMSD profiles of two adjacent monomers in this channel (see Fig. 6A(iii)) and high fluctuation in the corresponding RMSF profiles at that region. Though the CTDs in the Korkin model move inside the hydrophobic core, the conformational changes are not as pronounced as those in the Feig model as reflected in the corresponding RMSD profiles. Significant conformational changes occur within the first microsecond of simulation in both cases.

The spatial arrangement of each model induces different lipid rearrangement around the proteins. The hydrophobic TMs of the channel are in direct contact with the bilayer core in the Feig model, whereas the CTDs shield the TMs of the Korkin model inside the bilayer. The CTDs have charged and polar residues that prefer to interact with charged regions of the lipids around them. As a result, the CTDs in the Korkin model attract charged lipid headgroups around them and pull

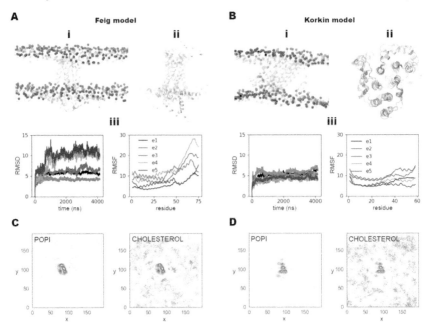

Fig. 6   Protein motions of the E channel in a model bilayer. Final snapshots showing the (i) membrane deformation around each channel, (ii) overlay of the initial (faded) and final channel structures, and the (iii) RMSD and RMSF per protein unit for the (A) Feig and (B) Korkin models; the purple spheres show the relative position of the phosphate group of membrane lipids. (C and D) Lipid density plots for inositol and cholesterol lipids around the channel, the protein density is also included in black for reference.

them deeper into the hydrophobic core, shown in darker spots around the protein density in Fig. 6C and D(i). On the other hand, the TMs in the Feig model, formed by non-polar and bulky residues, attract more cholesterol to the vicinity of the channel (see Fig. 6C(ii)). Cholesterol has a condensing effect, reducing the surface area per lipid and rendering the local environment more ordered and rigid. The opposite occurs around the Korkin model, which has a cholesterol-depleted region around the protein, where the local environment is more flexible. The channel itself, *i.e.*, the relative position of the TMs with respect to each other, does not fluctuate much as shown in the overlaid snapshots in Fig. 6A and B, but there are pronounced shifts of the CTD in both models as these actively interact and sort the lipids around the protein.

The difference in local environment around each model allows or restricts further protein motions. There are three phenylalanine rings in the TMs of the channel: PHE20, PHE23 and PHE26; only the last one is completely inside the channel during the entire trajectory and is also suggested as key modulator between the open and closed conformations of the channel.[14] Fig. 7A and B show 2D histograms for PHE26 residues in the TM of each model (the equivalent residue in the Korkin model is PHE19, as listed in Fig. 1C). The vertical axis corresponds to the rotation of the PHE ring plane, denoted by the cosine of the angle between the vector formed by the adjacent carbons to the ring tip (C_{E1} and C_{E2} in the protein topology file) and the bilayer normal; a value of one indicates

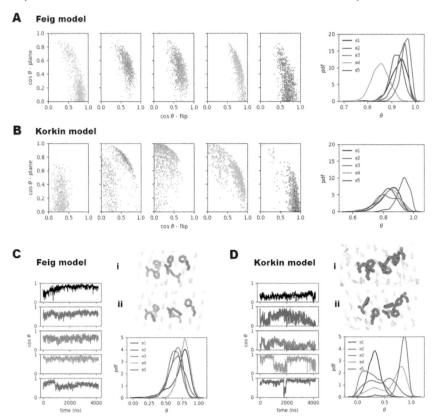

Fig. 7 Rotation maps to quantify the motions of the PHE ring plane as it rotates and flips inside the channel for the (A) Feig and (B) Korkin models, and the corresponding probability distribution of the TMs with respect to the bilayer normal in each model. (C and D) Time series and probability distribution of the cosine of the PHE26 angle with respect to the bilayer during the trajectory for the Feig and Korkin models, respectively; also shown are the (i) initial and (ii) final orientations of PHE26 inside each channel.

the PHE ring plane is perfectly aligned to the bilayer normal or z-axis in our simulation systems, whereas a value of zero indicates the ring plane is parallel to the bilayer center plane. The horizontal axis corresponds to the PHE flip inside the channel, denoted by the cosine of the angle between the vector formed by the base and tip carbons of the ring (C_β and C_Z in the topology) and the bilayer normal; a value higher than 0.5 indicates PHE points upwards in our simulation coordinates, towards the luminal side or the interior of the virion, and values lower than 0.5 indicate the PHE ring is nearly flat aligned with the bilayer center plane. The PHE residues in the TM in the Korkin model have higher incidence, in number of TMs, and frequency, in number of flips, of orientation changes. Even the probability density profiles show two distinct peaks for PHE rings in the TMs in the Korkin model (Fig. 7D).

We observed a more symmetric response in the membrane leaflets surrounding the Feig model in terms of the position of the P atoms. Though there is a change in the topology of the membrane surface from the initial to the final set of frames in the simulation, the change is coordinated between the leaflets,

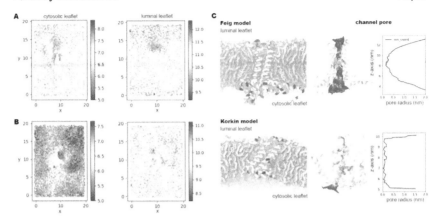

Fig. 8 Membrane deformation around the E channel: (A) Feig and (B) Korkin models. Each map shows the average location of the lipid phosphate groups during the last 50 ns of trajectory in the cytosolic (left) and luminal (right) leaflets, the top and bottom leaflets in the simulation box. (C) The Feig and Korkin models for the E channel shown as reference in the bilayer with a corresponding snapshot of the channel pore at the end of simulation computed on MDAnalysis implementing the HOLE[56] routine; the blue regions can accommodate two water molecules, green regions have enough space for one water molecule, and red regions are too narrow to allow the passage of a water molecule.

the same regions in the *xy*-plane are elevated or depressed. As shown in Fig. 8, the bilayer is pulled towards the cytosolic leaflet at the region where the CTDs of the Feig model are in contact with the membrane surface (shown in red in panel 8A). On the other hand, the leaflets surrounding the Korkin channel vary quite differently; the cytosolic leaflet remains pretty much at lower values, there is barely a small elevation in blue right at the edge of the pore created by the protein. Whereas the luminal leaflet, initially at a uniform elevation, reshapes to nearly half of its surface at 1.5–2 angstroms higher than the other half. Due to the CTDs inside the bilayer core, the cross-sectional area of the Korkin model is much more irregular than the highly conserved shape of the Feig model.

It is clear that the protein–lipid interplay around the E-channel is very complex. An additional homology model[14] has been proposed since the start of our study, simulations of an E monomer to characterize the orientation of the helices in the sequence,[57] an experimental structure for the TMs (7K3G).[11] It remains to be determined if the preferred E channel conformation resembles the Korkin model, or if a conformation like the one proposed by the Feig model is biologically relevant at a given stage of the viral life cycle. From previous studies, it seems the Korkin-like conformation is prevalent when the E channel functions as an ion channel, yet the channeling activity is highly modulated by local lipid environment.[14,15] Our results also favor that premise in the Korkin model could presumably let ions pass through with some selectivity, based on the size of the channel pore identified and shown in Fig. 8C. The Feig model, instead, remains wide enough throughout the 4 μs trajectory to accommodate up to two water molecules in its widest regions, identified by the blue coloring of the channel pore. Latest studies on the function of the E channel suggest it allows the passage of both monovalent and divalent cations modulated by the local lipid composition.[15] Furthermore, simulation studies suggest the glycosylation of the CTD helices could also determine their orientation with respect to the lipid bilayer.[57] There is a narrow area right at the location of PHE26, which our results also point as potentially key modulator for the ion channeling activity of the complex.

Since both E and M are implicated in viral assembly and budding, further modeling and simulation studies that examine their coordinated dynamics in the bilayer as well as in higher individual concentration are needed. Enhanced free energy sampling methods that enable us to quantify the energetics of protein–protein (M–M) and intra-protein (M–E) interactions would also serve to further understand the function of these integral membrane proteins in the assembly and budding of SARS-CoV-2 and related viruses. All-atom MD simulations offer an advantage to understand specific protein–lipid interactions and motions; however, a more multiscale approach is likely needed to gain better understanding of this intricate system.[58]

Conclusions

This work summarizes our initial efforts to understand specific protein–lipid interactions of two structural proteins of SARS-CoV-2. Specifically, we examined differences between two proposed homology models for a dimer of the M protein and the viroporin channel formed by five monomers of the E protein. We used all-atom MD and a complex membrane model to determine protein motions and interactions in a lipid environment that closely resembles the place where viral particles co-localize during early stages of viral assembly, the ERGIC. We propose that both of the two M dimer conformations suggested from homology models are present during viral assembly, potentially in different ratios depending on the stage of assembly and maturation. We discuss the effect of both the M and E viral proteins on lipid–lipid interactions and sorting around the proteins, and the potential impact of these interactions on membrane deformation during viral assembly. Finally, we identified PHE26, a residue located on the inside of the TMs of the E channel, as possibly involved in the ion gating function of this protein complex. Data from experiments also identified PHE26 pointing towards the E channel, inside the pore.[11] PHE 26 is one of the three PHE residues in the TM region of the channel that greatly shifts in the Korkin model. From experimental observations, all three residues are also identified as relevant modulators of the channel's function along with local lipid composition, which largely modulates the channel's gating activity.[11,15] Taken together, these observations seem to identify the Korkin model as the prevalent conformation when E serves as an ion channel, and the Feig model as a potential alternate conformation during other functions of the E viroporin. All the protein conformations we have examined in this work have a different effect on the immediate lipid environment, which is a key feature during viral assembly and membrane remodeling.

We continue to refine our simulation models in collaboration with experimental partners to provide further insights into the complex relationship between viral proteins and lipids in the cell. In addition to providing an atomistic view of these molecular interactions, data from this work can be used to systematically refine our multiscale coarse-grained model of the SARS-CoV-2 virion to better understand the collective dynamics of the entire viral particle.[58] Our simulation framework also enables us to update our models as new structures and insights from experiment become available. There is certainly pressing need to obtain experimental structures of the M and E proteins that will allow us to examine more closely their interactions and impact on lipid and viral dynamics in the cell. Gaining clear understanding of the viral mechanisms of SARS-CoV-2 is furthermore critical to the development of novel treatments, repurpose existing ones more efficiently, and anticipate strategies to combat natural mutations of this and related viruses.

Author contributions

G. A. V. provided funding, resources, and supervision. V. M. worked on conceptualization, data curation, formal analysis, and writing of the initial draft. G. A. V. and V. M. reviewed and edited the final manuscript. The authors thank the Voth Group, especially Dr Alex Pak, for insightful discussions on the simulation systems at the beginning of this work. We also thank Prof. Rommie Amaro and her team for initial discussions about interactions among structural membrane proteins of the virus.

Conflicts of interest

There are no conflicts to declare.

Acknowledgements

This work was supported in part by the National Science Foundation through NSF RAPID grant CHE-2029092 and in part by the National Institute of General Medical Sciences of the National Institutes of Health through grant R01 GM063796. Computational resources were provided by the COVID-19 HPC Consortium (project MCB20037); specifically, Frontera at the Texas Advanced Computer Center (TACC) and the Anton 2 machine at the Pittsburgh Super Computing Center (PSC) (available under NIH Grant R01GM116961). The Anton 2 machine at PSC was generously made available by D. E. Shaw Research. Initial equilibration of some of the systems in this study was computed on Midway2 at the Research Computing Center (RCC) at the University of Chicago.

References

1 S. K. M. Haque, O. Ashwaq, A. Sarief and A. K. Azad John Mohamed, *Future Virol.*, 2020, **15**, 625–648.

2 X. Y. Lim, L. Y. Ng, P. J. Tam and X. D. Liu, *Diseases*, 2016, **4**(3), 26.

3 S. G. Cynthia, M. T. Kathleen, G. K. Thomas, E. R. Pierre, A. C. James, W. L. William, A. R. Paul, B. Bettina, J. B. William and R. Z. Sherif, *Emerg. Infect. Dis.*, 2004, **10**, 320.

4 V. Corradi, B. I. Sejdiu, H. Mesa-Galloso, H. Abdizadeh, S. Y. Noskov, S. J. Marrink and D. P. Tieleman, *Chem. Rev.*, 2019, **119**, 5775–5848.

5 R. B. Chan, L. Tanner and M. R. Wenk, *Chem. Phys. Lipids*, 2010, **163**, 449–459.

6 W. Zeng, G. Liu, H. Ma, D. Zhao, Y. Yang, M. Liu, A. Mohammed, C. Zhao, Y. Yang, J. Xie, C. Ding, X. Ma, J. Weng, Y. Gao, H. He and T. Jin, *Biochem. Biophys. Res. Commun.*, 2020, **527**, 618–623.

7 S. Kang, M. Yang, Z. Hong, L. Zhang, Z. Huang, X. Chen, S. He, Z. Zhou, Z. Zhou, Q. Chen, Y. Yan, C. Zhang, H. Shan and S. Chen, *Acta Pharm. Sin. B*, 2020, **10**(7), 1228–1238.

8 C. E. McBride, J. Li and C. E. Machamer, *J. Virol.*, 2007, **81**, 2418.

9 Y.-C. Hsieh, H.-C. Li, S.-C. Chen and S.-Y. Lo, *J. Biomed. Sci.*, 2008, **15**, 707–717.

10 Y.-T. Tseng, C.-H. Chang, S.-M. Wang, K.-J. Huang and C.-T. Wang, *PLoS One*, 2013, **8**, e64013.

11 V. S. Mandala, M. J. McKay, A. A. Shcherbakov, A. J. Dregni, A. Kolocouris and M. Hong, *Nat. Struct. Mol. Biol.*, 2020, **27**, 1202–1208.

12 B. W. Neuman, G. Kiss, A. H. Kunding, D. Bhella, M. F. Baksh, S. Connelly, B. Droese, J. P. Klaus, S. Makino, S. G. Sawicki, S. G. Siddell, D. G. Stamou, I. A. Wilson, P. Kuhn and M. J. Buchmeier, *J. Struct. Biol.*, 2011, **174**, 11–22.

13 V. Navratil, L. Lionnard, S. Longhi, J. M. Hardwick, C. Combet and A. Aouacheria, *bioRxiv*, 2020, DOI: 10.1101/2020.04.09.033522.

14 M. Sarkar and S. Saha, *PLoS One*, 2020, **15**, e0237300.

15 C. Verdiá-Báguena, V. M. Aguilella, M. Queralt-Martín and A. Alcaraz, *Biochim. Biophys. Acta, Biomembr.*, 2021, **1863**, 183590.

16 Y.-Z. Fu, S.-Y. Wang, Z.-Q. Zheng, H. Yi, W.-W. Li, Z.-S. Xu and Y.-Y. Wang, *Cell. Mol. Immunol.*, 2021, **18**, 613–620.

17 B. Boson, V. Legros, B. Zhou, E. Siret, C. Mathieu, F.-L. Cosset, D. Lavillette and S. Denolly, *J. Biol. Chem.*, 2021, **296**, 100111.

18 Y. Dong, T. Dai, Y. Wei, L. Zhang, M. Zheng and F. Zhou, *Signal Transduction Targeted Ther.*, 2020, **5**, 237.

19 D. E. Gordon, G. M. Jang, M. Bouhaddou, J. Xu, K. Obernier, K. M. White, M. J. O'Meara, V. V. Rezelj, J. Z. Guo, D. L. Swaney, T. A. Tummino, R. Hüttenhain, R. M. Kaake, A. L. Richards, B. Tutuncuoglu, H. Foussard, J. Batra, K. Haas, M. Modak, M. Kim, P. Haas, B. J. Polacco, H. Braberg, J. M. Fabius, M. Eckhardt, M. Soucheray, M. J. Bennett, M. Cakir, M. J. McGregor, Q. Li, B. Meyer, F. Roesch, T. Vallet, A. Mac Kain, L. Miorin, E. Moreno, Z. Z. C. Naing, Y. Zhou, S. Peng, Y. Shi, Z. Zhang, W. Shen, I. T. Kirby, J. E. Melnyk, J. S. Chorba, K. Lou, S. A. Dai, I. Barrio-Hernandez, D. Memon, C. Hernandez-Armenta, J. Lyu, C. J. P. Mathy, T. Perica, K. B. Pilla, S. J. Ganesan, D. J. Saltzberg, R. Rakesh, X. Liu, S. B. Rosenthal, L. Calviello, S. Venkataramanan, J. Liboy-Lugo, Y. Lin, X.-P. Huang, Y. Liu, S. A. Wankowicz, M. Bohn, M. Safari, F. S. Ugur, C. Koh, N. S. Savar, Q. D. Tran, D. Shengjuler, S. J. Fletcher, M. C. O'Neal, Y. Cai, J. C. J. Chang, D. J. Broadhurst, S. Klippsten, P. P. Sharp, N. A. Wenzell, D. Kuzuoglu-Ozturk, H.-Y. Wang, R. Trenker, J. M. Young, D. A. Cavero, J. Hiatt, T. L. Roth, U. Rathore, A. Subramanian, J. Noack, M. Hubert, R. M. Stroud, A. D. Frankel, O. S. Rosenberg, K. A. Verba, D. A. Agard, M. Ott, M. Emerman, N. Jura, M. von Zastrow, E. Verdin, A. Ashworth, O. Schwartz, C. d'Enfert, S. Mukherjee, M. Jacobson, H. S. Malik, D. G. Fujimori, T. Ideker, C. S. Craik, S. N. Floor, J. S. Fraser, J. D. Gross, A. Sali, B. L. Roth, D. Ruggero, J. Taunton, T. Kortemme, P. Beltrao, M. Vignuzzi, A. García-Sastre, K. M. Shokat, B. K. Shoichet and N. J. Krogan, *Nature*, 2020, **583**, 459–468.

20 Y. Liu, K. Wang, T. F. Massoud and R. Paulmurugan, *ACS Pharmacol. Transl. Sci.*, 2020, **3**, 844–858.

21 E. Ketter and G. Randall, *Annu. Rev. Virol.*, 2019, **6**, 319–340.

22 S. Stertz, M. Reichelt, M. Spiegel, T. Kuri, L. Martínez-Sobrido, A. García-Sastre, F. Weber and G. Kochs, *Virology*, 2007, **361**, 304–315.

23 C. Appenzeller-Herzog and H.-P. Hauri, *J. Cell Sci.*, 2006, **119**, 2173.

24 L. Heo and M. Feig, *bioRxiv*, 2020, DOI: 10.1101/2020.03.25.008904.

25 S. Srinivasan, H. Cui, Z. Gao, M. Liu, S. Lu, W. Mkandawire, O. Narykov, M. Sun and D. Korkin, *Viruses*, 2020, **12**, 360.

26 C. B. Plescia, E. A. David, D. Patra, R. Sengupta, S. Amiar, Y. Su and R. V. Stahelin, *J. Biol. Chem.*, 2021, **296**, 100103.

27 R. Nardacci, F. Colavita, C. Castilletti, D. Lapa, G. Matusali, S. Meschi, F. Del Nonno, D. Colombo, M. R. Capobianchi, A. Zumla, G. Ippolito, M. Piacentini and L. Falasca, *Cell Death Dis.*, 2021, **12**, 263.

28 T. Harayama and H. Riezman, *Nat. Rev. Mol. Cell Biol.*, 2018, **19**, 281–296.

29 V. Monje-Galvan and J. B. Klauda, *Biochemistry*, 2015, **54**, 6852–6861.

30 G. van Meer, D. R. Voelker and G. W. Feigenson, *Nat. Rev. Mol. Cell Biol.*, 2008, **9**, 112–124.

31 S. Jo, T. Kim and W. Im, *PLoS One*, 2007, **2**, e880.

32 S. Jo, T. Kim, V. G. Iyer and W. Im, *J. Comput. Chem.*, 2008, **29**, 1859–1865.

33 B. R. Brooks, C. L. Brooks III, A. D. Mackerell Jr, L. Nilsson, R. J. Petrella, B. Roux, Y. Won, G. Archontis, C. Bartels, S. Boresch, A. Caflisch, L. Caves, Q. Cui, A. R. Dinner, M. Feig, S. Fischer, J. Gao, M. Hodoscek, W. Im, K. Kuczera, T. Lazaridis, J. Ma, V. Ovchinnikov, E. Paci, R. W. Pastor, C. B. Post, J. Z. Pu, M. Schaefer, B. Tidor, R. M. Venable, H. L. Woodcock, X. Wu, W. Yang, D. M. York and M. Karplus, *J. Comput. Chem.*, 2009, **30**, 1545–1614.

34 E. L. Wu, X. Cheng, S. Jo, H. Rui, K. C. Song, E. M. Davila-Contreras, Y. Qi, J. Lee, V. Monje-Galvan, R. M. Venable, J. B. Klauda and W. Im, *J. Comput. Chem.*, 2014, **35**(27), 1997–2004.

35 J. Lee, X. Cheng, J. M. Swails, M. S. Yeom, P. K. Eastman, J. A. Lemkul, S. Wei, J. Buckner, J. C. Jeong, Y. Qi, S. Jo, V. S. Pande, D. A. Case, C. L. Brooks, A. D. MacKerell, J. B. Klauda and W. Im, *J. Chem. Theory Comput.*, 2016, **12**, 405–413.

36 M. J. Abraham, T. Murtola, R. Schulz, S. Páll, J. C. Smith, B. Hess and E. Lindahl, *SoftwareX*, 2015, **1–2**, 19–25.

37 J. Huang, S. Rauscher, G. Nawrocki, T. Ran, M. Feig, B. L. de Groot, H. Grubmüller and A. D. MacKerell, *Nat. Methods*, 2017, **14**, 71–73.

38 D. E. Shaw, J. P. Grossman, J. A. Bank, B. Batson, J. A. Butts, J. C. Chao, M. M. Deneroff, R. O. Dror, A. Even, C. H. Fenton, A. Forte, J. Gagliardo, G. Gill, B. Greskamp, C. R. Ho, D. J. Ierardi, L. Iserovich, J. S. Kuskin, R. H. Larson, T. Layman, L. Lee, A. K. Lerer, C. Li, D. Killebrew, K. M. Mackenzie, S. Y. Mok, M. A. Moraes, R. Mueller, L. J. Nociolo, J. L. Peticolas, T. Quan, D. Ramot, J. K. Salmon, D. P. Scarpazza, U. B. Schafer, N. Siddique, C. W. Snyder, J. Spengler, P. T. P. Tang, M. Theobald, H. Toma, B. Towles, B. Vitale, S. C. Wang and C. Young, Anton 2: Raising the Bar for Performance and Programmability in a Special-Purpose Molecular Dynamics Supercomputer, *SC '14: Proceedings of the International Conference for High Performance Computing, Networking, Storage and Analysis*, 2014, pp. 41–53, DOI: 10.1109/SC.2014.9.

39 S. Jo, J. B. Lim, J. B. Klauda and W. Im, *Biophys. J.*, 2009, **97**, 50–58.

40 J. J. Madsen, J. M. A. Grime, J. S. Rossman and G. A. Voth, *Proc. Natl. Acad. Sci. U. S. A.*, 2018, **115**, E8595.

41 R. Liang, J. M. J. Swanson, J. J. Madsen, M. Hong, W. F. DeGrado and G. A. Voth, *Proc. Natl. Acad. Sci. U. S. A.*, 2016, **113**, E6955, DOI: 10.1073/pnas.1615471113.

42 W. G. Hoover, *Phys. Rev. A: At., Mol., Opt. Phys.*, 1985, **31**, 1695–1697.

43 S. Nosé, *Mol. Phys.*, 1984, **52**, 255–268.

44 H. J. C. Berendsen, J. P. M. Postma, W. F. van Gunsteren, A. DiNola and J. R. Haak, *J. Chem. Phys.*, 1984, **81**, 3684–3690.

45 S. Nosé and M. L. Klein, *Mol. Phys.*, 1983, **50**, 1055–1076.

46 M. Parrinello and A. Rahman, *J. Appl. Phys.*, 1981, **52**, 7182–7190.

47 T. Darden, D. York and L. Pedersen, *J. Chem. Phys.*, 1993, **98**, 10089–10092.

48 B. Hess, H. Bekker, H. J. C. Berendsen and J. G. E. M. Fraaije, *J. Comput. Chem.*, 1997, **18**, 1463–1472.

49 Y. Shan, J. L. Klepeis, M. P. Eastwood, R. O. Dror and D. E. Shaw, *J. Chem. Phys.*, 2005, **122**, 54101.

50 J.-P. Ryckaert, G. Ciccotti and H. J. C. Berendsen, *J. Comput. Phys.*, 1977, **23**, 327–341.

51 R. A. Lippert, C. Predescu, D. J. Ierardi, K. M. Mackenzie, M. P. Eastwood, R. O. Dror and D. E. Shaw, *J. Chem. Phys.*, 2013, **139**, 164106.

52 W. Humphrey, A. Dalke and K. Schulten, *J. Mol. Graphics*, 1996, **14**(33–38), 27–38.

53 R. J. Gowers, M. Linke, J. Barnoud, T. J. Reddy, M. N. Melo, S. L. Seyler, J. D. Domański, D. L. Dotson, S. Buchoux, I. M. Kenney and O. Beckstein, *Proceedings of the 15th Python in Science Conference*, 2016, 98–105, DOI: 10.25080/Majora-629e541a-00e.

54 N. Michaud-Agrawal, E. J. Denning, T. B. Woolf and O. Beckstein, *J. Comput. Chem.*, 2011, **32**, 2319–2327.

55 R. T. McGibbon, K. A. Beauchamp, M. P. Harrigan, C. Klein, J. M. Swails, C. X. Hernández, C. R. Schwantes, L.-P. Wang, T. J. Lane and V. S. Pande, *Biophys. J.*, 2015, **109**, 1528–1532.

56 O. S. Smart, J. G. Neduvelil, X. Wang, B. A. Wallace and M. S. P. Sansom, *J. Mol. Graphics*, 1996, **14**, 354–360.

57 A. Kuzmin, P. Orekhov, R. Astashkin, V. Gordeliy and I. Gushchin, *bioRxiv*, 2021, DOI: 10.1101/2021.03.10.434722.

58 A. Yu, A. J. Pak, P. He, V. Monje-Galvan, L. Casalino, Z. Gaieb, A. C. Dommer, R. E. Amaro and G. A. Voth, *Biophys. J.*, 2021, **120**, 1097–1104.

Faraday Discussions

PAPER

Structural changes in the model of the outer cell membrane of Gram-negative bacteria interacting with melittin: an *in situ* spectroelectrochemical study

Izabella Brand ⓘ * and Bishoy Khairalla ⓘ

Received 9th April 2020, Accepted 24th June 2020

DOI: 10.1039/d0fd00039f

The cell membrane of Gram-negative bacteria interacting with an antimicrobial peptide presents a complex supramolecular assembly. Fabrication of models of bacterial cell membranes remains a large experimental challenge. Langmuir–Blodgett and Langmuir–Schaefer (LS–LB) transfer makes possible the deposition of multicomponent asymmetric lipid bilayers onto a gold surface. Two lipids: 1-palmitoyl-2-oleoyl-*sn*-glycero-3-phosphoethanolamine (**POPE**) and di[3-deoxy-D-manno-octulosonyl]-lipid A (**KLA**) were used to deposit a model of the outer membrane of Gram-negative bacteria on the Au(111) substrate. The use of gold as the solid substrate enables control of the membrane potential. Molecular scale changes in the model membrane exposed to physiological electric fields and interacting with melittin antimicrobial peptide are discussed in this paper. The interaction of the outer membrane with melittin leads to an increase in the membrane capacitance and permeability to ions and water. The stability of the outer membrane with bound melittin decreases at positive membrane potentials. *In situ* polarization modulation infrared reflection absorption spectroscopy is used to investigate membrane potential-dependent changes in the structure of the outer membrane interacting with melittin. The hydration of the ester carbonyl groups is not affected by the interaction with melittin. However, the orientation and hydrogen bond network with the carboxylate groups in **KLA** changes drastically after **POPE**–**KLA** bilayer interacts with melittin. We propose that the positively charged groups in the amino acids present at the C-terminus of the peptide interact directly with the polar head group of **KLA**. Simultaneously, the packing order in hydrocarbon chains in the membrane with bound melittin increases. A hydrophobic match between the chains in the lipids and the peptide, which spans the membrane, seems to be responsible for the ordering of the hydrocarbon chains region of the bilayer. The N-terminus enters into the hydrophobic region of the membrane and forms a channel to the hydrophilic head groups in **POPE**.

Department of Chemistry, University of Oldenburg, 26111 Oldenburg, Germany. E-mail: Izabella.brand@uni-oldenburg.de; Tel: +497983973

Introduction

The cell envelope of bacteria differs in the composition and structure from the cell membranes of eukaryotes.[1] Gram-positive bacteria contain one phospholipid membrane which is surrounded by a 20–80 nm thick peptidoglycan film. The cell envelope of Gram-negative bacteria has a more complex structure. It is composed of two lipid bilayers. The inner cytoplasmic membrane contains phospholipids such as phosphatidylethanolamine (PE), phosphatidylglycerol (PG) and cardiolipin. The inner membrane is separated from the outer membrane (OM) by a thin (7–10 nm) peptidoglycan layer. The structure of the OM is highly asymmetric as schematically shown in Fig. 1. The inner leaflet contains the same phospholipids as the inner membrane. Lipopolysaccharides (LPS) are present in the outer leaflet. LPS have complex structures and display structural diversity between various bacterial species and strains.[1] The lipid A molecule consists of an amphiphilic fragment of each LPS. Lipid A contains 4 to 7 hydrocarbon chains which are bound to a phosphorylated 1–6-linked glucosamine dimer (Fig. 1b). The core fragment of LPS is bound to lipid A. It is subdivided into the inner and the outer core. The inner core has a highly preserved composition.[1] It is bound to lipid A *via* rare sugars: 2-keto-3-deoxyoctulosonate (Kdo) and L-*glycero*-D-mannoheptose. The outer core is composed of a few common sugar residues (hexoses and hexosamines). The polar head group of lipid A and the core regions are negatively charged. Divalent cations (Ca^{2+}, Mg^{2+}) are coordinated to the phosphate, carboxylate and hydroxide groups of this rigid fragment of LPS.[2–4] The most

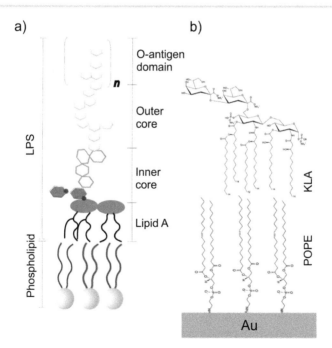

Fig. 1 (a) Schematic drawing of the structure of lipopolysaccharide and phospholipid molecules forming the OM of Gram-negative bacteria, (b) the structure of **KLA** and **POPE** molecules deposited by LB and LS transfer on the Au(111) surface.

outer fragment of LPS is called the O-antigen (Fig. 1a). It is a polysaccharide fragment composed of 3–6 monosaccharide repeating units, which form either linear or branched chains. The huge structural diversity of LPS combined with the complex supramolecular structure of the OM, have limited biomimetic studies of the cell membrane envelope of Gram-negative bacteria. Structural information about the packing of hydrocarbon chains and orientation of the polar head groups in the OM is predominantly available from computer simulation studies.[2,5–10] Experimental approaches include investigations of multilayer LPS-water systems[11–13] and monolayers at the air|water electrolyte interface.[4,14] Fabrication of a planar asymmetric lipid bilayer, containing LPS in the outer and phospholipids in the inner leaflet, is another experimental challenge. Here, the Langmuir–Blodgett (LB) and Langmuir–Schaefer (LS) transfer is used to deposit a planar supported bilayer containing 1-palmitoyl-2-oleoyl-*sn-glycero*-3-phosphoethanolamine (**POPE**) in the inner and di[3-deoxy-D-manno-octulosonyl]-lipid A (**KLA**) in the outer leaflet on the Au(111) surface (Fig. 1b). In this article we show that this asymmetric bilayer represents a simple but realistic model of the OM of Gram-negative bacteria.

The OM is the first contact surface for antimicrobial peptides (AMPs). AMPs are composed of short L-amino acid chains.[15] Their mode of action leads to direct lysis of the pathogenic cell membrane.[16–18] Melittin (Mel) is a 26-amino acid peptide which is the principal toxic component of bee venom. Mel has the net positive charge (+6) which except for Gln at the N-terminus is located at the C-terminus.[19] Mel has low specific activity. In other words it displays toxic activity toward bacterial and mammalian cells.[18,20] Numerous studies have demonstrated that Mel adsorbs on the surface of phospholipid bilayers.[21–25] After reaching a critical surface concentration, Mel inserts its hydrophobic N-terminus into the membrane forming pores and channels. The content of channels depends on the composition of the phospholipid bilayer, pH, and ionic strength of the electrolyte solution. With an increasing number of channels, solubilization of the model phospholipid bilayer is observed.[18,23] In addition, the formation of Mel channels depends on the potential drop across the lipid bilayer.[26,27] Numerous studies have been dedicated to the interaction of Mel with lipid membranes.[16,22–25] However, phospholipid bilayers have predominantly been used in previous investigations. Recent computational studies indicate however, that the structure, packing, surface properties, diffusion coefficients as well electric properties of the OM, differ significantly from the properties of phospholipid bilayers.[2,5,6,8,10] Due to challenges in the fabrication of realistic models of the OM, the results of theoretical studies still require experimental confirmation. We were able to deposit a stable planar asymmetric **POPE–KLA** lipid bilayer on a Au(111) surface (Fig. 1a). Deposition of the OM on a conductive surface, allows control of the potential applied to the Au(111) electrode, and therefore the membrane potential. This approach enabled us to follow electric potential dependent changes in the OM alone and during its interaction with Mel. To visualize the molecular scale changes taking place in the model OM interacting with Mel, a surface analysing technique: polarization modulation infrared reflection-absorption spectroscopy (PM IRRAS) with electrochemical control was used. We discuss potential dependent changes in the hydration and orientation of the polar head group region and the orientation and packing of the hydrocarbon chains in the hydrophobic fragment of the OM interacting with Mel.

Results and discussion

Electrochemical characterization of the model membrane interacting with melittin

Capacitance of a film adsorbed on an electrode surface represents an important macroscopic parameter, which provides information about the membrane compactness and permeability to water and ions. Fig. 2 shows the capacitance–potential plots of the pure **POPE–KLA** bilayer, and after interaction with Mel. The capacitance of the **POPE–KLA** bilayer is a function of potential applied to the Au(111) electrode (curve 1 in Fig. 2). The potential of zero charge (E_{pzc}) of the Au(111) electrode in 0.1 M KClO$_4$ and 5 mM Mg(ClO$_4$)$_2$ electrolyte solution, E_{pzc} = 0.305 V. The difference

$$E - E_{pzc} \approx E^m \tag{1}$$

is a good approximation of the membrane potential (E^m).[28,29]

In the negative potential scan, in the potential range $0.15 < E^m < -0.50$, the capacitance of the **POPE–KLA** bilayer is close to 3 µF cm^{-2}. A further negative potential shift leads to a fast increase in the capacitance (Fig. 2 and Table 1). At $E^m_{des} < -1.1$ V the capacitance of the **POPE–KLA** bilayer is the same as the capacitance of the unmodified Au(111) electrode, indicating desorption of the lipid bilayer from the metal surface.

In the reversed positive going potential scan, at $E^m > -0.90$ V a gradual decrease in the capacitance is observed. The **POPE–KLA** bilayer re-adsorbs on the electrode surface. In this potential scan a peak at $E^m_{tr} = 0.02$ V is seen in the capacitance–potential curve (curve 1, Fig. 2 and Table 1). This weak capacitance peak indicates potential-driven reorientations in the lipid bilayer. In the studied potential range, the surface charge density of the Au(111) surface changes from negative ($E < E_{pzc}$) to positive ($E > E_{pzc}$). This change may cause some rearrangement of the lipid molecules

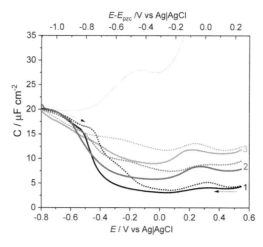

Fig. 2 Capacitance *vs.* potential applied to the electrode and membrane potential plots of the **POPE–KLA** bilayer on the Au(111) surface in 0.10 M KClO$_4$ and 5 mM Mg(ClO$_4$)$_2$ solution, curve (1) pure lipid bilayer, curve (2) after interaction with 1 µM for 15 min, and (3) 10 µM Mel solution for 15 min. Solid lines: negative, dotted lines: positive potential scans, thin line: capacitance of the unmodified Au electrode.

Table 1 The potential range of the capacitance minimum, membrane potential of the bilayer desorption (E_{des}^m) and phase transitions in the negative ($E_{tr,n}^m$) and positive ($E_{tr,p}^m$) going potential scans in the POPE–KLA bilayer interacting with 1 μM and 10 μM Mel solution for 15 min

$C_{melittin}$/μM, time/min	C_{min}/μF cm^{-2}	Potential window of C_{min}/V	E_{des}^m/V	$E_{tr,p}^m$/V	$E_{tr,n}^m$/V
0 μM, 0 min	3.1	$0.15 < E^m < -0.50$	−1.1	0.02	0.00
1 μM, 15 min	5.9	$-0.26 < E^m < -0.50$	−1.1	−0.01	−0.04
10 μM, 15 min	9.1	$-0.12 < E^m < -0.50$	−1.1	−0.06	−0.10

in the asymmetric OM. For example, in the asymmetric POPE–POPG:DPPG bilayer, in this potential range, an increase in the capacitance has been observed.[28] It was associated with the flow of the electrolyte solution in the polar head group region of the phospholipid membrane. In the studied OM, the POPE phospholipid molecules are located in the inner, electrode facing leaflet. Thus, rearrangements in the structure of the lipid membrane at $E^m \approx 0$ seem to be affected by the composition of the leaflet facing directly to the electrode surface. Molecular dynamics simulations of the electroporation process of the OM of *E. coli* indicated that only phospholipid molecules in the inner leaflet of the membrane undergo potential-driven reorientations.[8] From the electrochemical measurements we are able to provide information about the capacitance change in the entire membrane, not in each leaflet.

Electrochemical parameters, characterizing the POPE–KLA bilayer interacting with Mel are summarized in Table 1.

Exposure of the POPE–KLA bilayer for 15 min to the electrolyte solution containing Mel in the monomer form causes significant changes to the capacitance–potential plots (curves 2 and 3 Fig. 2). The capacitance minimum of the POPE–KLA bilayers with bound Mel increases (Fig. 2 and Table 1). Simultaneously, the potential window of the capacitance minimum decreases. In the bilayer with bound Mel the potential of the phase transition (E_{tr}^m) undergoes a negative shift, indicating a destabilization of the adsorbed state of the OM at $E_{tr}^m > 0.00$ V. Earlier electrochemical studies indicated that at negative membrane potentials the GlnNH$_2$ at the N-terminus of Mel enters into the lipid membrane forming a channel.[26,27] Under this condition, at positive membrane potentials, the electrostatic interactions between the N-terminus of Mel and the metal surface would be weakened. This, in turn, may lead to some potential-driven reorientations of lipid molecules in the POPE–KLA bilayer. However, the desorption potential of the POPE–KLA bilayer is not affected by the interaction with the AMP (Fig. 2 and Table 1). Thus, negative membrane potentials do not affect the stability of the OM with bound Mel. At negative membrane potentials the lipid molecules become permeable to the electrolyte solution causing the desorption of the bilayer from the metal surface.[28,30] Electrochemical results show that Mel affects the compactness and permeability to water and ions of the model OM. To present a molecular scale picture of potential-driven changes in the POPE–KLA bilayer interacting with Mel *in situ* PM IRRAS with electrochemical control is used.

Effect of melittin on the orientation and hydration of the polar head groups in the POPE–KLA bilayer

PM IRRA spectra in the 1820–1500 cm^{-1} spectral region of the pure POPE–KLA bilayer and after interaction with Mel are shown in Fig. 3. In this spectral region,

POPE and **KLA** lipids give three IR absorption modes. Fig. 3a shows the experimental PM IRRA spectrum of the lipid bilayer at $E^m = 0.00$ V and the calculated spectrum for randomly distributed molecules in a bilayer thick film. The asymmetric, intensive mode centred around 1733 cm^{-1} is assigned to the $\nu(C=O)$ in the ester groups in the **POPE** and **KLA** lipids. It is deconvoluted into three modes centred at 1741 ± 2, 1722 ± 2 and 1710 ± 2 cm^{-1}. These modes reflect differences in the hydration of the ester groups in both lipids present in the OM.[11,31,32] Bilayers of zwitterionic phospholipids give two $\nu(C=O)$ modes, which are centred around 1740 and 1725 cm^{-1}.[32-34] They reflect the asymmetry in the hydration of the lipid molecules in the two leaflets. The $\nu(C=O)$ mode in the ester carbonyl group of negatively charged lipids is more complex.[11,35] Three components at 1746–40, 1728–20 and $\sim$1710 cm^{-1} contribute to the $\nu(C=O)$ mode. Thus, the two high frequency bands of the $\nu(C=O)$ mode region originate from both **POPE** and **KLA** lipids in the OM. The absorption around 1710 cm^{-1} is assigned to the ester carbonyl groups in **KLA** and indicates that some ester groups are involved in the formation of strong hydrogen bonds to water.[11] Indeed, water content in the polar head group of the lipid A fragment is significantly larger than in the polar head group region of phospholipids.[2,10] Thus, the ester carbonyl groups in the intermediate region joining the polar head group in **KLA** with six hydrocarbon chains may be involved in the formation of hydrogen bonds to water.

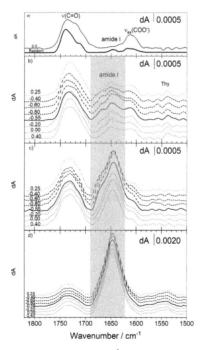

Fig. 3 PM IRRA spectra in the 1820–1500 cm^{-1} spectral region of the of the **POPE–KLA** bilayer on the Au(111) in 0.05 M KClO$_4$ and 5 mM Mg(ClO$_4$)$_2$ in D$_2$O at potentials marked in the figure. (a) Pure bilayer; thin line: at $E = 0.0$ V and thick line: calculated spectrum for randomly distributed molecules in the bilayer film; (b and c) after interaction with 1 μM for (b) 15 min, (c) 60 min; and (d) with 10 μM Mel for 15 min in the negative (dashed lines) and positive (solid line) going potential scans.

A weak mode in the PM IRRA spectra around 1633 cm^{-1} arises from the amide groups in **KLA** (Fig. 3a). The IR absorption mode around 1614–1600 cm^{-1} is assigned to the $\nu_{as}(COO^-)$ mode in the 2-keto-3-deoxyoctulosonate fragment of the polar head group of **KLA**.[11,36] In aqueous electrolyte solution (pH = 6.1) the carboxylic acid groups in **KLA** are deprotonated.[36] In addition, in the 1700–1610 cm^{-1} region the amide I mode of Mel is clearly seen in the spectra, indicating that the AMP is bound to the OM. The shape and intensity of this mode depend on the concentration and time of interaction between the OM and Mel (Fig. 3b–d).

First, the effect of Mel binding on the hydrophilic fragment of the **POPE–KLA** membrane is discussed. In the bilayer interacting with 1 μM Mel, the deconvolution of the ester carbonyl $\nu(C{=}O)$ mode gives three components (1740 and 1722–23 and 1712–1710 cm^{-1}). However, the deconvolution of the $\nu(C{=}O)$ mode in the **POPE–KLA** bilayer interacting with 10 μM Mel reveals two absorption modes centred at 1736 ± 2 and 1716 ± 3 cm^{-1}. This result indicates changes in the hydration of the ester carbonyl groups in the lipids present in the membrane. The high frequency mode undergoes a bathochromic shift. However, two $\nu(C{=}O)$ modes located in the pure bilayer at 1722 and 1710 merge, yielding one mode at 1717 cm^{-1}. This change indicates that in the membrane with bound Mel the strength of the hydrogen bonds in **KLA** decreases. This may be a result of direct electrostatic (charge–dipole and/or dipole–dipole) interactions between the carbonyl groups and the chain of the polypeptide.

Interaction of the **POPE–KLA** bilayer with Mel leads to small changes in the integral intensity of the $\nu(C{=}O)$ mode. In PM IRRAS, in an anisotropic film, the integral intensity of a given absorption mode is proportional to the surface concentration (constant) and the average orientation of the transition dipole vector of this fundamental mode.[37] Negligible changes in the intensity of the $\nu(C{=}O)$ mode indicate small potential- and Mel-dependent changes in the orientation of the ester groups. Order parameter (S) expresses the orientational and motional order in an anisotropic film,[38]

$$S = \left\{ \frac{1}{2}\langle 3(\cos^2 \theta) - 1 \rangle \right\} \tag{2}$$

where θ is the angle between the electric field vector (normal to the surface) and the direction of the transition dipole vector change in a given normal vibration. In the **POPE–KLA** bilayer adsorbed directly on the Au(111) surface the order parameter of the $\nu(C{=}O)$ mode $S_{\nu C{=}O} = 0.32 \pm 0.10$, giving an average tilt of ca. 38° between the direction of the C=O bonds and surface normal. Desorption of the bilayer leads to a decrease in $S_{\nu C{=}O}$ to 0.17 ± 0.06. In phospholipid bilayers the $S_{\nu C{=}O}$ is negative and indicates that the ester carbonyl groups adopt an almost parallel orientation to the bilayer plane.[28,34] Interestingly, in the asymmetric bilayer containing **KLA**, the ester carbonyl groups adopt an upwards orientation. Interaction with Mel introduces some orientation changes of the intermediate region of the ester carbonyl groups. The average $S_{\nu C{=}O}$ becomes independent of the membrane potential and decreases to 0.24 ± 0.12 indicating that the average tilt of the C=O groups increases to ca. 45° vs. surface normal.

Furthermore, interaction of the **POPE–KLA** bilayer with Mel causes large changes of the $\nu_{as}(COO^-)$ mode. This mode arises from the carboxylate groups in the 2-keto-3-deoxyoctulosonate residue in the core fragment of the **KLA** molecule

(Fig. 1b). In a solution with pH 6.0 the carboxylic acid groups in **KLA** are dissociated[36] and coordinated to the divalent cations (Mg^{2+}) present in the electrolyte solution.[2,4,6,8] In the pure **POPE–KLA** bilayer the $\nu_{as}(COO^-)$ mode is asymmetric and has two components (small points, Fig. 4). The position of the high wavenumber $\nu_{as}(COO^-)$ mode depends on the membrane potential and undergoes a shift from 1614.5 cm^{-1} at $E^m = 0.15$ V to 1611 cm^{-1} at $E^m = -1.00$ V. Monodentate binding of Mg^{2+} to carboxylate groups in the **KLA** in the outer leaflet causes a hypsochromic shift of the free $\nu_{as}(COO^-)$ mode.[39] Indeed, divalent cations are coordinated to the phosphate, carboxylate and hydroxyl groups in the polar head group region of **KLA**.[2,4,6,8,40] The second, weak $\nu_{as}(COO^-)$ mode is centred around 1602–1604 cm^{-1} and could be assigned to free carboxylate groups. Thus, a fraction of carboxylate groups in **KLA** is not coordinated to the metal ions, but makes hydrogen bonds to water. Interaction of the OM with Mel caused changes in this spectral region. In the **POPE–KLA** bilayer with bound Mel, the low-wavenumber $\nu_{as}(COO^-)$ mode disappears, while the high-wavenumber component undergoes a bathochromic shift (Fig. 4). Already after 15 minutes of interaction of the bilayer with 1 μM Mel solution, the main $\nu_{as}(COO^-)$ mode undergoes a 3 cm^{-1} bathochromic shift. Longer interaction time as well as increase in the concentration of the AMP leads to a further shift of the $\nu_{as}(COO^-)$ mode to 1607 ± 2 cm^{-1}. Clearly, the coordination and hydration of the carbonyl groups changes significantly after Mel binding to the OM. These results indicate that the binding of Mg^{2+} ions to the carboxylate residue is weakened. The position of the $\nu_{as}(COO^-)$ mode differs from the pure **POPE–KLA** bilayer, suggesting that carboxylate groups make hydrogen bonds to the amine groups in the C-terminus of Mel. Thus, Mel interacts directly with **KLA**. This direct interaction anchors the peptide to the OM surface.

Not only the position, but also intensity of the $\nu_{as}(COO^-)$ mode changes during the OM interaction with Mel. In other words, the OM–Mel interaction leads to

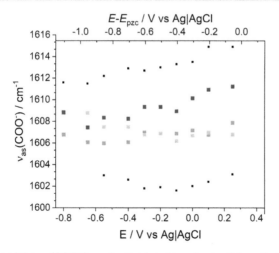

Fig. 4 Position of the $\nu_{as}(COO^-)$ mode as a function of potential applied to the Au(111) electrode and membrane potential of the pure **POPE–KLA** bilayer in the negative going potential scan: small black points, after interaction with 1 μM for 15 min, dark blue points; 60 min, blue points; with 10 μM Mel for 15 min; orange points in the negative going potential scan.

a decrease in the intensity of this mode (Fig. 4). The order parameter of the $\nu_{as}(COO^-)$ mode $[S_{\nu_{as}(COO^-)}]$ was calculated using eqn (2) and is shown in Fig. 5. In the pure **POPE–KLA** bilayer the $S_{\nu_{as}(COO^-)}$ depends on the membrane potential. It is equal to 0.65 ± 0.03 in the $0.2 < E^m < -0.40$ V potential range (Fig. 5a). These values of the $S_{\nu_{as}(COO^-)}$ indicate, that the transition dipole vector of the $\nu_{as}(COO^-)$ mode makes a small (*ca.* 20–25°) angle *versus* surface normal. The transition dipole vector of the $\nu_{as}(COO^-)$ mode is located along the line joining the two O atoms. The small average tilt angle indicates the upward orientation of this line. Desorption leads to an increase in the tilt of the carboxylate group. Thus, the carboxylate groups in the outer leaflet of the **POPE–KLA** bilayer have a well-defined orientation. As shown in Fig. 5, the interaction with Mel leads to a significant decrease in the $S_{\nu_{as}(COO^-)}$.

In the **POPE–KLA** bilayer interacting with 1 µM Mel for 15 and 60 minutes the $S_{\nu_{as}(COO^-)}$ is equal to 0.05 ± 0.09. In the OM incubated in 10 µM Mel solution and adsorbed directly on the Au(111) the $S_{\nu_{as}(COO^-)}$ decreases to -0.20. These results show that the interaction with Mel introduces large reorientations of the carboxylate groups in **KLA**. In the OM adsorbed directly on the Au surface the average tilt of the line joining two O atoms increases to *ca.* 55°–65° *versus* the

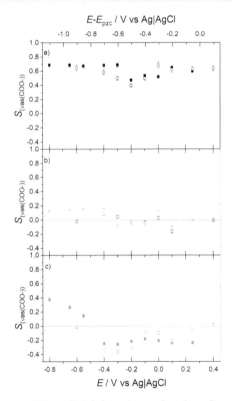

Fig. 5 Order parameter of the $\nu_{as}(COO^-)$ mode as a function of potential applied to the Au(111) electrode and membrane potential of (a) pure **POPE–KLA** bilayer and after its interaction with (b) 1 µM for 15 min, dark blue points; 60 min, blue points, and (c) 10 µM Mel for 15 min in the negative (filled points) and positive (open line) going potential scans. Dotted line at $S = 0$ indicates the random orientation.

surface normal. The line joining two O atoms in the carboxylate group tends to be oriented along the membrane plane. Clearly, the lipid–peptide interaction affects the orientation of carboxylate groups in **KLA** in the outer leaflet. The line joining two O atoms changes its direction from almost perpendicular to a parallel orientation with respect to the membrane surface. Probably this interaction stabilizes the membrane associated state of Mel. Interestingly, at negative membrane potentials, in the desorbed state, the $S_{\nu_{as}(COO^-)}$ increases to *ca.* 0.4 (Fig. 5c). At desorption membrane potentials the electrolyte flows between the electrode and the OM. A potential drop occurs across the water cushion layer. In consequence the membrane potential is lost.[41] Under this condition the orientation of the carboxylate groups in **KLA** changes, relaxing the interaction with Mel.

Effect of melittin on the orientation of the hydrophobic hydrocarbon chains in the POPE–KLA bilayer

Fig. 6 shows PM IRRA spectra of the **POPE–KLA** bilayer in the CH stretching modes region. Methyl and methylene groups are present in the hydrocarbon chains of both lipids. The hydrophobic fragment of the asymmetric lipid bilayer adsorbed on the Au(111) surface gives four well defined IR absorption modes: $\nu_{as}(CH_3)$ at 2956, $\nu_{as}(CH_2)$ at 2927, $\nu_s(CH_3)$ at 2872 and $\nu_s(CH_2)$ at 2855 cm^{-1}

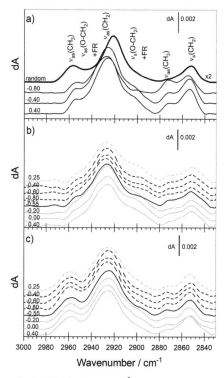

Fig. 6 PM IRRA spectra in the 3000–2820 cm^{-1} spectral region of the **POPE–KLA** bilayer on Au(111) in 0.05 M KClO$_4$ and 5 mM Mg(ClO$_4$)$_2$ in D$_2$O, at potentials marked in the figure; (a) pure bilayer, thick line shows a calculated spectrum for randomly distributed molecules in the bilayer film; (b) after interaction of the bilayer with 1 µM for 60 min and (c) with 10 µM Mel for 15 min in the negative (dashed lines) and positive (solid line) going potential scans.

(Fig. 6a). In the calculated spectrum of randomly distributed **POPE** and **KLA** molecules, in a bilayer thick film, the $\nu_{as}(CH_2)$ and $\nu_s(CH_2)$ modes appear at 2921 and 2853 cm^{-1}, respectively. Positions of the methylene stretching modes depend on the physical state of the hydrocarbon chain.[42,43] In the pure **POPE–KLA** bilayer supported on the Au(111) surface, the positions of the methylene stretching modes indicate that the hydrocarbon chains exist in a liquid disordered state. Interaction of the OM with Mel leads to a small (*ca.* 1–2 cm^{-1}) bathochromic shift of the methylene stretching modes (Fig. 6b and c). The $\nu_{as}(CH_2)$ mode is very broad and is overlapped with the Fermi resonance interactions between the $\nu_s(CH_2)$ fundamental and the overtones of the $\delta(CH_2)$ mode.[44] In addition, the saccharide moieties in the **KLA** possess methylene groups which are coordinated to the O atom (O–CH$_2$). The presence of the O atom adjacent to the methylene group leads to a hyposchromic shift of the methylene stretching modes.[45,46] Indeed the $\nu_{as}(O–CH_2)$ is overlapped with the FR mode around 2940 cm^{-1} while the $\nu_s(O–CH_2)$ mode appears around 2899 cm^{-1}. Since the IR absorption modes of the methylene groups in the hydrophobic hydrocarbon chains and saccharide residues are resolved, the deconvolution of this spectral region allows for the quantitative analysis of the order in the chain region of the **POPE–KLA** bilayer interacting with Mel.

In the spectra of randomly distributed molecules in the OM, the full width at half maximum (fwhm) of the $\nu_{as}(CH_2)$ is equal to 17.5 cm^{-1}, and for $\nu_s(CH_2)$, 13.5 cm^{-1}. In the pure **POPE–KLA** bilayer the fwhm increase to 22.3 ± 0.8 and 14.5 ± 0.3 cm^{-1}, respectively. Upon interaction with Mel, the fwhm of the $\nu_{as}(CH_2)$ increases further to 27.0 ± 1.2 cm^{-1}, and for the $\nu_s(CH_2)$ to 16.2 ± 0.8 cm^{-1}. An increase in fwhm of the methylene stretching modes indicates an increase in the motional rate of the hydrocarbon chains in the OM with bound Mel. In the **POPE–KLA** bilayer with bound Mel the population of hydrocarbon chains with different conformation and mobility increases. This is in line with previous studies of phospholipid bilayers which indicate differences in the conformation of hydrocarbon chains making direct contact to Mel, and those present in the bulk bilayer.[47]

In the pure **POPE–KLA** bilayer the integral intensities of the methylene stretching modes depend on the membrane potential. However, after the membrane interaction with Mel the intensities of the methylene stretching modes become independent of the membrane potential. Thus, Mel interacts not only with the polar head group in **KLA** but also with the hydrophobic region of the OM. Integral intensities of the methylene stretching modes were used to calculate the average angle between the transition dipole vector of the methylene stretching modes and the surface normal. These data were used to calculate the average tilt of the hydrocarbon chains in both lipid molecules in the OM and finally the chain order parameter (S_{chain}). The S_{chain} reflects the conformational and packing order of the hydrocarbon chains in an anisotropic film such as a lipid bilayer.[38] The S_{chain} *versus* potential plots are shown in Fig. 7.

The S_{chain}, and therefore the average orientation of the hydrocarbon chains in the **POPE–KLA** bilayer depend on the membrane potential. In the bilayer adsorbed directly on the Au(111) surface, the S_{chain} is close to 0.56 ± 0.04. In the desorbed state, at $E^m < -0.80$ V, the S_{chain} increases to 0.73 (Fig. 7a). Interestingly, the S_{chain} of the asymmetric **POPE–KLA** bilayer adsorbed directly on the Au(111) surface is lower than 0.65–0.80 as reported for supported phospholipid bilayers.[28,33,34,48] However, it is larger than the S_{chain} of 0.20–0.45 calculated for

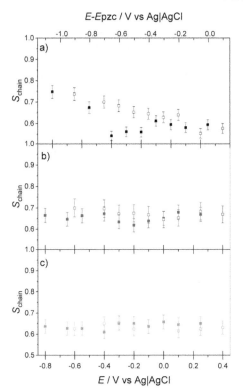

Fig. 7 Chain order parameter as a function of the potential applied to the Au(111) electrode and membrane potential of the (a) pure **POPE**–**KLA** bilayer, and bilayer interacting with (b) 1 μM Mel for 1 h, (c) 10 μM Mel for 15 min in the negative (filled points) and positive (hollow points) going potential scans.

hydrocarbon chains in mono- and bilayers of **LPS**.[2,49] The S_{chain} values determined in our study suggest asymmetry in the average orientation of the hydrocarbon chains in **POPE** and **KLA** molecules.

In the membrane with bound Mel, the S_{chain} becomes independent of the membrane potential. It is slightly larger than the S_{chain} of the pure **POPE**–**KLA** bilayer adsorbed directly in the Au(111) surface (0.63 ± 0.06, Fig. 7b and c). This result indicates a small increase in the order of packing of the hydrocarbon chains in the membrane with bound Mel. It is in line with previous studies which show that Mel introduces an ordering effect on the acyl chain fragment of phospholipid bilayers.[21,22] An increase in the order in packing of the hydrocarbon chains in the **POPE**–**KLA** bilayer is associated with the accommodation and incorporation of Mel into the membrane. The spanning of Mel through the hydrophobic fragment of the membrane may indeed lead to tighter packing of the lipids in the bilayer.

Experimental

Chemicals

1-palmitoyl-2-oleoyl-*sn-glycero*-3-phosphoethanolamine (**POPE**) and di[3-deoxy-D-manno-octulosonyl]-lipid A (ammonium salt) Kdo2-lipid A (**KLA**) were purchased from Avanti Polar lipids (Alabaster, USA). Lipids were used as received. Melittin (Mel, cat. no. M4171), $KClO_4$ (99.99%), $Mg(ClO_4)_2 \cdot 6H_2O$ (99%), acid-sodium salt dihydrate (EDTA), methanol and ethanol were purchased from Sigma Aldrich (Steinheim, Germany), NaCl (99.5%) was purchased from Carl Roth (Karlsruhe, Germany), 2-amino-2-(hydroxymethyl)-1,3-propanediol (Tris) from Fluka (Ulm, Germany) and D_2O from Eurisotop (Saarbrücken, Germany).

Langmuir–Blodgett transfer

Before each experiment, fresh lipid solutions were prepared. **POPE** was dissolved in $CHCl_3$. **KLA** was dissolved in $CHCl_3 : CH_3OH : H_2O$ in 13 : 6 : 1 volume ratio. The concentration of **POPE** was equal to 1 $\mu mol\ ml^{-1}$ and the concentration of **KLA** was equal to 1 $mg\ ml^{-1}$. Using a microsyringe (Hamilton, Reno, NV, USA) several microlitres of the lipid solution were placed at the liquid|air interface of the Langmuir trough (KSV, Helsinki, Finland). All aqueous solutions were prepared from ultrapure water [resistivity 18.2 MΩ cm (PureLab Classic, Elga LabWater, Celle, Germany)]. To evaporate the solvent, the lipid solution was left for 10 min. Surface pressure vs. area per molecule isotherms were recorded in the KSV LB mini trough (KSV Ltd., Helsinki, Finland) equipped with two hydrophilic barriers. Surface pressure was recorded as a function of the mean molecular area. The accuracy of measurements was $\pm 0.02\ nm^2$ for the mean molecular area and $\pm 0.1\ mN\ m^{-1}$ for surface pressure (π). Langmuir–Blodgett and Langmuir–Schaefer (LB–LS) transfers were used to prepare asymmetric supported planar lipid bilayers containing **POPE** in the inner (Au electrode oriented), and **KLA** in the outer (solution-oriented), leaflet on the gold surface. Prior to the LB–LS transfer, each monolayer was compressed to $\pi = 30\ mN\ m^{-1}$. First, the **POPE** monolayer was transferred from the aqueous subphase by a vertical LB withdrawing at a speed of 15 $mm\ min^{-1}$. The transfer ratio was 1.10 ± 0.10. After the transfer, the PE monolayer was left for 1 h for drying. Next, a monolayer of **KLA** on 0.1 M $KClO_4$ and 5 mM $Mg(ClO_4)_2$ aqueous subphase was compressed to $\pi = 30\ mN\ m^{-1}$ and a horizontal LS transfer was used to fabricate the second leaflet on the gold surface. Planar lipid bilayers were dried for at least 2 h before use in electrochemical and spectroelectrochemical experiments.

Interaction of melittin with lipid bilayers

The concentration of Mel in a buffer solution [20 mM Tris, 150 mM NaCl and 5 mM EDTA (pH = 7.3 ± 0.1)] was equal to either 1 or 10 μM. Under these experimental conditions Mel exists in a monomeric form.[50] Gold electrode modified with a lipid bilayer was incubated in a hanging meniscus configuration in the buffer solution containing the monomeric form of Mel. For the Mel concentration of 1 μM the incubation times were set to 15 min and 1 h. The incubation time from 10 μM Mel solution was set to 15 min. After this time the modified electrodes were carefully rinsed with water and transferred for further studies.

Electrochemistry

Electrochemical measurements were performed in an all-glass three-electrode cell using a disc of Au(111) single crystals (diam. 3 mm, MaTecK, Jülich, Germany) as the working electrode (WE) in a hanging meniscus configuration. The surface roughness of the Au electrode is below 0.01 μm per 1 cm^2 of the electrode surface. A gold wire was used as a counter electrode (CE) and a Ag|AgCl|sat. KCl (Ag|AgCl) as a reference electrode (RE). All potentials are referred against the Ag|AgCl|sat. KCl electrode. The electrolyte solution was 100 mM KClO$_4$ with 5 mM Mg(ClO$_4$)$_2$. A Metrohm Autolab potentiostat (Metrohm Autolab, Utrecht, Holland) was used to perform the electrochemical measurements. Prior to the experiment, the cell was purged with argon for 1 h. The cleanliness of the electrochemical cell was tested by recording the cyclic voltammograms in the electrolyte solution. Alternative current voltammetry (ACV) was used to measure the capacitance of the unmodified and lipid bilayer modified Au(111) electrode. AC voltammograms were recorded in negative and positive going potential scans at a rate of 5 mV s^{-1}, and the perturbation of the AC signal of 20 Hz and 10 mV amplitude. The differential capacitance *vs.* potential curves were calculated from the in-phase and out-of-phase components of the AC signal assuming the cell was equivalent to a resistor in series with a capacitor. Data from two independent measurements were averaged.

Polarization modulation infrared reflection absorption spectroscopy

The PM IRRA spectra were recorded using a Vertex 70 spectrometer with a photoelastic modulator ($f = 50$ kHz; PMA 50, Bruker, Germany) and demodulator (Hinds, Instruments, USA). A home-made spectroelectrochemical glass cell was washed in water, ethanol and placed in an oven (at 60 °C) for drying. CaF$_2$ prism was rinsed with water, ethanol and placed in a UV–ozone chamber (Bioforce Nanosciences, USA) for 10 min. The spectroelectrochemical cell has a built-in platinum CE. The RE was Ag|AgCl in 3 M KCl in D$_2$O. A disc of Au(111) single crystals (diam. 15 mm, MaTecK, Jülich, Germany) was used as the WE and mirror for IR radiation. The surface roughness of the Au electrode is below 0.01 μm per 1 cm of the electrode surface. A lipid bilayer was transferred on the WE surface using the LB–LS transfer. The electrolyte solution was 50 mM KClO$_4$ with 5 mM Mg(ClO$_4$)$_2$ in D$_2$O. The electrolyte solution was purged for 1 h with argon to remove oxygen. At each potential applied to the Au electrode 400 spectra with a resolution of 4 cm^{-1} were measured. In each experiment five negative and positive going potential scans were recorded. Each potential scan was analysed separately. In the negative going potential scan the following potentials were applied to the Au(111) electrode: 0.40, 0.25, 0.10, 0.00, −0.10, −0.20, −0.30, −0.40, −0.55, −0.65, −0.80 V while in the positive going potential scan: −0.80, −0.60, −0.40, −0.30, −0.20, −0.10, 0.00, 0.10, 0.25, 0.40 V. The thickness of the electrolyte layer between the Au(111) electrode and the prism varied between 3 and 5 μm in different experiments. The half wave retardation was set to 2900 cm^{-1} for the CH stretching, and at 1600 cm^{-1} for the amide I and C=O stretching modes. The angle of incidence of the incoming IR radiation was set to 55°. The spectra shown here correspond to the average obtained in two independent experiments. All the spectra are processed using the OPUS v5.5 software (Bruker, Ettlingen, Germany).

FT-IR transmittance measurements

Small unilamellar vesicles (SUVs) of the studied lipid mixture (**POPE–KLA**) (3.2 : 1 mole) were prepared fresh. A 10 mg ml^{-1} stock solution of each lipid in chloroform was prepared. Next, 250 μl of **POPE** and 250 μl of **KLA** stock solutions were mixed and dried in a flow of argon. To remove the remaining solvent, the vials were placed in a vacuum desiccator for at least 24 h. Next, 500 μl of 50 mM $KClO_4$ and 5 mM $Mg(ClO_4)_2$ in D_2O were added to the dry lipid and the mixture was sonicated (EMAG Technologies, Walldorf, Germany) at room temperature for 1 h. A clear solution containing lipid vesicles was used in transmission IRS measurements. The IR transmission spectra of 50 mM $KClO_4$ + 5 mM $Mg(ClO_4)_2$ in D_2O (background) and vesicles of **POPE–KLA** in the electrolyte solution (analyte), were recorded in a flow cell between two ZnSe windows (Aldrich, Steinheim, Germany) and a 50 μm Teflon spacer. A Vertex 70 IR spectrometer (Bruker, Ettlingen, Germany) was used to measure the spectra; 64 spectra with a resolution of 4 cm^{-1} were recorded for the background and analyte. The transmission spectra were calculated and used to extract the isotropic optical constants according to previously described procedures.[51] Optical constants were used to calculate a PM IRRA spectrum of randomly distributed molecules in a bilayer thick film.[37,51] The angle of incidence of the incoming IR radiation, the electrolyte layer thickness, and the stratified medium were the same as in *in situ* PM IRRAS experiments. The thickness of the lipid bilayer was 5.5 nm, which corresponds to the thickness of 2.1 nm of the **POPE** monolayer[52] and 3.4 nm of the **KLA**.[14] The surface coverage, calculated from the LB–LS transfer condition was equal to 0.82.

Conclusions

Our results show that the LB–LS transfer allows the deposition of bilayers with transverse asymmetry on a solid substrate. The asymmetric **POPE–KLA** bilayer, transferred using the LB–LS method, served as a model of the OM of Gram-negative bacteria. The orientation of the lipid molecules in the asymmetric OM depends on the membrane potential. Our results show that electric potentials cause significantly smaller changes in the packing and hydration of lipid molecules than in pure phospholipid bilayers.[28,33,34,48,53] Thus, in an electric field, the OM behaves differently to commonly studied phospholipid bilayers. Therefore, biomimetic studies of models of bacterial and viral membranes are highly required to recognize their response to external impulses such as temperature, electric field or addition of antibiotics.

The asymmetry in the composition and surface charge density on both sides of the OM may affect the membrane interactions with AMPs. Our results allow us to present a molecular scale picture of structural changes taking place in the OM interacting with Mel as a function of the membrane potential. As discussed here, spectroelectrochemical results show that lipids, in both leaflets of the OM, are involved in the interaction with Mel. First, Mel interacts electrostatically with carboxylate groups in the inner core of LPS (**KLA**). This interaction involves changes in the hydration and orientation of the carboxylate groups. Results of recent computational studies demonstrate that the sugar, carboxylate and phosphate groups in LPS are able to form hydrogen bonds to bulky amino acids such as Arg, Lys, Gln or Trp of membrane TonB-dependent transporter proteins.[6,9,54] The C-terminus in Mel contains Lys, Arg and Gln amino acids.[55] Therefore, it is highly probable that these amino acids in Mel are involved in the formation of hydrogen bonds with carboxylate groups in **KLA**. These results indicate that Mel is able to make hydrogen bonds with the carboxylate groups in **KLA**. These results raise the question if in general AMPs, bearing often a positive net charge, anchor in the same way to the OM surface. Moreover, LPS contain other negatively charged residues in the lipid A, core and antigen O regions. Thus, questions arise about the interaction of AMPs with other fragments of LPS.

Furthermore, the electrochemical results indicate that the interaction with Mel leads to an increase in membrane capacitance and permeability to ions and water. The electrostatic interactions between Mel and the OM are stabilized at negative membrane potentials and are weakened at positive membrane potentials. This could be connected to the insertion of Mel into the hydrophobic membrane fragment. Since the C-terminus of Mel interacts directly with the polar head group of **KLA**, the N terminus of the AMP is potentially able to penetrate through the hydrophobic fragment of the bilayer. The experimental evidence for the penetration of the N-terminus of Mel to the inner leaflet is not directly available from electrochemical measurements.

In situ PM IRRAS results show that the packing of the hydrocarbon chains fragment in the OM is affected by the interaction with Mel. The motional rate of the hydrocarbon chains in lipids of the OM interacting with Mel increases. Simultaneously, the order parameter of the hydrocarbon chain region increases. A hydrophobic match between the chains in lipids and the peptide, which spans the membrane, seems to be responsible for the ordering of the hydrocarbon chain region of the bilayer. The analysis of the PM IRRA spectra in the methylene stretching modes region corresponds to both leaflets of the OM. **POPE** and **KLA** have different structure of hydrophobic hydrocarbon chains, thus their packing and orientation in both leaflets of the OM interacting with Mel, may be different. The question of the interaction of Mel with the hydrocarbon chains in each leaflet separately, still remains unanswered.

Moreover, to find the mechanism of action of Mel on the OM, the orientation of Mel, interacting with the membrane exposed to changing electric fields has to be known. The amide I′ modes in Fig. 3 indicate that the membrane associated peptide undergoes conformation and orientation changes. The experimental challenge is how to distinguish these two processes and discuss the changes in the secondary structure and orientation of membrane associated Mel as a function of the membrane potential.

Mel is a well studied peptide and therefore, a comparison of the membrane attachment and penetration process of this peptide to the OM and phospholipid bilayers was possible. Substantial differences between the action of Mel on both kinds of model membranes were found. Our results indicate that use of realistic models of microbial cell membranes is essential to understand the molecular-level response of a microorganism to AMPs. Biomimetic studies of complex supramolecular assemblies of cell membrane envelopes of Gram-negative and Gram-positive bacteria may become an important and challenging future task.

Conflicts of interest

There are no conflicts to declare.

Acknowledgements

IB acknowledges support from DFG project number BR-3961-4.

References

1 C. Erridge, E. Bennett-Guerrero and I. R. Poxton, *Microbes Infect.*, 2002, **4**, 837–851.

2 E. L. Wu, O. Engstrom, S. Jo, D. Stuhlsatz, M. S. Yeom, J. B. Klauda, G. Widmalm and W. Im, *Biophys. J.*, 2013, **105**, 1444–1455.

3 E. Schneck, E. Papp-Szabo, B. E. Quinn, O. V. Konovalov, T. J. Beveridge, D. A. Pink and M. Tanaka, *J. R. Soc., Interface*, 2009, **6**, S671–S678.

4 E. Schneck, T. Schubert, O. V. Konovalov, B. E. Quinn, T. Gutsmann, K. Brandenburg, R. G. Oloviera, D. A. Pink and M. Tanaka, *Proc. Natl. Acad. Sci. U. S. A.*, 2010, **107**, 9147–9151.

5 P. C. Hus, F. Samsudin, J. Shearer and S. Khalid, *J. Phys. Chem. Lett.*, 2017, **8**, 5513–5518.

6 W. Im and S. Khalid, *Annu. Rev. Phys. Chem.*, 2020, **71**, 8.1–8.18.

7 S. Khalid, T. J. Piggot and F. Samsudin, *Acc. Chem. Res.*, 2019, **52**, 180–188.

8 T. J. Piggot, D. A. Holdbrook and S. Khalid, *J. Phys. Chem. B*, 2011, **115**, 13381–13388.

9 T. J. Piggot, D. A. Holdbrook and S. Khalid, *Biochim. Biophys. Acta*, 2013, **1828**, 284–293.

10 M. P. M. Lima, M. Nader, D. E. S. Santos and T. A. Soares, *J. Braz. Chem. Soc.*, 2019, **30**, 2219–2230.

11 K. Brandenburg, *Biophys. J.*, 1993, **64**, 1215–1231.

12 K. Brandenburg, H. Mayer, M. H. J. Koch, J. Weckesser, E. T. Rietschel and U. Seyder, *Eur. J. Biochem.*, 1993, **218**, 555–563.

13 K. Brandenburg and U. Seydel, *Eur. Biophys. J.*, 1988, **16**, 83–94.

14 A. P. Le Brun, L. A. Clifton, C. E. Halbert, B. Lin, M. Meron, P. J. Holden, J. H. Lakey and S. A. Holt, *Biomacromolecules*, 2013, **14**, 2014–2022.

15 F. Jean-Francois, J. Elezgaray, P. Berson, P. Vacher and E. J. Dufourc, *Biophys. J.*, 2008, **95**, 5748–5756.

16 H. Sato and J. B. Feix, *Biochim. Biophys. Acta*, 2006, **1758**, 1245–1256.

17 A. C. Rapson, M. A. Akhter Hossain, J. D. Wade, E. C. Nice, T. A. Smith, A. H. A. Clayton and M. L. Gee, *Biophys. J.*, 2011, **100**, 1353–1361.

18 G. Sessa, J. H. Freer, G. Colacicco and G. Weissmann, *J. Biol. Chem.*, 1969, **244**, 3575–3582.

19 T. C. Terwilliger, L. Weissman and D. Eisenberg, *Biophys. J.*, 1982, **37**, 353–361.

20 T. Katsu, M. Kuroko, T. Morikawa, K. Sanchika, Y. Fujita, H. Yamamura and M. Uda, *Biochim. Biophys. Acta*, 1989, **983**, 135–141.

21 C. R. Flach, F. G. Prendergast and R. Mendelsohn, *Biophys. J.*, 1996, **70**, 539–546.

22 S. Frey and L. K. Tamm, *Biophys. J.*, 1991, **60**, 922–930.

23 J. Juhaniewicz and S. Sek, *Electrochim. Acta*, 2015, **162**, 53–61.

24 S. Toraya, K. Nishimura and A. Naito, *Biophys. J.*, 2004, **87**, 3323–3335.

25 E. John and F. Jähnig, *Biophys. J.*, 1991, **60**, 319–328.

26 L. Beccucci and R. Guidelli, *Langmuir*, 2007, **23**, 5601–5608.

27 M. T. Tosteson and D. C. Tosteson, *Biophys. J.*, 1981, **36**, 109–116.

28 B. Khairalla, J. Juhaniewicz-Debinska, S. Sek and I. Brand, *Bioelectrochemistry*, 2020, **132**, 107443.

29 F. Abbasi, J. Alvarez-Malmagro, Z. F. Su, J. J. Leitch and J. Lipkowski, *Langmuir*, 2018, **34**, 13754–13765.

30 I. Burgess, M. Li, S. L. Horswell, G. Szymanski, J. Lipkowski, J. Majewski and S. Satija, *Biophys. J.*, 2004, **86**, 1763–1776.

31 R. N. A. H. Lewis and R. N. McElhaney, *Chem. Phys. Lipids*, 1986, **96**, 9–21.

32 A. Blume, W. Hübner and G. Messer, *Biochemistry*, 1988, **27**, 8239–8249.

33 E. Madrid and S. L. Horswell, *Langmuir*, 2013, **29**, 1695–1708.

34 I. Zawisza, X. Bin and J. Lipkowski, *Langmuir*, 2007, **23**, 5180–5194.

35 I. Zawisza, A. Lachenwitzer, V. Zamlynny, S. L. Horswell, J. D. Goddard and J. Lipkowski, *Biophys. J.*, 2003, **86**, 4055–4075.

36 A. Barkleit, H. Foerstendorf, B. Li, A. Rossberg, H. Moll and G. Bernhard, *Dalton Trans.*, 2011, **40**, 9868–9876.

37 V. Zamlynny and J. Lipkowski, in *Advances in electrochemical science and engineering*, ed. R. C. Alkire, D. M. Kolb, J. Lipkowski and P. N. Ross, Wiley-VCH, Weinheim, 2006, vol. 9, pp. 315–376.

38 A. Seelig and J. Seelig, *Biochemistry*, 1974, **13**, 4839–4845.

39 M. Nara, H. Torri and M. Tasumi, *J. Phys. Chem.*, 1996, **100**, 19812–19817.

40 R. D. Lins and T. P. Straatsma, *Biophys. J.*, 2001, **81**, 1037–1046.

41 Z. F. Su, M. Shodiev, J. J. Leitch, F. Abbasi and J. Lipkowski, *Langmuir*, 2018, **34**, 6249–6260.

42 R. A. MacPhail, H. L. Strauss, R. G. Snyder and C. A. Elliger, *J. Phys. Chem.*, 1984, **88**, 334–341.

43 R. G. Snyder, H. L. Strauss and C. A. Elliger, *J. Phys. Chem.*, 1982, **86**, 5145–5150.

44 P. T. T. Wong and H. H. Mantsch, *Chem. Phys. Lipids*, 1988, **46**, 213–224.

45 I. Brand, M. Nullmeier, T. Kluener, R. Jogireddy, J. Christoffers and G. Wittstock, *Langmuir*, 2010, **26**, 362–370.

46 T. Miyazawa, K. Fukushima and Y. Ideguchi, *J. Chem. Phys.*, 1962, **37**, 2764–2776.

47 I. W. Levin and F. Lavialle, *Biophys. J.*, 1982, **37**, 339–349.

48 N. Garcia-Araez, C. L. Brosseau, P. Rodriguez and J. Lipkowski, *Langmuir*, 2006, **22**, 10365–10371.

49 D. Jefferies, J. Shearer and S. Khalid, *J. Phys. Chem. B*, 2019, **123**, 3567–3575.

50 G. Schwarz and G. Beschiaschvili, *Biochemistry*, 1988, **27**, 7826–7831.

51 D. L. Allara and R. G. Nuzzo, *Langmuir*, 1985, **1**, 52–66.

52 K. Murzyn, T. Rog and M. Pasenkiewicz-Gierula, *Biophys. J.*, 2005, **88**, 1091–1103.

53 E. Madrid and S. L. Horswell, *Electrochim. Acta*, 2014, **146**, 850–860.

54 C. Balusek and J. C. Gumbart, *Biophys. J.*, 2016, **111**, 1409–1417.

55 T. C. Terwilligert and D. Eisenbergg, *J. Biol. Chem.*, 1982, **257**, 6016–6022.

Faraday Discussions

Bcl-xL inhibits tBid and Bax *via* distinct mechanisms†

Fabronia Murad[a] and Ana J. Garcia-Saez [ID] *[ab]

Received 25th April 2020, Accepted 7th July 2020

DOI: 10.1039/d0fd00045k

The proteins of the Bcl-2 family are key regulators of apoptosis. They form a complex interaction network in the cytosol and in cellular membranes, whose outcome determines mitochondrial permeabilization and commitment to death. However, we still do not understand how the action of the different family members is orchestrated to regulate apoptosis. Here, we combined quantitative analysis of the interactions and the localization dynamics of the family representatives Bcl-xL, Bax and tBid, in living cells. We discovered that Bax and tBid are able to constitutively shuttle between cytosol and mitochondria in the absence of other Bcl-2 proteins. Bcl-xL clearly stabilized tBid at mitochondria, where they formed tight complexes. In contrast, Bcl-xL promoted Bax retrotranslocation to the cytosol without affecting its shuttling rate, but by forming weak inhibitory mitochondrial complexes. Furthermore, analysis of phospho-mimetics of Bcl-xL suggested that phosphorylation regulates the function of Bcl-xL *via* multiple mechanisms. Altogether, our findings support a model in which the Bcl-2 network not only modulates protein/protein interactions among the family members, but also their respective intracellular localization dynamics, to regulate apoptosis.

Introduction

Apoptosis is the best studied form of programmed cell death, an essential, evolutionarily conserved mechanism that regulates development, tissue homeostasis and the defense against invasion of pathogens and genotoxic stress. Deregulation of apoptosis has been associated with the development of different diseases including cancer, stroke and neurodegeneration.[1] For these reasons, apoptosis regulation is not only a fundamental question in biology, but also a matter of medical relevance. In most cell types, irreversible apoptosis commitment occurs at the level of the mitochondrial outer membrane permeabilization (MOMP), which releases the apoptotic factors like Smac and

[a]Interfaculty Institute of Biochemistry, University of Tübingen, Tübingen, Germany. E-mail: ana.garcia@uni-koeln.de

[b]Institute for Genetics, CECAD Research Center, University of Cologne, Cologne, Germany

† Electronic supplementary information (ESI) available. See DOI: 10.1039/d0fd00045k

cytochrome c into the cytosol. This activates caspases, a family of proteases that cleaves a set of target proteins leading to the controlled dismantling of the cell.[2]

The proteins of the B-cell lymphoma 2 (Bcl-2) family are key regulators of MOMP.[3,4] They share one to four homology regions, known as Bcl-2 homology (BH) domains and can be classified into three groups regarding their function and the number of BH motifs they contain (ESI Fig. 1†). Pro-apoptotic effector members like Bax and Bak possess all four BH domains and directly mediate MOMP.[5,6] Their activation is promoted by the interaction with the BH3-only family members, like Bid and Bim, which are induced upon apoptotic stimuli. For example, during apoptosis Bid is cleaved into the truncated, active form named tBid. The inhibition of pro-apoptotic Bcl-2 proteins is mediated by the pro-survival members of the family, like Bcl-2 itself, Bcl-xL or Mcl-1, which also contain the four BH domains as well as a membrane anchor for insertion primarily into the mitochondrial outer membrane (MOM). The pro-survival proteins block the action of the pro-apoptotic effector proteins by forming with them inhibitory complexes.[7]

Furthermore, Bcl-xL inhibits apoptosis by promoting the retrotranslocation of mitochondrial Bax and Bak back into the cytosol.[8–10] Indeed, Bax exists in a dynamic equilibrium between the cytosol and mitochondria[11] and Bax activation not only depends on its interaction with other family members, but also on its residence time on the mitochondrial membrane.[12] Altogether, this yields a scenario where the many members of the Bcl-2 family form a complex interaction network in the cytosol and the mitochondrial membranes, whose outcome determines cell fate. However, how the interplay between the interaction network and the regulation of the localization dynamics of the Bcl-2 proteins controls apoptosis remains poorly understood.

Bcl-xL has been found upregulated and post translationally modified in different tumor types, which makes it an attractive drug target in the fight against cancer.[1] Several post translational modifications including phosphorylation and deamidation have been proposed to regulate Bcl-xL activity, leading to down- or upregulating of apoptosis.[13–21] Many of these modifications occur within a long unstructured domain (LUD) located between helices α1 and α2 (ref. 22 and 23) (Fig. 1a). For example, Bcl-xL phosphorylation at serine 62, located in the LUD, reverses the anti-apoptotic into an apoptotic activity.[13] Recently, Follis et al. reported that this modification promotes the interaction of the α1–α2 LUD with the folded core of Bcl-xL, which allosterically inhibits the pro-apoptotic activity of two types of targets, BH3-only proteins and p53.[24] However, we still do not understand how the different post translational modifications affect Bcl-xL function to modulate cellular apoptosis.

Here we combined two quantitative, live-cell microscopy approaches, Scanning Fluorescence Cross-Correlation Spectroscopy (SFCCS) and Fluorescence Loss In Photo-bleaching (FLIP), to compare the strength of the interactions and the shuttling between cytosol and mitochondria of representative Bcl-2 proteins, namely pro-survival Bcl-xL, and pro-apoptotic Bax and tBid. In parallel, we used suitable model membranes as chemically controlled systems to investigate the direct association of Bcl-xL with Bax and cBid and their effect on membrane permeabilization, in absence of other components and at different protein concentrations. We selected Giant- and Large Unilamellar Vesicles (GUV, LUV) for SFCCS and pore formation assays respectively, with a high cardiolipin

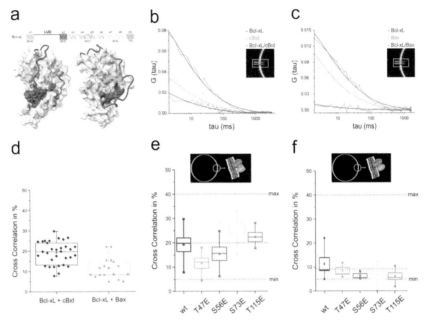

Fig. 1 Quantification of the interaction of Bcl-xL with cBid and with Bax in GUVs by SFCCS. (a) 2D and 3D structure schemes of Bcl-xL with its multiple Bcl-2 Homology (BH) regions and the Long Unordered Domain (LUD). (b–f) Analysis of the interactions between cBid, Bax and Bcl-xL on GUV membranes by SFCCS. (b) and (c) Representative SFCCS graphs for Bcl-xL with cBid (b) and with Bax (c) on GUV membranes. Red and green curves represent the auto-correlation of the fluorescence signal measured for each of the two labeled proteins. The blue curve represents the cross-correlation of the two signals. (d) Comparison of the association expressed in % cross-correlation (%CC) between Bcl-xL and cBid *versus* Bcl-xL and Bax, on GUV membranes. (e) and (f) Comparison of the association estimated from %CC between cBid (e) of Bax (f) with Bcl-xL-wt or phosphor-mimetic Bcl-xL variants on GUV membranes.

concentration as a simple model of the mitochondrial lipid composition that enables efficient binding of the proteins investigated. We could demonstrate that Bcl-xL modulates the localization dynamics of tBid and Bax in opposing ways *via* the formation of inhibitory complexes of different affinities. We also systematically evaluated the role of phosphorylation on the different activities of Bcl-xL using phospho-mimetic variants of sites previously reported in the literature. Our results suggest that regulation of apoptosis by the interplay between Bcl-2 proteins not only depends on the association between them, but also on the modulation of their respective redistribution dynamics.

Methods

Cell culture and transfection

The HCT116 allBcl-2 KO cell line was a gift from Prof. Xu Luo and was generated by knocking out all significant members of the Bcl-2 protein family. The cells were cultured in McCoy's 5A (modified) medium supplemented with 10% heat-activated fetal calf serum, 1% antibiotics (penicillin and streptomycin) at 37 °C and in the presence of 5% CO_2. All cells were regularly passaged at sub-confluence and plated at 4×10^4 cells per ml density. For confocal imaging experiments, the cells were seeded in LabTek chambers (Nunc) and for western blot analysis in 10 cm Petri dishes. Cells were transfected with the corresponding plasmids (8 ng for SFCCS, 10–25 ng for FLIP and 300 ng for western blot) and Lipofectamine 2000 (Life Technologies) according to manufacturer's specifications and protein overexpression was allowed to proceed between 6–72 hours depending on the experiment. For SFCCS, the medium was exchanged by air buffer (150 mM NaCl, 20 mM HEPES pH 7.4, 15 mM glucose, 150 µg ml^{-1} BSA, 20 mM trehalose, 5.4 MM KCl, 0.85 mM $MgSO_4$, 0.6 mM $CaCl_2$).

Protein expression, purification and labeling

The recombinant proteins full length human Bax and full length human Bid were expressed and purified as described in ref. 25. Purified Bid was cleaved by caspase 8 to cBid. Full length human Bcl-xL (Bcl-xL S4C, C151A) and phospho-mimetic Bcl-xL variants were over expressed in *E. coli* BL21/RIPL competent cells in a highly enriched medium at 37 °C and 180 rpm. At OD_{600} of 0.6 protein expression was induced with 1 mM IPTG for 4 hours at 20 °C. The bacteria were harvested by centrifugation at 6.000×g for 10 min. The bacterial pellets were flash-frozen in liquid nitrogen and stored at −80 °C. The bacterial pellets were thawed on ice, resuspended in buffer containing protease inhibitors (Merck, Darmstadt, Germany) and disrupted on ice by passing three times through the Emulsiflex C5 (Avestin, Mannheim, Germany). All the following steps were done at 4 °C and on ice. Immediately afterwards, ∼100 units of DNase I (Merck, Darmstadt, Germany) were added per liter bacterial culture and the lysates were incubated for 30 min on ice. Cellular debris were pelleted by centrifugation at 25 000 rpm (JA-25.50 Fixed-Angle Rotor and Beckman CoulterJ2 and Avanti Centrifuge) for 60 min at 4 °C.

Bax fused to intein-tag (plasmid pTYB1-Bax S4C, C62S, C126S) and Bcl-xL fused to intein-tag (plasmid pTYB2-Bcl-xL S4C, C151A) and phospho mimetic Bcl-xL variants (plasmids pTYB2-Bcl-xL S4C, C151A, T47E; pTYB2-Bcl-xL S4C, C151A, S56E; pTYB2-Bcl-xL S4C, C151A, S73E; pTYB2-Bcl-xL S4C, C151A, T115E) were purified following the manufacturer instructions. The intein-cleavage reaction was done overnight at 4 °C and the buffer was exchanged to 20 mM TRIS, pH 8 by dialysis for 4 hours at 4 °C. Afterwards the samples were further purified using an anion exchange column (HiTrap Q column from GE healthcare on an ÄKTA purifier FPLC system). The proteins were washed with ∼20 column volumes (CV) of buffer and then eluted with a gradient (8 CV) of high salt buffer (1 M NaCl, 20 mM TRIS, pH 8). The elution fractions were analyzed by SDS-PAGE and Coomassie staining. The buffer of the purest fraction was exchanged using dialysis (to 150 mM NaCl, 20 mM TRIS, pH 7.5 or 150 mM NaCL, 20 mM HEPES, pH 7.5).

Bid fused to His-tag (plasmid pET23-His-Bid) was purified using Ni-NTA Agarose beads (Qiagen, Hilden, Germany). Buffers and purification steps were done following the manufacturer instructions and the protein elution was done by a gradient adding 10, 25, and 250 nM imidazol to the elution buffer. Protein purity was investigated by SDS-PAGE and Coomassie staining. The buffer of the purest fraction was replaced by the caspase 8 cleavage buffer (50 mM NaCl, 5 mM DTT, 0.5 mM EDTA, 25 mM HEPES, 5% sucrose; pH 7.4) by dialysis. The mixture was incubated for 4 hours at RT to allow the cleavage of Bid to cBid (Bid/caspase 8 ratio

~1000 : 1; caspase 8 was a kind gift from Jean-Claude Martinou) and subsequently purified again based on Ni-NTA Agarose to remove caspase 8. Protein purity was tested by SDS PAGE. The protein purity of all the proteins was >95% (ESI Fig. 3†). A fraction of the proteins were aliquoted and flash frozen in liquid nitrogen and stored at −80 °C for further investigations by calcein release assays.

cBid was labeled with Alexa Fluor™ 647 C5 maleimide, Bax with Alexa Fluor™ 633 C5 maleimide and Bcl-xL with Alexa Fluor™ 488 C5 maleimide. For labeling, the proteins were incubated with 3× excess of TCEP for 30 min on ice, then with 10× excess of the label for 2 hours on ice, and another 5× excess of the label overnight at 4 °C. The proteins were separated from the free label using desalting columns. The concentration and the degree of labeling were calculated using a combination of Bradford assay and UV-vis spectroscopy. All the proteins were aliquoted in 10 μl portions and flash frozen in liquid nitrogen and stored at −80 °C.

Lipid mixture preparation, GUV formation and sample preparation

The lipid mixture composed of 80% L-α phosphatidylcholine (Egg, Chicken) (PC) and 20% cardiolipin (Heart, Bovine) (CL), or 70% PC and 30% CL (mol/mol) were used. All lipids were purchased from Avanti polar lipids (Alabaster, AL) and prepared in chloroform. Afterwards the chloroform was evaporated for 2 hours under vacuum and flash frozen in liquid nitrogen and stored at −80 °C.

The GUVs were formed by electro formation as described in ref. 26. In brief, 5 μg of the lipid mixture were distributed on the surface of the Pt electrodes of the electro formation system and incubated until the chloroform was evaporated, before immersion in 300 mM sucrose solution. Electro formation proceeded for 2 hours at 10 Hz, followed by 1 hour at 2 Hz.

For an assay with a final volume of 300 μL, 80 μL of the GUV suspension was added to a solution of 1× PBS buffer mixed with the proteins of interest in Lab-Tek 8-well chambered cover glasses (Nunc Lab-Tek Chamber Slide System, ThermoFischer Scientific) blocked by saturated casein suspension. The preparation of samples containing Bcl-xL/Bax or Bcl-xL/cBid or Bax/cBid was performed as described in ref. 26.

Calcein release assay

The LUVs were formed and the experiments done as described in ref. 27. LUVs were incubated with 20 nM cBid and varying concentrations of Bax (0–800 nM). The increase in fluorescence was monitored at room temperature for 2 hours in a Tecan Infinite M200 microplate reader (Tecan, Mannedorf, Switzerland). For the investigation of Bcl-xL activity in inhibiting the calcein release, varying concentrations of Bcl-xL-wt or phospho-mimetic variants (0–800 nM) were incubated with 20 nM cBid and 70 nM Bax in the presence of LUVs and the fluorescence intensity was measured as described for Bax. As a positive control, the LUVs were incubated with triton X-100 and, as negative controls, with the single proteins. The percentage of calcein released was calculated as reported in ref. 27.

Microscopy, scanning fluorescence cross-correlation spectroscopy (SFCCS) and fluorescence loss in photo-bleaching (FLIP)

All imaging experiments were performed using a Zeiss confocal Laser Scanning Microscope LSM 710 with ConfoCor 3, 40× N.A. water immersion and 63× N.A.

oil objectives, and excitation laser 488 and 561. SFCCS measurements on GUVs and the OMM as well as the quantification of the association strength were performed as described in ref. 26 and 28.

For FLIP analysis, the HCT116 allBcl-2 KO were seeded and transfected as described above. A region of 0.5 µM in the cytoplasm of a transfected cell was repeatedly bleached with two iterations of 100% power of the 488 nm laser. The time frame between the pulses was 30 s. Each bleach pulse was followed by two images. Next to the bleaching area, regions of interest (ROI) on mitochondria within a short and a long distance to the bleaching area, as well as ROI in a neighboring cell were also monitored for 600 s. The degree of loss in fluorescence in the ROIs was calculated from fluorescence intensity measurements using the Zeiss LSM software. The fluorescence intensities were normalized by setting the fluorescence in the beginning of the experiment to 100%. Only cells with comparable fluorescence intensities in the beginning of the measurement and within short distances (±20%) to the bleaching area were considered in the analysis.

Results

Bcl-xL interacts more strongly with tBid than with Bax in model membranes and in mitochondria of living cells

Fluorescence cross-correlation spectroscopy (FCCS) is a powerful method with single molecule sensitivity, where the fluorescence fluctuations due to the (co-) diffusion of individual proteins (tagged with spectrally different fluorophores) through the detection volume of a confocal microscope are auto- and cross-correlated to calculate diffusion coefficients, local concentrations and extent of complex formation.[29] In its scanning version, SFCCS, the focal volume is continuously scanned across the GUV membrane, which allows estimation of these parameters for labeled molecules embedded in membranes.[30,31]

By using SFCCS in giant unilamellar vesicles (GUVs), we have previously shown that the membrane plays an active role in modulating the strength of interactions between Bcl-xL and other Bcl-2 proteins.[32,33] This *in vitro* reconstituted model membrane system has the advantage of providing a chemically controlled environment where interactions between proteins embedded in membranes can be accurately quantified under different conditions. We therefore took advantage of this approach to systematically and quantitatively compare the interaction between Bcl-xL and the effector Bax, and between Bcl-xL and the BH3-only protein cBid.

We used a version of full length Bcl-xL (S4C, C151A) with a single cysteine for defined covalent labeling with maleimide fluorophores. We have shown in previous work that this mutant behaves like wild type Bcl-xL, and we subsequently refer here to this variant as Bcl-xL-wt. Additionally, we produced recombinant and fluorescently labeled Bax and cBid as reported before.[32–34] Here, we use the term cBid for recombinant Bid cleaved *in vitro* with caspase 8 and containing the two fragments p7 and p15. As a membrane model system, we used GUVs composed of 80 : 20 or 70 : 30 egg PC : bovine CL for Bcl-xL/cBid and Bcl-xL/Bax association, respectively.

To make the measurements comparable with each other, we used the same amounts of Bcl-xL-wt-488 for both BcL-xL/cBid and Bcl-xL/Bax association experiments. Additionally, only GUVs with a radius between 10 and 20 μm were considered. In the case of Bcl-xL/cBid, we incubated the GUVs with 40 nM Bcl-xL-wt-Al488 and 20 nM cBid-Al647 for 30 minutes at room temperature, followed by microscopy imaging to control the quality of the GUVs and subsequently SFCCS measurements. For quantification of Bcl-xL/Bax complexes, we incubated the GUVs with 40 nM Bcl-xL-wt-Al488 and 80 nM Bax-Al633 for 40 min at 43 °C in order to auto-activate Bax in absence of cBid.[33,34] From the analysis of the SFCCS measurements, we estimated the extent of association based on the average percentage of cross-correlation (%CC). In agreement with our previous studies,[33] we could demonstrate that in membranes the association of Bcl-xL with cBid (19 %CC) was stronger than that of Bcl-xL with Bax (8 %CC) (Fig. 1b–d).

We recently extended SFCCS to measure interactions between proteins in the more physiological settings of the intact MOM of living cells. We implemented suitable data acquisition protocols optimized for the complex structure and dynamic nature of mitochondria, which also consider the relatively slow diffusion of molecules in the mitochondrial membranes.[28] This gives us the opportunity to quantitatively compare the binding affinity between Bcl-xL and other Bcl-2 proteins in chemically controlled systems *versus* the complex native environment in the cell.

To quantify the association of Bcl-xL with tBid and with Bax without the influence of binding to other endogenous Bcl-2 proteins, we used a HCT 116 cell line lacking all Bcl-2 proteins (HCT116 allBcl-2 KO).[35] Accordingly, we transiently expressed mCherry-Bcl-xL and tBid-GFP or mCherry-Bcl-xL and GFP-Bax in HCT116 allBcl-2 KO cells. We visualized the mitochondrial localization of the transfected proteins in living cells with confocal microscopy (Fig. 2a), followed by SFCCS acquisition in the MOM of single cells. Only cells with similarly low expression levels of the mCherry- and GFP-tagged proteins were considered for SFFCS measurements. In agreement with the results obtained in GUVs, we found that mCherry-Bcl-xL associated more strongly with tBid-GFP (average 22 %CC) than with GFP-Bax (average 8 %CC) (Fig. 2b–d).

Bcl-xL stabilizes tBid, but not Bax in the mitochondrial membrane

It is known that the pro-survival protein Bcl-xL inhibits apoptosis not only by direct complex formation with the pro-apoptotic family members, but also by constant retrotranslocation of mitochondrial Bax back into the cytosol.[8,11]

To investigate the interplay of the mitochondrial/cytosolic shuttling of Bcl-xL on Bax and tBid, we applied Fluorescence Loss In Photo-bleaching (FLIP) as previously reported.[8,11] These measurements provide an estimation of the ability of these proteins to move between the mitochondria and cytoplasm compartments when they are expressed alone and in combination. For the same reasons mentioned above, we used the HCT116 allBcl-2 KO cell line for expression of GFP-Bax or tBid-GFP individually or in combination with mCherry-Bcl-xL. Since GFP-Bax overexpression directly kills HCT116 allBcl-2 KO cells, we optimized the expression conditions with low DNA amounts and short transfection times, and selected for our analysis cells that were clearly alive. A region in the cytoplasm of a transfected cell (white square) was repeatedly bleached for the whole time of the measurement (ESI Fig. 4†). The GFP- and mCherry-fluorescence intensities in the targeted cell were measured over the time of measurement in selected regions of

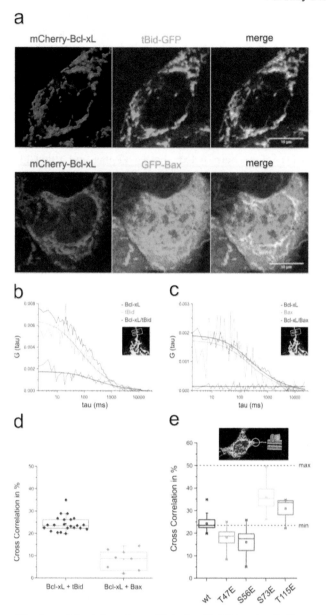

Fig. 2 Quantification of the interaction of Bcl-xL with tBid and with Bax on the Mitochondrial Outer Membrane (MOM) of intact cells by SFCCS. (a) Representative confocal images of HCT116 allBcl-2 KO cells transiently expressing mCherry-Bcl-xL-wt and tBid-GFP (upper panel) or mcherry-Bcl-xL-wt and GFP-Bax (lower panel). (b) and (c) Representative SFCCS graphs for mCherry-Bcl-xL-wt with cBid-GFP (b) and with GFP-Bax (c) at the MOM of living cells. Red and green curves represent the auto-correlation of the fluorescence signal measured for each of the two labeled proteins. The blue curve represents the cross-correlation of the two signals. (d) Comparison of the association in % CC between mCherry-Bcl-xL-wt and tBid-GFP and between mCherry-Bcl-xL-wt and GFP-Bax at the MOM of single cells. (e) Comparison of the association in %CC between tBid-GFP and mCherry-Bcl-xL-wt or phospho-mimetic mCherry-Bcl-xL variants at the MOM.

interest in the cytoplasm as well as on mitochondria within a short and a long distance to the bleaching area (red, green and blue circles respectively, ESI Fig. 4†). To check that the measurement itself does not affect the fluorescence intensity and does not induce photo-bleaching, the GFP- and mCherry-fluorescence intensities in a region of interest of a neighboring cell were also measured (black circle, ESI Fig. 4†). We then analyzed the decay in fluorescence intensity considering only regions of interest with comparable fluorescence intensities in the beginning of the measurement and within short distances (±20%) to the bleaching area (Fig. 3).

In the absence of mCherry-Bcl-xL, both GFP-Bax and tBid-GFP presented a high exchange rate between the mitochondria and cytoplasm compartments and a low immobilized fraction (Fig. 3 and 4d, e). Under our experimental conditions, the overexpression of mCherry-Bcl-xL did not significantly alter GFP-Bax exchange rate between the mitochondria and cytoplasm compartments (55% in the absence of Bcl-xL and 41% in the presence of Bcl-xL), or its immobilized fraction at the mitochondria (21% in the absence of Bcl-xL and 23% in the presence of Bcl-xL) (Fig. 3b, d and 4e). To our surprise, mCherry-Bcl-xL over-expression caused a much lower exchange rate (12%) and a higher immobilized fraction of tBid-GFP at the mitochondria (54%) (Fig. 3a, c and 4d). These data

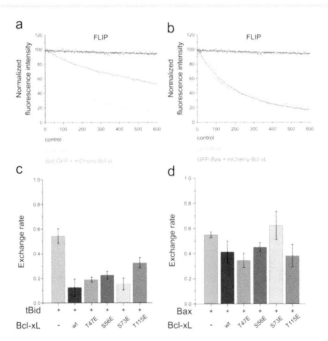

Fig. 3 Mitochondrial/cytosolic shuttling of Bax and tBid measured by FLIP is affected by Bcl-xL. (a) and (b) Normalized loss of fluorescence intensity over time during photo-bleaching for tBid-GFP (a) and GFP-Bax (b) at mitochondria in the absence (blue/green) and presence (purple/red) of overexpressed mCherry-Bcl-xL-wt. Fluorescence of the neighboring cell is shown as control (black). (c) and (d) The exchange rates between the mitochondria and cytoplasm compartments calculated for tBid–GFP (c) and GFP–Bax (d) in the absence and presence of overexpressed GFP-Bcl-xL-wt and its phospho-mimetic variants.

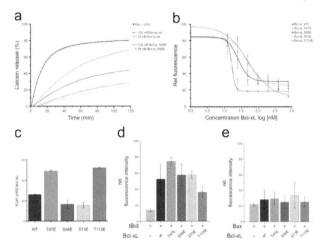

Fig. 4 Anti-apoptotic activity of recombinant Bcl-xL and phospho-mimetic variants. (a) Calcein release from LUVs incubated with recombinant Bax and cBid in the absence (positive control, curve in dark brown) and presence of recombinant Bcl-xL-wt and phospho-mimetic variant S56E. (b) Dose response analysis of Bcl-xL effect on the inhibition of the tBid/Bax-induced liposome permeabilization. (c) Inhibitory Concentration for half-maximal values (IC50) of Bcl-xL-wt and its phospho-mimetic variants in the calcein release assay. (d) and (e) The relative (Rel.) fluorescence intensities represent the immobile fractions of tBid-GFP (d) and GFP-Bax (e) in the absence and presence of overexpressed GFP-Bcl-xL-wt and phospho-mimetic variants measured by FLIP.

indicate that, in contrast to Bax, Bcl-xL stabilizes tBid at the mitochondria and reduces its exchange with the cytosol.

Phospho-mimetic variants of Bcl-xL show a pronounced potency to inhibit Bax pore formation

Several phosphorylation sites have been identified in Bcl-xL, but their functional effect remains unclear for most of them. To explore the role of Bcl-xL phosphorylation on its interplay with other Bcl-2 proteins, we selected Bcl-xL phospho-mimetic versions of sites reported in the literature that would affect its ability to inhibit cBid/Bax pore formation *in vitro*, as described below. Only the variants with altered activity are shown here.[36–38]

We produced different recombinant phospho-mimetic Bcl-xL variants in which single serine or threonine residues were exchanged by glutamic acid, which mimics both structurally and in charge the phosphorylated form of these residues. Three of the described phosphorylation sites, T47, S56 and S73 are located within the flexible LUD between α1 and α2 and only one, the T115 is located in the loop between α3 and α4 (Fig. 1a).

We used an *in vitro* assay of liposome permeabilization by tBid-activated Bax.[25,27,39] This strategy has the advantage that it is based on a chemically controlled system without unknown components, enabling us to uniquely test the inhibitory activity of Bcl-xL at different concentrations and perform a quantitative comparison. We measured the release of calcein from large unilamellar vesicles (LUVs) composed of egg PC and bovine CL (80 : 20). First we determined the lowest concentration of Bax required to permeabilize 80% of the liposomes in the presence of cBid, which served as a positive control (ESI Fig. 5a†). As negative controls to verify that the recombinant Bcl-xL variants do not have permeabilizing activity themselves, we tested the individual proteins for their ability to permeabilize LUVs (ESI Fig. 5b†).

Representative curves of cBid/Bax-induced calcein release over time in absence and presence of Bcl-xL-wt and Bcl-xL-S56E are shown in Fig. 4a, where the stronger inhibitory capacity of the mutant form can be observed. In order to systematically compare the efficacy of the different phospho-mimetic Bcl-xL variants to inhibit cBid/Bax-induced pore formation and to test whether their activity is concentration dependent, we quantified their dose response behavior over a concentration range of 12.5–800 nM at constant Bax and cBid concentrations and using Bcl-xL-wt as a reference. From the normalized calcein release we calculated the IC50 of each variant compared to Bcl-xL-wt (Fig. 4b). Our data show that Bcl-xL variants T47E and T115E had a lower inhibitory activity, with IC50 of 40 nM and 55 nM, respectively, while variants S56E and S73E were more potent inhibitors, with IC50 of 17 nM and 15 nM, respectively, compared to Bcl-xL-wt with IC50 of 26 nM (Fig. 4b and c). This comparative analysis reveals that phosphorylation at different sites in Bcl-xL can lead to either an increase or to a loss of its anti-apoptotic activity.

Phospho-mimetic variants of Bcl-xL exhibit altered affinity to activated Bid but not Bax

To compare whether Bcl-xL-phosphorylation affects its association with cBid and Bax in model membranes, we performed SFCCS of the different phospho-mimetic Bcl-xL variants with cBid or Bax on GUVs. Interestingly, phospho-mimetic Bcl-xL variants S73E and T115E showed a higher association with cBid compared to the wild type protein, with average cross-correlation values of 26% and 22%, respectively (Fig. 1e). In contrast, phospho-mimetic Bcl-xL variants T47E and S56E presented lower association to cBid compared to Bcl-xL-wt, with average cross-correlation values of 12% and 16%, respectively (Fig. 1e). These findings indicate that depending on the site, phospho-mimetics of Bcl-xL have opposing effects with respect to the strength of the interactions with cBid.

Unlike the association with cBid, we could not detect significant changes in the association of Bax with the different phospho-mimetic variants of Bcl-xL (Fig. 1f). These results suggest that the phosphorylation sites tested do not have the ability to significantly strengthen the affinity between these two proteins. However, the low %CC in our experimental system for the binding between Bax and wild type Bcl-xL in membranes provides a very small window to lower values and we cannot exclude that phosphorylation could have a negative impact on their interaction.

Next, we transiently expressed mCherry-Bcl-xL-wt or phospho-mimetic mCherry-Bcl-xL variants and tBid-GFP in HCT116 allBcl-2 KO cells. By imaging the distribution of the fluorescently tagged proteins using live cell confocal microscopy, we found that phospho-mimetic variants of Bcl-xL maintained predominant mitochondrial localization. We then performed SFCCS measurements in the OMM of single cells and calculated the corresponding diffusion coefficients and %CC. Importantly, the results of the quantitative analysis of the interactions between the Bcl-xL variants and tBid-GFP were in excellent agreement with the results obtained in GUVs. Mutations T47E and S56E in mCherry-Bcl-xL decreased the association with tBid-GFP, with %CC of 18 and 16, while mutations S73E and T115 increased the affinity between both proteins to 36 and 31 %CC, respectively (Fig. 2e).

Given the low %CC between mCherry-Bcl-xL and GFP-Bax and the higher error of SFCCS in mitochondria of living cells, we could not analyze the effect of Bcl-xL phospho-mimetic mutations on the binding to GFP-Bax, which in any case did not increase (not shown).

We reasoned that, beyond altering the affinity for other Bcl-2 proteins, phosphorylation of Bcl-xL might also affect apoptosis by modifying its retrotranslocation activity. To explore this possibility, we evaluated the ability of the phospho-mimetic variants of mCherry-Bcl-xL to retrotranslocate GFP-Bax and tBid-GFP. As shown in Fig. 3d, the mitochondrial shuttling of GFP-Bax was not largely affected by the co-expression of different mCherry-Bcl-xL phosphomimetics. Despite the changes in binding affinities of the different Bcl-xL mutants detected by SFFCS, all variants still maintained the ability to retain tBid-GFP in mitochondria. Only mCherry-Bcl-xL-T115E presented significantly decreased mitochondrial stabilization of GFP-tBid compared to the wild type protein. These findings indicate that the binding affinity between tBid and Bcl-xL phospho-mimetic variants does not directly correlate with the ability of Bcl-xL to stabilize tBid in mitochondria.

In addition to changes in the binding affinity to other Bcl-2 proteins and in its retrotranslocation activity, phosphorylation of Bcl-xL could also play a regulatory role at the level of protein stability in the cell. Furthermore, the differences in binding and retrotranslocation detected so far could be due to altered lifetime of the phospho-mimetic Bcl-xL variants. To test for this possibility, we analyzed protein expression levels under our experimental conditions. HCT116 allBcl-2 KO were transfected transiently with Bcl-xL-wt or the phospho-mimetic Bc-xL variants and the proteins were expressed for 24, 48 and 72 hours. The expression levels of the proteins analyzed by western blot (ESI Fig. 2†) confirmed that the protein levels of all the proteins changed in a similar way. These results suggest that phosphorylation of Bcl-xL does not modulate apoptosis by changing the lifetime of the protein.

Discussion

In previous work we quantified a reconstituted minimal Bcl-2 interactome in solution and in model membranes, where we could demonstrate that Bcl-xL prefers to interact with cBid than with Bax.[33] Now, for the first time, we quantitatively compared the interactions between these proteins directly in the mitochondria of living cells using a variant of SFCCS we recently implemented for tubular organelles.[28] Our findings show that complex formation between mCherry-Bcl-xL and GFP-Bax in mitochondria occurs to a lesser extent than between Bc-xL and tBid under the same experimental conditions. This is in absence of other endogenous Bcl-2 proteins that might compete for the interaction, although we cannot exclude the role of other cellular components. However, these results are in good agreement with the hierarchy of interactions obtained in our previous measurements *in vitro*, thereby supporting the "integrated" model for the Bcl-2 network.[33]

Yet, despite the low levels of binding between GFP-Bax and mCherry-Bcl-xL, co-expression of the two proteins rescued the large extent of apoptosis induced by GFP-Bax in the HCT116 allBcl-2 KO cell line.[35] This demonstrates that mCherry-Bcl-xL exerted its inhibitory function without strong binding to GFP-Bax. Instead, we consistently detected a largely cytosolic distribution and high retro-translocation rate of GFP-Bax in most cells co-expressing both proteins. These results suggest that the mechanism of inhibition *via* retrotranslocation promoted by Bcl-xL[8,11] dominates over inhibition of Bax *via* complex formation at mito-chondria.[7] The quantitative nature of the analysis performed here allowed us for the first time to compare the relative importance of these two modes of inhibition by Bcl-xL.

Importantly, we show that both Bax and tBid have the ability to constitutively shuttle between cytosol and mitochondria in the absence of other Bcl-2 proteins. While Bcl-xL did not extensively affect the retrotranslocation rate of inactive Bax, it prevented its accumulation in mitochondria where Bcl-xL and Bax formed complexes to a lower extent. In contrast, Bcl-xL clearly reduced the shuttling rate of tBid and stabilized it on mitochondria, where Bcl-xL and tBid formed tight complexes. This is surprising, since it indicates that Bcl-xL inhibits apoptosis *via* opposite effects on Bax and tBid when it comes to complex formation and mitochondrial accumulation. Yet, we still lack the necessary structural under-standing to explain these differences at the atomistic level.

It is important to note that we measured the shuttling rate of GFP-Bax between the cytosol and the mitochondria in healthy HCT116 allBcl-2 KO cells. In these cells most if not all GFP-Bax molecules were in their inactive form. This could only be achieved in cells with low GFP-Bax levels and short expression times. Based on this, we propose that the retrotranslocation rate calculated here corresponds to inactive Bax populations. According to such a model, Bcl-xL would then not affect the retrotranslocation rate *per se*, but instead the proportion of Bax molecules in an inactive conformation, in agreement with Gilmore and coworkers.[11]

Our previous data in *in vitro* reconstituted systems clearly demonstrate that Bcl-xL reduces the amount of Bax associated to the membrane in the absence of any other component, thereby constituting the minimal machinery that shifts the equilibrium of Bax conformational and activation states towards its dissociation from the membrane in a soluble form,[33,34] as observed in cells.[8] Since Bax acti-vation is a multi-step process[40] that can happen spontaneously,[35,41] our data support a model in which Bax alone shuttles between a soluble and a membrane-bound, inactive state. Contact with the membrane would catalyze the probability of transition to the active Bax conformation, which would be blocked by complex formation with Bcl-xL. This would effectively shift the equilibrium to the membrane-bound, inactive Bax state that can spontaneously dissociate from the membrane due to the low affinity for Bcl-xL. In the absence of pro-survival Bcl-2 proteins, continued exposure to the membrane would eventually lead to the spontaneous activation of a few Bax molecules, which would self-amplify *via* auto-activation of additional Bax molecules,[33,42] leading to the opening of membrane pores, to MOMP and to apoptotic cell death.[35]

Traditionally, most attention has been paid to the effect of phosphorylation on the binding affinity to other Bcl-2 proteins,[24] and other activities of Bcl-xL have been neglected. Here we took a more comprehensive and quantitative approach aimed at evaluating the effect of Bcl-xL phosphorylation on its potential pro-survival activities using mimetics of the post-translational modification. While we could demonstrate that Bcl-xL phospho-mimetic variants resulted in altered complex formation with tBid or in a different tendency to stabilize tBid in mitochondria, these effects did not correlate with each other, neither showed a clear trend on the ability of Bcl-xL to inhibit cBid/Bax-mediated pore formation. For example, mutation T115E increased the extent of binding to tBid, yet it stabilized it less at mitochondria and was less capable of blocking Bax compared to the wild type protein. Mutation T47E also reduced the inhibitory activity of Bcl-xL, but it instead reduced complex formation with tBid and did not have a strong effect on its shuttling rate. None of these mutations had a strong effect on Bax retrotranslocation or binding. Phosphorylation at these two sites has been linked with apoptosis induction.[36] Mutation S73E increased the inhibitory power of Bcl-xL, which correlated with higher affinity for tBid without affecting its shuttling rate, while it increased the retrotranslocation of Bax to the levels measured in absence of Bcl-xL. However, this phospho-mimetic in cells is cytotoxic.[37] In the case of S56E, Bcl-xL also was a more potent inhibitor of pore formation, but it interacted less with tBid without affecting much the shuttling of tBid nor Bax. These results reveal a complex scenario still lacking a general basis for the regulation of Bcl-xL by post-translational modifications that will deserve further investigation. Additional activities of Bcl-xL, like its effect on mitochondrial structure and function, on metabolism or on calcium signaling,[3] may also need to be taken into account.

Conclusions

Taken together, our study combining quantifications of interactions and cytosol/mitochondria shuttling of Bcl-2 proteins has further implications for our understanding of how these proteins regulate apoptosis. It supports a new paradigm beyond the Bcl-2 interactome as a single determinant of MOMP, where the network of Bcl-2 proteins regulates not only which complexes are formed, but also their shuttling rate between cellular compartments and their stabilization at mitochondria.

Conflicts of interest

There are no conflicts to declare.

Acknowledgements

We thank Hector Flores Romero, John SH Danial and Joseph Unsay for helpful discussions and Carolin Stegmüller for technical assistance. This work was funded by the Deutsche Forschungsgemeinschaft (DFG, German Research Foundation) – 232935877/FOR2036.

References

1 J. M. Adams and S. Cory, *Cell Death Differ.*, 2018, **25**, 27–36.
2 L. Galluzzi, I. Vitale, S. A. Aaronson, J. M. Abrams, D. Adam, P. Agostinis, E. S. Alnemri, L. Altucci, I. Amelio, D. W. Andrews, M. Annicchiarico-

Petruzzelli, A. V. Antonov, E. Arama, E. H. Baehrecke, N. A. Barlev, N. G. Bazan, F. Bernassola, M. J. M. Bertrand, K. Bianchi, M. V. Blagosklonny, K. Blomgren, C. Borner, P. Boya, C. Brenner, M. Campanella, E. Candi, D. Carmona-Guticrrcz, F. Cccc005i, F. K. Chan, N. S. Chandel, E. H. Cheng, J. E. Chipuk, J. A. Cidlowski, A. Ciechanover, G. M. Cohen, M. Conrad, J. R. Cubillos-Ruiz, P. E. Czabotar, V. D'Angiolella, T. M. Dawson, V. L. Dawson, V. De Laurenzi, R. De Maria, K. M. Debatin, R. J. DeBerardinis, M. Deshmukh, N. Di Daniele, F. Di Virgilio, V. M. Dixit, S. J. Dixon, C. S. Duckett, B. D. Dynlacht, W. S. El-Deiry, J. W. Elrod, G. M. Fimia, S. Fulda, A. J. Garcia-Saez, A. D. Garg, C. Garrido, E. Gavathiotis, P. Golstein, E. Gottlieb, D. R. Green, L. A. Greene, H. Gronemeyer, A. Gross, G. Hajnoczky, J. M. Hardwick, I. S. Harris, M. O. Hengartner, C. Hetz, H. Ichijo, M. Jaattela, B. Joseph, P. J. Jost, P. P. Juin, W. J. Kaiser, M. Karin, T. Kaufmann, O. Kepp, A. Kimchi, R. N. Kitsis, D. J. Klionsky, R. A. Knight, S. Kumar, S. W. Lee, J. J. Lemasters, B. Levine, A. Linkermann, S. A. Lipton, R. A. Lockshin, C. Lopez-Otin, S. W. Lowe, T. Luedde, E. Lugli, M. MacFarlane, F. Madeo, M. Malewicz, W. Malorni, G. Manic, J. C. Marine, S. J. Martin, J. C. Martinou, J. P. Medema, P. Mehlen, P. Meier, S. Melino, E. A. Miao, J. D. Molkentin, U. M. Moll, C. Munoz-Pinedo, S. Nagata, G. Nunez, A. Oberst, M. Oren, M. Overholtzer, M. Pagano, T. Panaretakis, M. Pasparakis, J. M. Penninger, D. M. Pereira, S. Pervaiz, M. E. Peter, M. Piacentini, P. Pinton, J. H. M. Prehn, H. Puthalakath, G. A. Rabinovich, M. Rehm, R. Rizzuto, C. M. P. Rodrigues, D. C. Rubinsztein, T. Rudel, K. M. Ryan, E. Sayan, L. Scorrano, F. Shao, Y. Shi, J. Silke, H. U. Simon, A. Sistigu, B. R. Stockwell, A. Strasser, G. Szabadkai, S. W. G. Tait, D. Tang, N. Tavernarakis, A. Thorburn, Y. Tsujimoto, B. Turk, T. Vanden Berghe, P. Vandenabeele, M. G. Vander Heiden, A. Villunger, H. W. Virgin, K. H. Vousden, D. Vucic, E. F. Wagner, H. Walczak, D. Wallach, Y. Wang, J. A. Wells, W. Wood, J. Yuan, Z. Zakeri, B. Zhivotovsky, L. Zitvogel, G. Melino and G. Kroemer, *Cell Death Differ.*, 2018, **25**, 486–541.

3 A. Pena-Blanco and A. J. Garcia-Saez, *FEBS J.*, 2018, **285**, 416–431.

4 A. J. Garcia-Saez, *Cell Death Differ.*, 2012, **19**, 1733–1740.

5 K. Cosentino and A. J. Garcia-Saez, *Trends Cell Biol.*, 2017, **27**, 266–275.

6 R. Salvador-Gallego, M. Mund, K. Cosentino, J. Schneider, J. Unsay, U. Schraermeyer, J. Engelhardt, J. Ries and A. J. Garcia-Saez, *EMBO J.*, 2016, **35**, 389–401.

7 F. Llambi, T. Moldoveanu, S. W. Tait, L. Bouchier-Hayes, J. Temirov, L. L. McCormick, C. P. Dillon and D. R. Green, *Mol. Cell*, 2011, **44**, 517–531.

8 F. Edlich, S. Banerjee, M. Suzuki, M. M. Cleland, D. Arnoult, C. X. Wang, A. Neutzner, N. Tjandra and R. J. Youle, *Cell*, 2011, **145**, 104–116.

9 F. Todt, Z. Cakir, F. Reichenbach, R. J. Youle and F. Edlich, *Cell Death Differ.*, 2013, **20**, 333–342.

10 F. Todt, Z. Cakir, F. Reichenbach, F. Emschermann, J. Lauterwasser, A. Kaiser, G. Ichim, S. W. Tait, S. Frank, H. F. Langer and F. Edlich, *EMBO J.*, 2015, **34**, 67–80.

11 B. Schellenberg, P. Wang, J. A. Keeble, R. Rodriguez-Enriquez, S. Walker, T. W. Owens, F. Foster, J. Tanianis-Hughes, K. Brennan, C. H. Streuli and A. P. Gilmore, *Mol. Cell*, 2013, **49**, 959–971.

12 T. Kuwana, L. E. King, K. Cosentino, J. Suess, A. J. Garcia-Saez, A. P. Gilmore and D. D. Newmeyer, *J. Biol. Chem.*, 2020, **295**, 1623–1636.

13 A. Basu and S. Haldar, *FEBS Lett.*, 2003, **538**, 41–47.

14 B. S. Chang, A. J. Minn, S. W. Muchmore, S. W. Fesik and C. B. Thompson, *EMBO J.*, 1997, **16**, 968–977.

15 M. Upreti, E. N. Galitovskaya, R. Chu, A. J. Tackett, D. T. Terrano, S. Granell and T. C. Chambers, *J. Biol. Chem.*, 2008, **283**, 35517–35525.

16 D. T. Terrano, M. Upreti and T. C. Chambers, *Mol. Cell. Biol.*, 2010, **30**, 640–656.

17 R. J. Clem, E. H. Y. Cheng, C. L. Karp, D. G. Kirsch, K. Ueno, A. Takahashi, M. B. Kastan, D. E. Griffin, W. C. Earnshaw, M. A. Veliuona and J. M. Hardwick, *Proc. Natl. Acad. Sci. U. S. A.*, 1998, **95**, 554–559.

18 L. Du, C. S. Lyle and T. C. Chambers, *Oncogene*, 2005, **24**, 107–117.

19 L. Saraiva, R. D. Silva, G. Pereira, J. Goncalves and M. Corte-Real, *J. Cell Sci.*, 2006, **119**, 3171–3181.

20 J. Wang, M. Beauchemin and R. Bertrand, *Cell. Signalling*, 2011, **23**, 2030–2038.

21 S. H. Dho, B. E. Deverman, C. Lapid, S. R. Manson, L. Gan, J. J. Riehm, R. Aurora, K. S. Kwon and S. J. Weintraub, *PLoS Biol.*, 2013, **11**, e1001588.

22 R. J. Youle and A. Strasser, *Nat. Rev. Mol. Cell Biol.*, 2008, **9**, 47–59.

23 A. Shamas-Din, J. Kale, B. Leber and D. W. Andrews, *Cold Spring Harbor Perspect. Biol.*, 2013, **5**, a008714.

24 A. V. Follis, F. Llambi, H. Kalkavan, Y. Yao, A. H. Phillips, C. G. Park, F. M. Marassi, D. R. Green and R. W. Kriwacki, *Nat. Chem. Biol.*, 2018, **14**, 458–465.

25 S. Bleicken, C. Wagner and A. J. Garcia-Saez, *Biophys. J.*, 2013, **104**, 421–431.

26 F. Murad and A. J. Garcia-Saez, *Methods Mol. Biol.*, 2019, **1877**, 337–350.

27 A. J. Garcia-Saez, M. Coraiola, M. Dalla Serra, I. Mingarro, G. Menestrina and J. Salgado, *Biophys. J.*, 2005, **88**, 3976–3990.

28 J. D. Unsay, F. Murad, E. Hermann, J. Ries and A. J. Garcia-Saez, *ChemPhysChem*, 2018, **19**, 3273.

29 J. Ries, Z. Petrasek, A. J. Garcia-Saez and P. Schwille, *New J. Phys.*, 2010, **12**, 113009.

30 J. Ries and P. Schwille, *Biophys. J.*, 2006, **91**, 1915–1924.

31 A. J. Garcia-Saez and P. Schwille, *Methods*, 2008, **46**, 116–122.

32 A. J. Garcia-Saez, J. Ries, M. Orzaez, E. Perez-Paya and P. Schwille, *Nat. Struct. Mol. Biol.*, 2009, **16**, 1178–1185.

33 S. Bleicken, A. Hantusch, K. K. Das, T. Frickey and A. J. Garcia-Saez, *Nat. Commun.*, 2017, **8**, 73.

34 Y. Subburaj, K. Cosentino, M. Axmann, E. Pedrueza-Villalmanzo, E. Hermann, S. Bleicken, J. Spatz and A. J. Garcia-Saez, *Nat. Commun.*, 2015, **6**, 8042.

35 K. L. O'Neill, K. Huang, J. Zhang, Y. Chen and X. Luo, *Genes Dev.*, 2016, **30**, 973–988.

36 S. Kharbanda, S. Saxena, K. Yoshida, P. Pandey, M. Kaneki, Q. Wang, K. Cheng, Y. N. Chen, A. Campbell, T. Sudha, Z. M. Yuan, J. Narula, R. Weichselbaum, C. Nalin and D. Kufe, *J. Biol. Chem.*, 2000, **275**, 322–327.

37 J. Megyesi, A. Tarcsafalvi, N. Seng, R. Hodeify and P. M. Price, *Cell Death Discovery*, 2016, **2**, 15066.

38 Y. Tamura, S. Simizu, M. Muroi, S. Takagi, M. Kawatani, N. Watanabe and H. Osada, *Oncogene*, 2009, **28**, 107–116.

39 S. Bleicken, O. Landeta, A. Landajuela, G. Basanez and A. J. Garcia-Saez, *J. Biol. Chem.*, 2013, **288**, 33241–33252.

40 J. F. Lovell, L. P. Billen, S. Bindner, A. Shamas-Din, C. Fradin, B. Leber and D. W. Andrews, *Cell*, 2008, **135**, 1074–1084.

41 H. C. Chen, M. Kanai, A. Inoue-Yamauchi, H. C. Tu, Y. Huang, D. Ren, H. Kim, S. Takeda, D. E. Reyna, P. M. Chan, Y. T. Ganesan, C. P. Liao, E. Gavathiotis, J. J. Hsieh and E. H. Cheng, *Nat. Cell Biol.*, 2015, **17**, 1270–1281.

42 C. Tan, P. J. Dlugosz, J. Peng, Z. Zhang, S. M. Lapolla, S. M. Plafker, D. W. Andrews and J. Lin, *J. Biol. Chem.*, 2006, **281**, 14764–14775.

Faraday Discussions

PAPER

Lipid distributions and transleaflet cholesterol migration near heterogeneous surfaces in asymmetric bilayers†

Elio A. Cino,*[ab] Mariia Borbuliak,[a] Shangnong Hu[a] and D. Peter Tieleman *[a]

Received 4th January 2021, Accepted 21st April 2021

DOI: 10.1039/d1fd00003a

Specific and nonspecific protein–lipid interactions in cell membranes have important roles in an abundance of biological functions. We have used coarse-grained (CG) molecular dynamics (MD) simulations to assess lipid distributions and cholesterol flipping dynamics around surfaces in a model asymmetric plasma membrane containing one of six structurally distinct entities: aquaporin-1 (AQP1), the bacterial β-barrel outer membrane proteins OmpF and OmpX, the KcsA potassium channel, the WALP23 peptide and a carbon nanotube (CNT). Our findings revealed varied lipid partitioning and cholesterol flipping times around the different solutes and putative cholesterol binding sites in AQP1 and KcsA. The results suggest that protein–lipid interactions can be highly variable, and that surface-dependent lipid profiles are effectively manifested in CG simulations with the Martini force field.

Introduction

Biological membranes are fundamental to life as we know it. They regulate signaling, compartmentalization and the flow of ions and other substances. Proteins are an integral component of cellular membranes and have essential roles in conferring distinct functionalities, which are often mediated through interactions with lipid molecules.[1,2] Prominent examples of biochemical process that involve protein–lipid interactions include regulation of sperm–egg fusion by the lipid anchoring of CD9/CD81,[3] and the maintenance of lung homeostasis *via* the association of surfactant proteins with cholesterol and phospholipids.[4]

Although protein–lipid interactions are known to be involved in numerous biological processes, their experimental characterization can be a challenge and MD simulations with all-atom (AA) and CG approximations have become

[a]Centre for Molecular Simulation and Department of Biological Sciences, University of Calgary, 2500 University Drive NW, Calgary, Alberta, T2N 1N4, Canada. E-mail: tieleman@ucalgary.ca; elio.cino@ucalgary.ca

[b]Department of Biochemistry and Immunology, Federal University of Minas Gerais, Belo Horizonte, MG, Brazil

† Electronic supplementary information (ESI) available. See DOI: 10.1039/d1fd00003a

important tools for their characterization.[5] While AA models can resolve molecular details more precisely, CG models can provide improved sampling of rather slow processes, such as the equilibration of lipid distributions and cholesterol movement between leaflets.[6] The lipid profiles in the vicinities of integral membrane proteins can be important for their functionality and MD simulations have proven useful in studying the associations between proteins and specific lipid types.[7–9] Membrane lipids often organize into distinct microdomains, or lipid rafts, of varying compositions around different proteins.[10] Using CG simulations, it has been shown that the phospholipid and cholesterol distributions in a membrane can be specific to the GPCR type and conformational state.[11] In addition to the lateral movement within monolayers, lipids can also transition between leaflets, which can be induced by the presence of certain proteins.[12] For example, ABCA1 specifically promotes cholesterol translocation from the inner to outer leaflet for the regulation of cellular events without altering the total cholesterol content in the membrane.[13] Flipping between leaflets occurs relatively slowly for phospholipids, but sterols, such as cholesterol, can exchange between leaflets on the microsecond timescale, which can be studied using MD simulations at the CG level.[14]

Here, we performed a systematic comparison of lipid distributions and cholesterol flipping in an asymmetric bilayer around six distinct surfaces, including eukaryotic and bacterial proteins, as well as synthetic entities (Fig. 1). Though most bacteria do not synthesize cholesterol, some can obtain it from their environment and many can synthesize sterol-like lipids, which may act as functional analogs.[15] Nevertheless, the different entities included in the study were primarily selected to encompass a range of chemical and structural diversity to identify features of lipid distributions that have a generic basis *versus* a possible functional importance. The systems were designed to be potentially useful in the future as model systems to test the agreement (or lack thereof) between Martini simulations, the new Martini 3 version,[16] further development of Martini 3 and

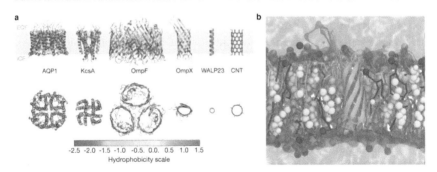

Fig. 1 An asymmetric model bilayer embedded with different membrane proteins and a carbon nanotube. (a) Membrane proteins and carbon nanotube models. AQP1, KcsA, OmpF, OmpX and the WALP23 peptide are shown in cartoon style, and CNT with ball and stick, colored from low to high Eisenberg hydrophobicity[17] with a blue-green-red gradient. The upper and lower panels show the side and top views, respectively. ECF: extracellular fluid; ICF: intracellular fluid. (b) Initial state of the OmpX system: OmpX (gray), cholesterol (yellow), POPC (purple), DPSM (cyan), POPE (blue), POPS (magenta) and water (translucent).

atomistic simulations. Variations in lipid localization and cholesterol flipping dynamics were observed around the different surfaces in the individual leaflets, suggesting that protein–lipid interactions can be highly variable and that unique lipid environments may be recruited by specific proteins.

Methods

System construction

Solvated asymmetric bilayer systems and topologies were built using the Martini Maker[18] module of CHARMM-GUI[19] and INSANE (INSert membrANE),[20] with the Martini v2.2 (ref. 21) parameter set and nonpolarizable water. The systems were composed of 1-palmitoyl-2-oleoyl-phosphatidylcholine (POPC) and N-stearoyl-D-erythro-sphingomyelin (DPSM) at 1 : 1 ratios in the upper leaflet, and 1-palmitoyl-2-oleoyl-phosphatidylethanolamine (POPE) and 1-palmitoyl-2-oleoyl-phosphatidylserine (POPS) at ~2 : 1 ratios in the lower leaflet, with equal numbers of cholesterol (CHOL) molecules in both leaflets, and one of AQP1-1J4N,[22] KcsA-1R3J,[23] OmpF-3POX,[24] OmpX-1QJ9,[25] the WALP23 peptide or a CNT (Table S1†). Two lipid only systems, one with half the number of cholesterol molecules, were also built (Table S1†). The lipid composition used was approximate to that of a simplified mammalian red blood cell membrane.[26] To conserve the protein structure, an elastic network was applied on the atom pairs within 0.9 nm with a 500 kJ mol^{-1} force constant.[27] A Martini-compatible CNT structure and topology was generated using cnt-martini.[28,29] The CNT consisted of 12 rings, with 8 beads per ring, giving a length of 4.5 nm and diameter of 1.2 nm. For the top and bottom rings, the CNP beads were substituted for more polar SNda beads. Atomistic AQP1 systems and topologies were generated by backmapping the final frame of each replica using CHARMM-GUI, the CHARMM36m force field[30] and the TIP3P water model (Table S1†).

MD simulations

Energy minimization, stepwise equilibration and final production runs at 310 K were carried out with GROMACS 2018 (ref. 31) using the default CHARMM-GUI generated run parameter files. Integration time steps of 20 and 2 fs were used for the CG and AA simulations, respectively. Each CG system was simulated in triplicate for 60 μs (3 × 20 μs), while 3 × 1 μs runs were conducted for the AA AQP1 systems.

Data analysis

Considering the periodicity of the simulation box, all systems were centered around the protein or CNT center of mass and aligned to remove rotation and translation. The gmx select tool was used to dynamically compute the number of lipids around the different solutes using the PO4 beads for POPC, DPSM, POPE and POPS, and ROH beads for CHOL. Cholesterol flipping between the upper and lower leaflets was assessed by calculating the angle between the C2-ROH vector of the cholesterol molecule and the bilayer normal. Successful flips were those that exhibited a stable vector sign change and angle difference of > 120° for at least 10 ns. The average flipping time for each cholesterol was measured as the time duration between flipping events and plotted as a function of the distance to the protein or CNT. Unless shown otherwise, data are presented as an average of the three replicas for each system over the second half (last 10 μs) of the trajectories. VMD was used for system visualization.[32] Numpy and Matplotlib were used for statistical analysis and data plotting.[33,34]

Results

To assess the system equilibration and reproducibility across the replicas, the number of contacts between the membrane lipids and embedded proteins or the CNT were calculated throughout the simulations (Fig. S1 and S2†). The lipid distributions around the different surfaces tended to stabilize early on in the simulations, within a few microseconds, and there was good consistency between replicates. Although the cholesterol molecules in each system were evenly distributed between the monolayers in the initial configurations, within a few microseconds of simulation, cholesterol became enriched in the upper leaflet, remaining stable thereafter at around a 1.3 : 1 ratio, and a 1.6 : 1 ratio for the lipid only system with a halved cholesterol concentration (Table S2†).

Next, the localization probability density of CHOL, POPC, DPSM, POPE and POPS was computed for each system (Fig. 2a). Enrichment of cholesterol was observed around the surface of AQP1 in both leaflets, deeper within the KcsA structure in the upper leaf, and around the OmpF surface in the upper leaflet (Fig. 2a and b). There was a higher localization probability of POPC around AQP1 and at the trimer interfaces of OmpF, which was not evident in the other systems. POPS was enriched around the outer surfaces of AQP1 and KcsA in the lower

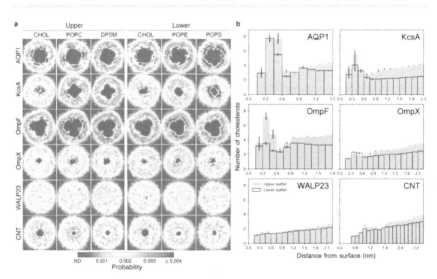

Fig. 2   Lipid distributions. (a) Probability density of CHOL, POPC, DPSM, POPE and POPS in the upper and lower leaflets. The areas in blue in the center of the 2D maps show the membrane regions displaced by the proteins or CNT. (b) Cholesterol counts as a function of distance from the different surfaces. The number of cholesterol molecules was calculated as a function of the nearest distance from the protein or CNT beads to the cholesterol ROH beads for the upper (shaded bars) and lower (empty bars) leaflets separately. Standard deviation is shown with black error bars.

leaflet, and increased POPE density around OmpF in the lower leaflet was noted (Fig. 2a). Aside from the aforementioned cases, lipid concentrations around the different surfaces tended to be approximate to their bulk values. The equilibrated lipid distributions around AQP1 were maintained in the atomistic continuation simulations (Fig. S3†).

Subsequently, the per-molecule and overall cholesterol flipping was assessed. The different systems presented similar bulk flip times of ~1 μs, but slower flipping dynamics were evident close to AQP1 and KcsA (Fig. 3a and b). Several cholesterol molecules around AQP1 and KcsA had flip times of > 10 μs (Fig. 3a), indicating that they did not flip during the simulation interval analyzed. In the lipid only system with half of the number of cholesterol molecules, flipping was faster, with an average rate of 1.4 flips per μs (Table S2†), which is also reflected in the flip time distributions (Fig. S5†). A few cholesterol molecules with flip times of > 1 μs were observed in the other systems at varying distances to the different surfaces, and also in the lipid only configuration, and was attributed to stochastic behavior (Fig. 3a and S4a†). No cholesterol flipping events were registered in the AQP1 AA systems.

Because areas of higher CHOL density colocalized with those of slow flipping in AQP1 and KcsA, these regions were inspected more closely to assess the possible cholesterol interaction sites. AQP1 and KcsA are similar in that they are both predominantly α-helical homotetramers. From analyzing the non-flipping cholesterol molecules from the AQP1 and KcsA trajectories, it was evident that they preferentially bind at the subunit interfaces (Fig. 4). Each intersubunit interface typically accommodated a single cholesterol molecule; however,

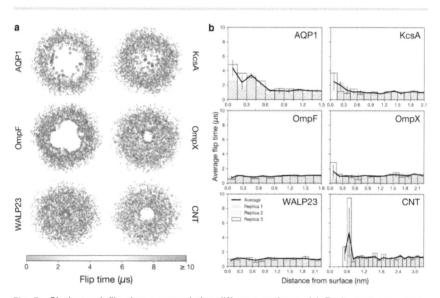

Fig. 3 Cholesterol flip times around the different surfaces. (a) Each spot represents a cholesterol flip event. The times between flip events are indicated by the spot color and size (larger spots reflect longer flip times). Each plot is centered around the protein's center of mass. (b) Average cholesterol flipping times as a function of the distance from the surfaces.

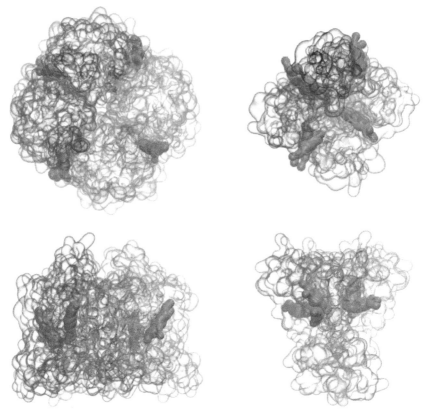

Fig. 4   Locations of the non-flipping cholesterol molecules around AQP1 and KcsA. Top and side views of non-flipping cholesterol molecules (red) from the AQP1 (left) and KcsA (right) trajectories. Individual subunits are colored in black, cyan, orange and magenta.

localization to the intersubunit grooves was reproducible across the different replicas, leading to the overlap of molecules in Fig. 4.

Discussion

Achieving a better understanding of lipid distributions and cholesterol dynamics around membrane proteins is crucial given the importance of protein–lipid interplay in biological processes.[35] We have taken a methodical approach to compare per-leaflet lipid distributions and transleaflet cholesterol flipping around six different surfaces in an asymmetric model membrane. The results are in accordance with previous reports of unique lipid environments around certain proteins. Here follows a contextualized discussion of our observations.

Though the simulations were initiated with an equal number of cholesterol molecules in the upper and lower leaflets, the reproducible migration of cholesterol from the lower to upper leaflet occurred in the early stages of the different simulations (Fig. S1, S2 and Table S2†). Similar behavior has been reported for asymmetric bilayers of comparable makeups, and possible explanations include cholesterol preference for certain lipid types and leaflet coupling to equilibrate membrane stress.[36] Allender and colleagues noted a similar degree of cholesterol enrichment in the outer leaflet as that observed in our simulations, which they attributed to a strong attraction between cholesterol and sphingomyelin.[37] They also predict that nearly 80% of the cholesterol could localize in the outer leaf if it were not for unfavorable bending energies associated with the spontaneous curvature induced by such a distribution. Our results showed that the saturated sphingomyelin lipid environment of the upper leaflet contained 25–40% more of the total cholesterol than the lower leaflet, with greater differences close to the protein surfaces in some cases (Table S2† and Fig. 2). In the systems with the same lipid composition, but half the number of cholesterol molecules, the upper leaflet contained an average of 60% more cholesterol than the lower leaflet (Table S2†). Based on the work of Allender *et al.*, it seems reasonable that by reducing the total amount of cholesterol in the bilayer, the upper leaflet is able to accommodate a higher fraction of cholesterol without inducing excessive curvature. The ability of cholesterol to exchange between monolayers has been shown to eliminate membrane stress when it solvates equally well in both leaflets, and it creates stress, resulting in membrane curvature, when partitioning to one leaflet is preferred.[38] It is evident that the intrinsic asymmetry of biomembranes, coupled with the ability of membrane components such as cholesterol to undergo transleaflet migration, leads to a complex but biologically relevant situation. For example, it has been suggested that caveolin-1 can induce membrane curvature through cholesterol recruitment, and that such caveolar complexes have central roles in cell signalling.[39]

As reported in other works,[1,11] the enrichment of specific lipid types around certain surfaces was detected in our simulations. One of the most evident observations was a high probability density of cholesterol around AQP1 and KcsA (Fig. 2 and 4). Cholesterol can affect membrane proteins specifically through direct binding, and non-specifically by altering bilayer properties.[40] Both AQP1 and KcsA exhibited cholesterol localization at distinct nonannular subunit interface sites (Fig. 4). Cholesterol has been shown to modulate the water permeability of aquaporins and specific interactions between AQP0 and cholesterol have been reported.[41,42] Although the mechanism by which regulation occurs has not been clarified, our findings suggest that the intersubunit grooves of aquaporin tetramers may represent putative cholesterol binding sites. Furthermore, the ion conductance of KcsA homologs can be regulated by the direct binding of cholesterol to transmembrane segments of the subunit interfaces.[43] Changes in bilayer thickness and tension induced by cholesterol and other sterols can impact the activity of aquaporins and KcsA, indicating that non-specific mechanisms likely play a role as well, though further studies are required to obtain a better understanding of the direct and indirect effects.[44,45] The association of membrane proteins with cholesterol can also affect their localization by promoting clustering in cholesterol-rich lipid raft domains, which has been demonstrated for AQP4.[46,47] Cholesterol may compete with anionic lipids for KcsA binding sites, and this idea is supported by our finding of POPS enrichment around KcsA in the lower leaflet. Similarly, the activity of aquaporins can be regulated by the density of negatively charged lipids in their immediate vicinity,[48] and we detected POPS clustering at the AQP1 surface in the inner leaflet (Fig. 2a). Additionally, POPC was found to associate at distinct sites around AQP1 in the upper leaflet, but the biological relevance of these interactions is currently unknown. A distinct lipid profile around OmpF was also noted (Fig. 2a), and our findings are consistent with the experimental demonstration of POPE and POPC binding to OmpF and regulating its function.[49,50] The OmpX, WALP23 and CNT systems presented considerably more homogeneous lipid distributions, with no clearly discernable patterns of enrichment around their surfaces (Fig. 2a). The absence of a particular lipid profile around the CNT was to be expected, considering that its surface was not functionalized such that especially strong interactions would occur with the other system components. However, in future studies it could be useful to evaluate if CNTs with different functional groups on their surfaces could recruit specific lipids, as such systems could serve as models for understanding the molecular details of lipid raft assembly. WALP peptides were developed as models for studying protein insertion in membranes, orientation and hydrophobic mismatch.[51] Although WALP23 did not present a distinct lipid profile in its vicinity, association with other lipid types has been demonstrated, suggesting that WALP and related KALP peptides could be useful models for studying protein–lipid interactions.[52]

Another property analyzed from our simulations was the cholesterol flipping dynamics between leaflets. As discussed earlier, the transleaflet migration of cholesterol can be influenced in a protein-dependent manner, regulating bilayer properties and biological functions. Computer simulations have shown that the rate of cholesterol flip-flop depends on several factors, such as the bilayer composition, temperature and force field.[14,53] Because our aim was to compare systems of the same lipid composition, the absolute rates of interleaflet diffusion are not highly important, though flipping was found to occur on the timescale of a few microseconds, which is in line with other Martini simulations.[54] It is also noteworthy that reducing the amount of cholesterol by half led to ~1.5× faster flipping (Table S2†). Like the lipid distribution profiles, the most obvious differences in cholesterol flipping times were seen in the vicinity of AQP1 and KcsA (Fig. 3). The slower transleaflet movement of cholesterol around these proteins was attributed to its tendency to associate at putative cholesterol binding sites at the subunit interfaces (Fig. 4).

Finally, we have shown that CG simulations can be beneficial for the equilibration of lipid profiles for subsequent atomistic studies. Due to the increased computational time needed to simulate atomistic *versus* coarse-grained systems,[55] it can be efficient to perform initial system equilibration at the CG level. To illustrate this point, the last frames of the AQP1 systems simulated at the CG level were backmapped to atomistic detail, and propagated for an additional microsecond. AQP1 was chosen because it presented a distinct lipid profile in its vicinity, which persisted in the atomistic simulations (Fig. S3†). It is worthwhile to mention that no cholesterol flipping events occurred in the three independent 1 μs atomistic simulations with AQP1. Though a lack of flipping was consistent with reported data of a comparable membrane setup simulated with the CHARMM36 force field,[36] it means that if one is interested in studying transleaflet cholesterol dynamics, CG simulations or atomistic force fields with different parameter sets, such as Slipids, are likely better options.[14]

In summary, our study represents an attempt to consistently evaluate and compare surface-dependent lipid profiles and cholesterol dynamics. To do so, a simple, yet physiologically representative, asymmetric biomembrane model was used, with different entities inserted. Our findings of protein- and leaflet-dependent lipid partitioning and cholesterol flip-flop suggest that CG simulations with Martini can be effective for studying protein–lipid fingerprints.

Conflicts of interest

There are no conflicts to declare.

Acknowledgements

Work in DPT's group is supported by the Natural Sciences and Engineering Research Council (Canada). Calculations were carried out on Compute Canada facilities, funded by the Canada Foundation of Innovation and partners. DPT acknowledges further support from the Canada Research Chairs program.

References

1 V. Corradi, B. I. Sejdiu, H. Mesa-Galloso, H. Abdizadeh, S. Y. Noskov, S. J. Marrink and D. P. Tieleman, *Chem. Rev.*, 2019, **119**, 5775–5848.

2 L. H. Wong, A. T. Gatta and T. P. Levine, *Nat. Rev. Mol. Cell Biol.*, 2019, **20**, 85–101.

3 R. Umeda, Y. Satouh, M. Takemoto, Y. Nakada-Nakura, K. Liu, T. Yokoyama, M. Shirouzu, S. Iwata, N. Nomura, K. Sato, M. Ikawa, T. Nishizawa and O. Nureki, *Nat. Commun.*, 2020, **11**, 1–11.

4 O. Cañadas, B. Olmeda, A. Alonso and J. Pérez-Gil, *Int. J. Mol. Sci.*, 2020, **21**, 3708.

5 J. Loschwitz, O. O. Olubiyi, J. S. Hub, B. Strodel and C. S. Poojari, *Prog. Mol. Biol. Transl. Sci.*, 2020, **170**, 273–403.

6 W. F. D. Bennett and D. P. Tieleman, *Biochim. Biophys. Acta, Biomembr.*, 2013, **1828**, 1765–1776.

7 R. Briones, C. Aponte-Santamaría and B. L. de Groot, *Front. Physiol.*, 2017, **8**, 124.

8 L. Domicevica, H. Koldsø and P. C. Biggin, *J. Mol. Graphics Modell.*, 2018, **80**, 147–156.

9 E. Barreto-Ojeda, V. Corradi, R. X. Gu and D. P. Tieleman, *J. Gen. Physiol.*, 2018, **150**, 417–429.

10 F. Mollinedo and C. Gajate, *J. Lipid Res.*, 2020, **61**, 611–635.

11 B. I. Sejdiu and D. P. Tieleman, *Biophys. J.*, 2020, **118**, 1887–1900.

12 M. S. Miettinen and R. Lipowsky, *Nano Lett.*, 2019, **19**, 5011–5016.

13 F. Ogasawara, F. Kano, M. Murata, Y. Kimura, N. Kioka and K. Ueda, *Sci. Rep.*, 2019, **9**, 1–10.

14 R. X. Gu, S. Baoukina and D. P. Tieleman, *J. Chem. Theory Comput.*, 2019, **15**, 2064–2070.

15 Z. Huang and E. London, *Chem. Phys. Lipids*, 2016, **199**, 11–16.

16 P. C. T. Souza, R. Alessandri, J. Barnoud, S. Thallmair, I. Faustino, F. Grünewald, I. Patmanidis, H. Abdizadeh, B. M. H. Bruininks, T. A. Wassenaar, P. C. Kroon, J. Melcr, V. Nieto, V. Corradi, H. M. Khan, J. Domański, M. Javanainen, H. Martinez-Seara, N. Reuter, R. B. Best, I. Vattulainen, L. Monticelli, X. Periole, D. P. Tieleman, A. H. de Vries and S. J. Marrink, *Nat. Methods*, 2021, **18**, 382–388.

17 D. Eisenberg, E. Schwarz, M. Komaromy and R. Wall, *J. Mol. Biol.*, 1984, **179**, 125–142.

18 P. C. Hsu, B. M. H. Bruininks, D. Jefferies, P. Cesar Telles de Souza, J. Lee, D. S. Patel, S. J. Marrink, Y. Qi, S. Khalid and W. Im, *J. Comput. Chem.*, 2017, **38**, 2354–2363.

19 S. Jo, T. Kim, V. G. Iyer and W. Im, *J. Comput. Chem.*, 2008, **29**, 1859–1865.

20 T. A. Wassenaar, H. I. Ingólfsson, R. A. Böckmann, D. P. Tieleman and S. J. Marrink, *J. Chem. Theory Comput.*, 2015, **11**, 2144–2155.

21 D. H. De Jong, G. Singh, W. F. D. Bennett, C. Arnarez, T. A. Wassenaar, L. V. Schäfer, X. Periole, D. P. Tieleman and S. J. Marrink, *J. Chem. Theory Comput.*, 2013, **9**, 687–697.

22 H. Sui, B. G. Han, J. K. Lee, P. Walian and B. K. Jap, *Nature*, 2001, **414**, 872–878.

23 Y. Zhou and R. MacKinnon, *J. Mol. Biol.*, 2003, **333**, 965–975.

24 R. G. Efremov and L. A. Sazanov, *J. Struct. Biol.*, 2012, **178**, 311–318.

25 J. Vogt and G. E. Schulz, *Structure*, 1999, **7**, 1301–1309.

26 J. H. Lorent, K. R. Levental, L. Ganesan, G. Rivera-Longsworth, E. Sezgin, M. Doktorova, E. Lyman and I. Levental, *Nat. Chem. Biol.*, 2020, **16**, 644–652.

27 X. Periole, M. Cavalli, S. J. Marrink and M. A. Ceruso, *J. Chem. Theory Comput.*, 2009, **5**, 2531–2543.

28 M. Vögele, J. Köfinger and G. Hummer, *Faraday Discuss.*, 2018, **209**, 341–358.

29 R. M. Bhaskara, S. M. Linker, M. Vögele, J. Köfinger and G. Hummer, *ACS Nano*, 2017, **11**, 1273–1280.

30 J. Huang, S. Rauscher, G. Nawrocki, T. Ran, M. Feig, B. L. De Groot, H. Grubmüller and A. D. MacKerell, *Nat. Methods*, 2017, **14**, 71–73.

31 M. J. Abraham, T. Murtola, R. Schulz, S. Páll, J. C. Smith, B. Hess and E. Lindah, *SoftwareX*, 2015, **1–2**, 19–25.

32 W. Humphrey, A. Dalke and K. Schulten, *J. Mol. Graphics*, 1996, **14**, 33–38.

33 S. Van Der Walt, S. C. Colbert and G. Varoquaux, *Comput. Sci. Eng.*, 2011, **13**, 22–30.

34 J. D. Hunter, *Comput. Sci. Eng.*, 2007, **9**, 90–95.

35 V. Corradi, E. Mendez-Villuendas, H. I. Ingólfsson, R. X. Gu, I. Siuda, M. N. Melo, A. Moussatova, L. J. Degagné, B. I. Sejdiu, G. Singh, T. A. Wassenaar, K. Delgado Magnero, S. J. Marrink and D. P. Tieleman, *ACS Cent. Sci.*, 2018, **4**, 709–717.

36 M. Blumer, S. Harris, M. Li, L. Martinez, M. Untereiner, P. N. Saeta, T. S. Carpenter, H. I. Ingólfsson and W. F. D. Bennett, *Front. Cell Dev. Biol.*, 2020, **8**, 575.

37 D. W. Allender, A. J. Sodt and M. Schick, *Biophys. J.*, 2019, **116**, 2356–2366.

38 A. Hossein and M. Deserno, *Biophys. J.*, 2020, **118**, 624–642.

39 S. Prakash, A. Krishna and D. Sengupta, *Faraday Discuss.*, 2020, DOI: 10.1039/d0fd00062k.

40 M. R. Elkins, I. V. Sergeyev and M. Hong, *J. Am. Chem. Soc.*, 2018, **140**, 15437–15449.

41 J. Tong, M. M. Briggs and T. J. McIntosh, *Biophys. J.*, 2012, **103**, 1899–1908.

42 J. W. O'Connor and J. B. Klauda, *J. Phys. Chem. B*, 2011, **115**, 6455–6464.

43 G. Hedger and M. S. P. Sansom, *Biochim. Biophys. Acta, Biomembr.*, 2016, **1858**, 2390–2400.

44 J. Tong, Z. Wu, M. M. Briggs, K. Schulten and T. J. McIntosh, *Biophys. J.*, 2016, **111**, 90–99.

45 M. Iwamoto and S. Oiki, *Proc. Natl. Acad. Sci. U. S. A.*, 2018, **115**, 13117–13122.

46 D. Sviridov, N. Mukhamedova and Y. I. Miller, *J. Lipid Res.*, 2020, **61**, 687–695.

47 K. Asakura, A. Ueda, S. Shima, T. Ishikawa, C. Hikichi, S. Hirota, T. Fukui, S. Ito and T. Mutoh, *Brain Res.*, 2014, **1583**, 237–244.

48 N. Klein, N. Hellmann and D. Schneider, *Biophys. J.*, 2015, **109**, 722–731.

49 I. Liko, M. T. Degiacomi, S. Lee, T. D. Newport, J. Gault, E. Reading, J. T. S. Hopper, N. G. Housden, P. White, M. Colledge, A. Sula, B. A. Wallace, C. Kleanthous, P. J. Stansfeld, H. Bayley, J. L. P. Benesch, T. M. Allison and C. V. Robinson, *Proc. Natl. Acad. Sci. U. S. A.*, 2018, **115**, 6691–6696.

50 A. H. O'Keeffe, J. M. East and A. G. Lee, *Biophys. J.*, 2000, **79**, 2066–2074.

51 J. A. Killian, *FEBS Lett.*, 2003, **555**, 134–138.

52 S. J. Marrink, V. Corradi, P. C. T. Souza, H. I. Ingólfsson, D. P. Tieleman and M. S. P. Sansom, *Chem. Rev.*, 2019, **119**, 6184–6226.

53 W. F. D. Bennett, J. L. MacCallum, M. J. Hinner, S. J. Marrink and D. P. Tieleman, *J. Am. Chem. Soc.*, 2009, **131**, 12714–12720.

54 S. Thallmair, H. I. Ingólfsson and S. J. Marrink, *J. Phys. Chem. Lett.*, 2018, **9**, 5527–5533.

55 I. Lima, A. Navalkar, S. K. Maji, J. L. Silva, G. A. P. de Oliveira and E. A. Cino, *Biochem. J.*, 2020, **477**, 111–120.

Faraday Discussions

ROYAL SOCIETY
OF CHEMISTRY

PAPER

Controllable membrane remodeling by a modified fragment of the apoptotic protein Bax

Katherine G. Schaefer,†[a] Brayan Grau,†[bc] Nicolas Moore,[b]
Ismael Mingarro, [c] Gavin M. King [★ad] and Francisco N. Barrera [★b]

Received 22nd May 2020, Accepted 1st September 2020
DOI: 10.1039/d0fd00070a

Intrinsic apoptosis is orchestrated by a group of proteins that mediate the coordinated disruption of mitochondrial membranes. Bax is a multi-domain protein that, upon activation, disrupts the integrity of the mitochondrial outer membrane by forming pores. We strategically introduced glutamic acids into a short sequence of the Bax protein that constitutively creates membrane pores. The resulting BaxE5 peptide efficiently permeabilizes membranes at acidic pH, showing low permeabilization at neutral pH. Atomic force microscopy (AFM) imaging showed that at acidic pH BaxE5 established several membrane remodeling modalities that progressively disturbed the integrity of the lipid bilayer. The AFM data offers vistas on the membrane disruption process, which starts with pore formation and progresses through localized exposure of membrane monolayers leading to stable and small (height $\sim$ 16 Å) lipid–peptide complexes. The different types of membrane morphology observed in the presence of BaxE5 suggest that the peptide can establish different types of membrane interactions. BaxE5 adopts a rare unstructured conformation when bound to membranes, which might facilitate the dynamic transition between those different states, and then promote membrane digestion.

Introduction

Apoptosis is a programmed cell death mechanism that plays an important role in multicellular organisms. The intrinsic or mitochondrial apoptosis process is triggered by the Bcl2 (B cell lymphoma-2)-family, which governs mitochondrial

[a]Department of Physics and Astronomy, University of Missouri, Columbia, MO 65211, USA. E-mail: kinggm@missouri.edu

[b]Department of Biochemistry & Cellular and Molecular Biology, University of Tennessee, Knoxville, 37996, USA. E-mail: fbarrera@utk.edu

[c]Departament de Bioquímica i Biologia Molecular, Estructura de Recerca Interdisciplinar en Biotecnologia i Biomedicina (ERI BioTecMed), Universitat de València, E-46100 Burjassot, Spain

[d]Department of Biochemistry, University of Missouri, Columbia, Missouri, 65211, USA

† These authors contributed equally.

outer membrane permeabilization. Within this family, pro- and anti-apoptotic members mediate cell survival through a complex network of interactions about which insufficient knowledge is available. Bax (Bcl2 associated protein X) is a pro-apoptotic multi-domain protein that is activated by tBid (truncated BH3 interacting-domain death agonist) to permeabilize the mitochondrial outer membrane. Specifically, Bax forms a membrane pore that releases pro-apoptotic factors from the mitochondrial intermembrane space.[1] It has been previously found out that $Bax^{\alpha5}$, a peptide comprising helix 5 plus a fraction of helix 6 of Bax, both of which are involved in dimerization and pore formation, can directly disrupt membranes, bypassing the need for activation for membrane pore formation.[2,3]

Controlled membrane pore formation has the potential to become a therapeutic tool, as leakage of contents out of the diseased cell could cause their demise. However, this possibility has been precluded by a lack of specificity, as pore formation cannot be efficiently prevented in healthy cells. Here, efforts are described towards controllable pore formation triggered by acidic pH. pH-dependency is a promising property for therapeutic applications, as diseased states characterized by extracellular acidosis include aggressive solid tumors,[4] arthritic inflammation,[5] and sepsis.[6]

The Barrera laboratory has previously used judicious introduction of acidic residues (aspartic or glutamic acid) to (i) design peptides that use the acidity of tumors to target the membranes of cancer cells,[7,8] and (ii) design membrane peptides that bind and activate the receptor tyrosine kinase EphA2.[9] Here, we modified $Bax^{\alpha5}$ to create a peptide that disrupts membranes selectively at acidic pH. To this end, glutamic acid residues were strategically introduced into the $Bax^{\alpha5}$ peptide sequence. The modified peptide exhibits pH-dependent membrane disruption, which results in major bilayer remodeling.

Materials and methods

Peptide preparation

The BaxE5 peptide was chemically synthesized using F-moc chemistry. Peptide purity was assessed by HPLC and MALDI-TOF. Peptide stocks were prepared by dissolving the powder using MilliQ water, where the pH was raised to pH 8 with an aliquot of NaOH. Peptide stock concentration was calculated by absorbance of the single Tyr residue in the sequence (Fig. 1), using a molar extinction coefficient of $1490~M^{-1}~cm^{-1}$.

Circular dichroism (CD)

CD samples were prepared by diluting concentrated stocks of peptide in 1 mM phosphate buffer at pH 7.5. Large unilamellar vesicles (LUVs) of the lipid POPC (1-palmitoyl-2-oleoyl-*glycero*-3-phosphocholine, from Avanti Polar Lipids) were formed by extrusion, using 100 nm pore filters. BaxE5 was incubated with LUVs to reach a peptide : lipid molar ratio of 1 : 200 in 1 mM sodium phosphate buffer (NaP$_i$) pH 7.5 for at least 45 min. After peptide–lipid incubation, pH was modified by adding sodium acetate buffer (pH 4.3), for 5 mM final buffer concentration and 5 μM final peptide concentration, followed by another 45 min incubation. CD measurements were performed on a Jasco J-815 spectropolarimeter at 25 °C in

(A)

BAX$^{\alpha 5}$ KGRVVALFYFASKLVLKALSTKVPELIRTK
              1   12   30
BaxE5   EGRVVALFYFAEKLVLKALSTKVPELIRTE

(B)

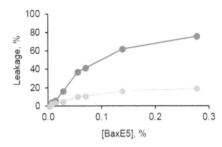

Fig. 1   BaxE5 selectively disrupts lipid integrity. (A) Sequence of the BaxE5 peptide. Three acidic residues were introduced into the sequence of the Bax$^{\alpha 5}$ peptide, at positions 1, 12 and 30. The acidic glutamic acid residues are shown in red, and basic residues in blue. (B) Leakage of sulforhodamine B encapsulated in POPC vesicles was measured at pH 4.0 (red) and pH 7.9 (grey) at different peptide concentrations, expressed as peptide : lipid molar percentage.

a 2 mm cuvette. The appropriate lipid-derived backgrounds were subtracted from data recorded. Values were normalized to molar ellipticity (millidegrees).

Leakage assay

Dried POPC lipids were rehydrated with 20 mM sulforhodamine B (SRB, Thermo) and extruded using 200 nm filters. Excess of free SRB was removed by using Sephadex G-25 desalting columns (GE Healthcare). The elution of SRB encapsulated in POPC vesicles was done using 1 mM NaP$_i$ buffer pH 7.5. Peptide dilutions for each assay were made from fresh 10 mM stock in 1 mM NaP$_i$ pH 7.5. Final vesicle concentration was 90 µM in 200 µL final volume. Measurements were performed using a BioTek Cytation V Imaging Reader at an excitation wavelength of 485 nm and an emission wavelength of 590 nm. Percentage of leakage was calculated using the following equation:

$$\% \text{ leakage} = \frac{\text{sample} - \text{control}}{\text{triton} - \text{control}} \times 100, \tag{1}$$

where Triton X-100 (10% v/v) was used to induce total leakage, and buffer in the absence of peptide was used as non-leakage control.

Atomic force microscopy

1,2-Dioleoyl-*sn-glycero*-3-phosphocholine (DOPC) (Avanti polar lipids) liposomes were prepared from a chloroform stock. Chloroform was first evaporated using argon gas to form lipid films which were then placed in a rough vacuum chamber (~0.1 Pa) overnight. An oil-free pump (XDS5, Edwards) was used to prevent backstreaming of oil, a potential contaminant. Samples were sealed with parafilm and stored at −20 °C. When required, a dry lipid stock was swelled in phosphate buffered saline solution (3 mM KCl, 8 mM Na_3PO_4, 140 mM NaCl, pH 7.5) for 60 minutes. After swelling, the lipid solution was extruded 30 times to form uni-lamellar liposomes (200 nm pore diameter, Liposofast, Avestin). This solution was aliquoted in 5 µL volumes and stored at −80 °C. Final DOPC concentration was approximately 5.5 mM. A volume (5 µL) of BaxE5 stock (31 µM or 33 µM) was added to DOPC at pH 7.5, such that the peptide : lipid molar ratio was either 1 : 100 or 1 : 170 and incubated for 5 minutes. The solution was subsequently diluted 1 : 11 in pH 4.0 buffer (10 mM Hepes, 100 mM KAc) or pH 7.5 buffer (10 mM Hepes, 100 mM Na_3PO_4) and incubated for an additional 10 minutes. At the time of delivery onto the supporting surface the final concentrations of lipid and peptide were, respectively, 1.4 µM and 250 µM for the 1 : 100 peptide to lipid (P : L) ratio, or 1.5 µM and 250 µM for the 1 : 170 P : L samples. One hundred µL of solution was deposited onto a freshly cleaved mica surface. This was incubated for 20–30 minutes to allow liposomes to rupture and form a planar bilayer on the surface.[10-12] The sample was then heavily rinsed by exchanging 100 µL of pH 4.0 buffer or pH 7.5 buffer 5–6 times and imaged. For samples originally prepared in pH 7.5 buffer, a series of images were taken. Then the tip was raised from the surface and the buffer exchanged for pH 4.0 imaging buffer (100 µL) 5–6 times and imaged. For control experiments performed with DOPC lipid in the absence of peptide, pH 4.0 buffer was added to DOPC to attain 250 µM, a similar concentration as the peptide samples. This solution was then deposited onto a freshly cleaved mica surface and incubated for 20–30 minutes at room temperature. These samples were rinsed 5–6 times with pH 4.0 buffer, then imaged. All images were obtained in fluid using a commercial instrument (Asylum Research Cypher) and tips (BL-AC40TS, Olympus). The images were acquired in tapping mode, with an estimated tip-sample force less than 100 pN. Analysis of pore-like feature topography was performed using custom software developed in Igor 7 (Wavemetrics, Portland, OR).[13] Smoothed histograms were produced in Igor 7 using Epanechnikov kernel density estimation.

Results

The sequence of the Bax^{α5} peptide contains a single acidic residue. Three additional glutamic acid residues were introduced at non-conserved positions (Fig. 1A). The resulting BaxE5 peptide is readily soluble in aqueous solution at close to neutral pH. The parental Bax^{α5} peptide permeabilizes lipid bilayers in a concentration-dependent manner. We studied the effect of BaxE5 on membrane integrity measuring its effect on the leakage of sulforhodamine B out of large unilamellar vesicles (LUVs) of the lipid POPC. Fig. 1B shows that at pH 4, BaxE5 can efficiently release sulforhodamine B at low concentrations. Interestingly, the degree of dye leakage was controlled by pH. At neutral pH inefficient leakage was observed. In contrast, BaxE5 efficiently disrupted membrane integrity at acidic pH, in agreement with the design principle we pursued. We used circular dichroism (CD) to determine the conformation of BaxE5 in the presence of POPC at acidic pH. Fig. 2A shows that the CD spectra had a minimum at 204 nm, a value typically observed for peptides and proteins in a largely unstructured conformation.[14,15] A shoulder was observed at 222 nm, which indicates that the peptide has

(A)

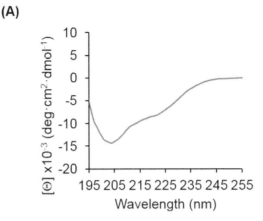

(B)

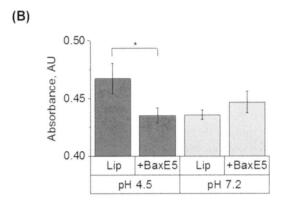

Fig. 2 BaxE5 disrupts membranes in a largely unstructured conformation. (A) Circular dichroism spectrum of BaxE5 in the presence of POPC vesicles at pH 4.3. The molar lipid to peptide ratio was 200 to 1. (B) Effect of BaxE5 in the absorbance at 250 nm in the presence of LUVs at pH 4.5 (red) or 7.2 (grey). Error bars are S.D. $n = 3$ and *, $p = 0.02$.

a weak α-helical character, in contrast to the original peptide that in aqueous solvent displayed values close to 40% total helical structure, considering both regular and distorted α-helix.[1] We observed that at pH 4.2 the presence of BaxE5 reduced the absorbance at 250 nm of the sample, but no effect was observed at neutral pH (Fig. 2B). The changes in absorbance possibly result from a reduction of the light the LUVs scatter, and suggest that BaxE5 induces a disruption in the vesicle structure.

    Atomic force microscopy (AFM) studies can shed direct light on the nature of peptide-induced membrane disruption.[16–18] Our initial AFM studies focused on samples prepared and deposited on the mica surface in acidic buffer (10 mM Hepes, 100 mM KAc, pH 4.0), corresponding to conditions of high vesicle leakage. Fig. 3 highlights the basic observations. In the presence of the BaxE5 peptide at pH 4, highly localized punctate depressions appeared in the DOPC membrane. Following our previous AFM work characterizing other pore forming peptides,[10–12] these features are likely to be pores or pore precursors, and we will refer to them generally as pore-like features. A control experiment was performed at the same

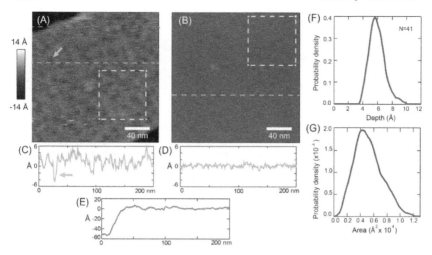

Fig. 3 BaxE5 induces pore-like features. Comparison of a DOPC bilayer imaged at pH 4 in the presence (A) and absence (B) of peptide. The z-scale is shown by the color bar on the left. Root-mean-square (RMS) roughness was measured inside 75 nm × 75 nm areas (white dashed boxes). Roughness values were 1.6 Å and 0.7 Å for (A) and (B), respectively. Line scans (C) and (D) are profiles through the image data (blue dashed lines in (A) and (B), respectively. A pore-like feature is indicated ((A) & (C), arrow). (E) An additional line scan, indicated by dashed red line in (A), reveals the approximately 50 Å height of the lipid bilayer. Hessian blob analysis was performed over (A), identifying pore-like features. Salient geometrical properties of the $N = 41$ pore-like features were collected into histograms, showing the distribution of apparent pore depth (F), which has a sharp peak around 6 Å, and area (G), which peaks at around 4000 Å^2.

pH without peptide (Fig. 3B). In this sample the bilayer surface showed no localized depressions. To quantitatively assess membrane disruption, measures of the root-mean-square (RMS) roughness were taken inside 75 nm × 75 nm areas. The RMS roughness was 0.7 Å for the bare membrane and 1.6 Å for the peptide-disrupted membrane. Line scans further show the dramatic change from a smooth bilayer surface in the absence of peptide (Fig. 3D) to numerous pore-like depressions in the presence of peptide (Fig. 3C). The presence of a DOPC lipid bilayer itself was verified by the expected ~50 Å step height from the supporting surface in both samples (Fig. 3E).[19–21]

We used the custom Hessian blob-detection software to analyze the AFM data and quantify the physical parameters defining the individual pore-like features. The depth histogram (Fig. 3F) has a single sharp peak at a depth of ~6 Å, with a slight shoulder extending out to 10 Å. However, convolution of the tip with the pore limits the depth measurement, as previously discussed.[10,11] A histogram of area (Fig. 3G) peaks around 4000 Å^2. If an approximately circular pore geometry is assumed, this gives an effective pore radius of about 36 Å. Single pore-like features demonstrated substantial conformational dynamics over time. Fig. 4 shows an example. Hessian blob analysis was used to measure the radius and depth of the pore-like feature. We observed that the dimensions of the feature fluctuated substantially from image to image. The average radius in the image sequence was 50 ± 6 Å (mean ± standard deviation) and the average depth was 7 ± 2 Å. We note that the precision of the vertical (depth) dimension in AFM is

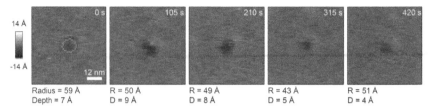

Fig. 4 Conformational dynamics of a single pore-like feature. Image sequence of DOPC in the presence of BaxE5 at pH 4. The data show a single pore-like feature undergoing structural variations over 7 minutes. Topographical characteristics were determined in each image using the Hessian blob algorithm, which outputs feature depth, area, and perimeter. One such perimeter is shown ($t = 0$ panel, red line). Assuming circularity, areas can be converted to feature radius. The calculated radius (R) and depth (D) are shown for each image.

typically ~1 Å.[22] Hence, these observations are not likely to be dominated by noise, but rather genuine topographic changes of the local membrane/lipid structure. Further, duplicate images of the same area (not shown), but with the tip scanning in the opposite direction, reproduced similar dimensional changes. For all 10 images in the sequence (trace and retrace), the average radius was 48 ± 8 Å and the average depth was 7 ± 2 Å.

To evaluate pH-dependent lipid remodeling, samples were deposited and incubated at higher pH, using a buffer comprising 10 mM Hepes, 100 mM Na_3PO_4, pH 7.5. Fig. 5A shows a representative AFM image of the membrane at pH 7.5, where the lipid remained undisturbed. The RMS roughness measurement was 0.8 Å, a similar value (within the estimated error of 0.3 Å) to bilayers in the absence of peptide at pH 4, which exhibited an RMS roughness of 0.7 Å (Fig. 3B).

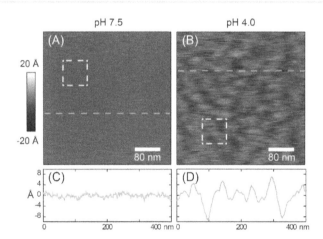

Fig. 5 Membrane disruption is only observed at acidic pH. A DOPC bilayer was first imaged in the presence of BaxE5 at pH 7.5 (A). After approximately 45 minutes, the pH was lowered to 4.0 by washing with acidic buffer. This pH change caused substantial remodeling of the lipid (B). The RMS roughnesses measured inside a 75 nm × 75 nm area (white dashed boxes) were 0.8 Å and 2.8 Å in (A) and (B), respectively. Line scans are shown for pH 7.5 (C) and pH 4.0 (D).

Forty-five minutes later, the pH was lowered by washing multiple times with acidic buffer. Fig. 5B shows the same sample after lowering the pH to 4. The lipid experienced a large-scale disruption, indicating that BaxE5 remained bound to the membrane. Pore-like structures were less common in favor of more widespread disruption. The roughness measurement was 2.8 Å, larger than the roughness shown in Fig. 3A. The change from distinct pores to widespread disruption appeared progressive. The nature of membrane disruption exhibited several modalities, from single pore-like features to disruption of the membrane and finally, removal of the bilayer from the mica surface. Fig. 6A shows an area where the membrane was remodeled and eventually started to detach over the course of imaging. Initial images show pore-like features, which increase in number and size over time. Lace-like structures thinner than the bilayer eventually form until the bilayer has completely disappeared, presumably resuspended into the imaging buffer solution. An interesting phenomenon was that this large-scale destruction was non-uniform on a small scale. Fig. 6B shows another instance of large membrane disruption. Over time the lipid bilayer in the lower left corner of the image is nearly completely lost, whereas a patch ~100 nm away in the upper right of the images only shows initial signs of disruption (small, punctate features appearing in the membrane in panel C). This may suggest that after rinsing there is a non-uniform BaxE5 density at the surface. Because both patches were subject to the same imaging conditions, this data acts as an internal control, suggesting that the observed bilayer remodeling is not tip-induced.

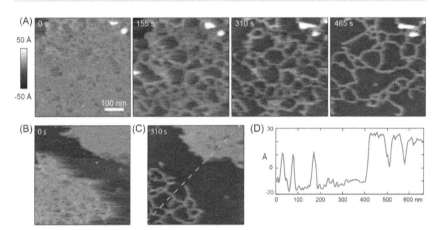

Fig. 6 Severe membrane remodeling by BaxE5 over time with spatially heterogeneous behavior. (A) A time sequence of images shows a bilayer at $t = 0$ s initially continuous, but with punctate and localized depressions, indicative of pore-like features. At times $t \geq 155$ s lace-like structures emerged. Over time, these features appeared more widespread and the continuous lipid bilayer was lost. Panel (B) shows that this behavior was spatially inhomogeneous. While the lipid patch in the lower left is being disrupted in the presence of peptide, the lipid patch located in the upper right corner of the same image, remains essentially undisturbed. (C) Three hundred and ten seconds later the lower left patch has been completely disrupted, while the second patch (upper right) is only showing initial signs, in the form of pore-like features. (D) A line scan through the image in panel (C) shows that the lace-like structures are fairly uniform in height (about 26 Å) and lower than the bilayer. The scale bar of 100 nm applies to all panels. All experiments were performed at pH 4.

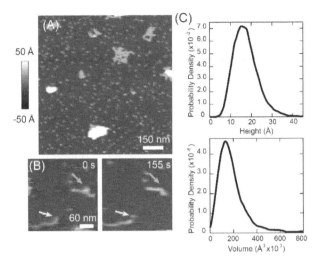

Fig. 7 Final stages of bilayer digestion. The leftover lace-like structures further degraded into smaller fragments of roughly uniform size (A). Evidence that these features originate from the lace structures is shown by zoomed in images taken over time, where features are seen to directly divide. Some of the fragments appear to dissociate from the surface ((B), green arrow); whereas many others remain surface-bound upon division ((B), yellow arrow). (C) Histograms of the heights and volumes of these fragments ($N = 1051$) as determined *via* Hessian blob analysis. A single peak is located in each distribution at ~16 Å and ~1.3×10^5 Å^3 for height and volume, respectively.

The lace-like structures that remained after lipid bilayer digestion continued to undergo degradation. Upon further imaging, these structures began to divide into smaller fragments. Fig. 7 shows the wide-scale spread of these fragments on the surface. A series of images (Fig. 7B) indicates that these fragments originate from the earlier lace structures. While some fragments seem to detach from the surface (green arrow), others stay bound to the surface (yellow arrow) for the remainder of the experiment. Hessian blob analysis showed that these fragments were homogeneous, and had similar height (~16 Å) by the single peak in a height histogram (Fig. 7C). This height is much smaller than the expected bilayer height of ~50 Å.[21] The volume histogram of these fragments is also sharp, with a peak at 1.3×10^5 Å^3.

Discussion

Peptides that insert into membranes triggered by a pH change have attracted significant attention from the standpoint of basic scientific knowledge as well as for potential therapeutic applications.[23-26] We applied biochemical and biophysical techniques in conjunction with imaging by single molecule AFM to provide quantitative characterization of lipid bilayer remodeling induced by a novel Bax-based peptide that was rationally designed to target membranes at low pH. The BaxE5 peptide disturbs the integrity of lipid membranes, as assessed by leakage experiments. The effect is strongly dependent on pH, and efficient leakage is only observed at pH 4. Severe membrane disruption is observed at nanomolar concentrations of BaxE5, as show in Fig. 1B, where the highest concentration used was 500 nM. BaxE5 also induced a change in the absorbance reading (Fig. 2B), likely resulting from changes in vesicle scattering. While small absorbance changes were observed in the absence of peptide at different pH, these might be merely caused by the different buffers used. Regarding the absorbance decrease observed at acidic pH, since scatter is proportional to vesicle size, the absorbance reduction suggests that BaxE5 disrupts the integrity of lipid vesicles.

AFM provides a direct vista into diverse nanoscale structures induced in supported lipid bilayers by peptides and detergents.[18] Examples range from the pores co-localized with membrane-thinned regions induced by a melittin derivative,[11] striations or ribbons formed by the WALP peptides,[27] to large scale cylindrical structures produced by a peptide from the simian immunodeficiency virus.[28] Detergent micelles and other structures, including highly ordered detergent domains within lipid bilayers have also been observed.[29-31] AFM has also been applied to study Bax, and large irregular pores (or membrane voids) with an ~50 nm radius have been observed.[1,32] These pores are significantly larger than those formed by $Bax^{\alpha 5}$.[3] We carried out AFM imaging to investigate how the BaxE5 peptide remodels a lipid bilayer. While no membrane disturbances were observed at neutral pH (Fig. 5A), under acidic conditions, significant distortion of the otherwise flat supported DOPC lipid bilayers was observed. The bilayer exhibited several discernable stages of remodeling: pore-like features, connected pore-like features (trenches), a large network of lace-like ribbons, and finally, highly localized fragments or membrane bits (see Fig. 8). This process resulted in most of the lipid bilayer being "digested", and as a consequence being lost into solution. The digestion of the membrane that BaxE5 carries out appears to have important differences to the mechanism detergents use to solubilize lipid bilayers. AFM imaging of detergent solubilization generally shows an all-or-none mechanism, where the bilayer desorbs from the mica surface without clear intermediates.[31,33] This observation contrasts with the several different intermediary membrane morphologies observed for BaxE5.

Pore-like features were observed under acidic conditions but were absent at neutral pH. This pH-dependent AFM observation is reminiscent of another recently studied peptide–lipid system.[10] A relatively narrow distribution of pore-like features were observed, with an average radius of ~36 Å, though individual pores exhibited $r > 50$ Å. We note that the AFM tip geometry is convoluted in these measurements. As a result, the measured pore depths, which varied between approximately 4 and 10 Å, do not reflect true depth due to the finite aspect ratio of the tip (80 Å nominal tip radius). Steric clash between the tip sidewall and the pore perimeter will lead to artificially shallow pore depths.[10,11] Our pore depth measurements therefore represent a lower limit of the real pore depth. The pore-like features often remained static in the minute time-scale (Fig. 4), suggesting they were coupled to the lower bilayer leaflet, which suppresses lateral diffusion *via* friction from the underlying supporting surface.[34] Based on this observation, we hypothesize that the pore-like features represent membrane indentations that reach at least the center of the bilayer. However, these features could indeed span the entire thickness of the bilayer, and be *bona fide* pores, as the AFM depth measurements correspond to a lower limit.

The depth and area distributions of the pore-like features show a shoulder at higher values (Fig. 3F and G), which might reveal the transition into larger membrane disruptions. Indeed, in addition to individual pore-like features,

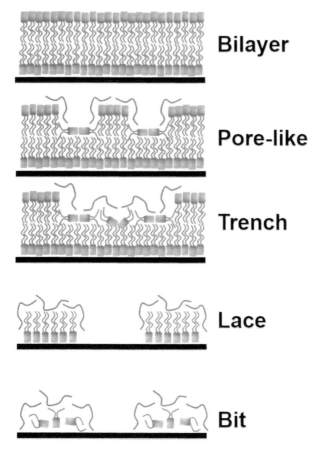

Fig. 8   Cartoon model describing the different stages of membrane remodeling caused by BaxE5. At acidic pH, the peptide initially induces the formation of pore-like, circular membrane depressions. These membrane features appear to converge, creating membrane trenches. At later times, or at higher BaxE5 concentrations, thin lace-like ribbons with the approximate thickness of a monolayer are observed. Finally, membrane ribbons are digested into thin membrane bits. Our data cannot exclude that the pore-like features correspond to *bona fide* pores. BaxE5 is shown in red, lipid in blue, and the mica surface in black.

larger scale structures were also apparent in the AFM image data. Trenches or joined pore-like features were prevalent, which we propose result from the fusion of independent pore-like features, which might originate at the initial points of attachment of BaxE5 molecules to the bilayer (Fig. 8). For example, a few trenches can be seen in Fig. 3A and many such structures are present in Fig. 5B, some of which span more than 50 nm in lateral extent. Over time, a striking stage of membrane remodeling became apparent, involving large scale lace-like ribbon formations. This result suggests that BaxE5 might assemble into linear oligomers, maybe similar to the line-type aggregates formed by gramicidin A.[35] Lace structures, many of which were over 100 nm in length (Fig. 6) had a height of ~26 Å above the mica surface. This value is close to half the full bilayer thickness,[36] which implies that the lace structures cannot correspond to bilayer tubes, or a single bilayer. One possibility is that these lace-like structures could be a lipid monolayer passivated by peptides. A less likely possibility is that they could be co-joined cylindrical reverse micelles or similar arrangements.[28]

The final stage of membrane remodeling resulted in small, well-dispersed fragments (membrane bits) on the supporting surface. These appeared to be similar to small-scale structures that have been observed in other membrane systems.[18,28] However, in those cases the partially solubilized particles appeared by re-deposition of mixed micelles that lead to bilayer formation. Since the height of the bits (16 Å) is less than half of the 50 Å bilayer thickness, the membrane bits observed here probably correspond to a lipo-protein complex. Because of their highly uniform size, these bits likely contain a lipid/peptide mixture of a well-defined stoichiometry. As with the small pores, tip geometry also convolves into volume measurements in AFM. To estimate the magnitude of this effect, a simulated image was created as previously described.[37,38] The AFM tip was modeled as a truncated cone shape with a hemispherical tip (cone angle = 15°, radius = 80 Å). The test feature was a delta function with the same height as the histogram peak (16 Å). Morphological dilation of the tip with the test feature created a simulated image whose volume was 6.0×10^4 Å^3. Subtracting this value from the experimentally measured volume roughly deconvolves the effect of tip geometry, giving a more realistic peak volume population of 7×10^4 Å^3. We note that this corrected volume is roughly an order of magnitude larger than that of a single BaxE5 monomer, which we estimate to be 4×10^3 Å^3 from the molecular weight and partial specific volume ($\sim$0.73 cm^3 g^{-1}). Based on the measured volume (7×10^4 Å^3), each bit likely contains less than 20 lipid or peptide molecules in total. One possibility is that the bits represent peptide aggregates absent of lipid. We disfavor this option based on the observation of Fig. 7B, as the bits originate from the membrane laces, and therefore likely contain lipid, probably at its core. We speculate that the molecular structure of such bits corresponds to a small stretch of lipid monolayer, lined by peptides (Fig. 8). However, since the thickness is less than the length of a fully extended DOPC acyl chain, and in fact $\sim$10 Å thinner than the lace structures (compare Fig. 6D and 7C) we propose that the presence of BaxE5 deforms the lipid acyl chains. This would result in a reduction in the van der Waals interactions that pack the acyl chains in a monolayer. The weakening of this stabilizing interaction will reduce the tendency of lipids to interact laterally. This might be a potential mechanism for the observed membrane digestion by BaxE5. However, further studies are required to validate this model.

The CD data indicates that BaxE5 disrupts membranes in a largely unstructured conformation, with a small helical content. This suggests that the addition of the three E residues (Fig. 1) destabilizes helix formation in BaxE5. These mutations are expected to slightly reduce helical content in solution.[39] The majority of peptides that permeabilize membranes are well-structured in the membrane-bound conformation.[23] In most cases these peptides adopt a helical structure, as is the case of the un-modified Bax$^{\alpha5}$ peptide,[2] while most of the rest adopt beta structure. Indolicidin is one of the rare instances of membrane-disrupting peptides that do not adopt stable secondary structure. Indolicidin, of only 13-aa, is secreted by neutrophils to disrupt the membrane integrity of viruses and bacteria.[40] This peptide is rich in tryptophan residues, and proline (25% of the sequence), in contrast to BaxE5, which contains no tryptophan residues and only one proline. However, a feature shared between indolicidin and BaxE5 is the presence of positively charged residues, as the former has a net charge of +4, while BaxE5 is expected to have a similar +5 charge at acidic pH. However, at neutral pH the unprotonated glutamic acids will yield a close to net neutral peptide. The membrane insertion of other acidic peptides where a drop in pH triggers membrane insertion is thought to be driven by the increase in overall peptide hydrophobicity,[7–9,26,41,42] resulting from the protonation of the side chain of acidic residues, and the resulting loss of negative charge. The molecular mechanism of the membrane insertion of BaxE5, on the other hand, might not be purely driven by hydrophobicity, but instead rely on the gain of a net positive charge. Only at acidic pH would BaxE5 be a poly-basic peptide, being similar in nature to other membrane-active peptides such as cell-penetrating peptides.[43,44]

While BaxE5 is a mutant of a single helix of the Bax protein, it might be useful to draw comparisons between the mechanisms the two use to disrupt membranes. Bax is monomeric in solution, and dimerizes in the membrane *en route* to further assemble into oligomers, which are thought to form a toroidal pore.[1] In the pore conformation, the Bax structure contains the latch domain and the core domain, comprising helix 5, which largely corresponds to the BaxE5 sequence. However, pore formation requires Bax to be activated by forming a transient complex with BH3-only proteins. In contrast, for BaxE5 to efficiently disrupt membranes requires "activation" by proton binding to the side chain of glutamic acids, as discussed above. Hence, Bax and BaxE5 use different mechanisms to disrupt membranes. Bax forms large and heterogeneous lipid-lined toroidal pores, after attaching to the membrane in a process that has been suggested to be analogous to a carpet model mechanism.[45,46] BaxE5 might also use a carpet model to initially attach to membranes, which leads to widespread membrane remodeling and a loss of bilayer integrity.

Why does BaxE5 induce such a massive membrane remodeling, resulting in digestion of the bilayer? We suspect that the unstructured conformation of BaxE5 is key for this behavior. Membrane peptides and proteins fold to avoid the steep energetic penalty of unsatisfied hydrogen bonds and unpaired salt bridges in the membrane.[47–49] Our CD data indicates that BaxE5 bound to the membrane does not form a structure that is significantly stabilized by hydrogen bonds. We hypothesize that BaxE5 remodels the bilayer in search of more energetically stable overall conformation. The linear aggregates of BaxE5 hypothesized in the lace structure would be more stable, but only marginally, and further evolve into the bit structure (Fig. 7 and 8). The relative homogeneous properties of the membrane bits suggest a stable conformation.

We show here how the BaxE5 peptide disrupts membranes in a specific acidic pH-triggered fashion. Therefore, BaxE5 has the potential to be developed into a therapeutic tool for cancer, among other diseases. Specifically, aggressive solid tumors are characterized by an acidic extracellular pH.[4,50,51] In fact, a slightly modified version of the $Bax^{\alpha5}$ peptide causes apoptosis in cell culture, and produces tumor regression in a mammary carcinoma xenograft model.[52] We do not know yet if BaxE5 might kill cells at the mildly acidic pH found in tumors (pH 6.5–7.0). If that occurred, it might digest the membrane of cancer cells, and also microenvironment cells that promote tumor growth and malignancy, including immuno-modulatory cells. If membrane disruption could be made to be specific, peptides causing this effect might be more efficient than peptides that form small membrane pores in acidic conditions, which might be more easily repaired by the cell.[10] While membrane-disrupting peptides have the potential to kill cancer cells, lack of specificity and toxicity have limited their use. In fact, clinical trials of cell-penetrating peptides have often yielded disappointing results.[53] Membrane insertion triggered by acidity might provide the basis for specificity, and prevent damage to healthy tissues. The use of nanocarriers might further reduce off-target effects.[8]

Conflicts of interest

There are no conflicts to declare.

Acknowledgements

This work was partially supported by NIH grant R01GM120642 (to F. N. B.), NSF grant 1709792 (to G. M. K.) and Generalitat Valenciana grant PROMETEU/2019/065 (to I. M.). B. G. was recipient of a predoctoral fellowship from the University of Valencia (Atracció de Talent program and Research staff stays program). We thank Kanokporn Chattrakun, Boomer Russell and Justin Westerfield for comments on the manuscript.

References

1 K. Cosentino and A. J. Garcia-Saez, Bax and Bak Pores: Are We Closing the Circle?, *Trends Cell Biol.*, 2017, **27**, 266–275.

2 A. J. Garcia-Saez, *et al.*, Peptides derived from apoptotic Bax and Bid reproduce the poration activity of the parent full-length proteins, *Biophys. J.*, 2005, **88**, 3976–3990.

3 A. J. Garcia-Saez, *et al.*, Peptides corresponding to helices 5 and 6 of Bax can independently form large lipid pores, *FEBS J.*, 2006, **273**, 971–981.

4 V. Estrella, *et al.*, Acidity generated by the tumor microenvironment drives local invasion, *Cancer Res.*, 2013, **73**, 1524–1535.

5 O. A. Andreev, *et al.*, Mechanism and uses of a membrane peptide that targets tumors and other acidic tissues in vivo, *Proc. Natl. Acad. Sci. U. S. A.*, 2007, **104**, 7893–7898.

6 K. E. Henry, *et al.*, Demarcation of Sepsis-Induced Peripheral and Central Acidosis with pH-Low Insertion Cyclic (pHLIC) Peptide, *J. Nucl. Med.*, 2020, **61**, 1361–1368.

7 V. P. Nguyen, D. S. Alves, H. L. Scott, F. L. Davis and F. N. Barrera, A Novel Soluble Peptide with pH-Responsive Membrane Insertion, *Biochemistry*, 2015, **54**, 6567–6575.

8 V. P. Nguyen, *et al.*, Mechanistic insights into the pH-dependent membrane peptide ATRAM, *J. Controlled Release*, 2019, **298**, 142–153.

9 D. S. Alves, *et al.*, A novel pH-dependent membrane peptide that binds to EphA2 and inhibits cell migration, *eLife*, 2018, **7**, e36645.

10 S. Y. Kim, *et al.*, Mechanism of Action of Peptides That Cause the pH-Triggered Macromolecular Poration of Lipid Bilayers, *J. Am. Chem. Soc.*, 2019, **141**, 6706–6718.

11 A. E. Pittman, B. P. Marsh and G. M. King, Conformations and Dynamic Transitions of a Melittin Derivative That Forms Macromolecule-Sized Pores in Lipid Bilayers, *Langmuir*, 2018, **34**, 8393–8399, DOI: 10.1021/acs.langmuir.8b00804.

12 S. J. Li, *et al.*, Potent Macromolecule-Sized Poration of Lipid Bilayers by the Macrolittins, A Synthetically Evolved Family of Pore-Forming Peptides, *J. Am. Chem. Soc.*, 2018, **140**, 6441–6447, DOI: 10.1021/jacs.8b03026.

13 B. P. Marsh, N. Chada, R. R. Sanganna Gari, K. P. Sigdel and G. M. King, The Hessian Blob Algorithm: Precise Particle Detection in Atomic Force Microscopy Imagery, *Sci. Rep.*, 2018, **8**, 978, DOI: 10.1038/s41598-018-19379-x.

14 S. M. Kelly, T. J. Jess and N. C. Price, How to study proteins by circular dichroism, *Biochim. Biophys. Acta, Proteins Proteomics*, 2005, **1751**, 119–139.

15 S. M. Kelly and N. Price, The use of Circular Dichroism in the investigation of protein structure and function, *Curr. Protein Pept. Sci.*, 2000, **1**, 349–384.

16 H. A. Rinia and B. de Kruijff, Imaging domains in model membranes with atomic force microscopy, *FEBS Lett.*, 2001, **504**, 194–199, DOI: 10.1016/s0014-5793(01)02704-1.

17 M. Meincken, D. L. Holroyd and M. Rautenbach, Atomic force microscopy study of the effect of antimicrobial peptides on the cell envelope of Escherichia coli, *Antimicrob. Agents Chemother.*, 2005, **49**, 4085–4092, DOI: 10.1128/AAC.49.10.4085-4092.2005.

18 K. El Kirat, S. Morandat and Y. F. Dufrene, Nanoscale analysis of supported lipid bilayers using atomic force microscopy, *Biochim. Biophys. Acta, Biomembr.*, 2010, **1798**, 750–765, DOI: 10.1016/j.bbamem.2009.07.026.

19 D. J. Muller and A. Engel, The height of biomolecules measured with the atomic force microscope depends on electrostatic interactions, *Biophys. J.*, 1997, **73**, 1633–1644.

20 R. R. Sanganna Gari, N. C. Frey, C. Mao, L. L. Randall and G. M. King, Dynamic structure of the translocon SecYEG in membrane: direct single molecule observations, *J. Biol. Chem.*, 2013, **288**, 16848–16854, DOI: 10.1074/jbc.M113.471870.

21 S. J. Attwood, Y. Choi and Z. Leonenko, Preparation of DOPC and DPPC Supported Planar Lipid Bilayers for Atomic Force Microscopy and Atomic Force Spectroscopy, *Int. J. Mol. Sci.*, 2013, **14**, 3514–3539.

22 C. A. Bippes and D. J. Muller, High-resolution atomic force microscopy and spectroscopy of native membrane proteins, *Rep. Prog. Phys.*, 2011, **74**, DOI: 10.1088/0034-4885/74/8/086601.

23 S. Guha, J. Ghimire, E. Wu and W. C. Wimley, Mechanistic Landscape of Membrane-Permeabilizing Peptides, *Chem. Rev.*, 2019, **119**, 6040–6085.

24 B. Bechinger, Towards membrane protein design: pH-sensitive topology of histidine-containing polypeptides, *J. Mol. Biol.*, 1996, **263**, 768–775.

25 W. Li, F. Nicol and F. C. Szoka, Jr, GALA: a designed synthetic pH-responsive amphipathic peptide with applications in drug and gene delivery, *Adv. Drug Delivery Rev.*, 2004, **56**, 967–985.

26 J. C. Deacon, D. M. Engelman and F. N. Barrera, Targeting acidity in diseased tissues: mechanism and applications of the membrane-inserting peptide, pHLIP, *Arch. Biochem. Biophys.*, 2015, **565**, 40–48, DOI: 10.1016/j.abb.2014.11.002.

27 H. A. Rinia, *et al.*, Domain formation in phosphatidylcholine bilayers containing transmembrane peptides: specific effects of flanking residues, *Biochemistry*, 2002, **41**, 2814–2824, DOI: 10.1021/bi011796x.

28 K. El Kirat, Y. F. Dufrene, L. Lins and R. Brasseur, The SIV tilted peptide induces cylindrical reverse micelles in supported lipid bilayers, *Biochemistry*, 2006, **45**, 9336–9341, DOI: 10.1021/bi060317x.

29 J. F. Liu, G. Min and W. A. Ducker, AFM study of adsorption of cationic surfactants and cationic polyelectrolytes at the silica-water interface, *Langmuir*, 2001, **17**, 4895–4903, DOI: 10.1021/la0017936.

30 H. A. Rinia, M. M. E. Snel, J. P. J. M. van der Eerden and B. de Kruijff, Visualizing detergent resistant domains in model membranes with atomic force microscopy, *FEBS Lett.*, 2001, **501**, 92–96, DOI: 10.1016/S0014-5793(01) 02636-9.

31 K. El Kirat, A. Pardo-Jacques and S. Morandat, Interaction of non-ionic detergents with biomembranes at the nanoscale observed by atomic force microscopy, *Int. J. Nanotechnol.*, 2008, **5**, 769–783.

32 R. F. Epand, J.-C. Martinou, S. Montessuit, R. M. Epand and C. M. Yip, Direct evidence for membrane pore formation by the apoptotic protein Bax, *Biochem. Biophys. Res. Commun.*, 2002, **298**, 744–749.

33 K. El Kirat, S. Morandat and Y. F. Dufrene, Nanoscale analysis of supported lipid bilayers using atomic force microscopy, *Biochim. Biophys. Acta*, 2010, **1798**, 750–765.

34 M. Przybylo, *et al.*, Lipid diffusion in giant unilamellar vesicles is more than 2 times faster than in supported phospholipid bilayers under identical conditions, *Langmuir*, 2006, **22**, 9096–9099, DOI: 10.1021/la061934p.

35 J. Mou, D. M. Czajkowsky and Z. Shao, Gramicidin A aggregation in supported gel state phosphatidylcholine bilayers, *Biochemistry*, 1996, **35**, 3222–3226.

36 N. Kucerka, *et al.*, Lipid bilayer structure determined by the simultaneous analysis of neutron and X-ray scattering data, *Biophys. J.*, 2008, **95**, 2356–2367.

37 R. R. Sanganna Gari, *et al.*, Direct visualization of the E. coli Sec translocase engaging precursor proteins in lipid bilayers, *Sci. Adv.*, 2019, **5**, eaav9404, DOI: 10.1126/sciadv.aav9404.

38 K. P. Sigdel, L. A. Wilt, B. P. Marsh, A. G. Roberts and G. M. King, The conformation and dynamics of P-glycoprotein in a lipid bilayer investigated by atomic force microscopy, *Biochem. Pharmacol.*, 2018, **156**, 302–311, DOI: 10.1016/j.bcp.2018.08.017.

39 C. N. Pace and J. M. Scholtz, A helix propensity scale based on experimental studies of peptides and proteins, *Biophys. J.*, 1998, **75**, 422–427.

40 D. I. Chan, E. J. Prenner and H. J. Vogel, Tryptophan- and arginine-rich antimicrobial peptides: structures and mechanisms of action, *Biochim. Biophys. Acta*, 2006, **1758**, 1184–1202.

41 H. L. Scott, J. M. Westerfield and F. N. Barrera, Determination of the Membrane Translocation pK of the pH-Low Insertion Peptide, *Biophys. J.*, 2017, **113**, 869–879.

42 M. Bano-Polo, L. Martinez-Gil, F. N. Barrera and I. Mingarro, Insertion of Bacteriorhodopsin Helix C Variants into Biological Membranes, *ACS Omega*, 2020, **5**, 556–560.

43 K. Fosgerau and T. Hoffmann, Peptide therapeutics: current status and future directions, *Drug Discovery Today*, 2015, **20**, 122–128.

44 R. M. Johnson, S. D. Harrison and D. Maclean, Therapeutic applications of cell-penetrating peptides, *Methods Mol. Biol.*, 2011, **683**, 535–551.

45 R. T. Uren, S. Iyer and R. M. Kluck, Pore formation by dimeric Bak and Bax: an unusual pore?, *Philos. Trans. R. Soc. London, Ser. B*, 2017, **372**, DOI: 10.1098/rstb.2016.0218.

46 Y. Shai, Mechanism of the binding, insertion and destabilization of phospholipid bilayer membranes by alpha-helical antimicrobial and cell non-selective membrane-lytic peptides, *Biochim. Biophys. Acta*, 1999, **1462**, 55–70.

47 J. Westerfield, *et al.*, Ions Modulate Key Interactions between pHLIP and Lipid Membranes, *Biophys. J.*, 2019, **117**, 920–929.

48 S. H. White, W. C. Wimley, A. S. Ladokhin and K. Hristova, Protein folding in membranes: determining energetics of peptide-bilayer interactions, *Methods Enzymol.*, 1998, **295**, 62–87.

49 M. Bano-Polo, *et al.*, Charge pair interactions in transmembrane helices and turn propensity of the connecting sequence promote helical hairpin insertion, *J. Mol. Biol.*, 2013, **425**, 830–840.

50 M. Xu, X. Ma, T. Wei, Z.-X. Lu and B. Ren, In Situ Imaging of Live-Cell Extracellular pH during Cell Apoptosis with Surface-Enhanced Raman Spectroscopy, *Anal. Chem.*, 2018, **90**, 13922–13928.

51 M. Anderson, A. Moshnikova, D. M. Engelman, Y. K. Reshetnyak and O. A. Andreev, Probe for the measurement of cell surface pH in vivo and ex vivo, *Proc. Natl. Acad. Sci. U. S. A.*, 2016, **113**, 8177–8181.

52 J. G. Valero, *et al.*, Bax-derived membrane-active peptides act as potent and direct inducers of apoptosis in cancer cells, *J. Cell Sci.*, 2011, **124**, 556–564.

53 P. P. Tripathi, H. Arami, I. Banga, J. Gupta and S. Gandhi, Cell penetrating peptides in preclinical and clinical cancer diagnosis and therapy, *Oncotarget*, 2018, **9**, 37252–37267.

Faraday Discussions

Estimating the accuracy of the MARTINI model towards the investigation of peripheral protein–membrane interactions†

Sriraksha Srinivasan, [ID] Valeria Zoni [ID] and Stefano Vanni [ID] *

Received 12th May 2020, Accepted 1st September 2020

DOI: 10.1039/d0fd00058b

Peripheral membrane proteins play a major role in numerous biological processes by transiently associating with cellular membranes, often with extreme membrane specificity. Because of the short-lived nature of these interactions, molecular dynamics (MD) simulations have emerged as an appealing tool to characterize at the structural level the molecular details of the protein–membrane interface. Transferable coarse-grained (CG) MD simulations, in particular, offer the possibility to investigate the spontaneous association of peripheral proteins with lipid bilayers of different compositions at limited computational cost, but they are hampered by the lack of a reliable *a priori* estimation of their accuracy and thus typically require *a posteriori* experimental validation. In this article, we investigate the ability of the MARTINI CG force field, specifically the 3 open-beta version, to reproduce known experimental observations regarding the membrane binding behavior of 12 peripheral membrane proteins and peptides. Based on observations of multiple binding and unbinding events in several independent replicas, we found that, despite the presence of false positives and false negatives, this model is mostly able to correctly characterize the membrane binding behavior of peripheral proteins, and to identify key residues found to disrupt membrane binding in mutagenesis experiments. While preliminary, our investigations suggest that transferable chemical-specific CG force fields have enormous potential in the characterization of the membrane binding process by peripheral proteins, and that the identification of negative results could help drive future force field development efforts.

Introduction

The interaction of peripheral proteins with cellular membranes plays a pivotal role in numerous biological processes including lipid metabolism and transport,[1–3] membrane trafficking, and signal transduction.[4,5] The interaction between

Department of Biology, University of Fribourg, Switzerland. E-mail: stefano.vanni@unifr.ch

† Electronic supplementary information (ESI) available. See DOI: 10.1039/d0fd00058b

proteins and the membrane surface is often reversible and transient,[6] and it is tightly modulated by numerous factors, including protein–protein interactions, lipid post-translational modifications, specific binding to signaling lipids, or recognition of bulk membrane properties such as electrostatics or membrane packing.[7]

However, due to the transient nature of protein–membrane interactions, determination of the three-dimensional structure of peripheral proteins is generally achieved in their soluble, membrane-free conformation. Furthermore, dynamic phenomena such as large-scale conformational changes upon membrane binding, and the molecular and energetic details of these interactions, are extremely difficult to elucidate *via* experimental methods.

Molecular dynamics (MD) simulations provide an alternative route by allowing a detailed investigation of protein–membrane complexes, thus enabling the study of their dynamic behavior and the molecular characterization of the membrane binding process.[8] In particular, coarse-grained (CG) MD simulations have emerged in the last few years as a powerful tool to investigate the interactions between proteins and membranes, and they have been shown to provide a powerful and cost-effective alternative to fully atomistic MD simulations.[9,10]

In detail, one of the most widely used CG models in this context is the MARTINI model,[11] which has fully compatible parameters available for several biomolecules such as proteins,[12,13] lipids,[11,14] nucleic acids,[15,16] and carbohydrates.[17] In the MARTINI model, four non-hydrogen atoms are generally represented as a single interaction site, called a bead. As a consequence, simulations can access much longer time- and size-scales, at a reasonable computational cost.

An intrinsic drawback of this CG strategy, however, is the almost complete lack of a reliable *a priori* estimation of its accuracy. This is related to the inevitable presence of systematic errors associated with the development of force field parameters, which generally occurs in simple molecular systems, and to the lack of a detailed understanding of whether these errors cancel or add up in more complex molecular systems,[18] such as those involving protein–membrane interactions. Thus, corroboration of the quality of a CG model is generally achieved *a posteriori*, based on comparison with experimental data.

For investigations pertaining to the binding of peripheral proteins to lipid bilayers, the MARTINI CG model has been shown in the last few years to be a very promising tool,[19] especially with respect to strong or irreversible membrane binding, such as in the case of the specific recognition of signaling lipids such as phosphoinositides[20–24] or in the case of lipidated proteins.[25,26] On the other hand, the ability of this model to accurately describe more transient ("on/off") membrane binding by peripheral proteins, as well as to reproduce the experimentally-determined ability of proteins to sense bulk membrane properties, such as electrostatics[4,27] or lipid packing density,[28–30] remains mostly unexplored.

A further complication of such investigations is the potential dependence of membrane binding on protein conformational changes or structural fluctuations. This is particularly challenging for the MARTINI force field as while the parameters for lipids and amino acids have been derived using a self-consistent strategy, a faithful representation of protein secondary structure in this model requires the *ad hoc* use of elastic network models (ENMs).[31] These approaches preserve the structure of the protein by generating additional harmonic bonds between the backbone beads of the protein and are thus characterized by supplementary parameters, such as the force constant of the harmonic bonds (fc) and the cut-off distance (R_C). Tuning of these parameters can affect the propensity of proteins to interact with lipid bilayers, and while recent studies suggest a potential role of the elastic network in protein–protein interactions and clustering,[32] its influence on interactions between peripheral proteins and membranes has not yet been thoroughly investigated.

In this work, we investigate the ability of the MARTINI model (specifically the 3 open-beta version) to accurately describe and predict the transient binding of 12 peripheral proteins or peptides to lipid bilayers of different compositions. We design a simple protocol to extract information from membrane binding/ unbinding time traces and we specifically focus on its ability to reproduce the sensing of bulk membrane properties displayed by certain protein families. Our data suggests that despite significant protein-to-protein variability and the presence of false positives and false negatives, this model provides, for the most part, meaningful structural information on the protein–membrane interface. In addition, we find that elastic network parameters have marginal influence on membrane binding for globular proteins, but they affect *in silico* binding for those proteins that have been shown experimentally to undergo conformational transitions upon binding.

Methods

Software details

All MD simulations of the protein–membrane systems were performed with the GROMACS (v 2018x)[33] package, using the open beta version of the Martini 3 force field.[19] All molecular images were rendered using Visual Molecular Dynamics (VMD).[34]

System setup

A total of 12 peripheral proteins were studied. They are listed in Table 1 along with their PDB IDs, the bilayer compositions they were set up with, and the duration of the simulations. The protein structures were obtained from RCSB PDB,[35] with the exception of ArfGAP1 amphipathic lipid packing sensor (ALPS) helix (residues 199 to 223) which was modeled as an alpha helix. For the Arf1 protein, the structures of two different conformations (GDP-bound and GTP-bound) are present in the RCSB PDB. Since the two conformations are markedly different, particularly in the N-terminal region which is responsible for membrane binding,[36] we removed the ligands from the structures in our MD simulations. This approach is particularly justified for CG simulations since the elastic network employed retains the secondary structure of the protein. The atomistic structures were converted to CG models with an additional elastic network using the martinize[12] script. The lower and upper elastic bond cut-offs were set to 0.5 nm and 0.9 nm respectively. For each protein system, independent simulations were performed with two elastic bond force constants: 500 and 1000 kJ mol^{-1} nm^{-2}. Side chain dihedral corrections were applied to the CG models using the addDihedral.tcl and bbsc.sh scripts, as necessitated for the 3 open beta version of the Martini force field.

Table 1 CG simulation details of the proteins and peptides tested

Protein	PDB ID	fc (kJ mol^{-1} nm^{-2})	Bilayer composition	Simulation time (µs)	Number of replicas	Total simulation time (µs)
Ubiquitin	1UBQ	500, 1000	DOPC	2	5	10
			DOPC/DOPS (80 : 20)	2	5	10
Hen egg white lysozyme	1AKI	500, 1000	DOPC	2	5	10
			DOPC/DOPS (80 : 20)	2	5	10
PDK1 PH domain (residues 407–549)	1W1D	500, 1000	DOPC	4	6	24
			DOPC/DOPS (90 : 10)	4	6	24
			DOPC/DOPS (80 : 20)	4	6	24
			DOPC/DOPS (70 : 30)	4	6	24
Evectin-2 PH domain (residues 1–112)	3VIA	500, 1000	DOPC	2	6	12
			DOPC/DOPS (80 : 20)	2	6	12
			DOPC/DOPS (70 : 30)	2	6	12
Lactadherin C2 domain (residues 1–158)	3BN6	500, 1000	DOPC	2	6	12
			DOPC/DOPS (90 : 10)	2	6	12
			DOPC/DOPS (80 : 20)	2	6	12
FVa C2 domain (residues 1–160)	1CZT	500, 1000	DOPC	3	6	18
			DOPC/DOPS (80 : 20)	3	6	18
PLA$_2$	1POA	500, 1000	DMPC	2	5	10
			DOPC	2	5	10
			DOPC/DOPS (80 : 20)	2	5	10
Arf1-GTP	2KSQ	500, 1000	DOPC	2	6	12
Arf1-GDP	2K5U	500, 1000	DOPC	2	6	12
Ricin	1BR5	500, 1000	DOPC	4	6	24
			DOPC/DOPS (70 : 30)	4	6	24
Mastoparan	1D7N	500, 1000	DMPC	1	6	6
			POPC	1	6	6
			DOPC	1	6	6
Osh4 ALPS peptide (residues 1–29)	1ZHY	500, 1000	DMPC	2	5	10
			POPC	2	5	10
			DOPC	2	5	10
ArfGAP1 ALPS peptide (residues 199–223)	Helical model	500, 1000	DMPC	2	6	12
			POPC	2	6	12
			DOPC	2	6	12
Total simulation time (µs)						860

Lipid bilayers of different compositions (as indicated in Table 1) with lateral dimensions of 20 nm × 20 nm were constructed using the CHARMM-GUI Membrane Builder,[37] and were equilibrated according to the standard six step equilibration protocol provided by CHARMM-GUI. The bilayers were then strip-ped of water molecules and ions, and the protein was placed such that the minimum distance between any bead of the protein and any bead of the lipid molecules was at least 2.5 nm. The initial orientation of the protein was such that its principal axes were aligned with the x, y, and z directions of the system, with the longer dimension of the protein aligned along the z direction. The systems were then solvated and ionized with 0.12 M sodium and chloride ions.

Simulation details

Initial equilibration was carried out by performing energy minimization using the steepest descent algorithm, followed by a short MD run of 250 ps with the protein backbone beads restrained. Production runs were performed at 310 K using a velocity-rescale thermostat,[38] with separate temperature coupling for the protein, lipids, and solvent particles. The pressure was maintained at 1 bar using the Parrinello–Rahman barostat,[39] along with a semi-isotropic pressure coupling scheme. The non-bonded interactions were calculated by generating a pair-list using the Verlet scheme with a buffer tolerance of 0.005. The coulombic terms were calculated using the reaction field method and a cut-off distance of 1.1 nm. A cut-off scheme was used for the vdW terms, with a cut-off distance of 1.1 nm, and the Verlet cut-off scheme was used for the potential-shift.[40] The md integrator was used, with a time step of 20 fs. The first 100 ns of the production runs were not considered for analyses.

Analysis details

Membrane binding events were characterized by analyzing the time trace of the minimum distance between the protein and the lipid bilayer (Fig. S1–S11†). This distance was computed using the *gmx mindist* tool in GROMACS.[33]

Membrane binding was subsequently assessed by generating probability density distributions from these time traces using the kernel density estimation (KDE) method, and bound states were defined as those instances when the minimum distance was lower than or equal to 0.7 nm. The percentage of binding, which is the ratio between the area below the curve for the distribution up to 0.7 nm and the total area under the curve, was computed in each case. Statistical errors for these values are reported as standard errors with respect to the computed values for each individual replica.

Membrane-interacting residues were determined using the following protocol: a residue was considered to interact with the membrane if the distance between any bead of the residue and any lipid-bead was lower than or equal to 0.5 nm. For each residue, the number of instances of its interaction with the membrane during the trajectory was counted and summed over all the replicas, and a cor-responding normalized value was computed.

Results

To investigate the binding of peripheral proteins to membranes, we initially positioned the proteins at a distance of at least 2.5 nm from equilibrated lipid bilayers of different compositions (Fig. 1A). After MD simulations for a few microseconds, as indicated in Table 1, binding events were identified and

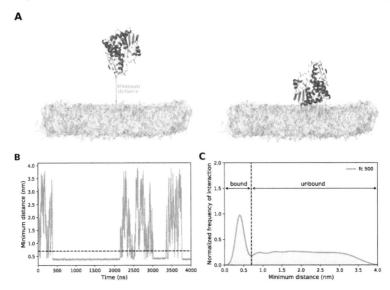

Fig. 1 Protocol used in this work to characterize membrane binding. (A) Initial system set-up for a representative peripheral protein. The protein is shown as a cartoon representation; the membrane is shown in gray. Water and ion beads are omitted for clarity. (B) Time trace of the minimum distance between the protein and the bilayer. (C) Probability density distribution of the time series in (B). Values are obtained by averaging over several independent replicas.

quantified using the protocol depicted in Fig. 1. First, the time trace of the minimum distance between the protein and the lipid bilayer was computed for all replicas (Fig. 1B, only one replica is shown), and the probability density distribution of these values (Fig. 1C) was generated by averaging the corresponding probability density distribution for all individual replicas. Membrane-bound states were defined as those instances when the minimum distance was lower than or equal to 0.7 nm (Fig. 1C), roughly corresponding to the first minimum observed in the probability density distribution.

Soluble proteins

As a first test to evaluate the ability of the MARTINI model to estimate membrane binding by peripheral proteins, we investigated, as a negative control, whether the model is able to accurately reproduce the lack of binding by proteins that are known to remain soluble in the cytosol and *in vitro* assays.[41–43] To this end, we tested the binding to pure DOPC lipid bilayers of two well-known soluble proteins, ubiquitin (Fig. 2A) and hen egg white (HEW) lysozyme (Fig. 2D), at two different force constants for the elastic network model, fc = 500 kJ mol^{-1} nm^{-2} and fc = 1000 kJ mol^{-1} nm^{-2}. In both cases, we observed no significant binding of the proteins to the bilayer, as indicated by the very low maxima at values below 0.7 nm in Fig. 2B and E. Rather, the two proteins adopted an unbound (protein–membrane minimum distance > 0.7 nm) state in solution for most (90%) of the trajectory (Fig. 2C and F and S1†). To further stress-test the methodology, we next added 20% DOPS lipids to our membrane composition, to potentially promote

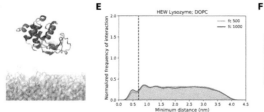

Fig. 2 Binding of soluble proteins to lipid bilayers. (A, D) Representative structures of ubiquitin (A) and HEW lysozyme (D). (B, E) Probability density distributions of minimum distance values for ubiquitin (B) and HEW lysozyme (E), for two values of the force constant, in DOPC bilayers. No membrane binding to DOPC bilayers was observed for both proteins. (C, F) Percentage of binding at different membrane compositions. Addition of DOPS to the bilayer did not enhance binding significantly for both proteins.

binding by means of electrostatic interactions. Even in these conditions, no significant binding was observed (Fig. 2C and F). Of note, no influence of the elastic network force constant was observed (Fig. 2B, C, E and F).

Globular proteins

Next, we evaluated whether the MARTINI model is able to reproduce the experimentally observed sensitivity to PS lipids that is a characteristic feature of specific protein families. To this end, we first tested two pleckstrin homology (PH) domains, belonging to 3-phosphoinositide-dependent kinase-1 (PDK1) (Fig. 3A) and evectin-2 (Fig. 3D), as well as two C2 domains – those of lactadherin (Fig. 5A) and human coagulation factor V (FVa) (Fig. 5D) – which have all been shown to bind to membranes in the presence of PS lipids.[44–48]

In the case of the PH domain of PDK1, we observe significant membrane binding in the presence of PS lipids (Fig. 3B), in agreement with experimental observations.[44] On increasing the concentration of PS lipids, membrane binding increased almost monotonically, with the protein showing almost no binding in the absence of PS lipids and binding for more than 50% of the time during the trajectory in the presence of 30% PS (Fig. 3C). On the other hand, for the second PH domain we tested, that of evectin-2, we observed only marginal binding in the mixed PC/PS bilayers (Fig. 3E). In the absence of PS lipids, membrane binding further, albeit slightly, decreased (Fig. 3F), suggesting that the MARTINI model possibly underestimates membrane binding affinity for this protein. In both cases, no effect of the elastic network force constant was observed (Fig. 3C and F).

We next analyzed the membrane binding interface for both PH domains (Fig. 4A and C). To do so, we counted the instances of residue–membrane interactions in the simulations performed at high PS concentration and collected the

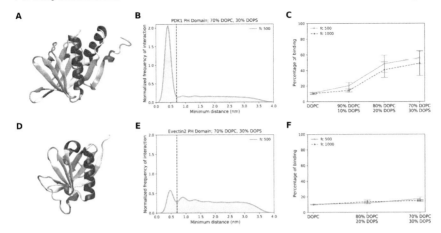

Fig. 3 Binding of PH domains to lipid bilayers. (A, D) Representative structures of the PH domains of PDK1 (A) and evectin-2 (D). (B, E) Probability density distributions of protein–membrane minimum distances in PC/PS lipid bilayers. (C, F) Percentage of binding of the two proteins at different membrane compositions. Addition of PS to the bilayer enhanced binding significantly for the PH domain of PDK1 but not for that of evectin-2.

corresponding frequencies (Fig. 4B and D). For PDK1, we found that residues R466 and K467 have the highest probability of binding the bilayer (Fig. 4B), in agreement with the binding mode that has been previously characterized

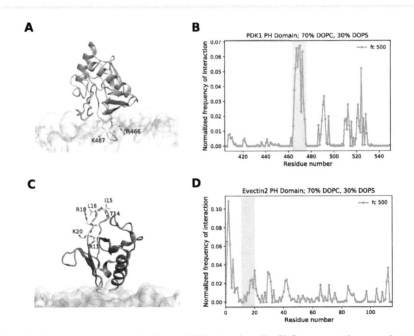

Fig. 4 Protein–membrane interface of PH domains. (A, C) Representative membrane binding modes for the PH domains of PDK1 (A) and evectin-2 (C). Experimentally determined residues responsible for membrane binding are shown explicitly in licorice representation. (B, D) Normalized frequency of contacts for protein residues for membranes with 30% PS. The shaded regions represent the experimentally observed binding regions.

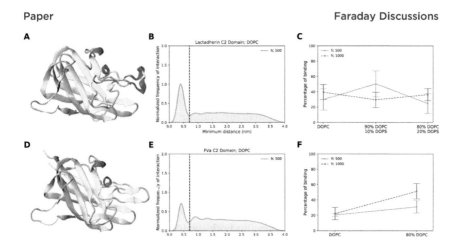

Fig. 5 Binding of C2 domains to lipid bilayers. (A, D) Representative structures of the C2 domains of lactadherin (A) and FVa (D). (B, E) Probability density distributions of protein–membrane minimum distances in PC lipid bilayers. (C, F) Percentage of binding of the two proteins at different membrane compositions. Addition of PS to the bilayer did not enhance binding significantly for both the proteins.

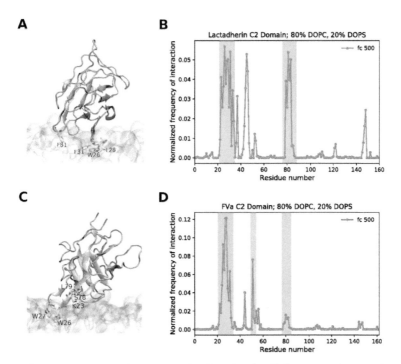

Fig. 6 Protein–membrane interface of C2 domains. (A, C) Representative membrane binding modes for the C2 domains of lactadherin (A) and FVa (C). Experimentally determined residues responsible for membrane binding are shown explicitly in licorice representation. (B, D) Normalized frequency of contacts for protein residues for membranes with 20% PS. The shaded regions represent the experimentally observed binding regions.

experimentally.[44] For evectin-2, on the other hand, the protein is known to bind to PS *via* a pocket made by three basic residues (R11, R18, K20) and *via* the backbone nitrogen atoms of T14, I15 and L16.[45,49] Analysis of the membrane-interacting residues, however, indicates no particular preference, and the model does not identify the correct binding interface (Fig. 4C and D).

Next, we performed similar analyses for the C2 domains of lactadherin (Fig. 5A) and FVa (Fig. 5D). Unlike the PH domains, in our simulations, the C2 domains displayed membrane binding despite the absence of PS lipids (Fig. 5B and E), and PS lipids did not significantly increase membrane binding (Fig. 5C and F), unlike in experimental observations.[45,46]

In both cases, however, analysis of the membrane binding interface (Fig. 6A and C) reveals that the model is capable of correctly identifying the experimentally known membrane binding interface for both the C2 domains (Fig. 6B and D). In detail, the lactadherin C2 domain is predicted to interact with membranes *via* its W26, G27, L28, F31 and F81 regions[47] while the FVa domain is predicted to do so *via* the residues K23, W26, W27, Q48, S78 and L79.[48]

Next, we investigated the binding of the enzyme phospholipase A_2 (PLA_2) (Fig. 7A) to lipid bilayers consisting of PC only or PC–PS mixtures (Fig. 7). The membrane binding of PLA_2 is a particularly challenging test case for the MARTINI force field, as it has been shown to depend on cation–pi interactions between lipid polar heads and protein aromatic residues.[50–52] We found that our model is not able to reproduce the experimentally observed binding to pure PC bilayers, regardless of acyl chain composition (Fig. 7B and S6†). On the other hand, adding PS lipids significantly increases binding (Fig. 7C), particularly *via* the binding interface proposed by a combination of atomistic and experimental assays using single point mutations, involving several aromatic residues (Y3, W18, W19, W61, F64, Y110)[50,53] (Fig. 7D and E).

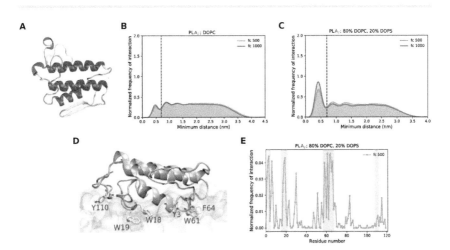

Fig. 7 Binding of PLA_2 to lipid bilayers. (A) Representative structure of PLA_2. (B, C) Probability density distributions of protein–membrane minimum distances in PC, and PC/PS lipid bilayers. (D) Representative membrane binding mode for PLA_2. Experimentally determined residues responsible for membrane binding are shown explicitly in licorice representation. (E) Normalized frequency of contacts for protein residues for membranes with 20% PS. The shaded regions represent the experimentally observed binding regions.

In summary, our data suggest that the model is able, with the sole exception of the PH domain of evectin-2, to identify most of the residues that have been experimentally identified to drive membrane association. On the other hand, the model is generally not well-suited to identifying the correct membrane conditions for binding, including the experimentally observed dependency of membrane binding upon membrane PS levels for the PH and C2 domains tested in this study, with the exception of that of PDK1. In all cases, varying the elastic network parameters has no significant effect on the observed binding behavior of globular peripheral proteins.

Peripheral proteins adopting different conformations

We next investigated whether this approach is able to discriminate between conformations that have been experimentally shown to promote or prevent membrane binding. To test this approach, we first focused on the protein Arf1, which has been shown to bind to membranes upon GDP-to-GTP exchange *via* its N-terminal amphipathic helix (AH).[36] As both GDP- and GTP-bound conformations are available in the Protein Data Bank (Fig. 8A and D),[54,55] we tested their binding to pure DOPC bilayers. In agreement with experimental observations,[55] we found that the GDP-bound conformer of the protein does not bind to DOPC bilayers (Fig. 8B), while the GTP-bound conformer does (Fig. 8E). Analysis of the protein residues that interact with the lipid bilayer in the GTP-bound form correctly identified the N-terminal AH as the main sequence responsible for membrane binding (Fig. 8F). Again, no major effect of the elastic network parameters was observed (Fig. 8B and E).

Next, we investigated whether our protocol is able to reproduce, at least at the qualitative level, membrane binding events in which protein conformational

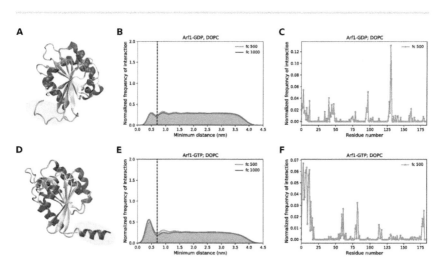

Fig. 8 Binding of Arf1 to lipid bilayers. (A, D) Representative structures of the Arf1 protein when bound to GDP (A) and GTP (D). GDP and GTP are omitted for clarity. The shaded region indicates the difference in the two structures at the N-terminus. (B, E) Probability density distributions of protein–membrane minimum distances in PC lipid bilayers. (C, F) Normalized frequency of contacts for protein residues for membranes with PC lipid bilayers. The shaded region represents the experimentally observed binding region.

changes have been proposed to occur. As a test system, we studied the ricin A chain (RTA) (Fig. 9A), a cytotoxic protein known to bind to PS containing vesicles, possibly upon binding-driven conformational changes.[56,57] In agreement with experimental observations,[56] we observed significant membrane binding in mixed PC/PS bilayers (Fig. 9B), but almost no binding of ricin to pure PC bilayers (Fig. 9C). Interestingly, membrane binding in the presence of PS lipids was larger when using a low fc of 500 kJ mol^{-1} nm^{-2} (Fig. 9B). On the other hand, when a high fc of 1000 kJ mol^{-1} nm^{-2} was used, the protein could initiate binding with the lipid bilayers, but this did not result in stable binding events (Fig. S8†). Analysis of the membrane binding residues for the simulations with PS-containing bilayers revealed that the binding interface is mostly consistent with what has been experimentally determined[56] (Fig. 9D and E).

Amphipathic helices

Finally, we investigated the ability of the model to reproduce the binding of AHs to lipid bilayers. AHs are unfolded in solution and adopt a helical conformation upon membrane binding, thus posing a unique challenge to our modeling strategy that requires the *a priori* determination of the protein secondary structure. As a consequence, modeling them in their folded, membrane-bound, helical conformation could also intrinsically favor membrane binding.

To investigate membrane binding of AHs to lipid bilayers we studied the binding of three distinct peptides, mastoparan (Fig. 10A) – an antimicrobial peptide known to stimulate the release of histamine from mast cells[58] – and the Amphipathic Lipid Packing Sensor (ALPS) motifs of two distinct proteins, the

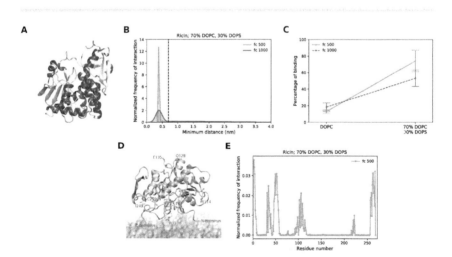

Fig. 9 Ricin binding to lipid bilayers. (A) Representative structure of ricin. (B) Probability density distribution of protein–membrane minimum distances in PC/PS lipid bilayers at two values of fc. (C) Percentage of binding of ricin at different membrane compositions. Addition of PS to the bilayer enhanced binding significantly, especially for fc = 500 kJ mol^{-1}nm^{-2}. (D) Representative membrane binding mode of ricin. Experimentally determined residues responsible for membrane binding are shown explicitly in licorice representation. (E) Normalized frequency of contacts for protein residues for membranes with 30% PS. The shaded regions represent the experimentally observed binding regions.

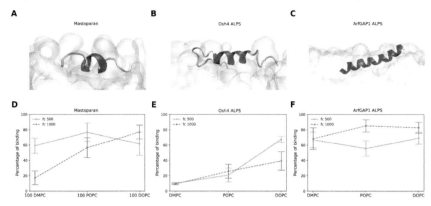

Fig. 10 AH binding to lipid bilayers. (A–C) Representative structures of the three AHs investigated in this study – mastoparan (A), ALPS motif of Osh4 (B), and ALPS motif of ArfGAP1 (C). (D–F) Percentage of binding of the three AHs at different membrane compositions.

lipid transporter Osh4 (Fig. 10B) and the GTPase-activating protein (GAP) for Arf1 (ArfGAP1)[59] (Fig. 10C). We studied the binding of these three peptides to bilayers composed of PC lipids with different acyl chains, which were shown to mimic an increase in lipid packing defects,[60] a membrane property that ALPS motifs are sensitive to,[30,59] and that substantially increases on going from saturated DMPC lipids to doubly unsaturated dioleoyl chains.[30,61]

Notably, the membrane binding behavior of mastoparan is affected by the choice of the elastic network parameters (Fig. 10D). In particular, when using a fc of 500 kJ mol^{-1} nm^{-2}, the peptide shows consistent binding to all lipid compositions, in agreement with experimental observations.[62] On the other hand, using a fc of 1000 kJ mol^{-1} nm^{-2}, the protein appears to exhibit sensitivity to lipid packing defects (Fig. 10A and D), even showing almost no binding in the presence of fully saturated DMPC lipids (Fig. 10D).

On the other hand, the ALPS motif of Osh4 displays the experimentally observed sensitivity for lipid packing defects regardless of the force constant used (Fig. 10E). Of note, this behavior is enhanced when using an fc of 500 kJ mol^{-1} nm^{-2}.

The ALPS motif of ArfGAP1, on the other hand, displays stable irreversible binding in all simulations, in disagreement with the experimentally observed sensitivity for lipid packing defects.[30] We attribute this negative result, in part, to the incorrect modeling of the ArfGAP1 AH as a fully folded helix, in disagreement with circular dichroism experiments showing a helical content of approximately 45% for the membrane bound conformation.[63] This also highlights a limitation of our protocol, which is able to discriminate between different proteins and/or membrane conditions only when multiple binding/unbinding events are observed in the MD simulations.

Taken together, these results suggest that the model is able, at least in some instances, to reproduce important properties of the binding of AHs to model membranes, including their sensitivity to lipid packing. At the same time, special care must be taken, as the choice of the protein conformation and of the elastic network force constant have a strong influence on the final results. Using a low fc of 500 kJ mol^{-1} nm^{-2} appears to provide a better agreement with experimental observations.

Discussion

In this article on peptide–membrane interactions, we investigated the ability of a widely-used CG model, the MARTINI force field, and more precisely its 3 open-beta release, to accurately describe the binding of peripheral proteins to model membranes. To this end, we have developed a protocol to characterize membrane binding from multiple replicas of unbiased MD trajectories for proteins and peptides belonging to different families and displaying different membrane binding mechanisms. Despite our limited dataset, our simulations allow us to draw some preliminary general conclusions about the performance of this model.

First, the model does not appear to overestimate protein–membrane binding affinity, as: (1) for most of the proteins in our dataset we observed multiple binding/unbinding events within the timescales of our simulations (microseconds), and (2) the model is able to correctly identify soluble proteins that do not bind to membranes, as we did not observe membrane binding for the two soluble proteins we selected as negative controls: hen egg white lysozyme and ubiquitin.

Second, the observed on/off behavior is paramount for the ability of our protocol to provide qualitative information on the binding process, and to quantify variations in membrane binding as a result of changes in membrane properties. For example, it allows for the investigation of protein sensitivity to specific membrane properties, such as electrostatics or lipid packing defects. To this end, we could correctly characterize the experimentally observed sensitivity to membrane properties, for example that of the PH domains of PDK1 to PS lipids,[44] and that of the ALPS motif of Osh4 to lipid packing defects.[59] On the other hand, we could not accurately reproduce the membrane sensitivity of different proteins, for example that of C2 domains for PS lipids, or that of ArfGAP1 for lipid packing defects. Interestingly, the membrane-binding interface of the C2 domains tested in this study is largely composed of aromatic residues, while that of PH domains is largely composed of positively-charged residues. It is thus possible that the negative results for C2 domains might originate from the previously established underestimation of aromatic residue–lipid interactions[52] by the MARTINI force field, an observation that could also explain why we could not observe the binding of PLA$_2$ to PC lipid bilayers that has been reported experimentally.[50] Taken together, these observations suggest that slight under- or overestimation of membrane binding affinity is sufficient to severely limit the ability of the model to correctly predict the ability to sense specific membrane properties displayed by some proteins.

Lastly, our protocol succeeds in correctly characterizing the protein–membrane interface for all but one protein in our dataset, as we were mostly able to identify the key residues found to disrupt membrane binding upon mutagenesis experiments. Since our approach lacks the ability to investigate protein conformational changes, it is possible that negative results might stem from the use of the crystal structure as the starting point, which might not represent the correct membrane-interacting conformation. Overall, however, when multiple independent binding events are observed, our findings indicate that MARTINI 3 can be used as a predictive tool to infer the protein–membrane interface and to rationally prioritize mutagenesis experiments for validation of the *in silico* results.

From a technical point of view, we found that varying the force constant in the elastic network model used to restrain the secondary structure of the protein has only minimal impact for folded globular proteins, but it becomes a relevant parameter in those cases where protein conformational flexibility plays a key role in membrane binding, as is the case, for example, for amphipathic helices. As the extent of conformational plasticity during binding is not generally known *a priori*, we suggest, as a practical rule, to keep the force constant low, to prevent unintentional biasing of the membrane binding affinity.

In summary, our preliminary investigations suggest that transferable chemical-specific CG force fields have enormous potential in the characterization of structural properties of the membrane binding process by peripheral proteins, even if room for improvement remains. We foresee that future investigations in this direction will shed further light on the capabilities and limitations of these approaches, possibly becoming part of the development and testing strategy for future releases of MARTINI or other CG force fields.

Conflicts of interest

There are no conflicts of interest to declare.

Acknowledgements

We thank Pablo Campomanes for useful discussions. This work was supported by the Swiss National Science Foundation (grant #PP00P3_163966). This project has received funding from the European Research Council (ERC) under the European Union's Horizon 2020 research and innovation programme (Grant agreement No. 803952). This work was supported by grants from the Swiss National Supercomputing Centre (CSCS) under project ID s842 and s980.

References

1 A. Chiapparino, K. Maeda, D. Turei, J. Saez-Rodriguez and A.-C. Gavin, *Prog. Lipid Res.*, 2016, **61**, 30–39.

2 L. H. Wong, A. Copic and T. P. Levine, *Trends Biochem. Sci.*, 2017, **42**, 516–530.

3 L. H. Wong, A. T. Gatta and T. P. Levine, *Nat. Rev. Mol. Cell Biol.*, 2019, **20**, 85–101.

4 W. Cho and R. V. Stahelin, *Annu. Rev. Biophys. Biomol. Struct.*, 2005, **34**, 119–151.

5 T. F. J. Martin, *Annu. Rev. Cell Dev. Biol.*, 1998, **14**, 231–264.

6 J. E. Johnson and R. B. Cornell, *Mol. Membr. Biol.*, 1999, **16**, 217–235.

7 B. Antonny, *Annu. Rev. Biochem.*, 2011, **80**, 101–123.

8 G. Enkavi, M. Javanainen, W. Kulig, T. Róg and I. Vattulainen, *Chem. Rev.*, 2019, **119**, 5607–5774.

9 H. I. Ingólfsson, C. Arnarez, X. Periole and S. J. Marrink, *J. Cell Sci.*, 2016, **129**, 257.

10 T. A. Soares, S. Vanni, G. Milano and M. Cascella, *J. Phys. Chem. Lett.*, 2017, **8**, 3586–3594.

11 S. J. Marrink, H. J. Risselada, S. Yefimov, D. P. Tieleman and A. H. de Vries, *J. Phys. Chem. B*, 2007, **111**, 7812–7824.

12 D. H. de Jong, G. Singh, W. F. D. Bennett, C. Arnarez, T. A. Wassenaar, L. V. Schäfer, X. Periole, D. P. Tieleman and S. J. Marrink, *J. Chem. Theory Comput.*, 2013, **9**, 687–697.

13 L. Monticelli, S. K. Kandasamy, X. Periole, R. G. Larson, D. P. Tieleman and S.-J. Marrink, *J. Chem. Theory Comput.*, 2008, **4**, 819–834.

14 S. J. Marrink, A. H. de Vries and A. E. Mark, *J. Phys. Chem. B*, 2004, **108**, 750–760.

15 J. J. Uusitalo, H. I. Ingólfsson, P. Akhshi, D. P. Tieleman and S. J. Marrink, *J. Chem. Theory Comput.*, 2015, **11**, 3932–3945.

16 J. J. Uusitalo, H. I. Ingólfsson, S. J. Marrink and I. Faustino, *Biophys. J.*, 2017, **113**, 246–256.

17 C. A. López, A. J. Rzepiela, A. H. de Vries, L. Dijkhuizen, P. H. Hünenberger and S. J. Marrink, *J. Chem. Theory Comput.*, 2009, **5**, 3195–3210.

18 R. Friedman, S. Khalid, C. Aponte-Santamaría, E. Arutyunova, M. Becker, K. J. Boyd, M. Christensen, J. T. S. Coimbra, S. Concilio, C. Daday, F. J. van Eerden, P. A. Fernandes, F. Gräter, D. Hakobyan, A. Heuer, K. Karathanou, F. Keller, M. J. Lemieux, S. J. Marrink, E. R. May, A. Mazumdar, R. Naftalin, M. Pickholz, S. Piotto, P. Pohl, P. Quinn, M. J. Ramos, B. Schiøtt, D. Sengupta, L. Sessa, S. Vanni, T. Zeppelin, V. Zoni, A.-N. Bondar and C. Domene, *J. Membr. Biol.*, 2018, **251**, 609–631.

19 S. J. Marrink and D. P. Tieleman, *Chem. Soc. Rev.*, 2013, **42**, 6801–6822.

20 E. Yamamoto, A. C. Kalli, T. Akimoto, K. Yasuoka and M. S. P. Sansom, *Sci. Rep.*, 2016, **5**, 18245.

21 E. Yamamoto, J. Domański, F. B. Naughton, R. B. Best, A. C. Kalli, P. J. Stansfeld and M. S. P. Sansom, *Sci. Adv.*, 2020, **6**, eaay5736.

22 L. Picas, J. Viaud, K. Schauer, S. Vanni, K. Hnia, V. Fraisier, A. Roux, P. Bassereau, F. Gaits-Iacovoni, B. Payrastre, J. Laporte, J.-B. Manneville and B. Goud, *Nat. Commun.*, 2014, **5**, 5647.

23 F. B. Naughton, A. C. Kalli and M. S. P. Sansom, *J. Phys. Chem. Lett.*, 2016, **7**, 1219–1224.

24 A. Buyan, A. C. Kalli and M. S. P. Sansom, *PLoS Comput. Biol.*, 2016, **12**, e1005028.

25 H. Li and A. A. Gorfe, *PLoS One*, 2013, **8**, e71018.

26 E. Jefferys, M. S. P. Sansom and P. W. Fowler, *Faraday Discuss.*, 2014, **169**, 209–223.

27 D. Murray, A. Arbuzova, B. Honig and S. McLaughlint, in *Current Topics in Membranes*, Academic Press, 2002, vol. 52, pp. 277–307.

28 B. Antonny, I. Huber, S. Paris, M. Chabre and D. Cassel, *J. Biol. Chem.*, 1997, **272**, 30848–30851.

29 J. Bigay and B. Antonny, *Dev. Cell*, 2012, **23**, 886–895.

30 S. Vanni, H. Hirose, H. Barelli, B. Antonny and R. Gautier, *Nat. Commun.*, 2014, **5**, 4916.

31 M. Cascella and S. Vanni, in *Chemical Modelling: Volume 12*, The Royal Society of Chemistry, 2016, vol. 12, pp. 1–52.

32 R. Alessandri, P. C. T. Souza, S. Thallmair, M. N. Melo, A. H. de Vries and S. J. Marrink, *J. Chem. Theory Comput.*, 2019, **15**, 5448–5460.

33 D. Van Der Spoel, E. Lindahl, B. Hess, G. Groenhof, A. E. Mark and H. J. C. Berendsen, *J. Comput. Chem.*, 2005, **26**, 1701–1718.

34 W. Humphrey, A. Dalke and K. Schulten, *J. Mol. Graphics*, 1996, **14**, 33–38.

35 H. M. Berman, J. Westbrook, Z. Feng, G. Gilliland, T. N. Bhat, H. Weissig, I. N. Shindyalov and P. E. Bourne, *Nucleic Acids Res.*, 2000, **28**, 235–242.

36 B. Antonny, S. Beraud-Dufour, P. Chardin and M. Chabre, *Biochemistry*, 1997, **36**, 4675–4684.

37 E. L. Wu, X. Cheng, S. Jo, H. Rui, K. C. Song, E. M. Davila-Contreras, Y. Qi, J. Lee, V. Monje-Galvan, R. M. Venable, J. B. Klauda and W. Im, *J. Comput. Chem.*, 2014, **35**, 1997–2004.

38 G. Bussi, D. Donadio and M. Parrinello, *J. Chem. Phys.*, 2007, **126**, 014101.

39 M. Parrinello and A. Rahman, *J. Appl. Phys.*, 1981, **52**, 7182–7190.

40 D. H. de Jong, S. Baoukina, H. I. Ingólfsson and S. J. Marrink, *Comput. Phys. Commun.*, 2016, **199**, 1–7.

41 B. P. Roscoe, K. M. Thayer, K. B. Zeldovich, D. Fushman and D. N. A. Bolon, *J. Mol. Biol.*, 2013, **425**, 1363–1377.

42 T. R. Butt, S. Jonnalagadda, B. P. Monia, E. J. Sternberg, J. A. Marsh, J. M. Stadel, D. J. Ecker and S. T. Crooke, *Proc. Natl. Acad. Sci. U. S. A.*, 1989, **86**, 2540.

43 S. B. Howard, P. J. Twigg, J. K. Baird and E. J. Meehan, *J. Cryst. Growth*, 1988, **90**, 94–104.

44 N. Lucas and W. Cho, *J. Biol. Chem.*, 2011, **286**, 41265–41272.

45 Y. Uchida, J. Hasegawa, D. Chinnapen, T. Inoue, S. Okazaki, R. Kato, S. Wakatsuki, R. Misaki, M. Koike, Y. Uchiyama, S.-i. Iemura, T. Natsume, R. Kuwahara, T. Nakagawa, K. Nishikawa, K. Mukai, E. Miyoshi, N. Taniguchi, D. Sheff, W. I. Lencer, T. Taguchi and H. Arai, *Proc. Natl. Acad. Sci. U. S. A.*, 2011, **108**, 15846–15851.

46 D. E. Otzen, K. Blans, H. Wang, G. E. Gilbert and J. T. Rasmussen, *Biochim. Biophys. Acta, Biomembr.*, 2012, **1818**, 1019–1027.

47 C. Shao, V. A. Novakovic, J. F. Head, B. A. Seaton and G. E. Gilbert, *J. Biol. Chem.*, 2008, **283**, 7230–7241.

48 S. Macedo-Ribeiro, W. Bode, R. Huber, M. A. Quinn-Allen, S. W. Kim, T. L. Ortel, G. P. Bourenkov, H. D. Bartunik, M. T. Stubbs, W. H. Kane and P. Fuentes-Prior, *Nature*, 1999, **402**, 434–439.

49 S. Okazaki, R. Kato, Y. Uchida, T. Taguchi, H. Arai and S. Wakatsuki, *Acta Crystallogr., Sect. D: Biol. Crystallogr.*, 2012, **68**, 117–123.

50 M. Sumandea, S. Das, C. Sumandea and W. Cho, *Biochemistry*, 2000, **39**, 4206.

51 R. V. Stahelin and W. Cho, *Biochemistry*, 2001, **40**, 4672–4678.

52 H. M. Khan, P. C. T. Souza, S. Thallmair, J. Barnoud, A. H. de Vries, S. J. Marrink and N. Reuter, *J. Chem. Theory Comput.*, 2020, **16**, 2550–2560.

53 Q. Waheed, H. M. Khan, T. He, M. Roberts, A. Gershenson and N. Reuter, *J. Phys. Chem. Lett.*, 2019, **10**, 3972–3977.

54 Y. Liu, R. A. Kahn and J. H. Prestegard, *Nat. Struct. Mol. Biol.*, 2010, **17**, 876–881.

55 Y. Liu, R. A. Kahn and J. H. Prestegard, *Structure*, 2009, **17**, 79–87.

56 P. U. Mayerhofer, J. P. Cook, J. Wahlman, T. T. J. Pinheiro, K. A. H. Moore, J. M. Lord, A. E. Johnson and L. M. Roberts, *J. Biol. Chem.*, 2009, **284**, 10232–10242.

57 T. S. Ramalingam, P. K. Das and S. K. Podder, *Biochemistry*, 1994, **33**, 12247–12254.

58 A. P. Watt, *Inflammopharmacology*, 2001, **9**, 421–434.

59 G. Drin, J.-F. Casclla, R. Gautier, T. Boehmer, T. U. Schwartz and B. Antonny, *Nat. Struct. Mol. Biol.*, 2007, **14**, 138–146.

60 S. Vanni, L. Riccardi, G. Palermo and M. De Vivo, *Acc. Chem. Res.*, 2019, **52**, 3087–3096.

61 S. Vanni, L. Vamparys, R. Gautier, G. Drin, C. Etchebest, P. F. J. Fuchs and B. Antonny, *Biophys. J.*, 2013, **104**, 575–584.

62 J. A. Whiles, R. Brasseur, K. J. Glover, G. Melacini, E. A. Komives and R. R. Vold, *Biophys. J.*, 2001, **80**, 280–293.

63 P. Gonzalez-Rubio, R. Gautier, C. Etchebest and P. F. Fuchs, *Biochim. Biophys. Acta*, 2011, **1808**, 2119–2127.

Faraday Discussions

Theoretical and experimental comparisons of simple peptide–membrane systems; towards defining the reaction space: general discussion

Marie-Isabel Aguilar, Kareem Al Nahas, Francisco N. Barrera, Patricia Bassereau, Burkhard Bechinger, Izabella Brand, Amitabha Chattopadhyay, Ronald J. Clarke, William F. DeGrado, Evelyne Deplazes, Marcus Fletcher, Franca Fraternali, Patrick Fuchs, Ana J. Garcia-Saez, Robert Gilbert, Bart W. Hoogenboom, Zack Jarin, Paul O'Shea, Georg Pabst, Sreetama Pal, John M. Sanderson, John M. Seddon, Durba Sengupta, David P. Siegel, Anand Srivastava, D. Peter Tieleman, Madhusmita Tripathy, Johanna Utterström, Robert Vácha, Stefano Vanni and Gregory A. Voth

DOI: 10.1039/d1fd90065j

Burkhard Bechinger opened discussion of the introductory lecture by William F. DeGrado: For the Influenza M2 transmembrane domain you mention the importance of the lipid composition, environmental and membrane details. Because X-ray (XR) investigations and structures in micelles have also been published, including in high impact journals, I would be interested what kind of details you still consider valid from such latter experiments? Which other details absolutely require a lipid bilayer and which peptide/lipid ratio do you consider acceptable?

William F. DeGrado responded: One really does need to look at whether a structure makes sense when all the biological and biophysical data are considered. My view is that crystal structures will provide energetically reasonable conformations, sometimes showing multiple states, even in a single unit cell. Often these states represent intermediates that are observed during function (*e.g.*, conductance in the case of M2 (see accompanying manuscript). For M2, we had companion data from Mei Hong, using SSNMR, which gave a very similar structure to our C-closed crystal structure. And it explained a large body of electrophysiological and pharmacological data. Another method we apply frequently now are unrestrained all-atom microsecond simulations of the protein embedded in a relevant lipid composition. Structures not compatible with this environment will drift away from the starting configuration over this time period. Overall, the more data we can gather the better. Ultimately, we need to connect high-resolution data from cryo-EM and crystallography to the functionally relevant ensemble of structures in membranes with compositions closely matching those of the cellular membrane. In the end, however, we always need to compare all structures to experiments conducted in cellular and organismic context.

Burkhard Bechinger enquired: The LS-peptides adopt amphipathic helical structures when membrane-associated and should be happy to sit on the membrane surface. What driving forces make them go transmembrane and what is the proportion of the transmembrane *versus* in-planar populations?

William F. DeGrado responded: Yes indeed! They are surface absorbed, but a small fraction is inserted and stabilized by a transmembrane voltage. The inserted state can then associate as reviewed in ref. 1.

1 K. S. Åkerfeldt, J. D. Lear, Z. R. Waserman, L. A. Chung and W. F. DeGrado, *Acc. Chem. Res.*, 1993, **26**, 191–197.

Patricia Bassereau queried: Would the dual faces of the β3 integrin allowing interaction with mixed alpha partners contribute to drive integrins clustering at adhesion sites? For example, engaging two binding partners simultaneously as one driving force may result in clusters of β3 integrins (maybe mixed with alpha partners).

William F. DeGrado replied: I love the model and what you suggest is intriguing, but we do not have data to support it (or refute it). It would be interesting to find out.

Sreetama Pal asked: Based on the present understanding of lipid–protein interactions, what factors play a decisive role in determining the optimum number of helices in a transmembrane protein?

William F. DeGrado responded: Good question, I expect that the number of helices would be more dependent on the function, rather than folding. A protein would need to be large enough to create, for example, a binding site for small molecules, and in many cases also signal.

David P. Siegel continued: Professor DeGrado made the interesting speculation that some synuclein peptides could stabilize fusion/fission pores by binding with high affinity to lipid membranes with negative Gaussian curvature. I think one could determine whether peptides bound more readily to bilayers that already have negative Gaussian curvature. Such an activity might be a way certain peptide sequences are recruited to membrane fission or fusion sites *in vivo*. One could also determine if certain sequences, bound to such surfaces, decreased Gaussian curvature even more (*i.e.*, made the structures smaller). That could be done by seeing if the bound peptides reduce the lattice constant of those phases. One might simultaneously monitor both the extent of binding of the peptide and effects on the lattice constant using SAXS, on the same samples. For example, it might be possible to make "plugs" of cubic phase in X-ray capillaries, by temperature cycling host lipids across the bilayer/non-bilayer phase transition temperature. One could add the peptides of interest by changing the column of fluid overlying the plugs. The extent of diffusion of the peptides into the cubic phase plug could be assayed by tracing the inherent fluorescence or Trp absorbance of the peptides under a light microscope. However, it is not clear how large a peptide can diffuse into cubic phase aggregates, either through the water channels in the bicontinuous structure, or along the bilayer–water interfaces. (The water channels are several nm in diameter, and the diameter can be changed as a function of temperature and lipid composition.) Obviously, this would have to be tested for the peptides of interest.

Durba Sengupta enquired: You mentioned lipid effects in M2 – but not in the helix dimers – where you mentioned structural motifs and polar residues. Wouldn't lipid effects be important there – especially with the polar residues embedded in the membranes? How could we probe it experimentally?

William F. DeGrado answered: M2 forms a tetramer, and it is important that the lipid chain length matches the length of the tetramer. In general, oligomerization also relies on matching of the hydrophobic thickness of the transmembrane domain (TMD) with that of the surrounding lipid. If the lipid bilayer is too thin or too thick, oligomerization can be reduced in model membranes. We have shown the effects of membrane on thermodynamic stability and functional dynamics in M2, and Dirk Schneider's group on glycophorin.[1–3]

1 A. L. Cristian, J. D. Lear and W. F. DeGrado, *Proc. Natl. Acad. Sci. U. S. A.*, 2003, **100**, 14772–14777.
2 C. W. Lin, B. Mensa, M. Barniol-Xicota, W. F. DeGrado and F. Gai, Activation pH and gating dynamics of influenza A M2 proton channel revealed by single-molecule spectroscopy, *Angew. Chem., Int. Ed.*, 2017, **56**(19), 5283–5287.
3 V. Anbazhagan and D. Schneider, *Biochim. Biophys. Acta*, 2010, **1798**, 1899–1907.

Durba Sengupta commented: The oncogenic mutations you mentioned in the growth factor receptors are at the head group region (or just below) so perhaps there are more lipid effects (than the simple hydrogen bonds between polar residues) that should be considered?

William F. DeGrado indicated that they have nothing to add.

Ana J. Garcia-Saez queried: Do you have an explanation of why up-regulation of chaperones is associated with membrane disruption? Is this happening in bacteria?

William F. DeGrado answered: Bacteria have two-component His kinases that sense and relay information about the environment to induce a transcriptional response of regulons beneficial to bacteria. Systems like Bae and Cox in *E. coli* sense capsule stress.

Paul O'Shea said: When you are looking at interactions of the lipids, you treat the lipid (such as cholesterol) that interacts as a single molecule with the proteinaceous species. I am thinking about dipole potentials and micro-viscosities which are more ensemble properties of the lipid and are there more specific interactions. So the question is are you treating the lipid as a homogeneous matrix together with very specific lipid molecule interactions or can you include the lipid environmental heterogeneity as a factor in the interactions?

William F. DeGrado replied: Good question Paul. In analysis of natural proteins, we and others look specifically at lipid effects, particularly cholesterol in M2. Of course, there is a massive literature on this for antimicrobial peptides (AMPs). For designed systems, we don't explicitly consider lipid effects in the design process (other than hydrophobic length and positions of Tyr, Trp, Arg, Lys…) until we run microsecond molecular dynamics (MD), once we have a potential sequence designed. At that point we need to choose a lipid composition.

Patricia Bassereau opened discussion of the paper by Gregory A. Voth: Does the place in the endoplasmic reticulum–Golgi intermediate compartment (ERGIC) where the virus is budding, possibly related to a local membrane composition or geometry, influence the M protein conformation, and thus budding?

Gregory A. Voth answered: It's pure speculation, but I would think so. We need to do coarse-grained simulations with many proteins to address this better. It would be wonderful to see you do some of your beautiful experiments on this protein binding to membrane tubules pulled from micro-pipettes like you have done for N-BARs (endophilin).[1]

1 M. Simunovic, E. Evergren, I. Golushko, C. Prévost, H.-F. Renard, L. Johannes, H. McMahon, V. Lorman, G. A. Voth and P. Bassereau, *Proc. Natl. Acad. Sci. U. S. A.*, 2016, **113**, 11226–11231.

Patricia Bassereau enquired: Since you show that the local lipid composition influences the conformation, what triggers one conformation *vs.* the other in the cell?

Gregory A. Voth responded: We do not yet know. Perhaps its lipid domain dynamics that triggers the conformational change, or perhaps it is a response to larger scale membrane morphological changes during budding.

Patricia Bassereau asked: How do changes in cholesterol level in the cell affect virus assembly?

Gregory A. Voth responded: I am not sure anyone knows the answer to this. There has really been relatively little work done on the virus assembly and budding. It's a critical need.

John M. Seddon queried: Are the trans-membrane segments of the proteins sensitive in terms of conformation and/or dynamics to the lateral pressure profile within the lipid bilayer, and do you calculate this?

Gregory A. Voth answered: We don't yet know an answer to that question. However, I have a feeling that the M protein conformational states we see in the simulations could be influenced by the lateral pressure profile. Yes, it should be possible to calculate it in the future.

Durba Sengupta asked: You showed high membrane deformation around the E channel – the viral envelope/membrane would be highly curved – would the high curvature alter the perturbation – and perhaps aid function?

Gregory A. Voth replied: It is possible. The pore in our model is pretty narrow so anything that increases its radius could affect ion conduction. On the other hand, the E channel appears to be non-selective.

Burkhard Bechinger enquired: From your simulations, can you decipher molecular details what kind of interactions make individual lipids (*e.g.* cholesterol) accumulate or deplete around the proteins? H-Bonds, van der Waals interactions, hydrophobic mismatch *etc.*?

Gregory A. Voth answered: Yes we can but we have not yet done so. One has to be a little careful about using local measures such as H-bonds though. N-BAR proteins are seen to aggregate in linear "strings" which are a result of membrane fluctuation-mediated forces between the proteins (like Casimir forces) and not local interactions.

Bart W. Hoogenboom asked: Could you speculate on how your findings with different lipids translate to real cell membranes?

Gregory A. Voth responded: Boy, that is a hard question. "Real" membranes are so complex yet many properties of membrane proteins have been elaborated pretty well in model membranes. We used a model for the ERGIC membrane.

Amitabha Chattopadhyay remarked: For HIV1, host cholesterol is known to be necessary for the virus to bud. It is not known whether this is true for SARS-CoV-2 virus.

Gregory A. Voth replied: I fully agree with you. However, the M protein which is believed to be important for SARS-CoV-2 budding shows a tendency in our simulations to sequester some cholesterol.

Izabella Brand asked: E protein is a channel protein. What triggers the opening and closure of the channel? Does the electric potential drop across the membrane affect the channel properties?

Gregory A. Voth responded: I'm afraid that no one really knows yet. Sorry! I don't think there is evidence the E channel is voltage gated though.

Georg Pabst opened discussion of the paper by Izabella Brand: The fabrication of lipid asymmetry on solid supports by the technique you are using has been shown to be challenging (see *e.g.* work by the groups of Lukas Tamm[1-3] and John Conboy[4]). How do you measure lipid asymmetry to make sure that your system is asymmetric to begin with?

1 J. M. Crane, V. Kiessling and L. K. Tamm, *Langmuir*, 2005, **21**(4), 1377–1388.
2 V. Kiessling, J. M. Crane and L. K. Tamm, *Biophys. J.*, 2006, **91**(9), 3313–3326.
3 V. Kiessling, C. Wan and L. K. Tamm, *Biochim. Biophys. Acta*, 2009, **1788**(1), 64–71.
4 T. C. Anglin, K. L. Brown and J. C. Conboy, *J. Struct. Biol.*, 2009, **168**, 37–52.

Izabella Brand replied: In general, fabrication of asymmetric lipid bilayers belongs to an experimental challenge. A freshly prepared bilayer is asymmetric, which is confirmed by electrochemical studies. The capacitance of a freshly prepared bilayer is ~ 3 μF cm^{-2}. If a KLA molecule with its large polar head group faced the electrode surface, due to presence of water and sugar residues on the Au surface, an increase in the capacitance, due to a higher value of the dielectric constant of the water/sugar system, would be observed. However, potential cycling for several times, did not lead to any increase in the membrane capacitance. This result suggests that the asymmetry, at least to a large extent is preserved. Possibly, application of more negative potentials (as tested in simulation studies[1]) would cause some irreversible changes in the bilayer structure. Indeed, in our previous studies of asymmetric lipid bilayers containing glycolipids (GM1 and Gd1a) a gradual increase in the membrane capacitance in following potential scans was observed.[2] In parallel, the intensities of IR spectra of the sugar residues in the asymmetric POPE–KLA bilayer are attenuated, indicating a uniform orientation of the polar head groups in KLA. In contrast in the previous studies the intensities of the IR absorption modes in sugar residues adapted a random orientation, indicating structural rearrangements in the bilayer. A flip–flop was suggested. Note that KLA molecules bind divalent ions. It is responsible for the formation of a rigid leaflet, which might be responsible for the stability of the asymmetric bilayer.

1 T. J. Piggot, D. A. Holdbrook and S. Khalid, *J. Phys. Chem. B*, 2011, **115**, 13381–13388, DOI: 10.1021/jp207013v.
2 M. Nullmeier, H. Koliwer-Brandl, S. Kelm, P. Zägel, K.-W. Koch and I. Brand, *ChemPhysChem*, 2011, **12**, 1066–1079, DOI: 10.1002/cphc.201100036.

Georg Pabst asked: As a follow up of the previous question, lipid flip–flop was reported to be fast in solid supported bilayers by the Conboy group, see, *e.g.*, ref. 1. We suggested in Marquardt *et al.*[2] that this is due to unavoidable defects in solid supported bilayers. Do you think that lipopolysaccharides (LPS) will increase the stability of your asymmetric systems?

1 T. C. Anglin, M. P. Cooper, H. Li, K. Chandler and J. C. Conboy, *J. Phys. Chem. B*, 2010, **114**, 1903–1914.
2 D. Marquardt, F. A. Heberle, T. Miti, B. Eicher, E. London, J. Katsaras and G. Pabst, *Langmuir*, 2017, **33**(15), 3731–3741.

Izabella Brand answered: Yes, I think that the LPS increase the stability of the asymmetric bilayer. I have discussed this problem in the answer to the previous question.

Paul O'Shea enquired: Are capacitance changes due to structural rearrangements of the lipid? Or is there a contribution from the dielectric properties if the peptide attached to the membrane?

Izabella Brand responded: Both aspects are very important. The electric potentials cause some changes to the lipid bilayer alone (Fig. 2, curve 1, in the paper; DOI: 10.1039/d0fd00039f). These changes are due to the electroporation and adsorption–desorption of the lipid bilayer. In the presence of a peptide attached to the membrane the dielectric constant of the outer membrane (OM) changes (increases), agreeing with the observed increase in the membrane capacitance in the presence of melittin (Fig. 2, curves 2, 3, in the paper; DOI: 10.1039/d0fd00039f). The dielectric properties of melittin may not only lead to an increase in the capacitance, but may affect the membrane electroporation potential too. Melittin insertion into the bilayer will also affect the potential of zero charge of the OM, a phenomenon which was experimentally observed. Both lipids and peptides respond to electric potentials and contribute to experimentally measured capacitance values.

Paul O'Shea asked: Can you correlate the kinetics of these changes? Kinetics of binding and insertion of peptides including melittin has been done by other groups, including my own early reports (see for example ref. 1), and with melittin particularly (ref. 2). Do any of these rate changes correlate?

1 C. Golding, S. Senior, M. T. Wilson and P. O'Shea, *Biochemistry*, 1996, **35**, 10931–10937.
2 J. Wall, C. A. Golding, M. V. Veen and P. O'Shea, *Mol. Membr. Biol.*, 1995, **12**, 183–192.

Izabella Brand replied: Kinetics of melittin interaction with lipid bilayers was intensively studied over the past decades. However, large differences in kinetic data were reported. Many aspects affect the kinetics of melittin interaction with lipid bilayers. Among the critical factors are: the lipid : peptide ratio, the lipid composition of the bilayer, the shape of the supramolecular aggregate (*e.g.* planar lipid bilayer, vesicle), the presence of cholesterol in the bilayer or temperature. Despite differences in kinetics (rate constants, association constants), some similarities could be observed between different reports. There is a general consensus that disordered melittin adsorbs on the membrane surface. After adsorption on the membrane surface a transition to α-helix structure and membrane incorporation take place (see for example ref. 1). Saturation of the membrane surface with a melittin layer depends on the peptide concentration. However, after *ca.* 60 minutes of the interaction, saturation was observed. In our experiments, high concentrations of melittin were used (1 and 10 μM). Our results showed that after 15 minutes of the interaction the peptide still had a disordered structure. A change to α-helical structure was observed after 1 h of the interaction. This result is in good agreement with literature. Next steps of interaction include peptide insertion into the membrane and its disruption. This process is very specific to the lipid composition of the membrane. Since phospholipids were usually used in kinetic studies, a direct comparison to a bilayer containing lipopolysaccharide cannot be done.

1 M. G. Burton, Q. M. Huang, M. A. Hossain, J. D. Wade, A. H. A. Clayton and M. L. Gee, *Langmuir*, 2013, **29**, 14613–14621, DOI: 10.1021/la403083m.

Sreetama Pal questioned: Is it possible to identify (from the IR-based readouts you mention) whether any specific residues of melittin interact with the bacterial membrane mimic in use?

Izabella Brand responded: IR is sensitive to the conformation of a peptide/protein. However, the measured IR absorption modes are not specific for any particular amino acids. Thus, the measured signal is an average answer of all amino acids in the studied molecule. Thus, we are not able to detect specific amino acids in melittin interacting with KLA and POPE lipids.

Sreetama Pal asked: What is the orientation of the peptide after it adsorbs onto the membrane? In the abstract (DOI: 10.1039/d0fd00039f) you mention that "the N-terminus enters into the hydrophobic region of the membrane and forms a channel to the hydrophilic head groups in POPE". If this implies a pseudo-transmembrane orientation of melittin, how are the lipid headgroups positioned with respect to the peptide?

Izabella Brand answered: In this paper the orientation of the lipid molecules was studied. We observed that the carboxylate groups in the inner core of the KLA lipid interact directly with the peptide. This interaction weakens the binding of Mg^{2+} ions, which may be responsible for a loss of the outer leaflet compactness. In the hydrocarbon chain region an increase in the order of their packing was observed. This result agrees with previous studies and indicates insertion of the peptide into the bilayer. Electrochemical results support this conclusion.

Of course, melittin gives rise to a strong amide I mode (see Fig. 3; DOI: 10.1039/d0fd00039f). This mode was not analysed in this paper. However, depending on the time of adsorption large changes in the intensity (concentration) and shape of the amide I mode were observed. Briefly, after 15 minutes of adsorption the amide I mode was centered around 1640 cm^{-1} indicating that a disordered peptide was adsorbed on the bilayer surface. However, after 1 h of interaction the amide I mode was centered at 1648 cm^{-1}, indicating its conformational transformation to α-helix. This structure is characteristic for membrane inserted melittin, thus we concluded that the peptide spans the membrane. Quantitative analysis of the amide I mode aiming at the determination of the average orientation of the long helix axis in the bilayer will be discussed in a separate paper.

John M. Sanderson enquired: In Fig. 3 of your article (DOI: 10.1039/d0fd00039f) you show polarization modulation infrared reflection-absorption (PM IRRA) data for the same system after 15 min and 60 min. There is a clear difference in the spectra, which indicates that the system had not reached equilibrium (or the endpoint of the binding process) after 15 min. Has equilibrium been reached after 60 min? I ask the question because in work we did with Alison Rodger to look at melittin binding to liposomes by dichroism methods, an endpoint was sometimes difficult to achieve.[1] We subsequently found that there was an acyl transfer reaction from the lipid to the peptide.[2]

1 A. Damianoglou, A. Rodger, C. Pridmore, T. R. Dafforn, J. A. Mosely, J. M. Sanderson and M. R. Hicks, *Protein Pept. Lett.*, 2010, **17**, 1351–1362, DOI: 10.2174/0929866511009011351.
2 C. J. Pridmore, J. A. Mosely, A. Rodger and J. M. Sanderson, *Chem. Commun.*, 2011, **47**, 1422–1424, DOI: 10.1039/c0cc04677a.

Izabella Brand replied: Our results show that we did not reach equilibrium, even after 1 h of interaction of the POPE–KLA bilayer with 10 µM melittin solution. In my opinion, this is a dynamic system, which will finally lead to the membrane dissolution, disruption. Our aim was to elucidate the changes in the lipid molecules taking place during interaction with melittin and the impact of this interaction on the membrane electric properties. Currently, we analyse the amide I mode of melittin to elucidate the changes in the secondary structure of the peptide interacting for different times with the membrane. After 1 h of the interaction, independently of the melittin concentration, the amide I mode indicated that melittin adapted an α-helix structure. The average tilt of the long helix axis will be estimated. After exposure of the membrane for a given period of time to melittin solution (15 or 60 min), the sample was transferred into an electrolyte solution without peptide. The spectral changes were monitored for up to 20 h. In this time the average structure of the bilayer was preserved. We cannot prove if the process of melittin acylation by acyl chains in KLA and/or POPE took place. Our technique is not sensitive to the progress of this reaction. It is a very interesting point, and will be considered in our future investigations.

John M. Sanderson queried: You see desorption of material from the electrode after longer time periods – have you considered collecting this material and analysing it to see whether there have been any chemical changes?

Izabella Brand responded: *In situ* PM IRRAS experiments took *ca.* 20 h. In this time several positive- and negative-going potential scans were recorded. Next, the spectra of each cycle are checked, compared and if no major deviations over time occurred, averaged. During 20 h of the experiment no significant spectral changes were observed. Since we work with a single bilayer of lipids adsorbed on a *ca.* 1 cm^2 large Au electrode, and the volume of the electrolyte solution in the *in situ* cell is large (*ca.* 60 mL) eventually desorbed material will have a very low concentration. For this reason, I did not undertake the analysis of the analyte. Now, knowing that acylation of melittin can be a reason for the lack of changes in the spectra, I am considering the analysis of the solution after reaction. Thank you very much, your questions, comments and fantastic talk (DOI: 10.1039/d1fd00030f) will have a large impact on the planning of future experiments.

Evelyne Deplazes asked: The membrane is made on a supported monolayer, and the lipids will therefore have limited lateral movement compared to a bilayer in the fluid state. How do you think this will impact the interactions with the peptide? Can the membrane still adapt its structure to the peptide, *i.e.* can the lipids rearrange themselves in response to membrane binding? And how does this reduced lateral movement of the lipids affect the measurements that are being taken?

Izabella Brand answered: The lateral mobility of POPE molecules in the inner, Au surface facing leaflet is lower than in fluid floating bilayers. The lateral mobility of KLA molecules present in the outer leaflet, facing the electrolyte solution is not affected by the substrate. The outer leaflet provides the first contact to melittin. Thus, I assume, the adsorption of melittin on the bilayer surface and interaction with a lipopolysaccharide is not affected by the substrate. However, the insertion of melittin into the second leaflet of the bilayer may be affected by the reduced lateral mobility of POPE molecules.

Anand Srivastava enquired: With POPE being a H-bond donor, can you please comment on how the pH of the environment would play a role in the process?

Izabella Brand responded: We did not investigate the pH effects on the membrane structure and stability. The pH of the used electrolyte was 6.5. We are aware that pH affects not only the H bonding in POPE but also the protonation of carboxylate groups in the KLA molecule and in general in lipopolysaccharides. Thus, pH changes would affect the structure of lipid molecules forming the bilayer, which may further affect its stability in changing electric fields.

Kareem Al Nahas queried: How to differentiate a capacitance shift between peptides present freely only in the aqueous phase near the outer layer of the membrane rather than bound/inserted in the lipid membrane?

Izabella Brand replied: The capacitance of the Au electrode modified by a molecular assembly depends on the surface density, packing and compactness of the molecules adsorbed on its surface. The capacitance of a defect-free film adsorbed on an electrode surface is given in eqn (1):

$$C = \varepsilon_0 \varepsilon_r A/d \tag{1}$$

where ε_0 is the permeability of vacuum, ε_r is the dielectric constant of the molecules present in the film, d is the thickness of the adsorbed film, and A is the surface area of the electrode. When a defect-free film is formed on the electrode surface the dielectric constant of the adsorbing molecule has a large impact on the capacitance value. Lipid molecules are amphiphilic and contain long acyl chains. The dielectric constant of a hydrocarbon chain is $\varepsilon = 2$. Thus, lipid molecules have lower values of dielectric constants than proteins or peptides. Experimentally measured capacitance has a contribution from the capacitance of the lipid membrane (C_{lipid}), capacitance of eventually adsorbed protein film ($C_{protein}$) present on top of the membrane and of diffuse layer capacitance (C_{dif}):

$$1/C = 1/C_{lipid} + 1/C_{protein} + 1/C_{diff} \tag{2}$$

According to eqn (2) the adsorption of a peptide on the bilayer surface, will not significantly affect the measured capacitance, because it is determined by the lower capacitance of the system (in this case by the compact lipid bilayer). Thus, if melittin adsorbed only on the membrane surface the measured capacitance would practically not change compared to the bilayer of the intact bilayer. An increase in the capacitance indicated insertion of the peptide into the membrane which leads to an increase in the dielectric constant of the film adsorbed on the Au surface, and possibly caused some changes in the membrane packing and compactness.

Marie-Isabel Aguilar asked: You used 1 and 10 μM melittin. What P : L ratio did this correspond to in the systems you studied? Did you observe any capacitance changes as you approached a lytic concentration?

Izabella Brand answered: We did not perform our studies for a longer time than 1 h of a direct exposure of the KLA–POPE bilayer to melittin solution. In this time no lysis of the membrane was observed. If the membrane became unstable, the measured IR signals would change during the experiment and it would not be possible to obtain any answer. The P : L ratio, considering the supramolecular assembly of the outer membrane, has changed during protein adsorption. We were unable to determine the surface concentration of melittin; therefore we do not know the P : L ratio.

Paul O'Shea opened discussion of the paper by Ana J. Garcia-Saez: Can you comment on the targeting on the mitochondrial surface of the initial binding reactions?

Ana J. Garcia-Saez replied: The targeting of Bax and tBid to mitochondria is mainly mediated by the sequence of the C-terminal anchor of Bax, and in the case of tBid by Mtch2 and the cardiolipin binding domain identified in tBid.

The role of lipids in the membrane targeting has been mostly explored in *in vitro* reconstituted systems with model membranes. Here, the presence of cardiolipin at about 20% concentration is usually sufficient for efficient binding of tBid to the membrane. In the absence of tBid or another activator, recombinant, inactive Bax does not spontaneously bind to membranes, even if they contain cardiolipin. The presence of tBid induces Bax binding to cardiolipin-containing membranes. Bax can also bind to cardiolipin-containing liposomes if activated *in vitro* by incubation at 42 °C. The requirement of cardiolipin for binding is mainly due to electrostatic interactions, as it can be replaced by phosphatidyl glycerol at a similar density of negative charges.

Paul O'Shea asked: Binding seems to be related to negative charge on the surface of the membrane. Would you know if there is any localisation of the binding to say cardiolipin which can be localised in the membrane. So I guess I'm asking is there any correlation with cardiolipin regions?

Ana J. Garcia-Saez answered: Yes, the *in vitro* experiments suggest that membrane binding is due to electrostatic interactions with negatively charged lipids. Now, the question of cardiolipin regions is more complicated. In healthy mitochondria, the concentration of cardiolipin in the outer membrane is believed to be around 4–6%, while it is around 16% in the inner mitochondrial membrane. Although it has been proposed that the amount of cardiolipin in the outer membrane may increase during apoptosis, it still remains unclear whether cardiolipin plays a fundamental role in the targeting of tBid and Bax to mitochondria during apoptosis. One possible explanation is the higher local concentration of cardiolipin in potential cardiolipin domains, which could be sufficient for driving the targeting. However, strong evidence for these domains in the outer membrane of healthy mitochondria is missing. Furthermore, GFP-Bax presents a homogeneous localization at the microscopic level when it binds to mitochondria in healthy cells and in the initial stages of apoptosis.

John M. Seddon queried: Your model system contains a significant amount of cardiolipin, which has a strong tendency to form inverse non-lamellar phases in the presence of binding cations. Did you see any sign of this in your systems?

Ana J. Garcia-Saez responded: Not under the experimental conditions that we used. We usually work with giant unilamellar vesicle (GUV) preparations that present a high number of good quality GUVs. We cannot discard the formation of non-lamellar phases in the lipid layers attached to the electrodes, but we do not image them. I am also not sure whether we could unambiguously detect non-lamellar phases with the resolution of optical microscopy.

Patricia Bassereau asked: What do you think is the benefit to get such an indirect way to repress death (a 3-body, rather than a 2-body process)?

Ana J. Garcia-Saez answered: It provides additional levels of regulation. The decision to undergo cell death is the ultimate irreversible decision for a cell, as it ends the cell's existence and there is no way back from it. It thus makes sense that cells have developed sophisticated regulation processes with multiple levels of fine tuning to decide whether to die or survive. The proteins of the BCL-2 family are a very good example in this regard, because the activator, effector and inhibitor functions overlap between several family members with small differences in their mechanisms and functions.

Durba Sengupta enquired: Is there anything known at the structural level – especially about the residue-level specificity for Bax and tBid interacting with cardiolipins?

Ana J. Garcia-Saez replied: There is nothing really known at the structural level about a specific interaction of these proteins with cardiolipin. In the case of tBid, a cardiolipin-interacting domain could be localized within helix 6, which is also important for the pro-apoptotic activity of the protein.

David P. Siegel opened discussion of the paper by **D. Peter Tieleman**: Sorry if I am missing something, but my impression is that CHOL, POPC, POPS and DPSM are all enriched around AQP1 & OmpF, according to the probability scale beneath the figure (Fig. 2 in the paper; DOI: 10.1039/d1fd00003a). This seems to reflect a fairly general phenomenon, relatively insensitive to lipid structure. In the case of CHOL you show that there are specific CHOL–protein interactions. But for the phospholipids, what do you think this more general effect may be? Perhaps a length mis-match effect? Perhaps I merely misunderstand the definition of relative probability. I do see a patchiness in the association of POPC and POPS with the protein periphery.

D. Peter Tieleman responded: Emphasis on CHOL-protein interactions was guided by the relatively discrete nature of CHOL density around AQP1 and KcsA seen in Fig. 2, potentially suggestive of specific binding sites. As noted, some of the phospholipids are also enriched around a few of the protein entities, namely AQP1, OmpF, and KcsA, albeit more diffusely compared to CHOL. While these findings are consistent with previous studies, as discussed in the paper, the causes and consequences of specific phospholipid enrichment are not clear in all cases. The comparably diffuse distribution of phospholipids compared to CHOL around some of the proteins suggests they don't form strong specific interactions in this annular shell. Though the molecular basis of annular lipid shells was not a focus of this investigation, they typically arise due to favorable, yet rather low affinity interactions between lipid head and tail groups and residues on the protein surface. We look forward to delving further into the atomistic details of such interactions in follow-up studies.

Durba Sengupta enquired: A more technical question. Did you calculate the potential of mean force (PMF) of cholesterol flip–flop in Martini 3 – and is there a higher penalty to flip–flop than in Martini 2 and what are the differences to the values calculated with atomistic force-fields?

D. Peter Tieleman replied: No, we haven't. There is a Martini 3 beta model at the moment but no finalized version. We have previously published PMFs with Martini, with the atomistic Berger/GROMOS87 combination and with the atomistic Slipids Force Field. There is quite a spread of values for the peak in the PMF. CHARMM37 seems to have much slower rates. Overall, however, the rates are still fast, in agreement with experimental difficulties to measure cholesterol flip flop rates and the notion that cholesterol equilibrates between leaflets quickly on macroscopic or biological time scales.

Robert Gilbert asked: It seems from your data that the smoother the structure of the membrane protein, the more likely the flip flop. In simple terms, if you have a protein that is "knobbly" on its intra-membrane surface, then it doesn't flip flop much. Could you comment on surface topography and the likelihood of flip flop. Also, is it possible to map the trajectory of the flip flop?

D. Peter Tieleman responded: I think that is a plausible hypothesis but it's hard to dis-entangle this from size and shape effects of the protein. To test it I'd consider artificial systems that look like walls (graphene sheet) with different levels of "knobbly-hood", but we haven't done that. You can certainly map the trajectory of flip flop. We have in the past, *e.g.* Gu *et al.*[1]

1 R.-X. Gu, S. Baoukina and D. P. Tieleman, *J. Am. Chem. Soc.*, 2020, **142**, 2844–2856.

Franca Fraternali questioned: Is Martini 3 clearly improving the description of protein flexibility and are the networks of contacts captured in the flexibility descriptors. How is the flexibility captured in this new force-field?

D. Peter Tieleman answered: Not in itself, in its current form, but there are a number of reasons why I'm optimistic that there is room for improvement. First, in addition to elastic networks the Go model approach to maintaining protein structure looks promising, and has been used a few times. This is not specific to Martini 3 but an alternative to elastic networks and in principle allows encoding more structural information because it's a potential function based on native contacts instead of an elastic network that just locks everything in place. Second, the choice of bead size in Martini 3 will allow redefining the protein backbone with more detail. The current Martini 2 protein force field is unusual in coarse-grained protein force fields in that it has very accurate (relatively speaking) side chains but a very simple backbone, while foldable coarse-grained protein force fields tend to have very complex backbones to parameterize backbone conformations but very simple side chains. In Martini 2 there is not much room to make the backbone more accurate. We tried, with dummies and extra particles, but with the smaller particles in Martini 3 there is much more room to work on this in the future.

Franca Fraternali queried: The elastic network can be made variable in terms of membrane positioning? The flexibility and the rigidity are factors that could impact the position in the membrane.

D. Peter Tieleman replied: Yes, the elastic network definition and parameters can be adjusted. People have used some flexibility in definition to include or exclude loops, or have separate networks for domains or subunits, but generally I haven't seen very compelling results from this. We have used position restraints to anchor proteins in a specific location while the lipids and solvent are free to move, which means you don't define or pre-determine the position in the membrane (although there is an impact on possible tilting). For many purposes with the current version of Martini the preferred approach probably is default elastic network settings and no position restraints, although for specific questions that of course can change.

Georg Pabst enquired: The diffusion of cholesterol to the opposing leaflet leads to a mass imbalance in your system. As a consequence the differential stress within your system will change (specifically also because of the intrinsic curvature of cholesterol). Do you expect that these changes in bulk membrane properties couple back to your observations?

D. Peter Tieleman responded: They might, although you need quite a large imbalance to have observable effects on pretty much anything in a simulation. This might not be true for stress profiles but these are hard to calculate accurately. I do think this is an interesting and important direction for the next few years.

Paul O'Shea queried: How is your new modeling approach handling polarisation – *i.e.* polarisable force fields? Or are you using *via* massive parameterisation? Or something else maybe in terms of hybrids with atomistic models?

We find that there is a contribution of the water and a contribution of the lipid to say the membrane dipole potential and their mutual contribution differs markedly depending on the type of lipid. When we and others have tried to study this atomistically, the problem of equilibrium occurs in which we don't think the system has reached equilibrium with say 30 mol% cholesterol. So we have the problem of not being able to address biologically-relevant cholesterol-rich nanodomains and makes things very different. So it seems to me that the atomistic approach is some way behind where you are. Is there a way forward for this with the new Martini model or with a hybrid?

D. Peter Tieleman replied: A force field embodies specific approximations. An atomistic force field has no electronic polarization but has polarization effects due to reorienting dipoles formed by fixed partial charges. A polarizable force field like Amoeba or a force field based on Drude oscillators does incorporate electronic polarization. Martini beads interact through parameterized potentials that have to incorporate polarization implicitly. Although it has explicit partial charges, on most beads these are zero because they represent a fragment with no net charge. The main advantage of Martini compared to atomistic simulations is its vastly increased sampling power, but if your scientific questions require both sampling of Martini and detail of an atomistic or polarizable force field you're out of luck. You'll have to wait for faster computers. Approaches to hybrid simulations for now focus on things like peptide or protein conformation described in detail with a Martini environment. Whether that is useful depends on the specific question of interest.

Paul O'Shea asked: We can download structures for atomistic modeling that are supposedly in some kind of equilibrium but many lipid structures are unavailable although they are observed in wet experiments to interact quite differently with peptides *etc* (see ref. 1). We observe in wet experiments that membranes containing all the ingredients to form micro-domains (cholesterol, sphingomyelin, *etc*) take a very long time to reach some kind of steady state structures and so are unlikely to satisfy equilibrium conditions suitable for atomistic modeling so is there information that coarse-grained (CG) is giving that atomistic cannot?

1 T. Asawakarn, J. Cladera and P. O'Shea, *J. Biol. Chem.*, 2001, **276**, 38457–38463.

D. Peter Tieleman responded: Structures in the protein data bank still have flexibility and a distribution of conformations at room temperature in solution or a membrane and proteins undergo changes in conformation as they go through their function. The notion of 'supposedly equilibrium' is problematic in this respect. The main power of CG force fields is that they allow better sampling, but at the cost of less chemical detail. It's a trade off you'll have to make for each specific research question.

Patrick Fuchs enquired: Now that MARTINI 3 has fixed the problem of "stickiness" between membrane proteins, have you tried to simulate more than one protein in the box? If yes, what is the lipid distribution like with more than one protein (in the case of multiple copies of the same protein or if there are different proteins)?

D. Peter Tieleman answered: We have published a number of simulations with multiple proteins but these were restrained in place (*e.g.* ref. 1 and 2). We have ongoing simulations on a larger number of proteins in a single bilayer but the time scales are so slow these systems are difficult to sample and we currently have no detailed comparisons between Martini 2 and 3.

1 V. Corradi, E. Mendez-Villuendas, H. I. Ingólfsson, R.-X. Gu, I. Siuda, M. N. Melo, A. Moussatova, L. J. DeGagné, B. I. Sejdiu, G. Singh, T. A. Wassenaar, K. Delgado Magnero, S. J. Marrink and D. P. Tieleman, *ACS Cent. Sci.*, 2018, **4**, 709–717.
2 B. I. Sejdiu and D. P. Tieleman, *Biophys. J.*, 2020, **118**, 1887–1900.

Izabella Brand queried: Can you please comment on the results of the incorporation of carbon nanotubes (CNTs) into lipid membranes? How did you select the length of the CNT? Did you detect any interactions (hydrophobic/hydrophilic preference) between CNTs and lipid molecules?

D. Peter Tieleman replied: The nanotube was just a model for a mostly featureless object, without the geometrical features of side chains and the diverse chemistry of side chains. It was restrained to not move in the simulations, and its length is unrealistically short compared to actual CNTs but just chosen to be long enough to span the membrane. We have previously looked at a range of chemically modified CNTs of different thicknesses and lengths and their interactions with lipids.[1]

1 S. Baoukina, L. Monticelli and D. P. Tieleman, *J. Phys. Chem. B*, 2013, **117**, 12113–12123.

Ronald J. Clarke asked: If cholesterol flipping occurs on the microsecond timescale, then on a timescale relevant for many membrane protein conformational changes, *e.g.* ion pumping or ion channel gating, which occur over the millisecond-to-second timescale, one would expect the cholesterol distribution across the membrane to be in equilibrium, *i.e.*, with no asymmetric distribution across the membrane. Can you envisage a situation where an asymmetric cholesterol distribution across the membrane could persist for long enough to be relevant for an ion pump or the gating of an ion channel?

D. Peter Tieleman responded: I do expect the cholesterol distribution to be in equilibrium, but that does not mean it is equal. Cholesterol flips but other lipids do not flip, except on timescales of seconds to hours, and their distribution is tightly controlled by proteins and cellular processes. Equilibrium for cholesterol means equal chemical potential in both leaflets but the chemical environment in both is different and therefore the concentration that results from equal chemical potential is different. Protein conformational changes might well be coupled to local cholesterol concentration near the protein. I think that is an interesting suggestion you could look at with Martini simulations fairly readily. I don't think it will play a role in the overall cholesterol distribution between leaflets at a larger scale.

Ana J. Garcia-Saez opened discussion of the paper by Francisco N. Barrera: Related to the presence of negatively charged ions in the anti-apoptotic proteins – did you consider the helix5 of Bcl-xL, which in contrast to Bax naturally contains negatively charged residues?

Secondly, you presented studies in supported bilayers. Do you know if BAX can be pH regulated in other systems, like in liposomes with encapsulated dyes?

Francisco N. Barrera answered: We focused for these studies on BAX, seeking to harness the natural ability of helix alpha 5 to participate in pore formation. While we cannot predict with certainty the results of testing helix5 of Bcl-xL, we reasoned that BAX could be a more interesting target as it might be possible to extract biologically significant data on BAX poration from studies on BaxE5.

Ana J. Garcia-Saez remarked: That is interesting because the wild type peptide does not disrupt/dissolve membranes unless it is at a very high concentration.

Francisco N. Barrera responded: The differences between the WT peptide (Bax alpha 5) and the pH-sensitive peptide (BaxE5) are indeed surprising. One might imagine that the replacement of positive charges in the WT for negative charges in BaxE5 would be behind these differences, as lysine residues often influence peptide–membrane interactions.

Marcus Fletcher asked: In Fig. 1B of your submitted paper (DOI: 10.1039/d0fd00070a) there looks to be some membrane integrity disruption at pH 7.9 that is dependent on BaxE5 concentration, but in the AFM there is no observed disruption to the supported lipid bilayer. What do you think is happening there in the leakage assay to cause disruption?

Francisco N. Barrera replied: While we cannot rule out that supported bilayers are more stable than bilayers in a liposome, the lack of effect of BaxE5 in supported bilayers could be merely a kinetic effect, since a shorter incubation time was used than in the leakage experiments. BaxE5 at neutral pH displays a slow effect on membranes, but indeed it can attack membranes to some degree at long incubation times.

Johanna Utterström queried: Could the reason for the differences between leakage assay and AFM depend on that different lipids were chosen? POPC or DOPC? And what was the reason for using different lipids for the two experiments?

Francisco N. Barrera answered: Thanks for this question. POPC and DOPC, while similar in some respects, have different biophysical properties (like fluidity), and therefore it would not be surprising to me that the differences in acyl chain modulated the BaxE5 effect. The two lipids were used for simple utilitarian reasons.

Robert Gilbert asked: It seems to me, from your data, that the effect of the binding of protein is a key factor, aside from downstream effects. Could you comment on the extent to which your observations can be explained by the degree of protein binding in different conditions (*e.g.* pH).

Francisco N. Barrera replied: It would reasonable to expect that a major factor in the observed pH-responsiveness of the BaxE5 peptide is lipid partitioning. Specifically, at neutral pH the glutamic acid residues in BaxE5 will be negatively charged, precluding efficient membrane interaction/insertion. A pH drop will protonate these residues to increase the overall hydrophobicity of BaxE5. This effect will certainly increase the degree of peptide partitioning into the membrane, and increase the peptide concentration at the membrane surface. It is expected that in these conditions the peptide will be favored to self-assemble and disrupt membrane integrity.

Sreetama Pal enquired: I am curious to know whether the behavior of the Bax fragment could be reversed on increasing pH? If you observe some kind of hysteresis (similar to your work with the pH low insertion peptide), that might provide some interesting information about how protonation influences peptide–membrane interactions.

Francisco N. Barrera responded: This is an interesting question. We expect a pH rise to detach BaxE5 from the membrane. Hysteresis would be certainly expected, in no small part due to the changes in membrane state occurring as a result of the BaxE5 effect.

Sreetama Pal asked: It would be interesting to know the rationale for choosing this specific BaxE5 analog. Would you expect the behavior of the Bax fragment to change if you vary the number and relative position of the introduced glutamate residues? If yes, how?

Francisco N. Barrera replied: We introduced the smallest number of glutamic acids that we expected would provide control over membrane disruption by pH. However, changes in the number and position of the glutamic acid residues would probably affect the pH midpoint (pH_{50}) of membrane insertion, based on prior observations (see ref. 1).

1 H. L. Scott, J. M. Westerfield and F. N. Barrera, *Biophys. J.*, 2017, **113**, 869–879, PMID: 28834723.

Sreetama Pal queried: Is it possible to isolate or stabilize the BaxE5–membrane complexes at some or all the stages of membrane remodeling represented in the cartoon in the paper (Fig. 8; DOI: 10.1039/d0fd00070a)? That could represent a potentially useful system to understand more about the basic biophysics of how proteins shape membranes.

Francisco N. Barrera responded: That is a wonderful suggestion. The stabilization of some of the membrane remodeling stages could be probably achieved by raising the pH to physiological levels. This should de-activate or at least slow down BaxE5 action.

Burkhard Bechinger enquired: I understood from working on Bcl-xL and Bax proteins many years ago that these proteins remain largely folded as determined in their XR and solution NMR structures and anchor in the membrane merely through their hydrophobic carboxy terminus. Helix 5 would then remain sandwiched between other layers of helical domains and not be in contact with the membrane. Therefore, it is interesting that this domain alone has such potent activities. What can we learn from studies of helix 5 alone? Can this domain be used independently as a functional unit in therapeutic or diagnostic approaches?

Francisco N. Barrera replied: Data with BaxE5 (a variant of helix 5 of Bax) suggest that this sequence can forcefully remodel the membrane. A salient feature of BaxE5 is that its membrano-lytic effect is largely in check at neutral pH, while it is unleashed at acidic pH. Such pH-triggered membrane destruction could be potentially used to specifically target cancer cells, as they are often embedded into an acidic environment (see ref. 1). BaxE5 could therefore be a lead molecule for anti-cancer therapeutics. However, a tighter control of pH-responsiveness (*e.g.*, ref. 2), and a higher pH_{50} would be probably required for clinical promise.

1 J. C. Deacon, D. M. Engelman and F. N. Barrera, *Arch. Biochem. Biophys.*, 2015, **565**, 40–48, PMID: 25444855.

2 H. L. Scott, J. M. Westerfield and F. N. Barrera, *Biophys. J.*, 2017, **113**, 869–879, PMID: 28834723.

Ana J. Garcia-Saez added: The existing structures of the soluble forms of Bax and Bcl-xL, as well as of other family members, do not recapitulate the structures of these proteins in the active state in the membrane, at least for some of them.

In the case of Bax (and Bak), both proteins can retain the globular fold while anchored to the membrane *via* the C-terminus helix in an inactive state. Upon activation, they undergo a large conformational change and the central hairpin of helices partially opens, which is a fundamental step for the activation process. We and others have shown that the N-terminal part of the protein up to helix 5 is then involved in a stable symmetric dimerization domain *via* Bh3-into-groove inter-actions. From helix 6 on, the flexibility of the protein in the active conformation in the membrane is very high and so far it could not be represented by a unique structure. There is still a lot of work to be done regarding the structure of the active forms of these proteins and their complexes in the membrane.

Additionally, we also know that helices 5 and 6 are directly involved in inter-actions with the membrane which are interfacial, meaning, not classically trans-membrane. It is well established that these helices in the protein are responsible for the pore forming activity. From the studies with the individual helices, we learned that indeed they can recapitulate the pore-forming activity of the entire protein, yet they miss the regulation of this activity provided by the other protein domains. There have been some efforts to use peptides derived from helix 5 for antimicrobial and anticancer purposes. Furthermore, understanding the mechanisms of the pore activity of Bax is important to develop small molecules that may directly activate it, which would be of interest for anticancer therapy, for example.

Ana J. Garcia-Saez asked: Did you test the pore activity of the modified Bax peptide in liposomes? Does it induce tubulation/curvature in giant unilamellar vesicles?

Francisco N. Barrera responded: The pore activity of the modified Bax peptide (BaxE5) was tested using a liposome cargo release assay. These results showed clearly enhanced cargo release at acidic pH, as shown in Fig. 1 of the manuscript (DOI: 10.1039/d0fd00070a).

Atomic force microscopy performed on supported bilayers did not hint at any effect of BaxE5 causing tubulation or curvature. However, we cannot rule these out, and additional fluorescence experiments using giant unilamellar vesicles could offer a clearer answer to this question.

Robert Gilbert enquired: Can you comment on how the purported role of dimerisation in Bax activity relates to your work with the isolated Bax-derived peptide you are working with?

Francisco N. Barrera replied: The BaxE5 peptide carves a membrane pattern (dubbed "lace"), which is characterized by a consistent width. We posit that laces are formed when a peptide oligomer, of well-defined dimensions, wraps around the lipid monolayer in the lace, transiently stabilizing it. This model implies that BaxE5 is stable with a particular self-assembly stoichiometry. Our results would be compatible with an scenario where helix 5 of Bax participates in self-assembly of the protein, particularly at the membrane insertion stage where the pore is formed.

Paul O'Shea opened discussion of the paper by Stefano Vanni: I noticed in one of your slides (not included in your conference paper I think) that you have noise profiles when considering the free and bound structures. I wonder is there information in that noise which is not being used as it seems to change, so is probably not systematic uncertainty in the 'measurement'.

Stefano Vanni answered: If I understand the question correctly, it would be inappropriate to refer to this data as noise. These plots (also provided in the supplementary information of the paper; DOI 10.1039/d0fd00058b) represent the minimum distance between the protein and the membrane over the course of the simulations. The proteins chosen are known to interact transiently with membranes and hence the minimum distance varies between 0.5 and 3 nm as they bind and unbind continuously. From these minimum distance curves we extract the relevant information that we discuss in the article.

Evelyne Deplazes enquired: During the binding/unbinding event can you see the peptide tumbling on the membrane surface and are you thus sampling different binding interfaces on the protein including sampling favourable and unfavourable interactions? We often work with proteins that have hydrophobic patches and thus a more 'sticky' side that shows a strong preference for the membrane over other sides/faces on the protein. We have been struggling to find methods that sample the favourable and unfavourable interactions for these peptides and this is needed for calculating the free energy of binding. Do you think if the simulations are run for long enough (dozens of microseconds) the trajectory can be used to obtain and estimate the free energy of binding for both weak and strong binding peptides?

Stefano Vanni responded: Yes the protein does sample unfavourable interactions although transiently (with the exception of Evectin-2 PH domain in our dataset that binds strongly in an unfavourable conformation). In principle, one could indeed determine binding free energy from CG simulations with reasonable statistical accuracy by running long-enough simulations. Comparison with experiments (albeit not straightforward) would provide further information on the quality of the used force field.

Durba Sengupta asked: I am surprised by the large number of binding/unbinding events! Did you analyse the very short transient bound states and do they differ from the longer binding events? Both in terms of the protein orientations and the lipids interacting with the protein. And in that context, events such as lipid diffusion that haven't been sampled (*e.g.* clustering of the PS lipids) could play a role.

Stefano Vanni replied: Yes, we did analyse the membrane binding interface in the case of short transient states (binding duration < 50 ns) *vs.* longer binding events. For some proteins, the binding interface in the case of very transient binding is not the one that has been experimentally determined. This also sheds light on the accuracy of the model as it is necessary to ensure that the protein does not stay trapped with a 'wrong' binding interface that has stronger electrostatics for example. As far as clustering of lipids is concerned, the timescale required for lipid redistribution to occur after protein binding exceeds the time for which some proteins stay bound to the membrane, making it hard to sample and analyse such events.

Madhusmita Tripathy said: The figures in the ESI are the crucial things we missed in the talk and paper (DOI: 10.1039/d0fd00058b). I find the binding trajectory of PDK1 PH domain at 30% DOPS to be quite interesting. In half of your simulations you see stable binding events and not in the other half. Why so?

Stefano Vanni answered: Our interpretation is that this is purely a sampling issue as all the binding events in the particular example are stable. Hence if the simulations were longer, one could have potentially seen unbinding and rebinding in the other half of the trajectories as well.

Madhusmita Tripathy enquired: Given that you do see the correct binding orientations at 30% DOPC, why are there no stable binding events over the 4 μs long trajectory? In other words, the proteins that approach the membrane in experimentally observed orientation, do they unbind less often?

Stefano Vanni replied: These proteins have an intrinsic tendency to bind and unbind/exhibit transient binding. We attribute this to the fact that they are not membrane residual proteins but rather membrane sensing proteins. The fact that we can reproduce this characteristic with the MARTINI model tells us that the free energy of binding is not over-estimated and the protein is not trapped in the membrane bound state.

Madhusmita Tripathy asked: I am curious because I see binding events to be stable in only half of the simulations with 30% POPS. Is it okay to see binding events that are not stable? Are there any other factors that take place? In the case where there are lower percentages, does the protein ever approach the correct membrane orientation or is it random?

Stefano Vanni responded: We consider all the binding events in PDK1 PH at 30% PS as stable binding events. The protein does approach in the correct orientation as can be seen with some proteins such as FVa-C2 where the membrane binding is only ~30% at 80 DOPC, 20 DOPS membrane composition.

Patrick Fuchs enquired: When you observe multiple binding/unbinding events, do you get the converged free energy of binding (or at least converged ratio of bound/unbound conformations)? If yes, do you get the experimental order of magnitude (for known cases)?

Stefano Vanni replied: We did not convert binding/unbinding events to free energy of binding, so we cannot comment on that. In principle, it is true that one could indeed determine binding free energy from CG simulations with reasonable statistical accuracy by running long-enough simulations. Comparison with experiments (albeit not straightforward) would provide further information on the quality of the used force field.

Patrick Fuchs asked: Regarding amphipathic helices, do you see also multiple binding/unbinding events or irreversible binding? For ALPS from ArfGAP1, what about the flexibility of the peptide (especially around the two glycines)?

Stefano Vanni answered: The amphipathic helices bind irreversibly almost all of the time and only few unbinding events can be observed towards the beginning of the simulation. We did not take into account the flexibility of the peptide at specific positions, but we rather changed the overall force constant as a mean to increase flexibility. If one is interested in a partially disordered protein, a careful modelling of its secondary structure is clearly very important.

Robert Vácha enquired: Have you simulated ALPS with different amount/level of helicity (*e.g.* from CD experiments) to see if the helicity affects ALPS binding to membranes with different compositions?

Stefano Vanni replied: We did not test ALPS with different level of helicity, but rather simply changed the elastic network force constant, thus allowing for more/ less flexibility for the peptide. We did so as we meant our analysis to be high-throughput and not system-dependent, but we expect that changing the level of helicity (also depending on the lipid bilayer composition) would affect the binding of the peptide to the bilayer.

Zack Jarin asked: Thank you again for the great paper and talk. If I recall correctly, you showed that some proteins did not reproduce the experimental behavior and did not bind more readily to PC/PS membranes than PC membranes. What were the net charges of the proteins you investigated? Could the truncated electrostatics in Martini model affect this result?

Stefano Vanni responded: The net charges on the proteins we studied vary significantly, from −3 (for PLA2) to +11 (for FVa C2). We can not exclude that the truncated/reaction field treatment of electrostatic interactions might partially affect the results. However, two observations seem to suggest that this shouldn't be the cause of the issue: (1) the proteins undergo fast tumbling motions close to the bilayers, and all regions of the proteins approach the membrane within the cutoff distance and (2) ultimately, the proteins bind to the bilayer with the correct interface.

Patricia Bassereau questioned: Do you add a particular interaction to keep your proteins next to the membranes and not let them diffuse away?

Stefano Vanni answered: No, we did not add any external potential or inter-action to the classical energetic terms of the force field to prevent the diffusion away from the bilayer.

Anand Srivastava commented: Wanted to share some thoughts and see if they make sense. Though SPR and ITC gives different K_d – they are not generally orders of magnitude different – they are in the ball park. Similarly, one of the things that can be tested here is "calculate" the K_d (using the kinetics off and on rates – like we do for SPR) and also do some umbrella sampling or other kind of advanced sampling and get PMF and from the PMF (after using standard state correction), calculate the K_d and compare the two K_d and see how the Martini3 simulations compare. That would be nice to see.

Stefano Vanni responded: Indeed, that would be very interesting. On the one hand, however, the lack of flexibility in protein secondary structure might compromise the ability of the model to reproduce the correct K_d. On the other hand, it would be interesting to investigate whether the performance of the model is similar for all proteins or if there are some (positive or negative) outliers, as this would indicate different binding mechanisms (or possibly the involvement of conformational changes only in some instances).

Anand Srivastava said: Related to the previous comment, it will be nice to see how much K_d calculated from PMF prescription and k_d calculated using k_{off}/K_{on} data are different.

Stefano Vanni replied: Agreed, but different enhanced sampling techniques could also lead to different K_d values, so this would require extensive testing.

Ronald J. Clarke communicated: Many peripheral membrane proteins are thought to interact with membrane surfaces *via* electrostatic interactions between negatively charged phosphatidylserine lipid headgroups and positively charged basic amino acid residues on the protein, *i.e.*, lysine and arginine, as described in your paper for the 3-phosphoinositide-dependent kinase-1 (PDK1). In your simulations you state in the Methods section (DOI: 10.1039/d0fd00058b) that you solvated the systems studied with 0.12 M NaCl. If a binding event is due to an electrostatic interaction one would expect the frequency of interaction to increase or the interaction distance to decrease as the ionic strength is decreased due to reduced electrostatic screening. Have you carried out any simulations at low salt concentrations or at zero ionic strength to check whether this is the case?

Stefano Vanni communicated in reply: We tested slightly different ionic strengths and we did observe small variations, but further studies would be required to properly characterise this behaviour. Indeed, the MARTINI developers indicate that the force field for ions is significantly improved in MARTINI 3 and we think our database could be a good test system.

Conflicts of interest

Faraday Discussions

ROYAL SOCIETY
OF CHEMISTRY

PAPER

Order-disorder transitions of cytoplasmic N-termini in the mechanisms of P-type ATPases†

Khondker R. Hossain, [*a] Daniel Clayton,[a] Sophia C. Goodchild, [b]
Alison Rodger, [b] Richard J. Payne, [a] Flemming Cornelius [c]
and Ronald J. Clarke [*ad]

Received 16th April 2020, Accepted 1st September 2020

DOI: 10.1039/d0fd00040j

Membrane protein structure and function are modulated *via* interactions with their lipid environment. This is particularly true for integral membrane pumps, the P-type ATPases. These ATPases play vital roles in cell physiology, where they are associated with the transport of cations and lipids, thereby generating and maintaining crucial (electro-) chemical potential gradients across the membrane. Several pumps (Na^+, K^+-ATPase, H^+, K^+-ATPase and the plasma membrane Ca^{2+}-ATPase) which are located in the asymmetric animal plasma membrane have been found to possess polybasic (lysine-rich) domains on their cytoplasmic surfaces, which are thought to act as phosphatidylserine (PS) binding domains. In contrast, the sarcoplasmic reticulum Ca^{2+}-ATPase, located within an intracellular organelle membrane, does not possess such a domain. Here we focus on the lysine-rich N-termini of the plasma-membrane-bound Na^+, K^+- and H^+, K^+-ATPases. Synthetic peptides corresponding to the N-termini of these proteins were found, *via* quartz crystal microbalance and circular dichroism measurements, to interact *via* an electrostatic interaction with PS-containing membranes, thereby undergoing an increase in helical or other secondary structure content. As well as influencing ion pumping activity, it is proposed that this interaction could provide a mechanism for sensing the lipid asymmetry of the plasma membrane, which changes drastically when a cell undergoes apoptosis, *i.e.* programmed cell death. Thus, polybasic regions of plasma membrane-bound ion pumps could potentially perform the function of a "death sensor", signalling to a cell to reduce pumping activity and save energy.

a School of Chemistry, University of Sydney, Sydney, NSW 2006, Australia. E-mail: ronald.clarke@sydney.edu. au; Fax: +61-2-93513329; Tel: +61-2-93514406

b Department of Molecular Sciences, Macquarie University, Sydney, NSW 2109, Australia

c Department of Biomedicine, University of Aarhus, DK-8000 Aarhus C, Denmark

d The University of Sydney Nano Institute, Sydney, NSW 2006, Australia

† Electronic supplementary information (ESI) available. See DOI: 10.1039/d0fd00040j

Introduction

P-type ATPases are integral membrane proteins found in all forms of life, where their primary function is to pump ions or lipids across the membrane in which they are situated. One simple but vital reason for the existence of ion pumps is the maintenance of osmotic balance across the animal plasma membrane which prevents cell swelling or bursting. This is one of the roles of the Na^+, K^+-ATPase. Another vital function is performed by the H^+, K^+-ATPase, which maintains the acidity of the stomach necessary for digestion. For these active transport processes the proteins derive their energy from the hydrolysis of ATP to ADP. Characteristic of all P-type ATPases is that the phosphate released by ATP hydrolysis is transferred to a conserved aspartic acid residue on the protein.[1,2] During their pumping cycles P-type ATPases oscillate between two major conformational states, termed E1 and E2, which can exist in both phosphorylated and nonphosphorylated forms. The generally accepted mechanistic model of their pumping activity, which describes the sequence of steps the proteins follow, is termed the Albers–Post, Post–Albers or simply E1–E2 model.[3,4] In Fig. 1 this model is shown for the Na^+, K^+-ATPase, the first ion-transporting membrane protein to be discovered.[5]

Ion-pumping ATPases (as opposed to lipid-pumping ATPases, *i.e.* flippases) possess ion-binding sites within the protein which allow the ions to be transported across the membrane. These sites can be open to one side of the membrane, to the other side of the membrane or to neither side of the membrane (*i.e.* occluded states), but never open to both sides of the membrane at the same time. Thus, ion-pumping ATPases act in a similar fashion to a river lock, allowing the movement of ions in a direction which is uphill in energy, just as a river lock allows a boat to move uphill between two different levels in a river or canal.

ATPases often transport two ions in opposite directions. Thus, the Na^+, K^+-ATPase transports 3 Na^+ ions out of the cell and 2 K^+ ions into the cell for each ATP molecule hydrolysed. To do this the protein must alternate its ion binding affinity. Thus, in the unphosphorylated E1 conformation the protein has a high Na^+ affinity and takes up Na^+ ions from the cytoplasm, whereas in the phosphorylated E2P conformation the protein has a high affinity for K^+ ions and takes up K^+ ions from the extracellular fluid. Other ion pumps of the P-type family operate in a similar fashion, but with binding affinities for different ions.

Changes in protein conformation are thus key to altering the ion binding affinity and driving the ion pumping process forward. In the case of the Na^+, K^+-ATPase, biochemical studies have indicated that the protein's lysine-rich cytoplasmic N-terminus undergoes significant movement during the transition from the K^+-specific E2 state to the Na^+-specific E1 state,[7–13] and N-terminal truncation of pig kidney Na^+, K^+-ATPase was found to approximately halve both the purified enzyme's rate of phosphorylation by ATP and its overall turnover.[12,14] Kinetic studies have shown that the E2–E1 transition, which is required for ATP hydrolysis to proceed, is a major rate-determining step of the entire pump cycle.[15,16] Thus, any factors which could modulate the rate of this reaction would be expected to have a major impact on the overall ion pumping activity. Recently it was shown[17] that an increase in ionic strength of the medium induces a shift from the E2 to the E1 conformation,

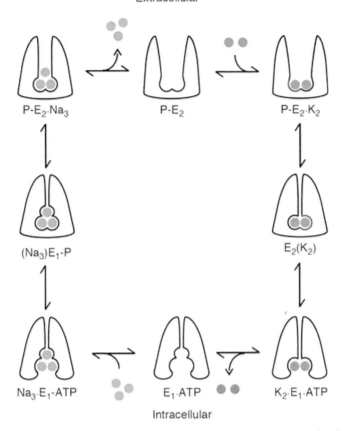

Extracellular

Intracellular

Fig. 1 Schematic diagram of the ion pumping reaction cycle of the Na^+, K^+-ATPase (adapted from Gadsby *et al.*, 2012).[6] Green and orange balls represent Na^+ and K^+ ions. The pump proceeds in a clockwise direction around this cycle, taking up 3 Na^+ ions from the cytoplasm and releasing them to the extracellular medium, and taking up 2 K^+ ions from the extracellular medium and releasing them to the cytoplasm for the hydrolysis of 1 ATP molecule. All other P-type ATPases operate similarly, but with variations in the ions transported and the stoichiometry.

suggesting that the E2 conformation is stabilized by an electrostatic interaction and that this interaction must be broken to allow the protein to convert to E1 for ion pumping to proceed. Considering the biochemical data implicating the involvement of the N-terminus in the E2–E1 transition, it appears likely that the N-terminus could be a key player in such an electrostatic interaction, which might occur between its positively charged lysine residues and negatively charged phosphatidylserine (PS) headgroups present on the cytoplasmic surface of the surrounding membrane.[18] This would be analogous to the situation found in many peripheral membrane proteins which are anchored transiently to the cytoplasmic surface of the plasma membrane *via* a polybasic peptide-anionic lipid headgroup interaction.[19–24]

In this paper we have investigated the possibility of N-terminus–membrane interaction(s) of the Na^+, K^+- and H^+, K^+-ATPases by chemically synthesising peptides corresponding to the N-terminus of both proteins, and studied their membrane interaction *via* quartz crystal microbalance measurements with dissipation monitoring (QCM-D) and circular dichroism (CD). The results obtained support the hypothesis of an electrostatic interaction, requiring the presence of PS within the membrane, with an increase in helical or other secondary structure content of the peptide occurring upon binding.

Materials and methods

Materials

Origins of the reagents used were as follows: NaCl (suprapure, Merck, Kilsyth, Australia), 2,2,2-trifluoroethanol (TFE) (99%, Sigma) EDTA (99%, Sigma), tris(hydroxymethyl)aminomethane (Tris) (99%, Alfa Aesar, Heysham, UK), NaOH (analytical grade, Merck), chloroform ($\geq$99.0%, Uvasol, Merck) and HCl (0.1 N Titrisol solution, Merck). Cholesterol, dioleoylphosphatidylcholine (DOPC) and dioleoylphosphatidylserine sodium salt (DOPS) were obtained from Avanti Polar Lipids (Alabaster, AL, USA). All Fmoc-amino acids were purchased from Mimotopes (Victoria, Australia) and used directly in peptide synthesis.

Peptide synthesis, HPLC purification and ESI-MS analysis

Peptides 1–40: GRDKYEPAAVSEHGDKKKAKKERDMDELKKEVSMDDHKLS (N-NKA) from the N-terminus of the α-subunit of pig kidney Na$^+$, K$^+$-ATPase, and 1–40: SKQDTYDMFEMGGEMDKKKKKKKMKKKEKLEGMKKEMDID (N-HKA) from the N-terminus of the α-subunit of gastric H$^+$, K$^+$-ATPase from *Siniperca chuatsi* (Chinese mandarin fish), were prepared at 100 µmol scale using Fmoc solid-phase peptide synthesis (SPPS) on Wang resin to yield the C-terminal acid. After the first residues were loaded using HBTU/DMAP chemistry the peptides were elongated (Na$^+$, K$^+$-ATPase, 1–40) on a CEM Liberty Blue automated microwave peptide synthesiser (USA, NC) [4 min coupling cycle: 2 min coupling (90 °C); 1 min deprotection (90 °C); 1 min associated washes and liquid handling] using a 5-fold excess of Fmoc amino acid, Oxyma and DIC, and (H$^+$, K$^+$-ATPase, 1–40) on a Biotage Syro I peptide synthesiser (Sweden, Uppsala) [30 min coupling cycle: 15 min coupling (75 °C); 2 × 2 min deprotection (RT); 11 min associated washes and liquid handling] using a 6-fold excess of reagents. 50 µmol of each peptide was cleaved from the resin and deprotected with a 20-fold volume to dry resin weight ratio of trifluoroacetic acid (TFA)-based cleavage solution (TFA/TIPS/H$_2$O, 95 : 2.5 : 2.5) for 4 h at room temperature, then worked up with diethyl ether, and lyophilised to yield the crude peptide. Both target ATPase peptides were purified using acetonitrile and H$_2$O buffer systems (buffer A; H$_2$O/0.1% TFA, buffer B; acetonitrile/0.1% TFA) using a Waters 2535 Quaternary gradient Module system and a 2489 UV/Vis detector (230/280 nm) with a Waters Sunfire, C18 OBD, 5 µm, 19 × 150 mm preparative column (20 mL min^{-1}) and at 0–30% B over 40 min. The identity of the peptides was confirmed on a Shimadzu UPLC-MS 2020 ESI instrument (LC-M20A pumps/SPD-M30A diode array detector) in positive mode (0.1% formic acid). The purity of the peptides was assessed on a Waters Acquity UPLC system (Acquity C18 BEH 1.7 µm 2.1 × 50 mm column) at 0–50% B, 5 min at 0.6 mL min^{-1}. Refer to ESI† for further synthesis detail and for characterisation, purity and yields.

Salt transfer

Prior to any CD or QCM-D measurements with the synthetic N-terminal peptide fragments, the TFA counter ion was transferred to the chloride ion by three cycles of lyophilisation from 1.5 mL of 5 mM HCl[25] with three further lyophilisation cycles performed from 50% acetonitrile/water to peptides as fluffy white solids. UPLC-MS analysis was performed on an aliquot of each final sample to confirm no acid-induced modification had occurred during the salt transfer procedure.

Vesicle preparation

For the preparation of unilamellar lipid vesicles, lipids were first dissolved in chloroform at a concentration of 3 mM. All solutions were prepared by weight, using the chloroform density of 1.48 g mL^{-1}. Appropriate volumes (again measured by weight) of the chloroform solutions of each lipid were mixed to obtain the required mol%, with a final total volume of 3 mL. Chloroform was then removed from each mixture by rotary evaporation at 40 °C at 474 mbar and maximum rotation speed. After no visible traces of chloroform could be detected in the flask, the resulting lipid film was dried for a further 30 min at 5 mbar. The lipid film was then hydrated by the addition of buffer (10 mM Tris, 0.3 mM EDTA, pH 7.2) with or without the presence of 150 mM NaCl, to obtain a final lipid concentration of 3 mM and sonicated in a sonic bath for 60 s to ensure complete lipid resuspension. The suspension was finally extruded 11 times through a 0.1 μm nucleopore polycarbonate membrane using an Avanti Mini-Extruder (Alabaster, AL, USA) at room temperature to break apart multilamellar vesicles and obtain a suspension of unilamellar vesicles. Vesicles were prepared with the following lipid compositions: pure DOPC; DOPC (60%) and cholesterol (40%); and DOPC (42%), DOPS (18%) and cholesterol (40%). All percentages are mol%. In the case of vesicles containing DOPS, the DOPS composition was chosen so that it was 30% of the total phospholipid content.

In order to study changes in secondary structure of the 1–40 N-terminal peptide fragments upon interaction with lipid vesicles *via* CD (see later), the peptides were encapsulated within lipid vesicles. In this case vesicles were prepared as described above except that a 500 μg mL^{-1} peptide solution (10 mM Tris, pH 7.2) was used to hydrate the lipid film. The vesicles were then centrifuged three times at 12 000 rpm for 10 min using an LSE high speed microcentrifuge (Corning, New York, USA) in order to remove any free peptides (unencapsulated) from the solution, and the lipid pellet was resuspended in 10 mM Tris, pH 7.2 buffer to obtain a final lipid concentration of 0.5 mM.

Quartz crystal microbalance measurements with dissipation monitoring (QCM-D)

A QCM-D E1 (QSense Analyzer, Biolin Scientific AB, Gothenburg, Sweden) was used to measure binding of the 1–40 N-terminal NKA and HKA fragments to supported lipid bilayers (SLBs) of varying compositions. Briefly, in QCM-D experiments, changes in the frequency (Δf) and the dissipation (ΔD) of a quartz crystal are derived from measurements of the piezoelectric properties of the crystal.[26] The change in frequency can be directly related to changes in mass, reflecting the adsorption of material onto the surface. The dissipation is attributed to energy losses of the material deposited on the oscillating crystal surface, and provides information on the viscoelastic properties of the attached molecules, *i.e.* a higher dissipation corresponds to a less rigid surface. Both the Δf and the ΔD values represent the changes relative to the bare quartz crystal.

The silicon dioxide quartz crystals (QSense) were cleaned before use for 30 min at room temperature using a 2% sodium dodecyl sulfate (SDS) solution, then washed thoroughly with Milli-Q water and dried using nitrogen gas. The chips were then treated in a plasma cleaner (Harrick Plasma, Ithaca, NY, USA) for 5 min at medium to high frequency, then immediately placed into the measurement chamber, and covered with ultrapure water (Merck-Millipore, Bayswater, Australia). A representative example of a QCM-D measurement is shown in the ESI Fig. 1.† The experiment was initiated by exchanging water in the flow cell with 50 mM NaCl, 10 mM Tris, 0.3 mM EDTA, at pH 7.2 for 5 min at a flow rate of 100 μL min^{-1}. Subsequently, a suspension of lipid vesicles (3 mM lipid) was introduced at the same flow rate to the crystal until the characteristic profile of a stable supported lipid bilayer (SLB) formation was observed.[27,28] After 5–10 min or until a stable baseline was observed, the sensor surface was rinsed with Milli-Q water for a period of 30 min at a flow rate of 100 μL min^{-1} to rinse away any salt present. Milli-Q water was then replaced with 10 mM Tris, 0.3 mM EDTA pH 7.2 buffer, for 5 min, followed by the addition of either 1–40 N-terminal NKA or HKA fragments (N-NKA or N-HKA) to the SLB surface at a concentration of 50 μg mL^{-1} and a flow rate of 50 μL min^{-1} to measure peptide binding. N-NKA and N-HKA were adsorbed for 15–20 min or until a stable baseline was observed. The quartz crystals containing the adsorbed peptide were then rinsed with 10 mM Tris, 0.3 mM EDTA pH 7.2, to study desorption from the SLB surfaces. The adsorption and desorption measurement cycles on each composition of SLB were performed in triplicate for both the N-NKA and N-HKA fragments. For experiments in the presence of NaCl, lipid vesicles prepared in 150 mM NaCl, 10 mM Tris, 0.3 mM EDTA pH 7.2, were used to make the SLB as previously described with the exception that there was no Milli-Q water wash step. Instead the peptides were added 5–10 min after SLB formation and the changes in frequency and dissipation were recorded. All measurements were performed at 24 °C.

Frequency and dissipation data were collected at the fundamental frequency (5 MHz) of the quartz crystal and at 6 additional overtones of the fundamental frequency. Overtones 5, 7, 9, and 11 were normalised and used in analysis. Traces of Δf and ΔD represent the normalised 5th overtone. Sauerbrey thickness and mass were calculated using the QTools analysis software (QSense) which incorporates the Sauerbrey equation, $\Delta m = -C(\Delta f/n)$, where C is a constant equal to 17.7 ng cm^{-2} Hz for a 5 MHz quartz crystal, and n is the overtone number. Since the mass of the initial SLBs varies due to change in lipid composition, the results are given as change in mass (Δ Mass) upon interaction of the peptides with the bilayer with time, *i.e.*, the initial mass of lipid deposited on the crystal has been subtracted. The adsorption and desorption measurement cycles on each composition of SLB were performed in triplicate for both the N-NKA and N-HKA fragments.

Circular dichroism (CD)

CD spectroscopic analysis of 1–40 N-terminal NKA and HKA fragments was performed using a J-1500 circular dichroism spectrophotometer (Jasco, Tokyo, Japan) at 25 °C using a Peltier thermostatted cell holder PTC-517 (Jasco). Spectra were obtained over a wavelength range of 180–250 nm, using a 1 mm path length quartz cell (Starna, Norwest, NSW, Australia) and continuous scanning mode with a response time of 2 s with 0.5 nm steps, a bandwidth of 2 nm, and a scan speed of 100 nm min^{-1}. Data were acquired after adding the peptides at a final concentration of 100 µg mL^{-1} directly to the appropriate buffer solutions in the 1 mm quartz cell. For experiments with trifluoroethanol (TFE), CD data were collected for the 1–40 N-terminal NKA and HKA fragments in various TFE concentrations ranging from 0–90%. CD data was also collected under varying pH conditions ranging from 4–12. No changes in the CD spectra of either N-NKA or N-HKA were observed over the pH range 4–7. Hence data is only shown over the basic pH range 9–12. For all samples, corresponding background samples without peptides were prepared and the backgrounds were subtracted from the sample spectra. Four spectra for each condition were averaged to achieve an appropriate signal-to-noise ratio.

Tris buffer (10 mM) was used for all measurements in aqueous solution due to its low absorbance above 190 nm.[29] In the case of measurements on vesicle-encapsulated peptides, spectra of lipid vesicles (0.5 mM lipid) in the absence of peptide were also collected as the background and subtracted from the sample spectra. All the buffers were filter-sterilized before use via 0.2 µm pore size Millipore syringe filters (Thermo Fisher Scientific, North Ryde, Australia).

Two web servers, CAPITO (CD Analysis and Plotting Tool; https://capito.uni-jena.de/)[30] and K2D3 (http://cbdm-01.zdv.uni-mainz.de/~andrade/k2d3/)[31] were used to evaluate the CD spectra and determine the percentages of different peptide secondary structures, e.g. α-helix and random-coil. CAPITO[30] utilises the molar ellipticity ratio, $[\theta_{200nm}]/[\theta_{222nm}]$, and compares the result with a databank of disordered protein model compounds. K2D3 is a publicly accessible server[31] which accepts entire CD spectra as input and estimates secondary structure content based on a reference set of theoretically derived spectra. For the peptides which we investigated, the use of either CAPITO or K2D3 resulted in essentially the same percentage values of each secondary structure. Therefore, only single percentage values are reported below.

Results and discussion

QCM-D studies of 1–40 N-terminal NKA and HKA peptides to lipid bilayers

To study the effect of membrane lipid composition on peptide interaction we carried out QCM-D measurements on SLBs formed using DOPC alone, 60 : 40 mol % DOPC : cholesterol and 42 : 18 : 40 mol% DOPC : DOPS : cholesterol. The interaction was measured with 50 µg mL^{-1} of each peptide. The final lipid composition was chosen so as to reproduce the native ratio between PC, PS and cholesterol in the cytoplasmic membrane leaflet surrounding the Na$^+$, K$^+$-ATPase.[18,32,33] Fig. 2 shows the averaged changes in frequency, Δf, and dissipation, ΔD, from three independent experiments of each of the different SLBs upon interaction with either N-NKA or N-HKA peptides. The Δf and ΔD data are shown five minutes prior to the addition of the peptide to stable SLBs at time = 5 min followed by adsorption for another 15–20 min, before being rinsed with 10 mM Tris, 0.3 mM EDTA pH 7.2, to study the desorption from the SLB surfaces.

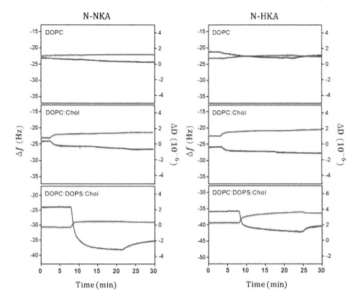

Fig. 2 Changes in frequency (Δf) and dissipation (ΔD) of various compositions of SLBs upon interaction with either N-NKA (left) or N-HKA (right) peptides. The Δf (blue line) and ΔD (red line) data of the stable SLBs are shown for 5 min prior to the addition of either N-NKA or N-HKA peptide (50 µg mL^{-1}) followed by adsorption for another 15–20 min before being rinsed with 10 mM Tris, 0.3 mM EDTA pH 7.2. The adsorption and desorption measurements on each SLB composition were performed in triplicate at 24 °C.

The addition of N-NKA and N-HKA to SLBs consisting of either pure DOPC or DOPC : cholesterol did not produce any significant change in Δf (Fig. 2). There were also no significant changes in ΔD observed for N-HKA in both the SLBs. However, N-NKA addition to DOPC : cholesterol membranes resulted in a small increase in ΔD from $(0.835 \pm 0.3) \times 10^{-6}$ to $(1.42 \pm 0.7) \times 10^{-6}$, indicating a thicker and softer membrane upon addition of the peptide. This may result from structural changes caused by the peptide upon interaction with the membrane. This is also reflected in a slight decrease in the Δf value. However, overall the results do not support any strong interaction of N-NKA or N-HKA with either pure DOPC lipids or DOPC : cholesterol.

Immediate changes in QCM-D signals (Δf and ΔD) occurred for N-NKA and N-HKA interactions with SLBs in the presence of anionic lipid DOPS at a concentration of 18 mol% (Fig. 2). When a DOPC : DOPS : cholesterol SLB was used, the Δf value decreased from −24 to −37.8 Hz for N-NKA and −35 to −42 Hz for N-HKA. This decrease in the frequency indicates a fast adsorption process of the peptides to the net negatively charged SLB. Steady-state conditions were reached within a few minutes of continuous flow of the peptides, where a minimum in Δf and a maximum in ΔD were observed prior to stabilization, suggesting either saturation of the SLB with peptide and/or reorganization of the lipid bilayer structure upon interaction with the peptides. The increase in ΔD values (Fig. 2) suggest the formation of a softer membrane upon interaction of the peptides with the DOPC : DOPS : cholesterol SLBs, further indicating the association of the peptides. After a 10 min period of equilibration without flow, rinsing with buffer resulted in desorption of the peptides from the DOPC : DOPS : cholesterol SLB. Interestingly, both N-NKA and N-HKA showed slight but not complete desorption from DOPC : DOPS : cholesterol SLBs upon washing with buffer. This is an indication of the relatively strong binding of the peptides to DOPC : DOPS : cholesterol SLB. The results, therefore, suggest that an electrostatic interaction between the positively charged amino acid residues of the peptides and the negatively charged DOPS lipid headgroups promote binding of these peptides with lipid membranes.

The effect of incorporating DOPS into the membrane becomes even more obvious when the raw frequency data are converted into mass of peptide adsorbed. When a DOPC : DOPS : cholesterol SLB was used, increases in SLB mass to 250 ± 10 and 97 ± 8 ng cm^{-2} for N-NKA and N-HKA, respectively, were observed (Fig. 3).

To understand better the nature of interaction of N-NKA and N-HKA with PS-containing lipid membranes, QCM-D experiments were carried out with DOPC : DOPS : cholesterol SLBs in the presence of buffer to which 150 mM NaCl had been added. Fig. 4 shows the average changes in frequency and dissipation from three independent experiments of DOPC : DOPS : cholesterol SLBs upon interaction with N-NKA and N-HKA in the presence of 150 mM NaCl. From the observed Δf and ΔD values shown in Fig. 4 it is clear that salt significantly weakens (or even abolishes) the interaction of the N-NKA and N-HKA peptides with DOPC : DOPS : cholesterol SLBs. This strongly suggests a salt-induced screening of an electrostatic interaction, *i.e.*, supporting the earlier suggestion of peptide binding being driven by an attraction between the positively charged basic residues (lysine and arginine) and the negatively charged phosphatidylserine headgroups of the lipid membrane.

Under the pH conditions of the experiments, *i.e.*, 7.2, the total charges of the N-NKA and N-HKA peptides would be expected to be −1 and +3, respectively. However, whereas the negatively charged aspartic acid (D) and glutamic acid (E) are

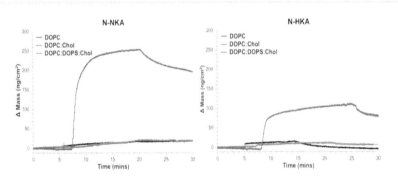

Fig. 3   The averaged changes in mass per cm^2 surface area of the quartz crystal, Δ Mass, of the different SLBs upon interaction with either N−NKA (left) or N−HKA (right) peptides. The SLBs were prepared with either DOPC alone (black line), 60 : 40 mol% DOPC : cholesterol (red line) or 42 : 18 : 40 mol% DOPC : DOPS : cholesterol (blue line) using a total lipid concentration of 3 mM. The Δ Mass data of the stable SLBs are shown for 5 minutes prior to the addition of the peptide. After addition of peptide, at concentration 50 µg mL^{-1}, peptide adsorption was followed for another 15−20 min before rinsing with 10 mM Tris, 0.3 mM EDTA pH 7.2, to study desorption from the SLB surfaces.

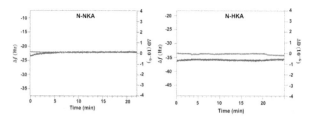

Fig. 4   Changes in frequency (Δf) and dissipation (ΔD) of DOPC : DOPS : cholesterol SLB upon interaction with N-NKA and N-HKA in the presence of 150 mM NaCl. All the measurements were performed at 24 °C in triplicate in 10 mM Tris, 0.3 mM EDTA, 150 mM NaCl at pH 7.2. The DOPC : DOPS : cholesterol lipid vesicles used were prepared at mole percentages of 42 : 18 : 40% in the same buffer solution.

distributed quite evenly along the length of both peptides, the positively charged lysine residues (K) are highly localized. Thus, the N-NKA peptide has a charge of +5 between residues 16–21 and the N-HKA peptide has a charge of +10 between residues 17–27. Therefore, we may conclude that the highly positively charged lysine clusters of both peptides dominate the interaction with lipid membranes.

The results shown in Fig. 2 and 3, however, appear to indicate a stronger interaction of the N-NKA peptide than the N-HKA peptide with DOPC : DOPS : cholesterol membranes. This is in spite of the N-HKA peptide possessing a more positive total charge than the N-NKA peptide. Therefore, it seems likely that the difference in binding strength could be due to other subtle differences in the amino acid sequences of the two peptides, not simply the number of lysine residues. A possible explanation could be that the hydrophobic effect also partially contributes to binding. If one inspects the sequences of the two peptides, it can be seen that the N-NKA peptide possesses 7 hydrophobic aliphatic residues (3 alanines, 2 valines and 2 leucines), whereas the N-HKA peptide only possesses 1 hydrophobic aliphatic residue (an isoleucine). One would, therefore, expect a stronger hydrophobic driving force for membrane binding of the N-NKA peptide. Nevertheless, the screening of the interaction by salt (see Fig. 4) indicates that the electrostatic interaction is decisive, and that any hydrophobic interactions, by themselves, are not sufficiently strong enough to hold the peptides on the membrane.

Another important class of amino acid residues are the acidic residues, aspartic acid and glutamic acid. These could also potentially explain differences in binding strength, by altering the total charge on the peptides or contributing to the alignment of the peptides when they approach the membrane. However, the total number of acidic residues are very similar for both peptides (11 for N-NKA and 10 for N-HKA). Therefore, it seems less likely that the acidic residues could explain different binding strengths of the peptides. However, a possibility could be that the spatial organisation of both positive and negative charges is different in the two peptides and that this could contribute to different binding strengths.

CD studies of 1–40 N-terminal NKA and HKA peptides

It is possible that N-terminal peptide binding to membranes could change its secondary structure, which could significantly affect the thermodynamics of ion pump conformational transitions.[14] No crystal structural data are currently available on either the entire Na$^+$, K$^+$-ATPase or H$^+$, K$^+$-ATPase which extends to the start of the N-terminus. Therefore, to investigate this, CD spectral measurements were conducted on the peptides under varying experimental conditions and in the absence and presence of lipid vesicles.

Fig. 5 shows the far-UV CD spectra of the N-NKA (left) and N-HKA peptides (right) dissolved in 0–90% v/v trifluoroethanol (TFE) (top panel) and at pH values ranging from 7–12 (bottom panel). The aim of the TFE titration was to investigate the peptides' propensity towards α-helix formation, with TFE known to promote folding into an α-helical conformation.[34–36] The spectra of both the N-NKA and N-HKA peptides at pH 7 and 25 °C in the absence of TFE show a major negative band centred at 200 nm and a much less intense negative band in the range 225–235 nm. The latter is probably due to aromatic residues (i.e. tyrosine (Y) in the case of N-NKA and, tyrosine and phenylalanine (F) in the case of N-HKA). This spectral behaviour is characteristic of disordered proteins.[37–39]

As shown in Fig. 5, both the N-NKA and N-HKA peptides display complex but similar changes in their CD spectra on increasing the TFE content of the solution. In both cases the negative maximum shifts from 200 nm to 213 nm with increasing TFE% and a positive maximum appears at 190 nm. Between 15 and 50% TFE, the N-HKA peptide displays a negative minimum (i.e. a dip between two negative maxima) at approximately 212 nm, which is characteristic of an α-helical structure. An increase in peptide secondary structure can be explained by a decrease in solvent polarity as the TFE% increases, which drives the peptides towards the formation of peptide intramolecular hydrogen bonding rather than intermolecular hydrogen bonding to solvating water molecules. From the

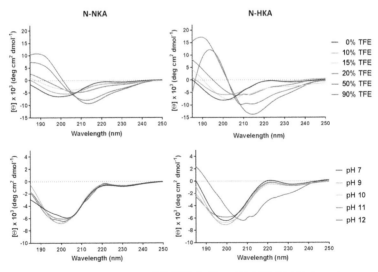

Fig. 5 Circular dichroism spectra of N-NKA (left) and N-HKA (right) at varying concentrations of TFE (top) and pH (bottom) at 25 °C. Data were acquired after adding the peptides in the appropriate buffers at a concentration of 100 µg mL^{-1}. For all samples, corresponding background samples without peptides were measured for spectral subtraction. Four spectra were averaged under each condition to obtain the spectra shown.

observed spectra at 0 and 90% TFE, the percentage of irregular structure was estimated *via* the CAPITO and K2D3 servers to drop from 85% to 34% for the N-NKA peptide and 86% to 13% for the N-HKA peptide. Note that in many of the spectra the negative minimum at around 212 nm is missing, suggesting under some conditions an increase in β-strand rather than helix structure.

The effect of pH (Fig. 5, bottom panel) was found to be very different between the two peptides. Between pH 7 and 12 no significant spectral changes of the N-NKA peptide were observed. However, the N-HKA peptide showed spectral changes indicative of a transition from a disordered state to a more ordered structure between pH 11 and 12 with the negative maximum shifting from 200 nm at pH 11 to 208 at pH 12, the appearance of positive ellipticity at <200 nm and a negative minimum at around 212 nm. Analysis of the N-NKA and N-HKA spectra at pH 12 using the CAPITO and K2D3 servers indicates that the N-NKA peptide is still 85% disordered, whereas the N-HKA is 94% structured. The increase in structure of the N-HKA peptide in the pH range 11 and 12 can presumably be attributed to deprotonation of basic amino acid side chains, most likely lysine. This would decrease the net positive charge on the peptide, reducing its strength of interaction with solvating water molecules, and, similar to the effect of TFE, promote intramolecular hydrogen bond formation and an α-helical structure. The absence of a pH effect on N-NKA can possibly be explained by the fact that it has fewer lysine residues than N-HKA and overall has a slight excess of acidic over basic residues. Thus, deprotonation of lysine residues of N-NKA would make the peptide more net negatively charged (*i.e.* moving away from its pI), whereas the N-HKA becomes less net positively charged (*i.e.* closer to its pI).

The experiments in which the % TFE or the solution pH were changed demonstrate the potential of the N-NKA and N-HKA peptides to undergo a transition from a disordered state to a more ordered state. Now we investigate whether such a transition occurs when peptide interacts with a biological membrane surface. To do this we encapsulated each peptide within lipid vesicles composed of DOPC : DOPS : cholesterol at the physiologically relevant percentages of 42 : 18 : 40 mol%. QCM-D measurements already provided strong evidence that membranes with this composition promote peptide binding (see Fig. 2 and 3). Encapsulation within the intravesicular volume increase the probability of the peptides coming into contact with the vesicle membrane, and more closely approximates the physiological situation where the peptides are attached to the rest of their respective ion pumps and hence cannot diffuse far from the membrane surface. Furthermore, encapsulation allows peptides in the bulk solution which would otherwise dominate the CD signal to be removed from the system.

CD spectra of DOPC : DOPS : cholesterol vesicles enclosing either the N-NKA or N-HKA peptides are shown in Fig. 6. In the absence of vesicles at neutral pH, both peptides showed a negative peak at 200 nm characteristic of a disordered peptide (see Fig. 5). In the presence of vesicles, the CD spectra have a lower intensity because of the much smaller amount of peptide being analysed, but there is a clear shift of the negative peak to a longer wavelength of around 207 nm, a shoulder in the range 215–220 nm and positive ellipticity at around 190 nm. Thus, the spectral form is comparable to that of the observed peptide spectra in 15–50% TFE, where a negative minimum characteristic of α-helical or β-sheet structures probably including significant turns to account for the relatively low

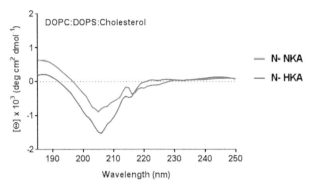

Fig. 6 Circular dichroism spectra of N–NKA (red) and N–HKA (blue) in DOPC : DOPS : cholesterol (42 : 18 : 40 mol%) vesicles. The peptides were encapsulated at a concentration of 500 µg mL^{-1} within lipid vesicles of 0.5 mM total lipid in 10 mM Tris, pH 7.2 buffer. Spectra of lipid vesicles at the same lipid concentration in the absence of the peptides were also collected as the background and subtracted from the sample spectra for each of the peptides. Four spectra for each condition were averaged to obtain the spectra shown. All other conditions were as in the caption of Fig. 5.

intensity at 222 nm (ref. 40), is observed for N-HKA. This implies that membrane interaction is occurring and causing the peptides to undergo a transition to an α-helical or other more folded conformation. Analysis of the spectra shown in Fig. 6 using either CAPITO or K2D3 results in an α-helical content of 94% for the N-NKA peptide and 81% for the N-HKA peptide. Though the structures adopted are not 100% helical in nature, they are certainly not random coils.

Theoretical predictions of secondary structure based on their amino acid sequences have previously suggested that both the Na$^+$, K$^+$-ATPase and the H$^+$, K$^+$-ATPase are likely to contain α-helical domains within the first 40 amino acids of their N-termini.[13,41] However, the present studies clearly indicate that the secondary structure of the N-termini of these ion pumps are strongly environmentally sensitive, with an unstructured conformation being preferred when they are free in solution.

Conclusions

The results obtained here *via* QCM-D and CD support the conclusion that the polybasic N-terminal sequence of the α-subunit of both the Na$^+$, K$^+$-ATPase and the H$^+$, K$^+$-ATPase, could interact with the surrounding lipid membrane *via* an electrostatic interaction with anionic lipid headgroups such as those of PS, and, furthermore, that the membrane-binding of these sequences is expected to cause a transition from a disordered conformation to a more structured one. The CD results, in particular the effects of TFE and pH, are similar to those obtained with an N-terminal polybasic (primarily lysine) peptide of yeast cytoplasmic aspartyl-tRNA synthetase.[42] This protein has a very different function to that of an ion pump, but it has been proposed that the polybasic tRNA synthetase domain allows it to interact electrostatically with polyanionic surfaces, in an analogous way to the N-termini of the Na$^+$, K$^+$- and H$^+$, K$^+$-ATPases. Likely binding partners of tRNA-synthetase N-termini are the polyanionic tails of α- and β-tubulins of which microtubules are composed,[43] thus providing a mechanism for compartmental-isation of the tRNA synthetase within the mRNA translation machinery. In the case of ion pumps which are embedded within the plasma membrane, it is hard to imagine a need for their polybasic N-termini as a means of compartmentali-sation. In this case a functional role would seem more likely.

The changes in peptide conformation observed here support a previous suggestion,[14] that differences in thermodynamic stability of some conformational states of the Na^+, K^+-ATPase and H^+, K^+-ATPase ion pumping cycles could involve significant entropic components due to binding and release of their N-termini from the membrane. Thus, although the overall thermodynamics of ion pump-ing are determined solely by the ion concentrations on each side of the membrane across which pumping occurs and the transmembrane electrical potential difference, the activation energy, and hence the kinetics of partial reactions contributing to the rate of ion pumping, could be controlled or regulated by protein segments such as the N-termini which are far from the proteins' ion transport sites.

It has been found that the proteins of hyperthermophilic organisms preferentially contain lysine residues (*i.e.*, over arginine), because of their greater number of rotameric conformations, which leads to an enhanced thermostability *via* an entropic stabilisation mechanism.[44] It seems feasible, therefore, that in ion pumps such as the Na^+, K^+-ATPase and the H^+, K^+-ATPase, the lysine residues of the N-termini could serve a dual function: *via* their charge they could provide an enthalpic stabilisation in certain enzyme conformational states where the N-terminus interacts electrostatically with the membrane, and in other states they could promote entropic stabilisation when the N-terminus is in a free unbound state.

By interacting electrostatically with the membrane, the N-termini of Na^+, K^+-ATPase and the H^+, K^+-ATPase could also provide a mechanism for detecting changes in lipid membrane composition, in particular PS content. It is known that when a cell is undergoing apoptosis a lipid scramblase within the plasma membrane is activated, which causes translocation of PS from the cytoplasmic leaflet to the extracellular leaflet, thus exposing PS to the surrounding cells. This provides an "eat-me" signal to neighbouring macrophages,[45] which then engulf the apoptotic cells. The consequent drop in PS content of the cytoplasmic membrane leaflet during apoptosis would be expected to affect N-terminal interaction of the Na^+, K^+- and H^+, K^+-ATPases with the surrounding membrane, leading to a change (presumably a drop) in their ion pumping activities. If a cell is about to die anyway, it would seem logical to reduce its energy expenditure and save available ATP. One could hypothesise, therefore, that the N-termini of the ion pumps might act as a "death sensor", signalling to pumps that apoptosis is underway. This hypothesis warrants further investigation.

Conflicts of interest

There are no conflicts to declare.

Acknowledgements

R. J. C. acknowledges financial support from the Australian Research Council (Discovery Grants DP121003548, DP150101112 and DP170101732). F. C.

acknowledges financial support from the Danish Medical Research Council and the Novo Nordisk Foundation.

References

1 M. Bublitz, J. P. Morth and P. Nissen, *J. Cell Sci.*, 2011, **124**, 2515–2519.
2 K. R. Hossain and R. J. Clarke, *Biophys. Rev.*, 2019, **11**, 353–364.
3 R. W. Albers, *Annu. Rev. Biochem.*, 1967, **36**, 727–756.
4 R. L. Post, C. Hegyvary and S. Kume, *J. Biol. Chem.*, 1972, **247**, 6530–6540.
5 J. C. Skou, *Biochim. Biophys. Acta*, 1957, **23**, 394–401.
6 D. C. Gadsby, F. Bezanilla, R. F. Rakowski, P. De Weer and M. Holmgren, *Nat. Commun.*, 2012, **3**, 669.
7 P. L. Jørgensen, *Biochim. Biophys. Acta*, 1975, **401**, 399–415.
8 P. L. Jørgensen, E. Skriver, H. Hebert and A. B. Maunsbach, *Ann. N. Y. Acad. Sci.*, 1982, **402**, 207–225.
9 P. L. Jørgensen and J. H. Collins, *Biochim. Biophys. Acta*, 1986, **860**, 570–576.
10 P. L. Jørgensen and J. P. Andersen, *J. Membr. Biol.*, 1988, **103**, 95–120.
11 C. H. Wu, L. A. Vasilets, K. Takeda, M. Kawamura and W. Schwarz, *Biochim. Biophys. Acta*, 2003, **1609**, 55–62.
12 F. Cornelius, Y. A. Mahmmoud, L. Meischke and G. Cramb, *Biochemistry*, 2005, **44**, 13051–13062.
13 R. Scanzano, L. Segall and R. Blostein, *J. Biol. Chem.*, 2007, **282**, 33691–33697.
14 K. R. Hossain, X. Li, T. Zhang, S. Paula, F. Cornelius and R. J. Clarke, *Biochim. Biophys. Acta*, 2020, **1862**, 183138.
15 C. Lüpfert, E. Grell, V. Pintschovius, H.-J. Apell, F. Cornelius and R. J. Clarke, *Biophys. J.*, 2001, **81**, 2069–2081.
16 P. A. Humphrey, C. Lüpfert, H.-J. Apell, F. Cornelius and R. J. Clarke, *Biochemistry*, 2002, **41**, 9496–9507.
17 Q. Jiang, A. Garcia, M. Han, F. Cornelius, H.-J. Apell, H. Khandelia and R. J. Clarke, *Biophys. J.*, 2017, **112**, 288–299.
18 K. Nguyen, A. Garcia, M.-A. Sani, D. Diaz, V. Dubey, D. Clayton, G. Dal Poggetto, F. Cornelius, R. J. Payne, F. Separovic, H. Khandelia and R. J. Clarke, *Biochim. Biophys. Acta*, 2018, **1860**, 1282–1291.
19 J. Kim, P. J. Blackshear, J. D. Johnson and S. McLaughlin, *Biophys. J.*, 1994, **67**, 227–237.
20 J. Kim, T. Shishido, X. Jiang, A. Aderem and S. McLaughlin, *J. Biol. Chem.*, 1994, **269**, 28214–28219.
21 C. T. Sigal, W. Zhou, C. A. Buser, S. McLaughlin and M. D. Resh, *Proc. Natl. Acad. Sci. U. S. A.*, 1994, **91**, 12253–12257.
22 T. J. Maures, H.-W. Su, L. S. Argetsinger, S. Grinstein and C. Carter-Su, *J. Cell Sci.*, 2011, **124**, 1542–1552.
23 F. Lei, J. Jin, C. Herrmann and W. Mothes, *J. Virol.*, 2013, **87**, 7113–7126.
24 M. A. Noguera-Salvà, F. Guardiola-Serrano, M. L. Martin, A. Marcilla-Etxenike, M. O. Bergo, X. Busquets and P. V. Escribá, *Biochim. Biophys. Acta*, 2017, **1859**, 1536–1547.
25 V. V. Andruschchenko, H. J. Vogel and E. J. Prenner, *J. Pept. Sci.*, 2007, **13**, 37–43.
26 M. Rodahl, F. Höök, A. Krozer and P. Brzezinski, *Rev. Sci. Instrum.*, 1995, **66**, 3924–3930.

27 C. A. Keller, K. Glasmästar, V. P. Zhdanov and B. Kasemo, *Phys. Rev. Lett.*, 2000, **84**, 5443–5446.

28 C. A. Keller and B. Kasemo, *Biophys. J.*, 1998, **75**, 1397–1402.

29 S. Kelly and N. Price, *Curr. Protein Pept. Sci.*, 2000, **1**, 349–384.

30 C. Weidemann, P. Bellstedt and M. Görlach, *Bioinformatics*, 2013, **29**, 1750–1757.

31 C. Louis-Jeune, M. A. Andrade-Navarro and C. Perz-Iratexta, *Proteins*, 2012, **80**, 374–381.

32 J. J. H. H. M. De Pont, A. Van Prooijen-Van Eeden and S. L. Bonting, *Biochim. Biophys. Acta*, 1978, **508**, 464–477.

33 W. H. M. Peters, A. M. M. Fleuren-Jakobs and J. J. H. H. M. De Pont, *Biochim. Biophys. Acta*, 1981, **649**, 541–549.

34 A. Jasanoff and A. R. Fersht, *Biochemistry*, 1994, **33**, 2129–2135.

35 P. Luo and R. L. Baldwin, *Biochemistry*, 1997, **36**, 8413–8421.

36 M. Vincenzi, F. A. Mercurio and M. Leone, *Curr. Protein Pept. Sci.*, 2019, **20**, 425–451.

37 A. F. Drake, G. Siligardi and W. A. Gibbons, *Biophys. Chem.*, 1988, **31**, 143–146.

38 R. W. Woody, *Adv. Biophys. Chem.*, 1992, **2**, 37–79.

39 S. H. Park, W. Shalongo and E. Stellwagen, *Protein Sci.*, 1997, **6**, 1694–1700.

40 B. Nordén, A. Rodger and T. Dafforn, *Linear Dichroism and Circular Dichroism: A Textbook on Polarized-Light Spectroscopy*, The Royal Society of Chemistry, Cambridge, UK, 2010.

41 D. Diaz and R. J. Clarke, *J. Membr. Biol.*, 2018, **251**, 653–666.

42 F. Agou, Y. Yang, J.-C. Gesquière, J.-P. Waller and E. Guittet, *Biochemistry*, 1995, **34**, 569–576.

43 R. Melki, P. Kerjan, J.-P. Waller, M.-F. Carlier and D. Pantoloni, *Biochemistry*, 1991, **30**, 11536–11545.

44 I. N. Berezovsky, W. W. Chen, P. J. Choi and E. I. Shakhnovich, *PLoS Comput. Biol.*, 2005, **1**, e47.

45 H. M. Hankins, R. D. Baldridge, P. Xu and T. R. Graham, *Traffic*, 2015, **16**, 35–47.

Faraday Discussions

PAPER

The influence of phosphatidylserine localisation and lipid phase on membrane remodelling by the ESCRT-II/ESCRT-III complex

Andrew Booth, [a] Christopher J. Marklew, [b] Barbara Ciani [b] and Paul A. Beales [*ac]

Received 21st April 2020, Accepted 14th July 2020

DOI: 10.1039/d0fd00042f

The endosomal sorting complex required for transport (ESCRT) organises in supramolecular structures on the surface of lipid bilayers to drive membrane invagination and scission of intraluminal vesicles (ILVs), a process also controlled by membrane mechanics. However, ESCRT association with the membrane is also mediated by electrostatic interactions with anionic phospholipids. Phospholipid distribution within natural biomembranes is inhomogeneous due to, for example, the formation of lipid rafts and curvature-driven lipid sorting. Here, we have used phase-separated giant unilamellar vesicles (GUVs) to investigate the link between phosphatidylserine (PS)-rich lipid domains and ESCRT activity. We employ GUVs composed of phase separating lipid mixtures, where unsaturated DOPS and saturated DPPS lipids are incorporated individually or simultaneously to enhance PS localisation in liquid disordered (L_d) and/or liquid ordered (L_o) domains, respectively. PS partitioning between the coexisting phases is confirmed by a fluorescent Annexin V probe. Ultimately, we find that ILV generation promoted by ESCRTs is significantly enhanced when PS lipids localise within L_d domains. However, the ILVs that form are rich in L_o lipids. We interpret this surprising observation as preferential recruitment of the L_o phase beneath the ESCRT complex due to its increased rigidity, where the L_d phase is favoured in the neck of the resultant buds to facilitate the high membrane curvature in these regions of the membrane during the ILV formation process. L_d domains offer lower resistance to membrane bending, demonstrating a mechanism by which the composition and mechanics of membranes can be coupled to regulate the location and efficiency of ESCRT activity.

[a]School of Chemistry, Astbury Centre for Structural Molecular Biology, University of Leeds, Leeds, LS2 9JT, UK. E-mail: p.a.beales@leeds.ac.uk

[b]Department of Chemistry, Centre for Membrane Interactions and Dynamics, University of Sheffield, Sheffield S3 7HF, UK

[c]Bragg Centre for Materials Research, University of Leeds, Leeds, LS2 9JT, UK

Introduction

The precise molecular composition of lipid membranes determines the physi-cochemical properties that regulate biological cell homeostasis,[1] such as membrane fluidity and phase behaviour. Lipid membrane fluidity and phase behaviour influence the mechanical properties of membranes (*e.g.*, bending modulus,[2] spontaneous curvature[3]) and phase separation imparts spatial ordering of lipids, which can play a role in biological signalling (in so-called 'lipid rafts').[4,5] This spatial ordering of lipids modulates the local mechanical properties of membranes[2,6] which plays a key role in signalling mechanisms comprising of membrane fusion and fission events.[4,5,7]

Membrane tension regulates the activity of membrane-remodelling protein complexes such as dynamin[8] and the endosomal sorting complex required for transport (ESCRT), which drives membrane budding and scission in multi-vesicular body biogenesis, viral budding, repair of plasma, nuclear and organelle membranes.[9-11] The core scission machinery of ESCRT is comprised of the ESCRT-III complex which initiates membrane invagination, and the adaptor complex ESCRT-II which binds specifically to phosphatidylinositide phosphates (PIPs).[12] In budding yeast, ESCRT-III assembles from four key subunits (Vps20, Snf7, Vps24 and Vps2) and the AAA+ ATPase Vps4 that regulates the formation of ESCRT-III three-dimensional spiral structures.[13] These structures are capable of stabilising membrane buds and constricting membrane necks in an ATP-dependent manner.[13]

The budding and scission activity of the ESCRT-II and ESCRT-III complexes can be reconstituted *in vitro* by following the generation of intralumenal vesicles (ILVs) within giant unilamellar vesicles (GUVs) after the addition of purified ESCRT components.[12,14] In this bulk-phase encapsulation assay, only ESCRT-generated ILV compartments contain fluorescent extravesicular media in their lumens allowing the quantification of ESCRT-mediated membrane-remodelling upon a change in experimental conditions. *In vitro* systems to study how the bulk phase is encapsulated into the membrane compartment by protein complexes such as ESCRT also represent a useful biomimetic tool to create multi-compartment architectures incorporating spatially segregated, chemically distinct environments, and possess significant promise in the construction of complex artificial cells.[15,16]

It is well established that specific lipid headgroups play key roles in signalling and protein recruitment processes.[17] Anionic PIPs and phosphatidylserine (PS) lipids are necessary to initiate ESCRT activity at the membrane.[18] Specifically, PS lipids are necessary for the binding of ESCRT-III subunits prior to complex assembly and membrane remodelling[9,19-21] whereas ESCRT-II subunits specifi-cally bind PI(3)P,[22] though phosphoinositides are not strictly required for scission activity.[14]

The affinity of ESCRTs for specific lipids therefore provides an opportunity to employ phase separated model membranes to investigate how spatially distinct lipid distribution between membrane phases of differing fluidity and mechanics influences their function and activity *in vitro*. Simple mixtures of saturated lipids, unsaturated lipids and cholesterol can be designed to phase separate into coex-isting fluid membrane phases that are thought to mimic the structural properties of native lipid rafts.[23-26] The liquid disordered (L_d) phase is enriched in unsaturated lipids and is characterised by higher fluidity and lower bending rigidity, while cholesterol and saturated lipids concentrate in more viscous and rigid liquid ordered (L_o) domains. Equilibration of phase separating GUV membranes results in membrane textures that coarsen into large microscale domains that minimise the interfacial tension between coexisting phases. These domains can easily be visualised by fluorescence microscopy techniques due to preferential partitioning of lipophilic trace fluorophores between distinct membrane phases.[27,28]

Here we show that ESCRT-II/ESCRT-III activity is increased in PS-rich L_d microdomains with the newly formed ILVs containing predominantly L_o phase membranes. Given the relationship between mechanical properties and membrane phases, our data provides further insight into the mechanism of mechanical deformation of membranes by the ESCRT complex. Furthermore, if specific lipids can indeed be used to 'target' ESCRT-driven ILV formation to membrane microdomains then we can envisage strategies for generating ILVs with physicochemical properties selectively distinct from their parent membrane.

Our data provides further insight into the *in vitro* mechanism of mechanical deformation of membranes by the ESCRT complex and might aid the understanding of ESCRT-related events at biological membranes.

Results and discussion

Strategy

We and others[14,15,29] have successfully used GUVs composed of homogeneous PC : cholesterol : PS mixtures to reconstitute the membrane remodelling activity of the ESCRT-II/ESCRT-III complex. Based on phase diagrams available in the literature for DOPC : DPPC : Chol mixtures,[30] we selected a baseline membrane composition of 35 : 35 : 30 as it is fairly central within the liquid ordered (L_o)–liquid disordered (L_d) coexistence region of the phase diagram. Thirteen mol% of PC lipids were exchanged for the corresponding PS lipid with the same acyl tails (*e.g.*: DOPC for DOPS) in each formulation, with either 13 mol% DOPS, 13 mol% DPPS or 6.5 mol% DOPS + 6.5 mol% DPPS. Fig. 1 depicts the expected distribution of PS lipids in our formulations due to preferential partitioning of these lipids between the coexisting phases and the hypothesised resulting ILV membrane compositions.

Controlling the phase distribution of PS-headgroup lipids

Confocal microscopy of GUVs treated with the PS-binding protein Annexin V, fluorescently labelled with AlexaFluor488 (AF-488-AxV), was used to assess PS localisation in each formulation with respect to membrane phase domain (Fig. 2). Phase character was indicated by the partitioning of naphtho[2,3-*a*]pyrene (Np) into L_o domains and of 1,2-dioleoyl-*sn-glycero*-3-phosphoethanolamine-*N*-(Lissamine rhodamine B sulfonyl) ammonium salt, ('Rh–PE') into L_d domains.[27] The great majority of GUVs were observed to have roughly hemispherical L_o/L_d distributions, although some striped and spotted domains were also sporadically observed. The expected phase-enriched distributions of PS lipids were obtained reliably (L_o PS enrichment with DPPS, L_d enrichment with

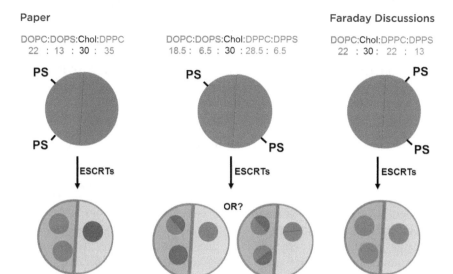

Fig. 1 Schematic of hypothesised outcomes from ESCRT-driven ILV formation activity against GUVs with a different distribution of PS-headgroup lipids. ILV membrane composition may reflect the relative distribution of the phase(s) with increased PS concentration. Red: L_d membrane, blue: L_o membrane.

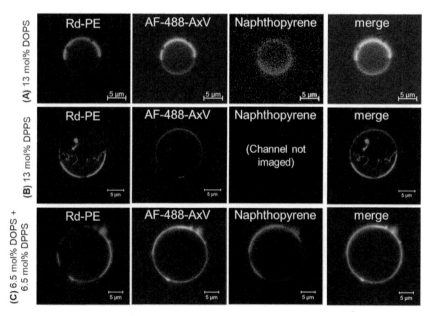

Fig. 2 Confocal microscopy of phase separated GUVs. (A) 28.5 : 28.5 : 13 : 30 DPPC : DOPC : DOPS : Chol, (B) 28.5 : 28.5 : 13 : 30 DPPC : DOPC : DPPS : Chol, (C) 28.5 : 28.5 : 6.5 : 6.5 : 30 DPPC : DOPC : DOPS : DPPS : Chol. Rh–PE preferentially partitions into L_d domains, napthopyrene (Np) preferentially partitions into L_o domains, AF488-AxV selectively binds to PS lipids to indicate distribution with respect to the domains of different membrane phases.

DOPS), with comparable distributions in the 6.5 mol% DPPS, 6.5 mol% DOPS '1 : 1' mixture. The distribution of AF-488-AxV within enriched phases was highly uniform other than some isolated cases where enhanced AF-488-AxV fluorescence was observed along phase boundaries in the '1 : 1' PS composition, but this was rare.

PS in the L_d phase enhances efficiency of ILV formation

ESCRT activity was assessed by ILV counting. Membrane-impermeable fluorescent dyes were added to GUV suspensions, followed by ESCRT proteins (50 nM Snf7, 10 nM each of ESCRT-II, Vps20, Vps24, Vps2 and Vps4, and 10 μM ATP.MgCl$_2$), all added simultaneously. The presence of fluorescent dye in the lumens of free-floating ILVs, after incubation with ESCRT proteins, is a strong indication that those ILVs were formed as a result of ESCRT activity.[15] The number of dye-containing ILVs were counted and then expressed as a number of ILVs per GUV volume equivalent, by taking the luminal volume observed during large area tile-scan imaging of GUV samples (cross sectional area × section thickness), divided by the volume of a typical 20 μm diameter GUV (Fig. 3). Automation of ILV counting was made possible using the calcein-Co^{2+} fluorescent dye–quencher pair in this assay, as fluorescence in the bulk medium could be quenched after the ESCRT incubation period by addition of the membrane-impermeable quencher

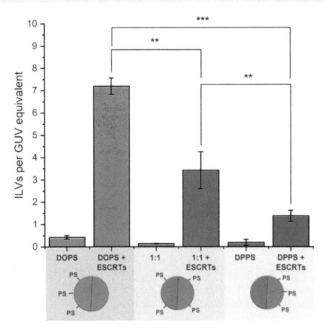

Fig. 3 The impact of PS phase localisation on the ILV-forming activity of ESCRTs. The graph shows the number of ILVs per GUV volume equivalent for GUVs incubated with the ESCRT-II/III complex (Snf7 50 nM, ESCRT-II, Vps20, Vps24, Vps2, Vps4 (10 nM) and ATP.MgCl$_2$ 1 μM). ILVs are determined to be formed post-ESCRT addition by the presence of calcein fluorescence. GUVs either contained DOPS localised to the L_d domains, DPPS localised to the L_o domains, or a 1 : 1 mix of DOPS and DPPS that is fairly evenly distributed between the coexisting phases. Diagram: red: L_d membrane, blue: L_o membrane.

(CoCl$_2$). This means that only ILVs containing calcein fluorescence had formed during the period of incubation with ESCRTs.

The results, shown in Fig. 3, indicate that the relative localisation of PS-headgroup lipids in phase-separated GUVs has a significant influence on the extent of ESCRT-induced ILV formation. The highest ILV formation activity was seen within GUVs containing 13 mol% DOPS, where PS lipids are predominantly located in the L$_d$ phase domains of the GUVs. In contrast, the lowest ILV formation activity was seen in GUVs where PS is preferentially located in the L$_o$-phase (13 mol% DPPS GUVs). Finally, where PS is found in both phases (1 : 1) an intermediate degree of ILV formation was observed.

This suggests that the concentration of PS lipids in L$_d$ domains also regulates the degree of ESCRT activity. We can rationalise these findings by taking into consideration the lower bending rigidity of each phase; the lower bending rigidity of an L$_d$ phase, offers lower mechanical resistance to ESCRT-induced membrane deformation than a L$_o$ domain.[2,31] ESCRT activity is negatively regulated by membrane tension, in a manner that is likely analogous to suppression of activity by higher bending rigidity.[15] We conclude that ESCRT activity must be higher at PS-containing L$_d$ domains where the membrane is more susceptible to deformation.

ESCRT-induced ILVs consist of L$_o$ phase membranes

Surprisingly, the ILVs that form mostly consist of L$_o$ membranes, despite local-isation of PS lipids in L$_d$ domains (Fig. 4, blue bars). This is determined by the observed depletion of the L$_d$-partitioning Rh–PE probe from the membranes of ILVs, indicating a lack of liquid disordered phase membrane. In contrast, the few ILVs present in the GUV lumen prior to addition of proteins, which lack encapsulated calcein, and form as a random by-product of the electroformation process (Fig. 4, red bars), showed greater L$_d$ content compared to those formed by ESCRT activity.

We therefore quantify the relative phase character of ILVs formed by ESCRTs and those that occur intrinsically, to ascertain the significance of this observation. We quantified the area mean fluorescence intensity of the Rh–PE for each individual ILV present before and after the addition of ESCRT proteins, taking this as a measure of the proportion of the L$_d$ phase in the volume of the ILV (Fig. 4). Analysis of Rh–PE fluorescence is more robust than analysis of the Np fluorescence, due to the weaker overall fluorescence and its minor overlap with the fluorescence emission of encapsulated calcein.

The 13% DOPS (Fig. 4a) and 1 : 1 DOPS : DPPS (Fig. 4b) GUV samples, show a distinctly lower mean Rh–PE fluorescence than intrinsic ILVs, we interpret this as a decrease in L$_d$ character of the membrane, or alternatively, a significant enrichment of the L$_o$ phase in ESCRT-induced ILVs.

This observation was surprising since it would be reasonable to anticipate that the ILVs form from the PS-enriched phase of the parent GUV membrane, and hence have a similar composition and phase character. However, this result is consistent with previous suggestions of localisation of ESCRT complexes at phase boundaries, where local membrane curvature may help to nucleate the assembly of ESCRT-III.[18]

The GUVs containing 13 mol% DPPS exhibit ILVs with a very broad range of mean fluorescence intensities with no obvious compositional preference (Fig. 4c). This lack of selectivity for L$_o$ or L$_d$ domains and significantly reduced ILV

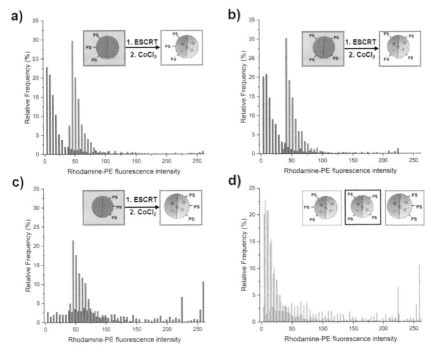

Fig. 4 ILV membranes are predominantly L_o phase. Histograms of mean Rh–PE fluorescence intensity in ILVs, a measure of the phase character of the membranes. Red plots: intrinsic ILVs (ESCRT-independent). Blue plots: ESCRT-induced ILVs. (a) 13 mol% DOPS GUVs, (b) 6.5 mol% DOPS, 6.5 mol% DPPS GUVs, (c) 13 mol% DPPS GUVs, (d) overlaid post-ESCRT ILV data for each composition (green = DOPS, grey = 1 : 1 mixture, yellow = DPPS). Inset diagrams: red, L_d membrane; blue, L_o membrane.

formation activity (Fig. 3) suggests ESCRT activity is dysregulated when PS is enriched in the L_o domains, the tighter packing of lipids in this phase may inhibit membrane insertion of the amphipathic helix of Snf7, thereby restricting its normal assembly and function.[32,33]

A representative image of DOPS-containing GUVs after incubation with ESCRTs is shown in Fig. 5A. Here, the significant reduction in Rh–PE ESCRT-induced fluorescence (those containing green calcein fluorescence), relative to intrinsic ILVs is strikingly illustrated. However, Fig. 5A also illustrates a common observation in GUV samples with compositions containing DOPS (13 mol% DOPS, 1 : 1 DOPS : DPPS); GUVs with a high abundance of ESCRT-induced ILVs tend to have a predominantly L_o parent membrane. This is a highly counterintuitive finding as the expectation would be that any mechanism that favours the formation of L_o ILVs necessarily leads to the removal of L_o lipids from the parent GUV membrane. Having already shown that PS-localisation in the L_o phase inhibits ESCRT activity, it is unlikely that ESCRTs have preference for L_o-only vesicles in the DOPS GUV samples. Indeed L_o-only GUVs were not observed prior to ESCRT addition, suggesting that such vesicles are ESCRT-induced and high ILV formation activity in a single vesicle likely results in shedding of the L_d phase from the parent membrane. We will show evidence for this in rare observations of stalled ESCRT-induced membrane buds.

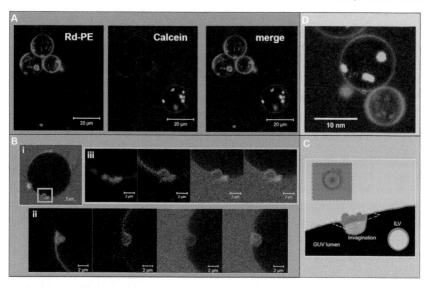

Fig. 5 Representative images of observed phase characteristics in GUVs incubated with ESCRTs. (A) High yields of ESCRT-induced ILVs in an individual GUV causes depletion of the L_d phase from the parent membrane. Confocal microscopy of GUVs (13 mol% DOPS) after incubation with ESCRT proteins and quenching of background calcein fluorescence with $CoCl_2$. Left, Rh–PE; centre, calcein; right, merge. (B) Confocal microscopy images of GUVs with apparent stalled ILV buds. (i) and (ii) are distinct GUVs, (iii) is an enlargement of the yellow box in (i), left to right; Rh–PE, Np, fluorescein labelled dextran, 70 kDa, both GUVs are '1 : 1' composition (6.5 mol% each of DPPS and DOPS). B(iii) is an enlargement of the area designated by the yellow square in (i): green, fluorescein labelled dextran (70 kDa); red, Rh–PE; blue, Np. (C) Schematic interpretation of ILV bud phase character distribution based on images (B)(i and ii). (D) L_o GUV with calcein containing ILVs with adherent L_d GUV, possibly a result of ESCRT-induced abscission.

Stalled ILV 'buds' reveal insights into the mechanism of L_o-enrichment in ILVs

We observed that a small number of GUVs with a 1 : 1 DOPS : DPPS composition contained stalled ILV buds (Fig. 5B). One mechanism by which these stalled ILV buds may appear is when the membrane tension of a particular GUV is sufficient to allow initial invagination of the ILV bud, but mechanical resistance stalls ILV formation prior to scission of the bud neck. We have previously observed a number of these stalled buds in homogeneous GUV membranes interacting with ESCRT proteins.[15] Alternatively, rare assembly errors in the formation of these complexes may have resulted in incompetent ESCRT machinery that is unable to complete the final neck closure and scission steps.

These nascent buds are yet to be closed off from the extravesicular medium given that we observe a continuity of fluorescein-labelled dextran ([70 kDa]) from the bulk phase into the bud volume. These buds have a neck of approximately 2 µm diameter, which correlates with the typical size range of ILVs observed in these experiments ($\sim$1–2 µm diameter). The membrane phases show an unusual distribution in these bud structures where buds occur at or near to a phase boundary, with the portion of the membrane that buds into the parent GUV lumen being predominantly L_o (Fig. 5B(i)). However, there is an accumulation of L_d membrane on the exterior of the parent membrane adjacent to the bud neck, which could be interpreted as a 'collar' of L_d membrane (see Fig. 5C). This contrasts somewhat with the structure observed in Fig. 5B(ii), where again, an L_o bud is observed at a phase boundary, with L_d accumulation around the bud neck, but there is also a protrusion of L_d material into the bud lumen. Despite the infrequent observation of this phenomenon, there is a clear correlation with the prevalence of L_o phase character in ESCRT-induced ILVs as shown in Fig. 4.

Proposed mechanism of L_o-enriched ILV formation

ILV formation is most efficient when PS lipids are localised within L_d domains of phase separated GUVs. Electrostatic association of the ESCRT-III proteins with PS lipids would be expected to concentrate initial binding onto L_d domains. Curvature generated at phase boundaries[34] may also preferentially nucleate ESCRT-III assembly at these sites.[18] Assembly of ESCRT complexes on the membrane likely increases the local membrane rigidity, suppressing thermal undulations. Suppression of membrane undulations in adhesion plaques has previously been shown to promote formation and localisation of more rigid membrane phases,[35] and similar observations have recently been reported *in vivo*.[36] Moreover, induced formation of L_o membrane beneath ESCRT-II assemblies has previously been reported on solid-supported lipid bilayers.[37]

Induced recruitment of L_o membrane beneath an ESCRT-II complex will create a line tension with the L_d phase at the domain boundaries. The energy cost of this phase boundary can be minimised by bending the membrane such that the circumference of the phase boundary decreases, promoting budding of the L_o domain.[38] The L_d phase at the L_o domain boundaries is likely required to facilitate the high curvature in the bud neck, while the line tension within the L_d–L_o phase boundary within the bud neck could enhance the efficiency of scission of the bud from the parent membrane, forming the ILV.[39–41] It should be noted that the curvature of the L_o domain bud (radius of curvature $\sim$ 1 μm) is much lower than in the bud neck (radius of curvature $\sim$ 10 s nm). These ESCRT-induced L_o buds may preferentially migrate to the macroscopic phase boundaries within the GUV, or preferentially nucleate in these regions of the vesicle, as observed in Fig. 5B(i) and (ii). Fig. 5B also shows strong evidence for involvement of L_d phase lipids in the neck of L_o-rich membrane buds.

Notably, there is evidence for lipid sorting during vesicle budding processes *in vivo*, in the generation of endosomes, multivesicular bodies and exosomes.[42] Interestingly, many lipidomics studies have reported enrichment of sphingomelin and cholesterol in endosomes and exosomes relative to their parent membrane (the plasma membrane), suggestive of a greater L_o phase character.[43–46] Furthermore, a recent detailed lipidomics analysis of membranes in erythrocytes reveals that PS lipids with highly unsaturated acyl chains tend to concentrate at the cytoplasmic-facing leaflet of the plasma membrane, thus implying they might preferentially localise within L_d regions of membranes where ESCRT activity is also engaged.[47] The parallels with the L_o-enrichment in phase separated GUV model membranes reported here are striking.

Importantly, ILV formation is strongly suppressed when PS lipids are enriched in L_o domains (Fig. 3). Therefore, we propose that ESCRT-II/ESCRT-III, preferentially recruited to these L_o domains through electrostatic association with DPPS, have suppressed functional activity at these membrane domains. This conclusion is supported by the following considerations. In the *in vivo* system, the anchoring of ESCRT-III *via* membrane-inserting motifs (myristoyl moiety in Vps20 (ref. 48) or the N-terminal amphipathic helix in Snf7 (ref. 33)) may be inhibited in the L_o phase, due to more ordered lipid packing and lower incidence of packing defects, resulting in lower surface concentration of bound protein.[49] Furthermore, the high membrane curvature required in the neck of nascent buds would have a much higher energy cost for L_o phase membrane than L_d phase membrane, where the energy required to bend the membrane is a factor of ~5–6 greater.[50] ESCRT assembly on the higher rigidity L_o membrane would also not favour recruitment of a different membrane phase beneath the complex, where line tension would no longer be generated to favour budding and scission as described above. These factors likely combine to strongly suppress ILV formation generated from ESCRT binding to L_o domains, where the few ILVs that do form are highly dysregulated in membrane composition as observed in Fig. 4C and D.

The observed depletion of L_d lipids from GUV membranes induced by ESCRT-driven ILV formation is harder to conclusively explain due to limited experimental evidence. We present hypotheses in Fig. 6B and C. In Fig. 6B some of the dense L_d phase membrane accumulation around ILV bud necks (as seen in Fig. 5B) is detached from the parent GUV membrane during the bud scission process. Following numerous ILV formation events, the L_d phase may be progressively depleted. Another source for L_o-rich GUVs may be ESCRT-driven abscission of entire L_d domains from GUVs, generating separate L_d-rich vesicles next to the L_o-rich vesicle as depicted in Fig. 6C. This is analogous to the role of ESCRT in cytokinetic abscission.[51] A small number of GUVs were observed in this

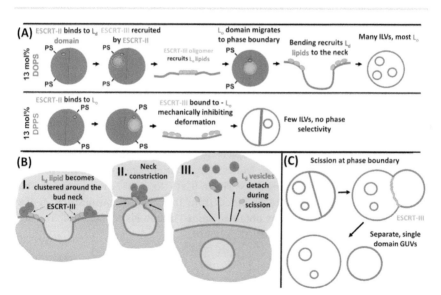

Fig. 6 Proposed mechanistic models of the phase behaviour observed in ESCRT-induced ILV formation in phase separated GUVs. (A) Model of ESCRT assembly and membrane remodelling activity in GUVs with PS lipids concentrated in L_d (13 mol% DOPS) or L_o (13 mol% DPPS) domains. (B) Depletion of GUV membrane L_d domain by ejection of clustered L_d vesicles from the ILV bud neck during scission. (C) Possible ESCRT-driven scission at GUV phase boundaries leading to distinct enriched or single domain GUVs as observed.

configuration in samples incubated with ESCRTs, a representative example is shown in Fig. 5D, but we currently have insufficient data to make any strong claims on the prevalence of this speculative mechanism. However, shedding of some membrane from GUVs caused by ESCRT activity should not be seen as too surprising since excess membrane shedding is required for the fission mechanisms at the basis of ESCRT activity *in vivo*.[52,53]

Conclusions

We have shown that ESCRT-driven membrane remodelling in phase separated model membranes is most efficient when PS lipids are preferentially located in the less rigid L_d domains. At first sight this result is intuitive since these domains offer lower mechanical resistance to the budding deformations required for ILV formation. However, surprisingly the ILVs that form are strongly enriched in L_o membrane. Combining rare observations of stalled ILV formation events and membrane biophysical concepts, we propose a mechanism by which this L_o phase enrichment occurs. Generation of ILV proto-organelles with a different membrane composition to the GUV parent membrane may have a useful role in generating artificial cells that mimic the variability of lipid composition between organelles and the plasma membrane of natural cells. Furthermore, the comparison of our observations with the enrichment of L_o-preferring lipids in endosomes and exosomes is compelling, suggestive of a similar phase-enrichment potentially occurring for ESCRT-generated ILVs *in vivo*.

Materials and methods

Materials

All lipids; DOPC (1,2-dioleoyl-*sn-glycero*-3-phosphocholine), DOPS (1,2-dioleoyl-*sn-glycero*-3-phospho-L-serine (sodium salt)), DPPC (1,2-dipalmitoyl-*sn-glycero*-3-phosphocholine), DPPS (1,2-dipalmitoyl-*sn-glycero*-3-phospho-L-serine (sodium salt)), cholesterol (ovine) and Rh–PE (1,2-dioleoyl-*sn-glycero*-3-phosphoethanolamine-*N*-(Lissamine rhodamine B sulfonyl) (ammonium salt)) were obtained from Avanti Polar Lipids Inc. (Alabaster, AL, USA). Np (naphtho[2,3-*a*]pyrene) was obtained from Tokyo Chemical Industries UK Ltd. Calcein, NaCl, Tris buffer, $CoCl_2$ (hexahydrate), $MgCl_2$, sucrose, ATP (adenosine 5′-triphosphate disodium salt hydrate) were all obtained from Sigma-Aldrich Company Ltd. (UK). 8-Chamber glass-bottomed microscope slides (Cat. No. 80827) were used for all confocal microscopy and were obtained from ibidi GmbH. Annexin V AF488 conjugate (Cat. No. A13201) was obtained from Thermo Fisher Scientific Inc. All proteins were expressed and purified in-house.

Protein expression

The plasmids for overexpression of *Saccharomyces cerevisiae* ESCRT-II (Vps22, Vps25, Vps36) and ESCRT-III subunits Vps20, Snf7, Vps2 and Vps4 are a gift from James Hurley (addgene plasmids #17633, #21490, #21492, #21494, #21495).[14,54] The *Saccharomyces cerevisiae* Vps24 gene was subcloned into a pRSET(A) expression vector, containing a His6-tag and enterokinase protease cleavage site. We have previously reported the detailed expression and purification methods for the

ESCRT-II and ESCRT-III proteins used in this the study, including character-isation and quality control data in ref. 15.

Electroformation of GUVs

Giant unilamellar vesicles (GUVs) were prepared using the electroformation method. Briefly, the desired lipid mixture is applied to indium tin-oxide coated glass slides (8–12 Ω sq^{-1}, Sigma-Aldrich) as a chloroform solution (15 µL, 0.7 mM), and dried under nitrogen to form a thin film. Lipid compositions were either: '13 mol% DOPS': 35 : 22 : 13 : 30 DPPC : DOPC : DOPS : Chol, '13 mol% DPPS': 22 : 35 : 13 : 30 DPPC : DOPC : DPPS : Chol, or '1 : 1' 28.5 : 28.5 : 6.5 : 6.5 : 30 DPPC : DOPC : DOPS : DPPS : Chol. An additional 0.5 mol% of Rh–PE and 1 mol% of Np were added to each stock solution.

Two slides, coated with the desired lipid mixture, were assembled into an electroformation chamber by separation of their conductive, lipid-coated surfaces by a silicone rubber gasket (~2 mm thickness). Electrical contacts were made to each slide using copper tape and these were connected to a function generator. The chamber was then filled with sucrose solution (300 mM) and sealed with a silicone rubber plug. The chambers were then placed into an oven at 60 °C. An AC field was applied to the chambers; 0.1 V$_{peak-to-peak}$, 10 Hz, sinusoidal waveform, ascending to 3 V$_{pp}$ over 10 min, and maintained at 3 V for 2 h, after which the frequency of the oscillation was reduced from 10 to 0.1 Hz over 10 min. At this point the GUV-containing sucrose solution was harvested using a syringe.

Osmotic relaxation of membrane tension was then performed. GUVs were diluted 1 : 4, vol/vol, GUV suspension : hypertonic buffer, with 10 mM Tris buffered saline (~150 mM NaCl; pH 7.4), adjusted to 10 mOsm kg^{-1} higher osmolarity than the sucrose solution and stored overnight at 4 °C prior to use. Solution osmolarity was measured using an Advanced Instruments 3320 osmometer.

Confocal microscopy

All experiments were performed on Zeiss LSM 700 or 880 systems, imaging with a Plan-Apochromat 40x/1.4 Oil DIC M27 objective lens, NA = 1.4, 405 nm, 488 nm, and 561 nm lasers. GUV suspensions were placed in chamber slides (µslide, 8 well, glass bottomed, ibidi GmbH). Glass surfaces in the chamber slides were passivated prior to use by addition of 5 wt% bovine serum albumin solution in deionised water, incubation for 5 min, followed by copious rinsing with Milli-Q water, and finally blown dry using a stream of nitrogen gas. Two-hundred µL aliquots of GUV suspension were added to each well.

Annexin V PS localisation assay

Prior to ESCRT assays, separate aliquots of GUVs from the same samples were treated with Annexin V, to confirm PS distribution was as expected. AlexaFluor488-labelled Annexin V (Thermo Fisher cat. no. A13201) (2 µL per 200 µL GUV suspension) was added to the desired wells and incubated for ~5 min before imaging.

ESCRT-induced ILV counting assay

Calcein (2 µL of a 4 mM stock solution in osmotically balanced saline buffer [10 mM Tris, pH 7.4]) was added to each well (200 µL GUV suspension), immediately followed by ESCRT proteins: ESCRT-II (Vps22, Vps25, Vps36), Vps20, Snf7, Vps24, Vps2 and Vps4 and ATP.MgCl$_2$, to give a final concentration of 10 nM for ESCRT-II, Vps20, Vps24, Vps2, Vps4, 50 nM for Snf7, and 10 μM ATP.MgCl$_2$. After a 20 min incubation period, CoCl$_2$ (2 μL of a 4 mM stock solution in osmotically balanced saline buffer [10 mM Tris, pH 7.4]), was added to quench calcein fluorescence in the extravesicular solution. Alternatively, fluorescently-labelled dextrans of varying molecular weights were added to assess uptake of larger species into ILV lumens. All imaging was performed with a pinhole setting that corresponded to a 3.1 μm section thickness, allowing for a greater volume of a GUV's lumen to be captured and to increase the likelihood of ILVs being entirely captured within the visible z-dimension, to improve the accuracy of membrane phase characterisation of both ILVs and GUVs. Large area tile-scanning was performed in order to capture sections through as high a number of GUVs as possible, to allow statistical analysis across the GUV population. Images were analysed using the Zeiss Zen software and Fiji. To improve comparability between results, ILV counts were expressed as 'ILVs per GUV volume equivalent', being the total GUV lumen observed (cross-sectional area × section thickness), divided into units of typically 20 μm diameter spherical GUV volumes.

ILV phase character analysis

The phase character of ESCRT-induced ILVs was assessed by calculating the mean fluorescence intensity of Rh–PE, over the area of the ILV-cross section, as a measure of the amount of L$_d$ membrane present. ESCRT-induced ILVs were identified as those which had calcein fluorescence in their lumen. This was compared to the same analysis of intrinsic ILVs occurring in GUV samples prior to the addition of ESCRT proteins (generated during electroformation of GUVs). These 'naturally occurring' ILVs were used as a benchmark for the typical average membrane composition of the parent GUV. Therefore a higher Rh–PE fluorescence than this benchmark would indicate that ESCRT-induced ILVs are enriched in L$_d$ phase, and a lower average Rh–PE fluorescence would indicate that ESCRT-induced ILVs are enriched in L$_o$ phase.

Conflicts of interest

There are no conflicts to declare.

Acknowledgements

This work was funded by the UK Engineering and Physical Sciences Research Council (EPSRC): EP/M027929/1 (PAB) and EP/M027821/1 (BC). We also thank Dr Daniel Mitchell for expression and purification of the Vps4 subunit.

References

1 G. van Meer, D. R. Voelker and G. W. Feigenson, *Nat. Rev. Mol. Cell Biol.*, 2008, **9**, 112–124.
2 H. P. Duwe and E. Sackmann, *Phys. A*, 1990, **163**, 410–428.
3 H. T. McMahon and E. Boucrot, *J. Cell Sci.*, 2015, **128**, 1065–1070.

4 M. A. Alonso and J. Millán, *J. Cell Sci.*, 2001, **114**, 3957–3965.

5 P. Varshney, V. Yadav and N. Saini, *Immunology*, 2016, **149**, 13–24.

6 W. Shinoda, *Biochim. Biophys. Acta, Biomembr.*, 2016, **1858**, 2254–2265.

7 S.-T. Yang, A. J. B. Kreutzberger, V. Kiessling, B. K. Ganser-Pornillos, J. M. White and L. K. Tamm, *Sci. Adv.*, 2017, **3**, e1700338.

8 A. Roux, K. Uyhazi, A. Frost and P. D. Camilli, *Nature*, 2006, **441**, 528–531.

9 L. Christ, C. Raiborg, E. M. Wenzel, C. Campsteijn and H. Stenmark, *Trends Biochem. Sci.*, 2017, **42**, 42–56.

10 D. J. Thaller and C. P. Lusk, *Biochem. Soc. Trans.*, 2018, **46**, 877–889.

11 J. H. Hurley, *EMBO J.*, 2015, **34**, 2398–2407.

12 J. H. Hurley and P. I. Hanson, *Nat. Rev. Mol. Cell Biol.*, 2010, **11**, 556–566.

13 S. Maity, C. Caillat, N. Miguet, G. Sulbaran, G. Effantin, G. Schoehn, W. H. Roos and W. Weissenhorn, *Sci. Adv.*, 2019, **5**, eaau7198.

14 T. Wollert, C. Wunder, J. Lippincott-Schwartz and J. H. Hurley, *Nature*, 2009, **458**, 172–177.

15 A. Booth, C. J. Marklew, B. Ciani and P. A. Beales, *iScience*, 2019, **15**, 173–184.

16 C. J. Marklew, A. Booth, P. A. Beales and B. Ciani, *Interface Focus*, 2018, **8**, 20180035.

17 J. H. Hurley, Y. Tsujishita and M. Pearson, *Curr. Opin. Struct. Biol.*, 2000, **10**, 737–743.

18 I.-H. Lee, H. Kai, L.-A. Carlson, J. T. Groves and J. H. Hurley, *Proc. Natl. Acad. Sci. U. S. A.*, 2015, **112**, 15892–15897.

19 W. M. Henne, N. J. Buchkovich and S. D. Emr, *Dev. Cell*, 2011, **21**, 77–91.

20 A. L. Schuh and A. Audhya, *Crit. Rev. Biochem. Mol. Biol.*, 2014, **49**, 242–261.

21 G. Bodon, R. Chassefeyre, K. Pernet-Gallay, N. Martinelli, G. Effantin, D. L. Hulsik, A. Belly, Y. Goldberg, C. Chatellard-Causse, B. Blot, G. Schoehn, W. Weissenhorn and R. Sadoul, *J. Biol. Chem.*, 2011, **286**, 40276–40286.

22 N. D. Franceschi, M. Alqabandi, N. Miguet, C. Caillat, S. Mangenot, W. Weissenhorn and P. Bassereau, *J. Cell Sci.*, 2019, **132**(4), jcs217968.

23 T. Baumgart, A. T. Hammond, P. Sengupta, S. T. Hess, D. A. Holowka, B. A. Baird and W. W. Webb, *Proc. Natl. Acad. Sci. U. S. A.*, 2007, **104**, 3165–3170.

24 N. Kahya, D. Scherfeld, K. Bacia, B. Poolman and P. Schwille, *J. Biol. Chem.*, 2003, **278**, 28109–28115.

25 S. L. Veatch and S. L. Keller, *Biochim. Biophys. Acta, Mol. Cell Res.*, 2005, **1746**, 172–185.

26 S. L. Veatch and S. L. Keller, *Phys. Rev. Lett.*, 2002, **89**, 268101.

27 T. Baumgart, G. Hunt, E. R. Farkas, W. W. Webb and G. W. Feigenson, *Biochim. Biophys. Acta*, 2007, **1768**, 2182–2194.

28 A. S. Klymchenko and R. Kreder, *Chem. Biol.*, 2014, **21**, 97–113.

29 Y. Avalos-Padilla, R. L. Knorr, R. Javier-Reyna, G. García-Rivera, R. Lipowsky, R. Dimova and E. Orozco, *Front. Cell. Infect. Microbiol.*, 2018, **8**, 53.

30 S. L. Veatch and S. L. Keller, *Biophys. J.*, 2003, **85**, 3074–3083.

31 R. D. Usery, T. A. Enoki, S. P. Wickramasinghe, V. P. Nguyen, D. G. Ackerman, D. V. Greathouse, R. E. Koeppe, F. N. Barrera and G. W. Feigenson, *Biophys. J.*, 2018, **114**, 2152–2164.

32 M. A. Zhukovsky, A. Filograna, A. Luini, D. Corda and C. Valente, *Front. Cell Dev. Biol.*, 2019, **7**, 291.

33 N. J. Buchkovich, W. M. Henne, S. Tang and S. D. Emr, *Dev. Cell*, 2013, **27**, 201–214.

34 T. Baumgart, S. T. Hess and W. W. Webb, *Nature*, 2003, **425**, 821–824.

35 V. D. Gordon, M. Deserno, C. M. J. Andrew, S. U. Egelhaaf and W. C. K. Poon, *Europhys. Lett.*, 2008, **84**, 48003.

36 C. King, P. Sengupta, A. Y. Seo and J. Lippincott-Schwartz, *Proc. Natl. Acad. Sci. U. S. A.*, 2020, **117**, 7225–7235.

37 E. Boura, V. Ivanov, L.-A. Carlson, K. Mizuuchi and J. H. Hurley, *J. Biol. Chem.*, 2012, **287**, 28144–28151.

38 R. Lipowsky and R. Dimova, *J. Phys.: Condens. Matter*, 2002, **15**, S31–S45.

39 B. Różycki, E. Boura, J. H. Hurley and G. Hummer, *PLoS Comput. Biol.*, 2012, **8**, e1002736.

40 J. Liu, M. Kaksonen, D. G. Drubin and G. Oster, *Proc. Natl. Acad. Sci. U. S. A.*, 2006, **103**, 10277–10282.

41 W. Römer, L.-L. Pontani, B. Sorre, C. Rentero, L. Berland, V. Chambon, C. Lamaze, P. Bassereau, C. Sykes, K. Gaus and L. Johannes, *Cell*, 2010, **140**, 540–553.

42 J. H. Hurley, E. Boura, L.-A. Carlson and B. Różycki, *Cell*, 2010, **143**, 875–887.

43 T. Skotland, K. Sandvig and A. Llorente, *Prog. Lipid Res.*, 2017, **66**, 30–41.

44 A. Llorente, T. Skotland, T. Sylvänne, D. Kauhanen, T. Róg, A. Orłowski, I. Vattulainen, K. Ekroos and K. Sandvig, *Biochim. Biophys. Acta, Mol. Cell Biol. Lipids*, 2013, **1831**, 1302–1309.

45 K. Laulagnier, C. Motta, S. Hamdi, S. Roy, F. Fauvelle, J.-F. Pageaux, T. Kobayashi, J.-P. Salles, B. Perret, C. Bonnerot and M. Record, *Biochem. J.*, 2004, **380**, 161–171.

46 K. Sobo, J. Chevallier, R. G. Parton, J. Gruenberg and F. G. van der Goot, *PLoS One*, 2007, **2**(4), e391.

47 J. H. Lorent, K. R. Levental, L. Ganesan, G. Rivera-Longsworth, E. Sezgin, M. Doktorova, E. Lyman and I. Levental, *Nat. Chem. Biol.*, 2020, **16**, 644–652.

48 M. Babst, D. J. Katzmann, E. J. Estepa-Sabal, T. Meerloo and S. D. Emr, *Dev. Cell*, 2002, **3**, 271–282.

49 J. Bigay and B. Antonny, *Dev. Cell*, 2012, **23**, 886–895.

50 R. S. Gracià, N. Bezlyepkina, R. L. Knorr, R. Lipowsky and R. Dimova, *Soft Matter*, 2010, **6**, 1472–1482.

51 J. Lafaurie-Janvore, P. Maiuri, I. Wang, M. Pinot, J.-B. Manneville, T. Betz, M. Balland and M. Piel, *Science*, 2013, **339**, 1625–1629.

52 A. J. Jimenez, P. Maiuri, J. Lafaurie-Janvore, S. Divoux, M. Piel and F. Perez, *Science*, 2014, **343**, 1247136.

53 Y.-N. Gong, C. Guy, H. Olauson, J. U. Becker, M. Yang, P. Fitzgerald, A. Linkermann and D. R. Green, *Cell*, 2017, **169**, 286–300.

54 T. Wollert and J. H. Hurley, *Nature*, 2010, **464**, 864–869.

Faraday Discussions

ROYAL SOCIETY
OF CHEMISTRY

PAPER

Impact of antimicrobial peptides on *E. coli*-mimicking lipid model membranes: correlating structural and dynamic effects using scattering methods†

Josefine Eilsø Nielsen, [a] Sylvain François Prévost, [b] Håvard Jenssen [c] and Reidar Lund [*a]

Received 27th April 2020, Accepted 11th September 2020

DOI: 10.1039/d0fd00046a

The mechanism of action of antimicrobial peptides (AMPs) has been debated over many years, and various models have been proposed. In this work we combine small angle X-ray/neutron scattering (SAXS/SANS) techniques to systematically study the effect of AMPs on the cytoplasmic membrane of *Escherichia coli* bacteria using a simplified model system of 4 : 1 DMPE : DMPG ([1,2-dimyristoyl-*sn-glycero*-3-phosphoethanolamine] : [1,2-dimyristoyl-*sn-glycero*-3-phospho-(10-*rac*-glycerol)]) phospholipid unilamellar vesicles. The studied antimicrobial peptides aurein 1.2, indolicidin, LL-37, lacticin Q and colistin vary in size, charge, degree of helicity and origin. The peptides insert into the bilayer to various degrees, and are found to accelerate the dynamics of phospholipids significantly as seen by time resolved SANS (TR-SANS) measurements, with the exception of colistin that is suggested to rather interact with lipopolysaccharides (LPS) on the outer membrane of *E. coli*. We compare these results with earlier published data on model systems based on PC-lipids (phosphatidylcholines), showing comparable effect with regards to peptide insertion and effect on dynamics. However, model systems based on PE-lipids (phosphatidylethanolamine) are more prone to destabilisation upon addition of peptides, with formation of multilamellar structures and morphological changes. These properties of PE-vesicles lead to less conclusive results regarding peptide effect on structure and dynamics of the membrane.

Introduction

Antimicrobial peptides (AMPs) are important agents in the first line of defence to kill pathogenic microorganisms in humans, animals, insects and are even

---

[a] *Department of Chemistry, University of Oslo, 0315 Oslo, Norway. E-mail: reidar.lund@kjemi.uio.no*

[b] *Institut Laue - Langevin, 38000 Grenoble, France*

[c] *Department of Science and Environment, Roskilde University, 4000 Roskilde, Denmark*

† Electronic supplementary information (ESI) available. See DOI: 10.1039/d0fd00046a

secreted by some bacteria. They defend the host against foreign infectious organisms including Gram-positive and Gram-negative bacteria, fungi, protozoa and viruses,[1–3] through host immune modulation and/or direct targeting of the infectious organisms. Their direct antimicrobial properties have been known for decades, yet the mode of action of AMPs are found to be quite complex, and many different theories have been presented. However, there is a general consensus that most AMPs in some way or another mainly target the cytoplasmic membrane of the microorganism.[4] Their ability to only attack foreign organisms and not their host is amongst others explained by the difference in lipid composition of the cytoplasmic membranes of eukaryote and prokaryote cells. Eukaryote cell membranes mostly consist of neutrally charged lipids (zwitterionic) and choles-terol,[5] while prokaryote membranes include a substantial amount of negative charged lipids in combination with zwitterionic lipids and cardiolipins.[6] Even though AMPs vary vastly in structure, the common feature of most of the peptides is their overall cationic (positive) charge. This enables electrostatic interactions with anionic (negative) lipids in the bacteria membrane, thus potentially playing a regulatory role in the target cell selectivity.[7]

To study peptide–lipid interactions in detail using biophysical and biochem-ical methodologies, the use of model membrane systems to mimic the cyto-plasmic membrane of bacteria is essential.[8] Real bacteria cells contain much more than the cytoplasmic membrane such as an outer membrane and intra-cellular ribosomes, a chromosome, and plasmids. Even the inner membrane itself is intricate with a diverse group of membrane proteins as well as different phospholipids as described above. The complexity will for many of these tech-niques obscure the results, complicating the interpretation with regards to the specific membrane–peptide interaction. Therefore, the model systems most frequently used are phospholipid membranes either as flat supported bilayers, tethered lipid bilayers, free-floating lipid micelles, or vesicles. The flat model membranes are used for surface sensitive techniques like neutron reflectometry (NR),[9–15] atomic force microscopy (AFM),[16–19] dual polarisation interferometry (DPI)[19] and surface plasmon resonance spectroscopy.[19,20] While the free-floating lipid micelles or vesicles can be studied by techniques like nuclear magnetic resonance (NMR),[11,21–23] small angle X-ray or neutron scattering (SAXS/SANS),[13,24–35] and fluorescence spectroscopy.[20,36] Recent efforts have gone into developing model systems more closely related to the real bacteria membrane. Clifton and co-workers recently presented a full NR characterization of a floating lipid membrane closely mimicking the inner and outer membrane of *E. coli* including the LPS layer.[10] Even some reports of nanoscale structural determina-tions of live bacterial cells have been reported recently. While Semeraro *et al.* have determined the ultrastructure of live *E. coli* using ultra-SAXS and detailed modelling,[37] Nickels and co-workers have used SANS and contrast variation to characterise the membrane heterogeneities of live *Bacillus subtilis*.[38] Although these findings reveal that it is possible to determine structural parameters of live cells they still justify the need for simplified model systems due to the high complexity in differentiating the specific effects of an added substrate.

The composition of the model systems can be adjusted depending on the specific cell type that one wishes to mimic. The most common composition of lipids used to mimic bacteria membranes is the combination of zwitterionic phospholipid with a phosphatidylcholine (PC) headgroup and negatively charged

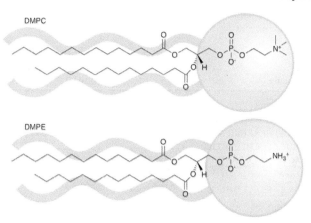

Fig. 1 Chemical structure of DMPC and DMPE phospholipids showing the difference in the head group from a choline group in DMPC to an amine in DMPE.

phospholipids with for example phosphatidylglycerol (PG) headgroups. These lipids are often chosen both because of commercial availability and because they are known to form relatively stable uniform lipid vesicles, and supported lipid bilayers with high coverage on for example mica or silica surfaces.[39] However, the use of PC to mimic the overall neutral part of the membrane, which in reality should be phosphatidylethanolamine (PE) in the case of most bacteria, has been debated in the literature.[40] The difference in the chemical structure of the PC and PE headgroup is the three methyl groups on the nitrogen forming a choline group in PC, which is exchanged for hydrogens in PE (Fig. 1). This seemingly small difference in the chemical structure results in a major increase in the phase transition temperature where the lipids in the membrane change from the gel phase to the liquid crystalline phase. While 14:0 PC (DMPC) exhibit a phase transition temperature of 24 °C, the equivalent 14:0 PE (DMPE) lipid shows transition at 50 °C.[41] This drastic increase may be explained by the PE head-group's ability to form hydrogen bonds in-between lipid in the membrane resulting in a more stable structure.

In this work, we have compared the structure of lipid bilayers consisting of 4 : 1 DMPC/DMPG or 4 : 1 DMPE/DMPG, and their interaction with the natural antimicrobial peptide, indolicidin.[42] Furthermore we present additional data for the lipid–peptide interaction with PE lipids for a wide range of natural AMPs: aurein 1.2,[43] LL-37,[44] lacticin Q[45] (details given in Table 1) and colistin[46,47] (negative control because of expected lack of interaction with cytoplasmic membranes) showing the effect of the peptides on the membrane structure using SAXS, and membrane dynamics using time resolved SANS.

Experimental

Materials and methods

Synthetic   DMPE   (1,2-dimyristoyl-sn-glycero-3-phosphoethanolamine, $C_{33}H_{66}NO_8P$), d54-DMPE (fully deuterated tail $(C_{13}D_{27})_2$, hydrogenated polar headgroup $C_7H_{12}NO_8P$), DMPG (1,2-dimyristoyl-sn-glycero-3-phospho-(10-rac-

Table 1 Overview of antimicrobial peptides

Peptide	Sequence	M_w (g mol^{-1})	Net charge at pH 7.0	Ratio of hydrophilic residues/total number of residues
Aurein 1.2	GLFDIIKKIAESF-CONH$_2$	1480	+1	38%
Indolicidin	ILPWKWPWWPWRR-CONH$_2$	1906	+4	23%
LL-37	LLGDFFRKSKEKIGKEFKRI VQRIKDFLRNLVPRTES	4493	+6	54%
Lacticin Q	MAGFLKVVQLLAKYGSKAVQWAWANKG KILDWLNAGQAIDWVVSKIKQILGIK	5898	+6	34%

glycerol) (sodium salt), $C_{34}H_{66}O_{10}P$), d54-DMPG (fully deuterated tail ($C_{13}D_{27}$)$_2$, hydrogenated polar headgroup $C_8H_{12}O_{10}P$) and DMPE–PEG (1,2-dimyristoyl-*sn*-*glycero*-3-phosphoethanolamine-*N*-[methoxy(polyethylene glycol)-2000], with M_w (PEG) = 2000 g mol^{-1}, *ca.* 45 CH_2CH_2O units) were purchased from Avanti Polar Lipids. The peptides indolicidin, aurein 1.2, LL-37 and lacticin Q were purchased from Schafen-N ApS, Copenhagen, while colistin was purchased from Sigma-Aldrich. The 50 mM Tris buffer was prepared by mixing Tris-base with Tris–HCl (Sigma Aldrich) to achieve a pH of 7.4 in either pure H_2O (MilliQ) for SAXS measurements or in 50% D_2O (Sigma-Aldrich) and 50% H_2O for TR-SANS measurements.

For DMPE/DMPG lipid vesicle preparation 75 mol% DMPE, 22.5 mol% DMPG and 2.5 mol% DMPE–PEG were dissolved in a 1 : 3 methanol : chloroform solution. The organic solvents were removed completely under vacuum using a Hei-dolph rotary evaporator with a Vacuubrand vacuum pump to prepare a thin lipid film. Then the film was hydrated with 50 mM Tris buffer, pH 7.4, for at least one hour at a temperature of 55 °C. After sonication using an ultrasonic bath for 10 minutes, the lipid dispersions were extruded through a 100 nm pore diameter Avanti polycarbonate filter (≥21 times) using an Avanti mini-extruder fitted with two 1 mL airtight syringes.

The antimicrobial peptides were dissolved in 50 mM Tris buffer, pH 7.4, to the desired concentration directly before the experiments.

Small angle X-ray scattering (SAXS)

The synchrotron SAXS data was collected at beamline P12 operated by EMBL Hamburg at the PETRA III storage ring (DESY, Hamburg, Germany).[48] The data was obtained using a radiation wavelength of 1.24 Å and a detector distance of 3.0 m, covering a Q range of 0.0032 Å^{-1} to 0.73 Å^{-1}, where Q is the magnitude of the scattering vector: $Q = \dfrac{4\pi}{\lambda}\sin\dfrac{\theta}{2}$, with θ the scattering angle. Data reduction was done automatically with the software available at the beam line and the 1D data were brought to absolute intensity scale using water as a primary standard. The SAXS results were analysed using the theoretical model described in the ESI.†[29,30] In short, the model provides a comprehensive description of the membrane by dividing it into probability functions for each component (lipid sub-units/peptide) across the bilayer. Error of scattering analysis is estimated to be ≤4%.

Time resolved small angle neutron scattering (TR-SANS)

TR-SANS experiments were performed on the D11 beamline at The Institut Laue-Langevin (ILL) facilities, Grenoble, France using detector distance of 20.5 m and a wavelength of 6 Å (fwhm 9%), covering a Q range of 0.002 Å^{-1} to 0.034 Å^{-1}. D-liposomes (vesicles consisting of lipids with deuterated tails) were mixed with H-liposomes (vesicles consisting of lipids with proteated tails) 1 : 1 directly before the first measurement using a Finntip micropipette and mixed with either pure buffer (to make sure concentration of the non-peptide samples compare to the peptide samples) or peptide solution 1 : 1. The samples were filled into Hellma quartz banjo-cells with a path length of 1 mm and kept in a temperature-controlled rack during the full experiment. The samples were measured with shorter measurement times and intervals in the beginning but increasing measurement times and longer intervals over time due to the descending contrast.

The TR-SANS data were evaluated by determining the relaxation function $R(t)$ according to:

$$R(t) = \sqrt{\frac{I(t) - I_\infty}{I(0) - I_\infty}} \tag{1}$$

where $I(t) = \int I(Q,t) \mathrm{d}Q$ is the integral intensity at a given time, I_∞ is the intensity of the premixed blend representing the final state and $I(0)$ is the averaged intensity of the H-vesicles and D-vesicles measured separately representing the initial state before exchange and flip-flop have taken place.

Results and discussion

Comparing PE and PC lipids in model systems

The SAXS results together with calculated electron density profiles of model membranes consisting of DMPE/DMPG (PE-vesicles) and DMPC/DMPG (PC-

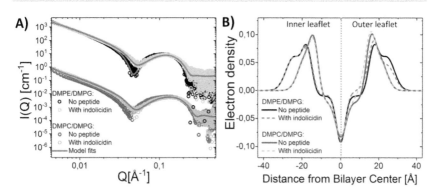

Fig. 2 (A) SAXS data comparing DMPE/DMPG and DMPC/DMPG[30] liposomes with and without addition of indolicidin (in a ratio of 1 : 20) at 37 °C plotted together with model fits. The data of DMPE/DMPG liposomes have been offset with a factor of 1000 to better display the data. (B) Electron density profiles calculated from the fit parameters of SAXS data shown in (A) (detailed fit parameters included in ESI Table S1† for DMPE/DMPG vesicles and in ref. 30 for DMPC/DMPG vesicles).

vesicles) (as previously published by Nielsen and co-workers[30]) at 37 °C are displayed in Fig. 2. From these data we can compare the structure of the pure lipid membrane as well as how the membrane is affected by adding an antimicrobial peptide. As seen from the scattering curves plotted in Fig. 2A comparing the scattering of the pure lipid vesicles, the shape of the scattering curve from the PE-vesicles differs from that of the PC-vesicles. This difference can be explained mainly by PE-vesicles having a thicker membrane than PC-vesicles at this temperature, as well as a contrast difference due to the smaller volume of the PE headgroup and incorporation of a slightly higher amount of PEGylated lipids to stabilise the system (5% DMPE–PEG in the case of PE vesicles contrary to only 2.5% DMPE–PEG in the case of PC-vesicles). To confirm that the increased amount of PEG does not affect the structure of the lipid bilayers, pure PE-vesicles with both 2.5 and 5% of PEGylated lipids have been analysed (see ESI Fig. S1†).

The estimated overall thickness of the DMPE/DMPG membrane based on model fits was found to be 45 Å (see Table S1 of the ESI†). Lee and co-workers have reported a membrane thickness for a bilayer with the same DMPE/DMPG mixture of 44 Å calculated from DPI measurements, but in these measurements the temperature was 28 °C (ref. 49) potentially explaining the slight difference. The increased thickness of the membrane of PE-vesicles when compared to the PC-vesicles is obvious in the electron density profile in Fig. 2B and can be explained by the difference in the phase transition temperature. As the lipids in the PE-vesicles are in the gel phase the membrane is more ordered resulting in tighter packing in the lateral direction, however upon melting to the liquid crystalline phase the lipids are more disordered resulting in a thinning of the membrane. Similar behaviour has been reported previously for PC lipids in studies where the thickness has been measured as a function of temperature.[50]

Overall the PE-vesicles were found to be less stable than the PC-vesicles upon addition of any positively charged substrates, with rapid fusion of vesicles and formation of multilamellar structures observed at higher concentrations. The difference in the molecular geometry of PE lipids in comparison to PC lipids provides an explanation of this behaviour. PE lipids prefer a slightly negative curvature resulting in a deformed membrane and formation of multilamellar structure.[40,51] Due to the incorporation of 5% PEGylated lipids (DMPE–PEG) in the membrane the unilamellar vesicles are more stable against fusion due to steric hindrance.[30] This allows us to quantitatively study the peptide–membrane interaction by analysing the individual bilayer structure using SAXS.

As seen, the scattering data for both lipid model systems (Fig. 2A) clearly show how addition of indolicidin in a lipid : peptide mol ratio of 20 : 1 results in a slight shift in the first minima to higher Q. This effect has previously been explained by Nielsen et al. as a change in the electron density of the membrane core upon insertion of the peptide in the bilayer.[13,29,30] Indolicidin is reported to insert in the outer leaflet of the membrane in the interface between the head and tail region of DMPC/DMPG membranes as seen by SAXS, SANS and NR,[13,30] an observation supported by molecular dynamics simulations.[52] As is evident from the electron density profile in Fig. 2B the same behaviour can be seen in the PE-vesicles where you see an increase in the electron density in the outer part of the tail region, close to the interface with the outer head group. However, contrary to the results reported for PC-vesicles[13,30] the volume of the head group is increased upon peptide addition. This is seen in the electron density as a change in the contrast of the outer headgroup. This change can be explained by the peptide disrupting the packing more for lipids in the gel phase, as well as potential destabilisation due to breakage of hydrogen bonds caused by the peptide insertion. Furthermore, the scattering profile from peptide–lipid mixtures in the case of PE-vesicles could not be explained solely by the insertion of peptide in the membrane as the case is for PC-vesicles. A slight solubilisation of the membrane was also observed with an estimate of 4% mixed peptide–lipid micelles present in the system after peptide addition. This may, as mentioned above, be related to the negative curvature strain on the membrane that favours destabilization and micelle formation.

Comparing data on a wide range of natural AMPs

To broaden the knowledge of the lipid interactions of AMPs in general, a range of natural peptides from various sources was mixed with the PE-vesicles mimicking the composition of E. coli membrane. The full results for varying amounts of peptide addition in the range from 1 : 20 to 1 : 100 peptide : lipid ratio are shown in Fig. 3 (fit for lacticin Q 1 : 20 is not shown as addition of this amount of peptide leads to formation of multilamellar structures preventing an accurate fit analysis with the model described in ref. 30). As seen from the scattering curves in Fig. 3A–D, all four peptides cause a similar slight shift in the first minima at intermediate Q, with increasing effects as a function of increasing concentration. This can as described above be explained by the shift in electron density of the hydrocarbon core region of the membrane upon insertion of peptide, due to the large contrast between the aliphatic tails and the peptides. In Fig. 3E–H the volume probability

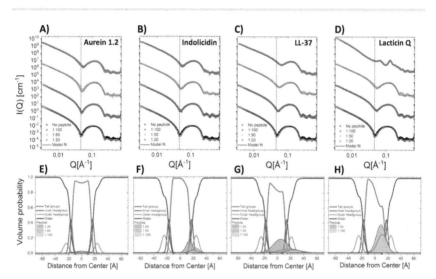

Fig. 3 SAXS data of peptide–lipid interactions of a wide range of natural AMPs: aurein 1.2 (A), indolicidin (B), LL-37 (C) and lacticin Q (D) mixed with E. coli mimicking lipid vesicles (DMPE/DMPG), together with model fits. The lipid : peptide ratio is indicated in the plot, and a line has been added at the minimum of the pure liposomes in order to visualise the shift upon peptide addition. Volume probability distribution for lipid membranes with addition of aurein 1.2 (E), indolicidin (F), LL-37 (G) and lacticin Q (H) showing the position of the peptide in the membrane.

plots calculated from the fit parameters of the SAXS data (parameters are included in Table S1 of the ESI†) are displayed. From these plots, the spatial distribution of the peptide within and at the membrane can be extracted.

Aurein 1.2 is the smallest α-helical peptide in this study with only 13 residues and the lowest overall cationic charge of +1 at pH 7. It is found to insert into the core of the hydrocarbon region of the membrane as seen in Fig. 3E. In addition to the deep insertion the peptide was found to have a pronounced detergent effect on the bilayer at the highest concentration (1 : 20) resulting in formation of around 48% mixed peptide–lipid micelles. Indications of a similar detergent effect of this peptide was reported by Fernandez and co-workers based on NMR and NR studies.[11] However, at lower concentrations at 1 : 50 and 1 : 100 ratios, only 5% and 0% mixed micelles were detected. This leaves room for a discussion on whether the solubilisation effects observed at the high peptide concentration provides a relevant explanation for the antimicrobial activity at physiological conditions which likely involve lower concentrations.

Indolicidin, is also a 13 residues peptide, but compared to aurein 1.2 it adopts an extended random coil structure in solution and a net charge of +4. As described in the section above, the indolicidin addition leads to a very slight solubilisation at the highest lipid : peptide ratio of 1 : 20. However, at all concentrations most of the peptide seems to insert into the membrane without any significant changes in the thickness or overall structure of the membrane. The position of the peptide in the membrane seems quite stable in the outer leaflet with only a slight deeper penetration with increasing concentration as seen in Fig. 3F. The differences in the penetration depth between the similarly sized aurein 1.2 and indolicidin can be explained by a variation in the number of charged and aromatic amino acids. Because of the high tryptophan content of indolicidin and the presence of the partially charged and bulky indole side group this peptide is more likely to position in the outer leaflet, in close proximity to the lipid–water interface.[53] For LL-37, a much bigger α-helical peptide with 37 residues a concentration dependent insertion is visible, with increasing penetration as a function of higher peptide : lipid ratio (Fig. 3G). Also, for this peptide a solubilisation effect was observed, with 1–5% formation of mixed micelles dependent on the concentration of peptide. The same effect has been reported in the past with PC-vesicles by both Sevcsik and co-workers[26,54] and Nielsen and co-workers[29] with almost full solubilisation of the membrane at high peptide concentrations.

Lacticin Q with 53 residues is the largest α-helical peptide and is according to reported literature much less studied than the other peptides included in the study. As seen in the volume probability plot in Fig. 3H, the peptide exhibits a similar insertion in the outer leaflet of the bilayer to the much smaller peptide indolicidin. However, the size of the peptide is much bigger resulting in a larger portion of the Gaussian distribution even at low peptide : lipid ratios. Interestingly even though the highest peptide ratio of 1 : 20 resulted in destabilisation of the lipids and eventually phase separation, the lower ratios shown here (1 : 50 and 1 : 100) did not lead to solubilisation of the membrane, and the changes in the scattering pattern can solely be explained by peptide insertion. As seen from the sequence presented in Table 1, lacticin Q has in the same way as indolicidin several aromatic tryptophans and tyrosines, as well as charged groups which support the similar peptide positioning in the outer head–tail interface as observed for indolicidin. The difference in the ability to solubilise the membrane between LL-37 and lacticin Q may be explained by LL-37 having fewer bulky aromatic groups and more charged amino acids (even though the net charge is the same) increasing the membrane solubilisation abilities.

As a negative control we used the natural cyclic peptide colistin, a commercially available peptide antibiotic that is known to have other targets than the cytoplasmic membrane, contrary to what is suggested for the other AMPs included in the study. Colistin is reported to rather affect the outer membrane through displacement of divalent cations and interaction with lipopolysaccharides (LPS), explaining its selectivity towards Gram-negative bacteria.[47,55,56] Alternatively, intracellular targets like peptide binding to ribosomes indicating passage through the outer and inner membrane has been presented.[57] However, although some scenarios have been presented where colistin was suggested to also target the inner membrane itself[58] our current results do not support this. The absence of any membrane interaction with colistin in the range of 1 : 10 to 1 : 100 was confirmed using SAXS. This can be seen by a perfect overlap of the measured peptide–lipid vesicle mixture and the calculated average in Fig. 4. These results confirm that our methodology is able to differentiate between the effect or lack thereof different cationic peptides.

The effect on exchange and flip-flop of phospholipids upon addition of peptides

In order to study the effect of the peptides on lipid dynamics, contrast variation TR-SANS measurements were performed. The method is illustrated in Fig. 5A and described in detail in ref. 29. As seen in Fig. 5B the reduced intensity, $R(t)$, proportional to the excess contrast (eqn (1)), decreases as a function of time for DMPE/DMPG lipid vesicles with and without addition of AMPs (see ESI Fig. S2† for raw SANS curves over time). The decrease in scattering intensity can be directly correlated to intervesicular exchange and intravesicular flip-flop of phospholipids.[29,59–61] The activation energy of lipid flip-flop of PE is estimated to be lower than for PC when both lipids are in the liquid crystalline phase due to a smaller lipid head group volume and a smaller hydration shell for PE lipids.[62] However, in our experiment the membrane is in the gel phase as the experiment is done at 37 °C and the phase transition temperature for the 4 : 1 DMPE/DMPG was estimated by DSC to be 44.7 °C (see ESI Fig. S3†). Trial experiments at 47 °C, above the

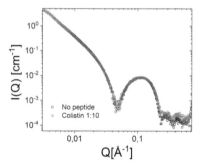

Fig. 4 Qualitative comparison of the measured data of DMPE/DMPG-vesicles mixed 10 : 1 with colistin to the calculated average where the scattering from the liposomes and the peptide have been measured separately and summed together.

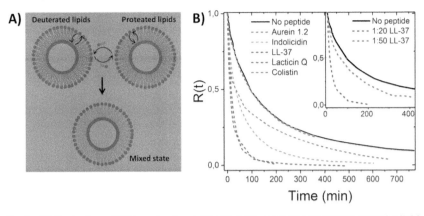

Fig. 5   (A) Illustration of H/D contrast variation technique used to study intravesicular lipid flip-flop and intervesicular lipid exchange by mixing deuterated and hydrogenated lipids in a zero-average contrast Tris buffer (mix of H_2O and D_2O to match the average contrast of the mixed liposomes). (B) Rate of contrast decay, $R(t)$, of pure DMPE/DMPG liposomes (solid line) and liposomes with added aurein 1.2 (1 : 50), indolicidin (1 : 20), LL-37 (1 : 20), lacticin Q (1 : 50) and colistin (1 : 10) (dotted lines) based on TR-SANS measurements (lipid : peptide ratios chosen based on the stability of liposome system). Inset graph highlights the concentration dependent effect of LL-37. All samples were measured at 37 °C.

phase transition temperature, were attempted but resulted in destabilization of the vesicles with formation of multilamellar structures over time both with and without addition of peptide. Nakano and co-workers have previously shown that the exchange and flip-flop rate constants can be extracted from TR-SANS data on pure PC-liposomes by fitting a double exponential decay function to the data.[60] However, this model was not able to explain the $R(t)$ of the DMPE/DMPG liposomes indicating that in this system there are more than two rates due to the mixture of the lipids.

As qualitatively seen from the $R(t)$ functions in Fig. 5B, all peptides except colistin were found to accelerate the lipid dynamics significantly. The same trend of lipid dynamics acceleration upon introduction to AMPs has previously been seen using PC based membranes.[27–29,63,64] Nielsen and co-workers previously reported lowering of the activation energy for lipid flip-flop upon addition of indolicidin after measuring 4 : 1 DMPC/DMPG lipid vesicles mixed with indolicidin 1 : 20 at four different temperatures.[29] For the DMPE/DMPG lipid vesicles, a combination of a lowering in the activation energy as well as partly solubilisation of the membrane as seen by SAXS provides an explanation of the observed effects. A potential peptide induced breakage of hydrogen bonding between PE lipids may play a significant role lowering the effective activation barrier.

As seen from Fig. 5B the largest peptide in the study, lacticin Q, seems to have the most pronounced acceleration of the peptides reaching the end state after only around 2 hours, even at the low peptide–lipid ratio of 1 : 50. While LL-37 shows the same degree of acceleration at 1 : 20, at 1 : 50 the effect on the dynamics is lower. A significant concentration dependence for LL-37 on the lipid dynamics was also observed for in PC-vesicles by Nielsen and co-workers,[29] and was related to the observed structural effect. The penetration of the peptide in the bilayer seems to increase progressively with increasing concentrations where the peptide at localized at the interface at low peptide addition whereas it assumes a more trans-membrane conformation at higher concentrations. Aurein 1.2 causes a dramatic acceleration of the lipid dynamics at higher concentrations than 1 : 50, reaching end-state outside the resolution of our experimental setup. This can be explained by partial solubilisation of the membrane as seen by SAXS. However, at 1 : 50 the effect is on the level of LL-37 at the same ratio showing a significant acceleration as compared to pure lipid vesicles. The same is seen for indolicidin, while colistin overlaps with the $R(t)$ curve of the pure membrane indicating no effect on the dynamics. When correlating these results with the absence of structural effects as seen by SAXS it is not surprising that colistin did not have any observed effect on the membrane dynamics. While all the peptides that were found to insert into the membrane by SAXS seem to also affect the lipid dynamics. Nguyen *et al.* found in their study on asymmetric lipid vesicles that peptides localized towards the surface of the membrane have a more pronounced effect on the dynamics than peptides that penetrate deeper into the membrane.[28] This supports the pronounced effect we see for lacticin Q and indolicidin, both found to position on the outer leaflet. However, this does not explain the dramatic effects seen for LL-37 and aurein 1.2 that seem to be able to penetrate deeper into the bilayer according to SAXS. Though, in the latter case the detergent effects observed specifically for PE-lipids may provide parts of the explanation. Due to these effects observed by SAXS, and the destabilisation effect seen upon increase in temperature, it becomes difficult to extract quantitative rates for exchange and flip-flop, and activation energy in the same way as done previously for PC-lipid systems using the same technique.[29,59–61] Thus, we can conclude by qualitative comparison that the same general trend of acceleration of lipid dynamics upon peptide addition can be observed also for lipid model membranes consisting of DMPE/DMPG lipids.

Conclusions

We have demonstrated how the effect of a range of different natural antimicrobial peptides on lipid membranes can be studied in detail using small angle scattering techniques. By using complex model systems, we can mimic the lipid composition of the cytoplasmic membrane of bacteria. However, SAXS/SANS results show how these systems with PE lipids are less stable upon peptide addition when comparing to frequently used model systems where the PE lipids are exchanged for PC lipids. Even though PC and PE both are zwitterionic and have a comparable structure, they vary considerably in phase transition temperature due to the PE lipids ability to form inter-molecular hydrogen bonds. This results in PE vesicles being more prone to undergo morphological transitions and formation of multilamellar structures. As seen using SAXS, addition of PEGylated PE lipids in the vesicles prevents the formation of multilamellar structures upon peptide addition, and in our case 5% of PEGylated PE enabled detailed studies of structure and dynamics of vesicles mixed with five different AMPs.

Based on modelling of SAXS data we were able to compare the insertion of aurein 1.2, indolicidin, LL-37, lacticin Q and colistin in the lipid membrane, and estimate the amount of solubilisation of the vesicles in the studied concentration range. The results show that four of the peptides insert into the membrane with a deeper insertion of aurein 1.2 and LL-37 (the latter also in a concentration dependent manner) than for indolicidin and lacticin Q. A fifth one, colistin, was not found to interact with the lipid vesicles. By using H/D contrast variation and TR-SANS we see that four out of five peptides have significant effect on the dynamics of the phospholipids, with a clear acceleration when comparing to the pure lipid system at 37 °C. This is not surprising as colistin is known to interact with the LPS outer membrane of Gram-negative bacteria. The data presented indicate that the larger peptides LL-37 and lacticin Q seem to have a more pronounced effect on the transport than the smaller peptides. However, the peptide–membrane interaction is very complex, and the effect on dynamics cannot easily be predicted from the structural interaction of the peptide or the peptide sequence independently, in support of previous data from Nielsen *et al.* on DMPC/DMPG lipids. The data do reveal that AMPs generally have profound impact on lipid dynamics that may have important consequences in the spatial distribution of lipids on the cytosolic and outer membrane side. In addition, the ion transport may be influenced by the enhanced lipid transport. These factors may in turn lead to lipid scrambling, signalling events or loss of net electrical potential which eventually lead to cell death.

Conflicts of interest

There are no conflicts to declare.

Acknowledgements

JEN, HJ and RL gratefully acknowledge NordForsk (project no. 82004) for financial support. We would like to thank DESY in Hamburg for allocated beam time at the P12 beam line, and Alexey Kikhney and Dmitry Molodenskiy for assistance in using the beamline. Further, the authors would like to thank ILL in Grenoble for allocation of beam time at D11 (DOI: 10.5291/ILL-DATA.8-02-869), and the PCSM lab for support during the SANS experiment. We acknowledge use of the Norwegian national infrastructure for X-ray diffraction and scattering (RECX). We would also like to thank Victoria Ariel Bjørnestad for her contribution to the SAXS experiments and Vitaliy Pipich for valuable discussions regarding the modelling of the SAXS data. We would also like to thank Abdullah Lone for input on choosing a representative selection of peptides for the study.

Notes and references

1 Y. Lai and R. L. Gallo, *Mod. Trends Immunol.*, 2009, **30**, 131–141.

2 M. Zasloff, *Nature*, 2002, **415**, 389.

3 H. Jenssen, P. Hamill and R. E. Hancock, *Clin. Microbiol. Rev.*, 2006, **19**, 491–511.

4 M. R. Yeaman and N. Y. Yount, *Pharmacol. Rev.*, 2003, **55**, 27–55.

5 A. A. Spector and M. A. Yorek, *J. Lipid Res.*, 1985, **26**, 1015–1035.

6 N. Shaw, *Adv. Appl. Microbiol.*, 1974, **17**, 63–108.

7 H. W. Huang, *Biochemistry*, 2000, **39**, 8347–8352.

8 D. Roversi, V. Luca, S. Aureli, Y. Park, M. L. Mangoni and L. Stella, *ACS Chem. Biol.*, 2014, **9**, 2003–2007.

9 L. A. Clifton, R. A. Campbell, F. Sebastiani, J. Campos-Terán, J. F. Gonzalez-Martinez, S. Björklund, J. Sotres and M. Cárdenas, *Adv. Colloid Interface Sci.*, 2020, **277**, 102118.

10 L. A. Clifton, S. A. Holt, A. V. Hughes, E. L. Daulton, W. Arunmanee, F. Heinrich, S. Khalid, D. Jefferies, T. R. Charlton and J. R. Webster, *Angew. Chem., Int. Ed.*, 2015, **54**, 11952–11955.

11 D. I. Fernandez, A. P. Le Brun, T. C. Whitwell, M.-A. Sani, M. James and F. Separovic, *Phys. Chem. Chem. Phys.*, 2012, **14**, 15739–15751.

12 D. I. Fernandez, A. P. Le Brun, T.-H. Lee, P. Bansal, M.-I. Aguilar, M. James and F. Separovic, *Eur. Biophys. J.*, 2013, **42**, 47–59.

13 J. E. Nielsen, T. K. Lind, A. Lone, Y. Gerelli, P. R. Hansen, H. Jenssen, M. Cárdenas and R. Lund, *Biochim. Biophys. Acta, Biomembr.*, 2019, **1861**, 1355–1364.

14 T. K. Lind, L. Darre, C. Domene, Z. Urbanczyk-Lipkowska, M. Cárdenas and H. Wacklin, *Biochim. Biophys. Acta, Biomembr.*, 2015, **1848**, 2075–2084.

15 T. K. Lind, M. W. Skoda and M. Cárdenas, *ACS Omega*, 2019, **4**, 10687–10694.

16 T. K. Lind, P. Zielinska, H. P. Wacklin, Z. Urbanczyk-Lipkowska and M. Cárdenas, *ACS Nano*, 2014, **8**, 396–408.

17 H. J. Askou, R. N. Jakobsen and P. Fojan, *J. Nanosci. Nanotechnol.*, 2008, **8**, 4360–4369.

18 J. E. Shaw, J.-R. Alattia, J. E. Verity, G. G. Privé and C. M. Yip, *J. Struct. Biol.*, 2006, **154**, 42–58.

19 K. Hall, T.-H. Lee, A. I. Mechler, M. J. Swann and M.-I. Aguilar, *Sci. Rep.*, 2014, **4**, 5479.

20 S. E. Blondelle, K. Lohner and M.-I. Aguilar, *Biochim. Biophys. Acta, Biomembr.*, 1999, **1462**, 89–108.

21 M.-A. Sani and F. Separovic, *Acc. Chem. Res.*, 2016, **49**, 1130–1138.

22 S. Zhu, M. A. Sani and F. Separovic, *Pept. Sci.*, 2018, **110**, e24061.

23 F. Porcelli, A. Ramamoorthy, G. Barany and G. Veglia, in *Membr. Protein.*, Springer, 2013, pp. 159–180.

24 G. Pabst, S. L. Grage, S. Danner-Pongratz, W. Jing, A. S. Ulrich, A. Watts, K. Lohner and A. Hickel, *Biophys. J.*, 2008, **95**, 5779–5788.

25 E. F. Semeraro, L. Marx, M. P. Frewein and G. Pabst, *Soft Matter*, 2021, **17**, 222–232.

26 E. Sevcsik, G. Pabst, W. Richter, S. Danner, H. Amenitsch and K. Lohner, *Biophys. J.*, 2008, **94**, 4688–4699.

27 M. Doktorova, F. A. Heberle, D. Marquardt, R. Rusinova, R. L. Sanford, T. A. Peyear, J. Katsaras, G. W. Feigenson, H. Weinstein and O. S. Andersen, *Biophys. J.*, 2019, **116**, 860–873.

28 M. H. Nguyen, M. DiPasquale, B. W. Rickeard, M. Doktorova, F. A. Heberle, H. L. Scott, F. N. Barrera, G. Taylor, C. P. Collier, C. B. Stanley, J. Katsaras and D. Marquardt, *Langmuir*, 2019, **35**, 11735–11744.

29 J. E. Nielsen, V. A. Bjørnestad, V. Pipich, H. Jenssen and R. Lund, *J. Colloid Interface Sci.*, 2021, **582**, 793–802.

30 J. E. Nielsen, V. A. Bjørnestad and R. Lund, *Soft Matter*, 2018, **14**, 8750–8763.

31 V. Castelletto, R. H. Barnes, K.-A. Karatzas, C. J. Edwards-Gayle, F. Greco, I. W. Hamley, R. Rambo, J. Seitsonen and J. Ruokolainen, *Biomacromolecules*, 2018, **19**, 2782–2794.

32 V. Castelletto, R. H. Barnes, K.-A. Karatzas, C. J. Edwards-Gayle, F. Greco, I. W. Hamley, J. Seitsonen and J. Ruokolainen, *Langmuir*, 2019, **35**, 1302–1311.

33 S. Qian and W. T. Heller, *J. Phys. Chem. B*, 2011, **115**, 9831–9837.

34 M. Pachler, I. Kabelka, M.-S. Appavou, K. Lohner, R. Vácha and G. Pabst, *Biophys. J.*, 2019, **117**, 1858–1869.

35 I. Kabelka, M. Pachler, S. Prévost, I. Letofsky-Papst, K. Lohner, G. Pabst and R. Vácha, *Biophys. J.*, 2020, **118**, 612–623.

36 A. K. Buck, D. E. Elmore and L. E. Darling, *Future Med. Chem.*, 2019, **11**, 2447–2460.

37 E. F. Semeraro, J. M. Devos, L. Porcar, V. T. Forsyth and T. Narayanan, *IUCrJ*, 2017, **4**, 751–757.

38 J. D. Nickels, S. Chatterjee, C. B. Stanley, S. Qian, X. Cheng, D. A. Myles, R. F. Standaert, J. G. Elkins and J. Katsaras, *PLoS Biol.*, 2017, **15**, e2002214.

39 A. Åkesson, T. Lind, N. Ehrlich, D. Stamou, H. Wacklin and M. Cárdenas, *Soft Matter*, 2012, **8**, 5658–5665.

40 K. Lohner and E. J. Prenner, *Biochim. Biophys. Acta, Biomembr.*, 1999, **1462**, 141–156.

41 J. R. Silvius, *Lipid-Protein Interact.*, 1982, **2**, 239–281.

42 M. E. Selsted, M. J. Novotny, W. L. Morris, Y.-Q. Tang, W. Smith and J. S. Cullor, *J. Biol. Chem.*, 1992, **267**, 4292–4295.

43 T. Rozek, K. L. Wegener, J. H. Bowie, I. N. Olver, J. A. Carver, J. C. Wallace and M. J. Tyler, *Eur. J. Biochem.*, 2000, **267**, 5330–5341.

44 J. Johansson, G. H. Gudmundsson, M. E. Rottenberg, K. D. Berndt and B. Agerberth, *J. Biol. Chem.*, 1998, **273**, 3718–3724.

45 K. Fujita, S. Ichimasa, T. Zendo, S. Koga, F. Yoneyama, J. Nakayama and K. Sonomoto, *Appl. Environ. Microbiol.*, 2007, **73**, 2871–2877.

46 P. Stansly and G. Brownlee, *Nature*, 1949, **163**, 611.

47 J. Li, R. L. Nation, R. W. Milne, J. D. Turnidge and K. Coulthard, *Int. J. Antimicrob. Agents*, 2005, **25**, 11–25.

48 C. E. Blanchet, A. Spilotros, F. Schwemmer, M. A. Graewert, A. Kikhney, C. M. Jeffries, D. Franke, D. Mark, R. Zengerle and F. Cipriani, *J. Appl. Crystallogr.*, 2015, **48**, 431–443.

49 T.-H. Lee, C. Heng, M. J. Swann, J. D. Gehman, F. Separovic and M.-I. Aguilar, *Biochim. Biophys. Acta, Biomembr.*, 2010, **1798**, 1977–1986.

50 N. Kučerka, M.-P. Nieh and J. Katsaras, *Biochim. Biophys. Acta, Biomembr.*, 2011, **1808**, 2761–2771.

51 R. M. Epand and R. F. Epand, *Biophys. J.*, 1994, **66**, 1450.

52 J. C. Hsu and C. M. Yip, *Biophys. J.*, 2007, **92**, L100–L102.

53 W.-M. Yau, W. C. Wimley, K. Gawrisch and S. H. White, *Biochemistry*, 1998, **37**, 14713–14718.

54 E. Sevcsik, G. Pabst, A. Jilek and K. Lohner, *Biochim. Biophys. Acta, Biomembr.*, 2007, **1768**, 2586–2595.

55 N. Paracini, L. A. Clifton, M. W. Skoda and J. H. Lakey, *Proc. Natl. Acad. Sci. U. S. A.*, 2018, **115**, E7587–E7594.

56 D. E. Santos, L. Pol-Fachin, R. D. Lins and T. A. Soares, *J. Chem. Inf. Model.*, 2017, **57**, 2181–2193.

57 M. J. Trimble, P. Mlynárčik, M. Kolář and R. E. Hancock, *Cold Spring Harbor Perspect. Med.*, 2016, **6**, a025288.

58 F. G. Dupuy, I. Pagano, K. Andenoro, M. F. Peralta, Y. Elhady, F. Heinrich and S. Tristram-Nagle, *Biophys. J.*, 2018, **114**, 919–928.

59 M. Nakano, *Chem. Pharm. Bull.*, 2019, **67**, 316–320.

60 M. Nakano, M. Fukuda, T. Kudo, H. Endo and T. Handa, *Phys. Rev. Lett.*, 2007, **98**, 238101.

61 M. Nakano, M. Fukuda, T. Kudo, N. Matsuzaki, T. Azuma, K. Sekine, H. Endo and T. Handa, *J. Phys. Chem. B*, 2009, **113**, 6745–6748.

62 R. Homan and H. J. Pownall, *Biochim. Biophys. Acta, Biomembr.*, 1988, **938**, 155–166.

63 M. Kaihara, H. Nakao, H. Yokoyama, H. Endo, Y. Ishihama, T. Handa and M. Nakano, *Chem. Phys.*, 2013, **419**, 78–83.

64 T. C. Anglin, K. L. Brown and J. C. Conboy, *J. Struct. Biol.*, 2009, **168**, 37–52.

Faraday Discussions

ROYAL SOCIETY
OF **CHEMISTRY**

PAPER

Caveolin induced membrane curvature and lipid clustering: two sides of the same coin?†

Shikha Prakash,‡ Anjali Krishna‡ and Durba Sengupta ⓘ *

Received 15th May 2020, Accepted 16th September 2020

DOI: 10.1039/d0fd00062k

Caveolin-1 (cav-1) is a multi-domain membrane protein that is a key player in cell signaling, endocytosis and mechanoprotection. It is the principle component of cholesterol-rich caveolar domains and has been reported to induce membrane curvature. The molecular mechanisms underlying the interactions of cav-1 with complex membranes, leading to modulation of membrane topology and the formation of cholesterol-rich domains, remain elusive. In this study, we aim to understand the effect of lipid composition by analyzing the interactions of cav-1 with complex membrane bilayers comprised of about sixty lipid types. We have performed a series of coarse-grain molecular dynamics simulations using the Martini force-field with a cav-1 protein construct (residue 82–136) that includes the membrane binding domains and a palmitoyl tail. We observe that cav-1 induces curvature in this complex membrane, though it is restricted to a nanometer length scale. Concurrently, we observe a clustering of cholesterol, sphingolipids and other lipid molecules leading to the formation of nanodomains. Direct microsecond timescale interactions are observed for specific lipids such as cholesterol, phosphatidylserine and phosphatidylethanolamine lipid types. The results indicate that there is an interplay between membrane topology and lipid species. Our work is a step toward understanding how lipid composition and organization regulate the formation of caveolae, in the context of endocytosis and cell signaling.

Introduction

Caveolae are highly curved membrane domains that are enriched in specific lipid components.[1,2] Several roles have been attributed to caveolar domains including lipid homeostasis, mechanoprotection and transport.[3,4] The primary lipid constituents of these nanodomains include cholesterol along with sphingomyelin

National Chemical Laboratory, Council of Scientific and Industrial Research, Dr. Homi Bhabha Road, Pune 411008, India. E-mail: d.sengupta@ncl.res.in

† Electronic supplementary information (ESI) available. See DOI: 10.1039/d0fd00062k

‡ Contributed equally to this work.

and saturated phospholipids.[5–7] Caveolin-1 (cav-1) is one of the main protein components of caveolae.[8] It is a dynamic, palmitoylated multi-domain protein that has been shown to co-localize with cholesterol.[9] Cav-1 has been implicated in cancer owing to its role in cell cycle regulation, as well as diseases such as lipodystrophy and pulmonary arterial hypertension.[2] Additionally, cav-1 has been attributed with an intrinsic propensity to interact with cholesterol[10] as well as generate curved membrane domains,[11] highlighting an integral role in forming caveolar domains.

A series of biochemical results have led to a consensus on the topological model of cav-1, mainly delineating four domains: the N-terminal domain (residues 1–81), the cav-1 scaffolding domain (CSD; residues 82–101), the intramembrane domain (IMD; 102–134), and the C-terminal domain (135–178).[12,13] Cav-1 is palmitoylated on three cysteine residues; one of which is toward the end of the IMD, and the remaining two at the C-terminal.[14] The full-length protein has not yet been structurally resolved due to the inherent problems of crystallizing lipidated membrane proteins and intrinsically disordered protein domains. However, circular dichroism (CD) and NMR (nuclear magnetic resonance) studies have characterized the IMD and CSD constructs of varying lengths and have the predicted structural propensity of the different domains.[12] Consensus reports validate that the central IMD domain has a helical hairpin structural motif which partially inserts into the membrane without completely traversing it.[15–18] The hairpin structure of the IMD has been shown to be stable by detailed molecular dynamics simulations.[17,19] The structural dynamics of the CSD appears to be membrane composition dependent with reports of both α-helical and β-sheet conformations. The CSD was shown to adopt a β-sheet conformation in cholesterol-containing membranes,[10,20] in contrast to an α-helix which has been observed in experiments involving membranes lacking cholesterol.[16–18,21,22] In molecular dynamics simulations, the helix in the CSD[19] has been reported to unfold in cholesterol-rich bilayers but not in simple phospholipid bilayers,[17] reiterating a cholesterol-dependent structure.

A high lipid sensitivity has been reported for cav-1 and it has been shown to insert into membranes in a cholesterol dependent manner. In fact, the scaffolding domain contains the putative cholesterol recognition CRAC motif (94–101),[23,24] which has been proposed to target proteins towards cholesterol-rich caveolar domains. However, it has been shown that CRAC containing cav-1 peptides of varying length differ in their potency to promote cholesterol-rich domains[10] suggesting a complex mechanism for cholesterol enrichment. We have previously shown using microsecond timescale coarse-grain simulations that cav-1 induces cholesterol clustering upon binding to 1,2-dipalmitoyl-sn-glycero-3-phosphocholine (DPPC)/cholesterol bilayers to stabilize cav-1 induced curvature.[25] Sphingolipids, especially sphingomyelins have been reported to co-localize in caveolar domains[6] and direct interactions have been reported.[26] Contradictory reports exist for glycosphingolipids, with some reports suggesting globotriaosylceramide (Gb3) is enriched in caveolae but not other glycosphingolipids, such as monosialotetrahexosylganglioside (GM1).[6,27] Previous simulations in sphingolipid-rich bilayers did not report any specific interactions with cav-1 [19] at low concentrations, although interactions were observed at high sphingomyelin concentrations.[28] Similarly, caveolae were reported to be enriched in saturated fatty acids (mainly palmitoyl C16:0 and stearoyl C18:0 chains).[29] In this context, we have previously shown that cav-1 interacts more favorably with DPPC bilayers, compared to 1-palmitoyl-2-oleoyl-*sn-glycero*-3-phosphocholine (POPC) and 1,2-dioleoyl-*sn-glycero*-3-phosphocholine (DOPC) bilayers.[30] Taken together, the data suggests that a unique interplay is seen between the lipid interactions of cav-1 and composition of caveolae, but molecular details are missing.

Here, we have performed coarse-grain simulations of a palmitoylated cav-1 construct with membrane bilayers representing mammalian plasma membranes. We show that cav-1 binding is a two-step process where the protein binding precedes the insertion of the palmitoyl tail. Upon binding, cav-1 interacts with specific lipids at a nanosecond to microsecond timescale. Interestingly, we observe that lipid species with negative intrinsic curvature such as cholesterol and sphingolipids cluster in the extracellular leaflet of the membrane. Increased interactions are observed with phosphatidylethanolamine (PE) lipids that cluster in the intracellular leaflet. Lipid clustering appears to be directly correlated to the small positive curvature induced in the intracellular leaflet by cav-1. Overall, our work is an important step towards understanding lipid-dependent membrane topology changes induced by cav-1.

Methods

Molecular dynamics simulations were performed with a palmitoylated cav-1 construct in the presence of a membrane bilayer, with the protein placed initially in the aqueous layer and not directly interacting with the membrane. The membrane and the protein were represented in the MARTINI coarse-grain representation (version 2). To better represent a eukaryotic cell membrane, we considered an asymmetric membrane bilayer with more than 60 lipid species. This bilayer composition is detailed below and has been previously considered as a general model of a realistic cell membrane.[31] The palmitoylated cav-1 protein construct used in the study encompasses residues 82 to 136 and has been characterized previously by coarse-grain simulations.[25] The details of the structural model are given below.

Cav-1 construct structural models

The cav-1 protein construct was taken from our previous study[25] and encompasses residues 82 to 136 (DGIWKASFTTFT VTKYWFYRLL SALFGIPMAL IWGIYFAIL SFLHIWAVVPCIRS). This stretch includes the entire membrane interacting domains (IMD and CSD). Due to the implementations of the Martini coarse-grain force-field (version 2), the secondary structure of the protein is required to assign the structure-dependent dihedral potentials between the backbone beads. Since the structure of this cav-1 construct has not been experimentally resolved, we considered a structural model based on domain-wise experimental data. The helical hairpin forms the core of the protein (IMD), and has been characterized by several experimental and computational studies. The maximum structural "uncertainty" is in the short stretch from 85 to 93, that has been reported to have both α-helical and β-sheet propensity by experimental CD and NMR studies. Since a β-sheet structure is predicted in experimental studies containing cholesterol, we consider only the β-sheet conformation for the CSD in this study. We have previously tested both the α-helical and β-sheet models and have shown that the β-sheet is important for membrane curvature and clustering in model membranes.

Structural modeling was performed in steps using in-house scripts to generate the coarse-grain topology of cav-1 by assigning the corresponding secondary structure and matching Martini bead types, using standard force constants and constraints. The IMD was individually modeled by the DAFT protocol[32] that assigns ideal helical stretches through geometric considerations. This construct has been previously validated and shown to reproduce experimental cholesterol-dependent binding to model membranes.[30] The CSD was similarly constructed coupling the ideal α-helix and ideal β-sheet geometries. The short and long elastic bonds (harmonic restraints) with a force constant of 2500 kJ mol^{-1} were used to model the β-sheet. The residues at the termini were modeled as having a coil structure without additional constraints to their backbone beads. The script is similar to the tool "Molmake.py" that is a part of the coarse-grain toolset and attempts to generate a structure from standard forcefield parameters, by mini-mizing initially semi random coordinates while slowly fading in nonbonded interactions. The different fragments were combined by modeling them in VMD,

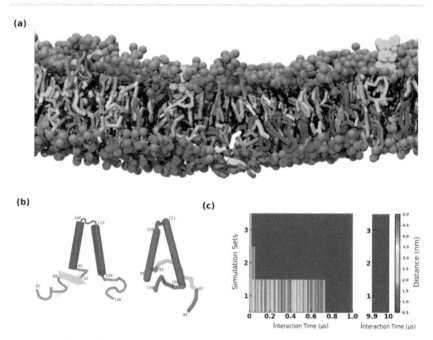

Fig. 1 A schematic representation of the cav-1 protein bound to multi-component cellular membranes. (a) The final binding mode of cav-1 in which it orients parallel to the bilayer surface, and the palmitoyl tail inserts into the bilayer. The cav-1 protein is shown in a ribbon representation in red (IMD) and green (CSD), and the palmitoyl tail is in white. The lipid components of the membrane are coloured in multiple colours and all head-groups are shown as blue spheres. The surrounding water is not shown for clarity. (b) The consensus atomistic structure from structural models of cav-1. (c) The minimum distance between cav-1 and the membrane during the course of the simulations. For clarity only the first 1 μs and the last 0.1 μs of the simulations are shown. The distance is color coded as shown in the scale bar.

and standard Martini backbone bonds were defined between them. The entire structure was minimized with the steepest descent algorithm, followed by short simulations in vacuum and then in water. The final structure of the cav-1 construct considered in the simulations is shown in Fig. 1.

To validate our coarse-grain Martini model of cav-1, we used the I-TASSER server[33] to predict the structure of cav-1 from its sequence spanning residues 82 to 136. The iterative threading assembly refinement (I-TASSER) server provides an automated pipeline of threading algorithms for predicting protein structure and function from amino acid sequences. We obtained five atomistic models, ranked by a scoring function (C-score). The helical hairpin motif was correctly predicted across three models. The secondary structure of the CSD varied and was predicted to be partially disordered with some β-sheet character in most models, and completely helical in others. The short N-terminal β-sheet in one of the predicted models overlaps with the amino acid stretch in the coarse-grain cav-1 CSD. We have shown the consensus structure of the I-TASSER cav-1 models as a snapshot with the varying CSD domains marked in Fig. 1.

Membrane composition

In this work, we have used a bilayer whose composition represents mammalian plasma membranes, and has been previously characterized by simulations.[31] A pictorial description is given in Fig. S1.† It consists of the major lipid types that have been reported in the plasma membrane, totalling 63 lipid types. The distribution and molar ratios of these lipids were matched to the idealized plasma membrane. Both bilayer leaflets, corresponding to the intracellular and extracellular leaflets, contain cholesterol, phosphatidylcholine (PC) lipids such as 1-palmitoyl-2-oleoyl-*sn-glycero*-3-phosphocholine (POPC) and 1,2-dioleoyl-*sn-glycero*-3-phosphocholine (DOPC) as well as phosphatidylethanolamine (PE) lipids *e.g.* 1,2-diarachidonoyl-*sn-glycero*-3-phosphoethanolamine (DAPE), 1,2-dioleoyl-*sn-glycero*-3-phosphoethanolamine (DOPE) in asymmetric ratios. The PC lipids were mostly present in outer leaflet (70%) and PE in the inner leaflet (80%). Several sphingomyelin (SM) lipid types were considered, such as *N*-stearoyl-D-*erythro*-sphingosylphosphorylcholine (DPSM) and XNSM (24:0, 24:1) that were predominantly present in the extracellular leaflet. In addition, the extracellular leaflet contains different glycosphingolipid (GM) *e.g.*, mono-sialotetrahexosylganglioside (GM1) lipids. The intracellular leaflet contains the charged lipids: phosphatidylserine (PS) *e.g.*, 1-palmitoyl-2-oleoyl phosphatidylserine (POPS), 1-stearoyl-2-arachidonoyl phosphatidylserine (PAPS), phosphatidic acid (PA), phosphatidylinositol (PI), and the PI-phosphate, -bisphosphate, and -trisphosphate (PIPs). The bilayer also consists of some important minor species: ceramide (CER), diacylglycerol (DAG) primarily in the extracellular leaflet (60–65%) and lysophosphatidylcholine (LPC) in the intracellular leaflet.

Simulation protocol

Molecular dynamics simulations were performed using the GROMACS program package (version Gromacs 2018). The Martini force-field (version 2.1) was used to describe the protein construct, lipids, and water with standard Martini parameters. The simulation box was of the order of 42 nm along the membrane surface and 18 nm normal to it, together with periodic boundary conditions. The temperature was coupled using the v-rescale algorithm (coupling time 0.1 ps) to a thermostat at 310 K. Replicate sets of simulations were performed with varying initial velocities randomly assigned from a Maxwell distribution at 300 K. The pressure was coupled using the Parrinello–Rahman semi-isotropic coupling algorithm at 1 bar pressure (coupling time 1.0 ps, compressibility 3×10^{-5} bar^{-1}). The nonbonded interactions were treated with a shift function from 0.0 to 1.2 nm for the Coulomb interactions, and 0.9 to 1.2 nm for the Lennard-Jones (LJ) interactions (pair-list update frequency of once per 10 steps). A time step of 20 fs was used for the simulations.

Analysis

Membrane interactions. To characterize the interactions between cav-1 and the membrane, we calculated the distance of closest approach (minimum distance) between the protein and the bilayer as a function of time. The "bound" regime is considered when the distance is below the cut-off distance of $\sim$0.5 nm. The depth of insertion was calculated as the distance from the bilayer center and the center of mass of cav-1. The orientation of cav-1 in the membrane was defined based on the angle between the membrane normal and each of its helices. The helix axis was defined as the axis joining the backbone atoms of residues 18 and 25 in helix 1 (IMD), and 34 and 41 in helix 2 (IMD). The inter-helical angle was calculated as the angle between the two helical axes.

Spatial distribution of lipids in the membrane plane. To calculate the spatial density distribution of cholesterol we used the *gmx spatial* routine in the GROMACS package. The cav-1 protein was centered in the membrane and the last 6 µs of each trajectory was analyzed. The bin size was set to 0.4 nm in the x and y directions. The values were averaged for each membrane leaflet, *i.e.* over the membrane leaflet directly interacting with cav-1 and the opposing (non-interacting) leaflet, and then averaged across the three replicate simulation sets.

Residue-wise total occupancy of lipids. A lipid molecule was defined to be bound to a particular amino acid residue if it were within the distance cut-off (0.55 nm) from that residue.[34] The total occupancy was calculated as the total time during which any lipid molecule was bound at a particular site. The occupancy was calculated over the "bound" regime and all values were normalized for simulation length, and averaged across replicate simulation sets.

Bilayer curvature. To calculate the mean curvature of a membrane leaflet we used the g_lomepro suite.[35] In this method, the curvatures are estimated by a Fourier transform of the lipid coordinates over time. In the current study, the cav-1 protein was centered and the last 1 µs of the trajectory was used to calculate the mean curvature individually for each set, and then averaged. The reference considered here is the center of the bilayer. In this definition, a membrane leaflet is defined as positive curvature which bends away from the mid plane (bilayer centre), and negative curvature implies it bends towards the bilayer centre. In regard to the full membrane, it is defined as positive curvature if it bends away from the mid plane towards the bound cav-1.

Results

To understand the molecular interactions between cav-1 and complex membranes, we performed coarse-grain molecular dynamics simulations of the cav-1 protein construct interacting with multicomponent bilayers. The membrane is comprised of more than 60 lipid types distributed asymmetrically in the extracellular and intracellular leaflets, and corresponds to a representative eukaryotic plasma membrane (Fig. S1†). During the simulations, the cav-1 protein that was initially placed in water diffused and interacted with the membrane interface within a nanosecond timescale. Representative snapshots of the membrane-bound state and the cav-1 protein construct are shown in Fig. 1a and b, respectively. The distance of closest approach between cav-1 and the membrane (minimum distance) was calculated to quantify the interactions. The values for the initial 1 μs and the last 0.1 μs are shown in Fig. 1c. Direct interactions between cav-1 and the bilayers (distance < 0.5 nm based on the coarse-grain force-field) are indicated by the dark blue stretches in the plot. In all three replicate simulation sets, the cav-1 protein interacted with the lower leaflet and a few close approaches were observed before the final bound regime was reached. The time to the final bound state was variable and ranged between 30–800 ns.

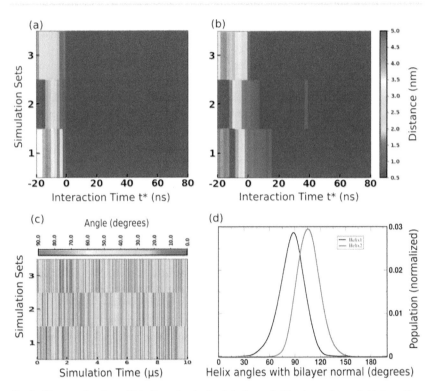

Fig. 2   The mechanism of binding of palmitoylated cav-1. We chose "pseudo" interaction time = 0 to represent the time that cav-1 first interacts with the bilayer (minimum distance < 0.5 as in Fig. 1). Panel (a) represents the minimum distance between the protein (not considering the lipid tail) and the membrane and (b) represents the minimum distance between palmitoyl tail of cav-1 and the membrane. The minimum distance has been calculated for 20 ns before and 80 ns after the binding of the peptide. The dark blue stretches (<0.5 nm) indicate a direct interaction. (c) Angle of the palmitoyl lipid tail with the membrane normal colour coded as in the scale bar. (d) Angle of the helices in IMD with the membrane normal.

Modes of cav-1-membrane binding

To delineate the mechanism of binding, we calculated the time evolution of interactions of the cav-1 protein and palmitoyl tail independently with the membrane. We consider a recalibrated time point ($t^* = 0$) as the time frame in which palmitoylated cav-1 binds to the membrane, as indicated by the distance of closest approach (Fig. 2a and b). The interaction time $t^* = 0$ differs from the actual simulation time between the sets but corresponds to the start of the bound regime in all sets. It is clear from the plots (Fig. 2a and b) that the protein interacts with the bilayer first (at $t^* = 0$), followed by the palmitoyl tail ($t^* > 0$) within a few nanoseconds. Again, a variable time of insertion is observed for the palmitoyl tail ranging from a few nanoseconds (set-3) to several tens of nanoseconds (set-2). The multiple close interactions that are observed initially are due to the protein contacts and the bound regime is observed only after insertion of the palmitoyl tail.

Upon binding, cav-1 orients parallel to the membrane interface, interacting mainly with the lipid headgroup moieties (Fig. 2c). The depth of insertion between cav-1 fluctuates but on at average lies about 2 nm (center of mass) from the bilayer center (Fig. S2†). Similarly, the angle between the two helices (from IMD) at the bilayer normal fluctuates around 90° (Fig. 2d). Although the IMD is parallel to the membrane surface, the palmitoyl tail is normal to the membrane surface. The palmitoyl tail appears to be quite flexible and samples several orientations in the membrane. In general the palmitoyl tail aligns parallel to the other acyl-chains (*i.e.* parallel to membrane normal) but may bend outwards (Fig. 2c). The helical hairpin domain of cav-1 samples several conformations in the bilayers. Both compact (angle between helical axes < 60°) and open (angle between helical axes > 90°) conformations were observed (Fig. S3†). The open conformations resemble the banana shape of several curvature-inducing proteins such as the BAR domain proteins. Overall, the cav-1 protein is surface aligned with the flexible palmitoyl tail inserting further into the membrane.

Distinct patterns of protein–lipid interactions

During the course of the simulations, multiple protein–lipid binding/unbinding events are observed and the interaction time of the different lipid species ranges from nanosecond to microsecond time scales. To analyze these direct interactions between cav-1 and the membrane components, we calculated the contacts between cav-1 and all lipid families (PC, PE, PS, SM *etc.*). Based on the preliminary data, we chose a few representative lipid species and calculated the occupancy time of these lipids (Fig. 3). The occupancy time is defined as the total time a lipid species is bound to the protein, normalized to the simulation time. At a residue-wise level, lipid-specific distinct patterns are observed for each of the residues. The highest occupancy time for all lipids is observed for the palmitoylated cysteine residue.

Cholesterol interacts with the protein almost throughout the simulation time and exhibits the highest occupancy time amongst all lipids. High occupancy times are observed at the IMD, especially near the turn motif of the helical hairpin structure. Interactions with the CSD are lower, but direct interactions are observed at the CRAC motif (residues 94–101 KYWFYRLL). Upon decomposing these interactions, we find that on an average four molecules of cholesterol from

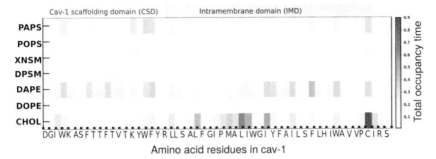

Fig. 3 The total occupancy time of representative lipid species at each residue of the cav-1 construct considered in the study. The occupancy time is defined as the total time where any lipid molecule is bound to a residue (minimum-distance cut-off of 0.55 nm). The values have been normalized for all simulation lengths, and averaged over the three replicate sets. The values are color coded as per the scale bar on the right. The panels correspond to cholesterol (CHOL), DOPE, DAPE, DPSM, XNSM, POPS and PAPS.

the lower leaflet interact directly with cholesterol at a given time. In addition, cholesterol molecules from the outer leaflet or one which is flipped in may also interact. High occupancy times were also observed for PE and PS lipids. Interestingly, a lipid dependence was observed amongst these lipid types. For instance in the PE lipids, DAPE interacts longer with cav-1 than DOPE. Similarly for the PS lipids, the total occupancy of PAPS lipids are much higher than that of POPS. Surprisingly, the total time of interactions with sphingomyelin molecules (represented by DPSM and XNSM lipids) is low although interactions are observed with the palmitoylated Cys residue. To further investigate, we calculated the occupancy times considering all SM lipid types and cav-1 as whole (Fig. S4†). We observe increased occupancy times suggesting that interactions of SM with cav-1 are dynamic with multiple binding and unbinding events, both in terms of SM species and residues involved. We were unable to identify prolonged interactions with the remaining lipids. Overall, we are able to discern specific lipid- and residue-dependent interactions between cav-1 and the constituents of the complex membrane.

Clustering and redistribution of lipids

The most important modulation of the membrane upon cav-1 binding is the phase separation or clustering of several membrane components. We calculated the time-averaged spatial distribution of the different lipid species in the two leaflets, centered around the cav-1 protein (Fig. 4 and 5). Two opposing effects are observed in the clustering behavior. The first set of lipid types, mainly cholesterol and sphingomyelins are enriched in the extracellular leaflet exactly opposing cav-1 that is bound in the intracellular leaflet. The second set of lipids, such as PE and PS are enriched in the intracellular leaflet around cav-1 but excluded from the opposing extracellular leaflet. Glycosphingolipid clusters are observed but are not correlated to cav-1 interactions. The remaining lipids do not appear to cluster and a uniform distribution was observed. We discuss below the detailed behavior of the representative lipids.

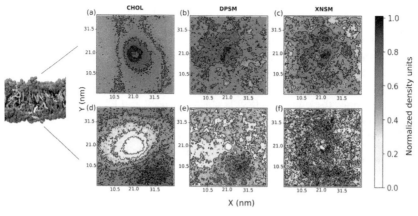

Fig. 4   The spatial density distribution of cholesterol and sphingomyelin molecules in cav-1 bound membrane bilayer. The densities were independently calculated for the leaflet bound to cav-1 (interacting) and the opposing leaflet. The average cholesterol density for (a–c) the opposing leaflet and (d–f) the interacting leaflet for (a and d) cholesterol (CHOL), (b and e) DPSM and (c and f) XNSM. The distribution corresponds to the membrane plane and the values are averaged over the bilayer normal independently for the two membrane leaflets, *i.e.* directly interacting with cav-1 and opposing it. The data shown here has been averaged across the three replicate simulation sets and color coded as shown in the scale bars. The left panel shows the representative snapshots of the membrane bound cav-1 for clarity.

The most striking effect observed is the redistribution of cholesterol molecules in the extracellular leaflet to form a high-cholesterol density domain. A distinct cholesterol nano-domain is observed centered around the protein in the opposing leaflet. The time-averaged density of cholesterol around cav-1 in the intracellular

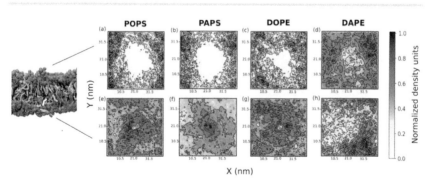

Fig. 5   The spatial density distribution of cholesterol and sphingomyelin molecules in cav-1 bound membrane bilayer. The densities were independently calculated for the leaflet bound to cav-1 (interacting) and the opposing leaflet. The average cholesterol density for (a–d) the opposing leaflet and (e–h) the interacting leaflet for (a and e) POPS, (b and f) PAPS and (c and g) DOPE (d and h) DAPE. The distribution corresponds to the membrane plane and the values are averaged over the bilayer normal independently for the two membrane leaflets, *i.e.* directly interacting with cav-1 and opposing it. The data shown here has been averaged across the three replicate simulation sets and color coded as shown in the scale bars. The left panel shows the representative snapshots of the membrane bound cav-1 for clarity.

leaflet is reduced. Similarly, a clustering of the sphingomyelins was observed mainly in the extracellular leaflet. Amongst the sphingomyelins those with saturated lipids such as DPSM (16:0 or 18:0) cluster predominantly in the extracellular leaflet. Clusters of sphingolipids such as XNSM (24:1) with unsaturated tails are observed in both leaflets, although it is more prominent in the extracellular leaflet.

Interestingly, the opposite effect is observed in PE and PS lipids. In the lower leaflet, we observed a redistribution of the charged PS lipids around cav-1. It should be noted that the construct of cav-1 considered in the study is positively charged due to the large number of lysine and arginine residues in the IMD and CSD. However, a lipid-dependent clustering is observed around cav-1 suggesting it is not just a simple electrostatic interaction. For instance, the clustering of PAPS is the maximum amongst all PS lipids. Lipid-dependent effects are observed in PE lipids as well. Lipids such as DOPE are observed to cluster in the vicinity of cav-1 in the intracellular leaflet, and are absent in the opposing extracellular leaflet. The distribution of polyunsaturated PE lipids such as DAPE does not seem to increase in the vicinity of cav-1 in either leaflet. The distribution of the other lipids appears to be independent of cav-1.

Cav-1 binding induces low curvature around the protein

Upon binding, cav-1 appears to induce a local curvature in the vicinity of the protein. We centered the trajectories around cav-1 and calculated a 2D profile of local mean curvature along the plane of the bilayer in the upper and lower leaflet (Fig. 6). A nanometer-size positive curvature domain is observed in the intracellular leaflet encompassing the bound cav-1 protein. An opposing negative curvature domain is observed in the extracellular leaflet opposing the bound cav-1. The curved domains are observed in all simulation sets and appear to persist during the course of the simulations. Several fluctuations are observed in the membrane further from cav-1 that randomly vary during the simulations. The values of curvature reported here appear rather small but need to be compared with systems of similar topology. As a comparison, the curvature observed for a completely spherical triolen lens in bilayer[36] was reported to be ± 0.1 nm^{-1}. Similarly, for SNARE mediated fusion,[37] curvature of a vesicle was reported to be ± 0.1 nm^{-1}. Our results indicate that a single cav-1 protein is able to generate curvature in large complex membranes but oligomerization and multiple copy numbers would be required for large-scale topology changes.

Discussion

Caveolar domains are critical for cellular processes such as mechanoprotection, lipid homeostatis and trafficking. Although critical for its function, the molecular interactions between cav-1 and the constituent lipids of these caveolar domains remains unclear. Here, we have delineated several direct and indirect interactions of cav-1 within complex membranes using coarse-grain molecular dynamics simulations. We show a striking redistribution of specific lipid species in the membrane that are correlated with the positive curvature induced in the intracellular leaflet in the vicinity of cav-1. We are also able to identify unique patterns

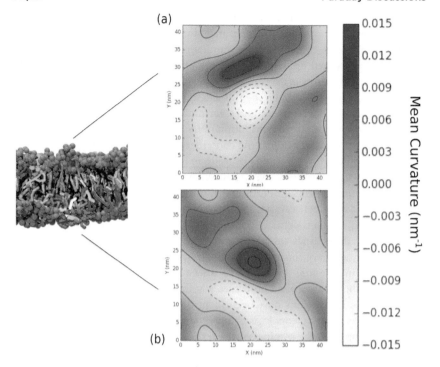

Fig. 6 The mean curvature of membrane upon cav-1 binding. The panels correspond to the (a) extracellular and (b) intracellular leaflets. The dark blue regions represent positive membrane curvature, and the yellow regions represent a negative curvature. The curvature is calculated along the membrane surface and color coded as indicated by the scale bar. The plots correspond to the mean curvature calculated across the last 1 μs of a single representative simulation set. The corresponding representative snapshot of the membrane-bound cav-1 is shown in leftmost side.

of lipid signatures on cav-1 and specific residue-wise interactions have been identified for a few lipid species.

One of our main findings is that cholesterol modulates cav-1 dynamics by both direct and indirect effects. In particular, we observe the clustering or redistribution of cholesterol in the extracellular leaflet opposing the bound cav-1. This clustering of cholesterol as well as sphingomyelin in the extracellular leaflet appears to help maintain the curvature induced upon binding of cav-1. For simpler systems we have shown that cholesterol clustering and curvature are intricately related.[25] In general, this interplay between membrane curvature and lipid distribution is still an active field of research[38-40] and it is not evident whether membrane curvature energetics can overcome mixing entropy leading to lipid sorting. For thin tubes pulled from the membrane, the mixing entropy is higher than the bending penalty and lipid sorting was not observed to be related to curvature.[41,42] However, the energy compensation could be system-dependent and it has been shown that certain bacterial membrane proteins induce lipid sorting while others do not.[43] Similarly, cardiolipin was reported to oligomerize and concentrate in negatively curved regions of membranes and it was suggested that membrane shape alone can modulate the distribution of cardiolipin without the involvement of any cellular protein machinery.[44]

Another interesting outcome of the simulations is the clustering of specific PS and PE lipids. Charged lipids such as PS are observed to cluster in the intracellular leaflet in the vicinity of cav-1 interacting with the positively charged IMD and CSD domains of cav-1. Although the full-length cav-1 has a lower charge than the construct considered here, the N- and C-terminal domains mainly lie outside the membrane. As a result, the redistribution of PS lipids in the vicinity of cav-1 may be a feature of cav-1, and not just the construct considered in this study. In fact, cav-1 interaction with PS lipids have been recently observed in cellular membranes.[45] Counterintuitively, a few PE lipids cluster on the intracellular leaflet despite their intrinsic negative curvature. In addition, there appears to be no direct correlation between occupancy, reporting on specific interactions and lipid clustering, a non-specific effect. A complex combination of the direct and indirect effects determines cav-1 interactions with membrane lipids.

In this study, we have chosen the Martini coarse-grain forcefield to analyze cav-1 membrane interactions owing to its successful implementation in understanding membrane associated processes. For instance, several protein–lipid and protein–protein interactions have been identified that have been experimentally validated.[46,47] Membrane topology changes such as in fusion have been well studied, starting from the seminal work on the mechanism of vesicle fusion that was shown to be in good agreement with experiments.[48] Since then the Martini force-field has been instrumental in providing molecular insights into membrane fusion[49,50] and the effect of fusogenic molecules[51] and proteins.[37,52,53] Likewise, the Martini force-field has been commonly used to probe membrane topology changes governed by membrane composition[54] and cholesterol distribution.[55] Phase separation of multicomponent membranes has been reproduced,[56] although with caveats and limitations due to the coarse-grain nature of the force-field.[57] Studies using Martini simulations have been able to reproduce experimental neutron scattering profiles on interactions between cholesterol and polyunsaturated fatty acid containing lipids.[58,59] The Martini force-field has also been used to study cav-1, and simulations have shown that cav-1 induces a positive curvature around the edges of the spontaneously aggregated triolein lens in a flat lipid bilayer.[36] Using similar methodologies, we have been able to reproduce experimental trends on depth of binding of a shorter cav-1 construct.[30] In contrast, super coarse-grain methods and meso-scale simulations have been seminal in understanding curvature induced by the BAR protein family.[60–62] Cav-1 does indeed exhibit the typical banana-shape protein and shallow membrane insertion. In fact, we have previously shown that such a conformation is required to maintain membrane curvature.[25] However, no lipid-dependence has been reported in these super coarse-grain approaches and these methods may lack the chemical resolution required. Overall, a coarse-grain model such as the Martini model is well suited to analyze caveolar domains, specifically the interplay between curvature and lipid clustering, while being computationally feasible.

Large strides have been made in our understanding of how the peripheral protein interactions affect membrane topology as well as the lateral distribution of membrane components, but general mechanisms are still missing.[63] In the specific case of caveolar domains, our work and others have identified unique lipid interactions, but how these lipids are sub-distributed between caveolae bulb and the funnel region still needs experimental testing. In fact, although topology and lipid de-mixing are related, we still need to identify the exact scenarios when bending modulus can be counteracted by mixing entropy, and when not. Another interesting problem is interleaflet coupling. Theoretical studies suggest that it coordinates the organization of molecules in two leaflets but further work is needed to quantify the effect.[64] An integrative approach with complementary theoretical, simulation and experimental studies could the provide crucial link for the dynamic, non-equilibrium membrane structure.

In conclusion, we have analyzed the unique interactions of cav-1 with specific lipids using comprehensive coarse-grain simulations. We are able to discern both direct cholesterol interactions as well indirect cholesterol clustering. Consequently, our results indicate that the enrichment of cholesterol in caveolae is driven by both cav-1 interactions and phase separation of domains with varying fluidity and topology. Further, our results identified indirect effects of sphingomyelin and PS clustering as well as direct interactions with PE and PS lipids that have not been reported earlier. This work is an important step towards a molecular picture of how cav-1 interacts with complex membranes and forms phase-separated highly curved nanodomains.

Author contributions

AK, SP and DS designed the research, performed simulations, analyzed and wrote the article.

Conflicts of interest

There are no conflicts of interest to declare.

Acknowledgements

AK and DS acknowledges the Dept. Science Technology (DST, (Govt. of India) Grant: EMR/2016/002294) and SP gratefully acknowledges the Dept. Biotechnology (DBT, Govt. of India) for the BioCARe grant for funding. We thank members of our research group for critically reading the manuscript.

References

1 E. J. Smart, G. A. Graf, M. A. Mcniven, W. C. Sessa, J. A. Engelman, P. E. Scherer, T. Okamoto and M. P. Lisanti, Caveolins, liquid-ordered domains and signal transduction, *Mol. Cell. Biol.*, 1999, **19**, 7289–7304.

2 R. G. Parton, M. A. del Pozo, S. Vassilopoulos, I. R. Nabi, S. Le Lay, R. Lundmark, A. K. Kenworthy, A. Camus, C. M. Blouin, W. C. Sessa, *et al.*, Caveolae: The FAQs, *Traffic*, 2020, **21**, 181–185.

3 R. G. Parton, Caveolae: Structure, function, and relationship to disease, *Annu. Rev. Cell Dev. Biol.*, 2018, **34**, 111–136.

4 C. A. Han, B. a. Copeland, A. Tiwari and A. K. Kenworthy, Assembly and turnover of caveolae: What do we really know?, *Front. Cell Dev. Biol.*, 2016, **4**, 68.

5 C. J. Fielding and P. E. Fielding, Cholesterol and caveolae: Structural and functional relationships, *Biochim. Biophys. Acta*, 2000, **1529**, 210–222.

6 S. Sonnino and A. Prinetti, Sphingolipids and membrane environments for caveolin, *FEBS Lett.*, 2009, **583**, 597–606.

7 U. Örtegren, M. Karlsson, N. Blazic, M. Blomqvist, F. H. Nystrom, J. Gustavsson, P. Fredman and P. Strålfors, Lipids and glycosphingolipids in caveolae and surrounding plasma membrane of primary rat adipocytes, *Eur. J. Biochem.*, 2004, **271**, 2028–2036.

8 T. M. Williams and M. P. Lisanti, The caveolin proteins, *Genome Biol.*, 2004, **5**, 214.

9 E. Ikonen, S. Heino and S. Lusa, Caveolins and membrane cholesterol, *Biochem. Soc. Trans.*, 2004, **32**, 121–123.

10 G. Yang, H. Xu, Z. Li and F. Li, Interactions of caveolin-1 scaffolding and intramembrane regions containing a CRAC motif with cholesterol in lipid bilayers, *Biochim. Biophys. Acta*, 2014, **1838**, 2588–2599.

11 N. Ariotti, J. Rae, N. Leneva, C. Ferguson, D. Loo, S. Okano, M. M. Hill, P. Walser, B. M. Collins and R. G. Parton, Molecular characterization of caveolin-induced membrane curvature, *J. Biol. Chem.*, 2015, **290**, 24875–24890.

12 K. T. Root, S. M. Plucinsky and K. J. Glover, Recent progress in the topology, structure, and oligomerization of caveolin: A building block of caveolae, *Curr. Top. Membr.*, 2015, **75**, 305–336.

13 S. Monier, R. G. Parton, F. Vogel, J. Behlke, A. Henske and T. V. Kurzchalia, VIP21-caveolin, a membrane protein constituent of the caveolar coat, oligomerizes in vivo and in vitro, *Mol. Biol. Cell*, 1995, **6**, 911–927.

14 T. Okamoto, A. Schlegel, P. E. Scherer and M. P. Lisanti, Caveolins, a family of scaffolding proteins for organizing "preassembled signaling complexes" at the plasma membrane, *J. Biol. Chem.*, 1998, **273**, 5419–5422.

15 S. Aoki, A. Thomas, M. Decaffmeyer, R. Brasseur and R. M. Epand, The role of proline in the membrane re-entrant helix of caveolin-1, *J. Biol. Chem.*, 2010, **285**, 33371–33380.

16 J. Lee and K. J. Glover, The transmembrane domain of caveolin-1 exhibits a helix-break-helix structure, *Biochim. Biophys. Acta*, 2012, **1818**, 1158–1164.

17 H. Rui, K. T. Root, J. Lee, K. J. Glover and W. Im, Probing the U-shaped conformation of caveolin-1 in a bilayer, *Biophys. J.*, 2014, **6**, 1371–1380.

18 G. Yang, Z. Dong, H. Xu, C. Wang, H. Li, Z. Li and F. Li, Structural study of caveolin-1 intramembrane domain by circular dichroism and nuclear magnetic resonance, *Biopolym. - Pept. Sci. Sect.*, 2015, **104**, 11–20.

19 H. Liu, L. Yang, Q. Zhang, L. Mao, H. Jiang and H. Yang, Probing the structure and dynamics of caveolin-1 in a caveolae-mimicking asymmetric lipid bilayer model, *Eur. Biophys. J.*, 2016, **45**, 511–521.

20 C. L. Hoop, V. N. Sivanandam, R. Kodali, M. N. Srnec and P. C. A. Van Der Wel, Structural characterization of the caveolin scaffolding domain in association with cholesterol-rich membranes, *Biochemistry*, 2012, **51**, 90–99.

21 C. Le Lan, J.-M. Neumann and N. Jamin, Role of the membrane interface on the conformation of the caveolin scaffolding domain: A CD and NMR study, *FEBS Lett.*, 2006, **580**, 5301–5305.

22 C. Le Lan, J. Gallay, M. Vincent, J. M. Neumann, B. De Foresta and N. Jamin, Structural and dynamic properties of juxta-membrane segments of caveolin-1 and caveolin-2 at the membrane interface, *Eur. Biophys. J.*, 2010, **39**, 307–325.

23 M. Murata, J. Peranen, R. Schreinert, F. Wielandt, T. V. Kurzchalia and K. Simons, VIP21/caveolin is a cholesterol-binding protein, *Proc. Natl. Acad. Sci. U. S. A.*, 1995, **92**, 10339–10343.

24 R. M. Epand, B. G. Sayer and R. F. Epand, Caveolin scaffolding region and cholesterol-rich domains in membranes, *J. Mol. Biol.*, 2005, **345**, 339–350.

25 A. Krishna and D. Sengupta, Interplay between membrane curvature and cholesterol: Role of palmitoylated caveolin-1, *Biophys. J.*, 2019, **116**, 69–78.

26 P. Haberkant, O. Schmitt, F. X. Contreras, C. Thiele, K. Hanada, H. Sprong, C. Reinhard, F. T. Wieland and B. Britta Brügger, Protein-sphingolipid interactions within cellular membranes, *J. Lipid Res.*, 2008, **49**, 251–262.

27 S. Sonnino, S. Prioni, C. Vanna and A. Prinetti, Interactions between caveolin-1 and sphingolipids, and their functional relevance, *Adv. Exp. Med. Biol.*, 2012, 97–115.

28 A. Krishna, S. Prakash and D. Sengupta, Sphingomyelin effects in caveolin-1 mediated membrane curvature, *J. Phys. Chem. B*, 2020, **124**, 5177–5185.

29 Q. Cai, L. Guo, H. Gao and X.-A. Li, Caveolar fatty acids and acylation of caveolin-1, *PLoS One*, 2013, **8**, e60884.

30 D. Sengupta, Cholesterol modulates the structure, binding modes, and energetics of caveolin-membrane interactions, *J. Phys. Chem. B*, 2012, **116**, 14556–14564.

31 H. I. Ingólfsson, M. N. Melo, F. J. Van Eerden, C. Arnarez, C. A. Lopez, T. A. Wassenaar, X. Periole, A. H. De Vries, D. P. Tieleman and S. J. Marrink, Lipid organization of the plasma membrane, *J. Am. Chem. Soc.*, 2014, **136**, 14554–14559.

32 T. A. Wassenaar, K. Pluhackova, A. Moussatova, D. Sengupta, S. J. Marrink, D. P. Tieleman and R. A. Böckmann, High-throughput simulations of dimer and trimer assembly of membrane proteins. The DAFT approach, *J. Chem. Theory Comput.*, 2015, **11**, 2278–2291.

33 J. Yang, R. Yan, A. Roy, D. Xu, J. Poisson and Y. Zhang, The I-TASSER suite: protein structure and function prediction, *Nat. Methods*, 2014, **12**, 7–8.

34 D. Sengupta and A. Chattopadhyay, Identification of cholesterol binding sites in the serotonin1A, *J. Phys. Chem. B*, 2012, **116**, 12991–12996.

35 V. Gapsys, B. L. De Groot and R. Briones, Computational analysis of local membrane properties, *J. Comput.-Aided Mol. Des.*, 2013, **27**, 845–858.

36 W. Pezeshkian, G. Chevrot and H. Khandelia, The role of caveolin-1 in lipid droplets and their biogenesis, *Chem. Phys. Lipids*, 2018, **211**, 93–99.

37 H. J. Risselada, C. Kutzner and H. Grubmuller, Caught in the act: Visualization of SNARE-mediated fusion events in molecular detail, *ChemBioChem*, 2011, **12**, 1049–1055.

38 S. Baoukina, H. I. Ingólfsson, M. J. Siewert and D. P. Tieleman, Curvature-induced sorting of lipids in plasma membrane tethers, *Adv. Theory Simul.*, 2020, **1**, 1800034.

39 X. Woodward, E. E. Stimpson and C. V. Kelly, Single-lipid tracking on nanoscale membrane buds: The effects of curvature on lipid diffusion and sorting, *Biochim. Biophys. Acta, Biomembr.*, 2018, **1860**, 2064–2075.

40 P. Sengupta and J. Lippincott-Schwartz, Revisiting membrane microdomains and phase separation: a viral perspective, *Viruses*, 2020, **12**, E745.

41 B. Sorre, A. Callan-Jones, J. B. Manneville, P. Nassoy, J.-F. Joanny, J. Prost, B. Goud and P. Bassereau, Curvature-driven lipid sorting needs proximity to

a demixing point and is aided by proteins, *Proc. Natl. Acad. Sci. U. S. A.*, 2009, **106**, 5622–5626.

42 A. Tian and T. Baumgart, Sorting of lipids and proteins in membrane curvature gradients, *Biophys. J.*, 2009, **96**, 2676–2688.

43 P. C. Hsu, M. F. Samsudin, J. Shearer and S. Khalid, It is complicated: curvature, diffusion and lipid sorting within the two membranes of Escherichia coli, *J. Phys. Chem. Lett.*, 2017, **8**, 5513–5518.

44 E. Beltrán-Heredia, F.-C. Tsai, S. Salinas-Almaguer, F. J. Cao, P. Bassereau and F. Monroy, Membrane curvature induces cardiolipin sorting, *Commun. Biol.*, 2019, **2**, s42003.

45 Y. Zhou, N. Ariotti, J. Rae, H. Liang, V. Tillu, S. Tee, M. Bastiani, A. T. Bademosi, B. M. Collins, F. A. Meunier, *et al.*, Dissecting the nanoscale lipid profile of caveolae, *bioRxiv*, 2020, 909408.

46 D. Sengupta, X. Prasanna, M. Mohole and A. Chattopadhyay, Exploring GPCR–lipid interactions by molecular dynamics simulations: excitements, challenges, and the way forward, *J. Phys. Chem. B*, 2018, **122**, 5727–5737.

47 S. A. Kharche and D. Sengupta, Dynamic protein interfaces and conformational landscapes of membrane protein complexes, *Curr. Opin. Struct. Biol.*, 2020, **61**, 191–197.

48 S. J. Marrink and A. E. Mark, The mechanism of vesicle fusion as revealed by molecular dynamics simulations, *J. Am. Chem. Soc.*, 2003, **125**, 11144–11145.

49 S. J. Marrink and A. E. Mark, Molecular view of hexagonal phase formation in phospholipid membranes, *Biophys. J.*, 2004, **87**, 3894–3900.

50 M. Fuhrmans, V. Knecht and S. J. Marrink, A single bicontinuous cubic phase induced by fusion peptides, *J. Am. Chem. Soc.*, 2009, **131**, 9166–9167.

51 M. Pannuzzo, D. H. De Jong, A. Raudino and S. J. Marrink, Simulation of polyethylene glycol and calcium-mediated membrane fusion, *J. Chem. Phys.*, 2014, **140**, 124905.

52 M. Fuhrmans and S. J. Marrink, Molecular view of the role of fusion peptides in promoting positive membrane curvature, *J. Am. Chem. Soc.*, 2012, **134**, 1543–1552.

53 H. J. Risselada, G. Marelli, M. Fuhrmans, Y. G. Smirnova, H. Grubmüller, S. J. Marrink and M. Müller, Line-tension controlled mechanism for influenza fusion, *PLoS One*, 2012, 7, 0038302.

54 H. J. Risselada and S. J. Marrink, Curvature effects on lipid packing and dynamics in liposomes revealed by coarse grained molecular dynamics simulations, *Phys. Chem. Chem. Phys.*, 2009, **11**, 2056–2067.

55 S. O. Yesylevskyy, A. P. Demchenko, S. Kraszewski and C. Ramseyer, Cholesterol induces uneven curvature of asymmetric lipid bilayers, *Sci. World J.*, 2013, **2013**, 965230.

56 H. J. Risselada and S. J. Marrink, The molecular face of lipid rafts in model membranes, *Proc. Natl. Acad. Sci. U. S. A.*, 2008, **105**, 17367–17372.

57 R. S. Davis, P. Sunil Kumar, M. M. Sperotto and M. Laradji, Predictions of phase separation in three-component lipid membranes by the MARTINI force field, *J. Phys. Chem. B*, 2013, **117**, 4072–4080.

58 S. J. Marrink, A. H. De Vries, T. A. Harroun, J. Katsaras and S. R. Wassall, Cholesterol shows preference for the interior of polyunsaturated lipid membranes, *J. Am. Chem. Soc.*, 2008, **130**, 10–11.

59 N. Kučerka, D. Marquardt, T. A. Harroun, M. P. Nieh, S. R. Wassall, D. H. De Jong, L. V. Schäfer, S. J. Marrink and J. Katsaras, Cholesterol in bilayers with PUFA chains: Doping with DMPC or POPC results in sterol reorientation and membrane-domain formation, *Biochemistry*, 2010, **49**, 7485–7493.

60 P. D. Blood, R. D. Swenson and G. A. Voth, Factors influencing local membrane curvature induction by N-BAR domains as revealed by molecular dynamics simulations, *Biophys. J.*, 2008, **95**, 1866–1876.

61 H. Cui, G. S. Ayton and G. A. Voth, Membrane binding by the endophilin N-BAR domain, *Biophys. J.*, 2009, **97**, 2746–2753.

62 M. Simunovic, E. Evergren, I. Golushko, C. Prévost, H.-F. Renard, L. Johannes, H. T. McMahon, V. Lorman, G. A. Voth and P. Bassereau, How curvature-generating proteins build scaffolds on membrane nanotubes, *Proc. Natl. Acad. Sci. U. S. A.*, 2016, **113**, 11226–11231.

63 P. Bassereau, R. Jin, T. Baumgart, M. Deserno, R. Dimova, V. A. Frolov, P. V. Bashkirov, H. Grubmüller, R. Jahn, H. J. Risselada, *et al.*, The 2018 biomembrane curvature and remodeling roadmap, *J. Phys. D: Appl. Phys.*, 2018, **51**, 343001.

64 J. B. d. l. Serna, G. J. Schütz, C. Eggeling and M. Cebecauer, There is no simple model of the plasma membrane organization, *Front. Cell Dev. Biol.*, 2009, **4**, 00106.

Faraday Discussions

PAPER

Lipid specificity of the immune effector perforin

Adrian W. Hodel, [ID] abc Jesse A. Rudd-Schmidt, [ID] ad
Joseph A. Trapani, [ID] de Ilia Voskoboinik [ID] *ad
and Bart W. Hoogenboom [ID] *bcf

Received 22nd April 2020, Accepted 13th August 2020

DOI: 10.1039/d0fd00043d

Perforin is a pore forming protein used by cytotoxic T lymphocytes to remove cancerous or virus-infected cells during the immune response. During the response, the lymphocyte membrane becomes refractory to perforin function by accumulating densely ordered lipid rafts and externalizing negatively charged lipid species. The dense membrane packing lowers the capacity of perforin to bind, and the negatively charged lipids scavenge any residual protein before pore formation. Using atomic force microscopy on model membrane systems, we here provide insight into the molecular basis of perforin lipid specificity.

Introduction

Killer T cells or cytotoxic T lymphocytes (CTLs) kill virus-infected and cancerous cells to maintain immune homeostasis. During the immune response, CTLs form a synapse with their target cells in which they secrete the pore forming protein perforin and pro-apoptotic granzymes.[1,2] Although both the CTL and the target cell plasma membrane are locally exposed at the synapse to perforin, perforin forms oligomeric pores in the target cell membrane but not in the CTL.[3] Through the pores, granzymes can enter and trigger apoptosis in the target cell (Fig. 1). By contrast, the CTLs remain impermeable to granzymes and thus remain viable, and can sequentially kill multiple target cells.[3,4] Without such resistance, CTLs (as well as natural killer cells) would be as vulnerable to perforin as the target cells.

[a]Killer Cell Biology Laboratory, Peter MacCallum Cancer Centre, 305 Grattan Street, Melbourne, VIC 3000, Australia. E-mail: ilia.voskoboinik@petermac.org

[b]London Centre for Nanotechnology, University College London, 19 Gordon Street, London WC1H 0AH, UK. E-mail: b.hoogenboom@ucl.ac.uk

[c]Institute of Structural and Molecular Biology, University College London, Gower Street, London WC1E 6BT, UK

[d]Sir Peter MacCallum Department of Oncology, University of Melbourne, Melbourne, VIC 3000, Australia

[e]Cancer Cell Death Laboratory, Peter MacCallum Cancer Centre, 305 Grattan Street, Melbourne, VIC 3000, Australia

[f]Department of Physics and Astronomy, University College London, Gower Street, London WC1E 6BT, UK

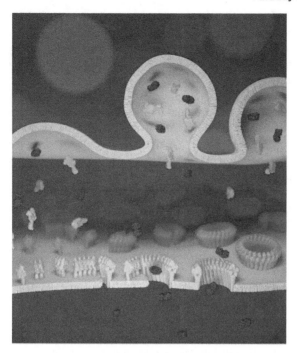

Fig. 1 Schematic illustration of perforin pore formation and granzyme delivery in the synapse. Perforin (blue) and granzymes (red) are transported to the pre-synaptic membrane (top) by cytotoxic granules and released into the synaptic cleft. The (monomeric) perforin subsequently binds to the target cell membrane (bottom). On the target membrane, from left to right, perforin first oligomerizes into short, non-lytic perforin prepores. These prepores can convert to the pore state by inserting into the membrane, and subsequently recruit further prepores to sequentially grow the pore size. Once the pore size is sufficiently large, granzymes can diffuse into the target cells to trigger apoptosis.

This would imply a (most costly) one-to-one ratio of killer cells to target cells and also prevent antigen experienced CTLs from differentiating into memory cells.

Perforin membrane binding – the first step in pore formation – is calcium-dependent and is mediated by its C2 domain.[5–9] It was initially thought that phosphocholine lipids were perforin receptors in the target membrane,[10] but it was later shown – by a comparison of lipids well above and just below their gel transition temperature – that lipid order was a more important factor in determining membrane sensitivity to perforin.[11] The relatively tight plasma membrane packing of CTLs thus served as an explanation of the resistance of CTLs to perforin lysis. In the context of unidirectional killing in the immune synapse however, this hypothesis failed to explain the capability of CTLs to target and kill other CTLs, nor did it explain the absence of a clear correlation between the membrane packing of target cells and their susceptibility to perforin lysis.[12] Following up on these early studies and on observations of perforin on model membranes by atomic force microscopy (AFM),[13] we have recently revealed a two-layered lipid-based mechanism that renders CTLs refractory to perforin pore formation:[14] firstly, increased lipid order and packing in the CTL membrane reduces perforin binding to the membrane, and secondly, perforin is sequestered and irreversibly inactivated by binding to the negative charge of externalised phosphatidylserine (PS) at the CTL surface. Importantly, these membrane changes are enhanced in the area of the CTL plasma membrane that is associated with the immune synapse.

This lipid specificity can be regarded in the context of other pore forming proteins in general[15] and of the membrane attack complex-perforin/cholesterol dependent cytolysin (MACPF/CDC) family of pore forming proteins, which perforin is part of.[16–19] The bacterial CDCs use – as implied by their name – cholesterol as a receptor on the membrane and only form pores in membranes above a rather sharp threshold of cholesterol contents, typically above $\sim$25–35%.[20–23] This cholesterol dependence defines the specificity of CDCs to eukaryotic target cells, as bacteria generally do not contain cholesterol in their membranes.[24,25] The mushroom derived MACPF pleurotolysin B utilizes the partner proteins ostreolysin A or pleurotolysin A to specifically bind and form pores in membranes containing sphingomyelin and cholesterol[26–28] or insect specific lipids.[29] In vertebrates, the membrane attack complex (MAC) is an immune effector that kills pathogenic bacteria. The formation of the MAC is facilitated on membranes that contain negatively charged lipids and show increased membrane tension, mimicking the surface of Gram-negative bacteria.[30] Other examples from the vertebrate immune system are the more recently discovered pore forming proteins of the gasdermin family, which share some structural elements with MACPF/CDCs and trigger cell death by perforating the membranes of infected cells from the inside out.[31] Gasdermin pore formation is related to negatively charged lipids that are in the inner leaflets of eukaryotic plasma membranes and mitochondria.

Besides cell-based assays, the lipid specificity of pore formation can be most conveniently studied on model lipid bilayers since these can be prepared from a wide selection of lipid components and therewith offer the ability to selectively alter biophysical properties. As a reference lipid, we used the dioleoyl derivative of phosphatidylcholine (PC), DOPC, which has a low liquid–gel transition, or melting, temperature (T_m, ca. $-17\,^\circ$C (ref. 32)), and is therefore present in a liquid disordered (L_d) state at physiological temperatures (37 °C). PC lipids in the L_d state (sometimes supplemented with cholesterol) are the most common components of model membranes used to visualize perforin assemblies,[7,13,14,33–35] and the addition of cholesterol to L_d membranes increases membrane order and, at sufficiently high concentrations, it can give rise to a liquid ordered (L_o) state.[25] Below their melting temperature T_m, lipids exist in a solid ordered (S_o) gel state. Different lipids have different T_m, e.g., the dipalmitoyl derivative of PC, DPPC, has a T_m of ca. 41 °C (ref. 36) and is thus in the S_o state at physiological temperature. Over time, a membrane containing a mix of lipids can phase separate and display domains of different states of membrane order. Commonly used mixtures to mimic eukaryotic membranes use a low T_m lipid species like DOPC, cholesterol and high T_m sphingomyelin (SM), e.g., egg SM. In such mixtures, one readily observes phase separation into PC-rich L_d domains and SM/cholesterol-rich L_o domains.[37,38] Similarly, mixtures of DOPC and excess DPPC can lead to L_d/S_o phase separation.[39] The lipid phase state is an important factor to consider when mixing different types of lipids. Thus, to retain the L_d state of a reference DOPC bilayer, we can use the dioleoyl derivatives of phospholipids, e.g. dioleoyl phosphatidylserine (DOPS, T_m *ca.* -11 °C (ref. 40)), ethanolamine (DOPE, T_m *ca.* -8 °C (ref. 41)), or glycerol (DOPG, T_m *ca.* -22 °C (ref. 42)).

Such model membranes can be prepared as supported lipid bilayers on a flat substrate, *e.g.*, mica or silica, facilitating their characterisation by in-liquid atomic force microscopy (AFM) experiments. AFM has become a popular tool to study the mechanisms of pore forming proteins,[43,44] in part because it allows a relatively straightforward distinction between prepore assemblies and membrane inserted pores. This can be achieved either by detecting a height change[45,46] or by the loss of mobility once the protein contacts the underlying substrate.[13,46] Another important feature of AFM is its ability to distinguish between different lipid domains *via* Ångström-sized differences in membrane thickness,[47] which allows us to simultaneously detect lipid phase boundaries and protein pores.

Here we use AFM-based experiments to expand upon our recent work[14] on establishing and elucidating how perforin function depends on the physico-chemical properties of the target membranes. Noting that perforin binding – and thus pore formation – is reduced on tightly packed membranes, we study and compare the effects of several properties that modulate membrane packing. In contrast to its response to changes in membrane order/packing, the interaction of perforin with the negatively charged DOPS is fundamentally different. The initial binding of perforin appears to be unaffected, but pore formation is disrupted. On model membranes, this effect is proportional to the amount of DOPS they contain. We therefore investigate how perforin interacts with pure DOPS membranes in the pursuit of understanding how perforin is deactivated by this lipid. Lastly, we describe the interaction of perforin with DOPE, as PE lipids are another major constituent of the plasma membrane.

Experimental

Recombinant proteins

Wild-type perforin (WT-PRF),[48] disulphide locked perforin (TMH1-PRF),[13] GFP fusion disulphide locked perforin (TMH1-GFP-PRF),[14] and C2 domain mutant perforin (D429A-PRF)[5] were expressed in baculovirus-infected Sf21 cells and purified from the supernatant as per the respective references provided. The CDC perfringolysin O (PFO) was kindly provided by Rana Lonnen and Peter Andrew (University of Leicester).

Preparation of lipid vesicles and AFM samples

1,2-Dioleoyl-*sn-glycero*-3-phosphocholine (DOPC), 1,2-dipalmitoyl-*sn-glycero*-3-phosphocholine (DPPC), 1,2-dioleoyl-*sn-glycero*-3-phosphoethanolamine (DOPE), 1,2-dioleoyl-*sn-glycero*-3-phospho-(1′-rac-glycerol) (DOPG), 1,2-dioleoyl-*sn-glycero*-3-phospho-L-serine (DOPS), egg sphingomyelin (egg SM) and cholesterol were purchased as powders from Avanti Polar Lipids (Alabaster, AL, USA). Where indicated, the lipids were mixed in the desired molar ratios ($\pm 5\%$ confidence intervals). Note that the provided mixing ratios do not necessarily represent the lipid concentration displayed on the final bilayer surface. The effective exposure of negatively charged lipids on the surface of supported lipid bilayers can be lower by a factor of 2 or more due to interactions with the substrate.[49] At a concentration of 0.5–1 mg mL^{-1}, unilamellar vesicles with a nominal diameter of 100 nm were prepared using the lipid extrusion method.[13,50]

4–8 µL of the unilamellar vesicles (containing 4 µg of lipid) were added to a freshly cleaved, ⌀ 10 mm mica disc (Agar Scientific, Stansted, UK) and topped up with 80 µL of adsorption buffer containing 150 mM NaCl, 25 mM Mg^{2+}, 5 mM Ca^{2+}, and 20 mM HEPES, pH 7.4. To form a pure DOPG bilayer, lower salt conditions were necessary[51] and the buffer was thus adjusted, instead containing no Mg^{2+} and 10 mM Ca^{2+}. The lipids were incubated for 30 minutes above the T_m of the constituent lipids to cover the mica substrate with an extended lipid bilayer. Excess vesicles were removed by washing the bilayer 6–12 times with 80 µL of the adsorption buffer.

Additional washes were applied to samples that contained DOPS, for which we found that Mg^{2+} in the buffer interfered with perforin binding, samples that contained DOPG to remove excess Ca^{2+}, or to control samples that required the removal of Mg^{2+}: these were washed an additional 6 times with 80 µL buffer containing 150 mM NaCl, 5 mM Ca^{2+}, and 20 mM HEPES, pH 7.4.

Wild-type perforin (WT-PRF), disulphide-locked perforin (TMH1-PRF), C2 domain mutant perforin (D429A-PRF) and perfringolysin O (PFO) were diluted up to *ca.* ten-fold to a volume of 40 µL in 150 mM NaCl and 20 mM HEPES, pH 7.4, and injected onto the sample, to reach concentrations of 150 nM or, where noted, *ca.* 400 nM above the model membrane. The protein was incubated for 2 (where noted) or 5 min at 37 °C. To unlock TMH1-PRF after its binding to the membrane, the engineered disulphide bond was reduced by addition of 2 mM DTT (Sigma-Aldrich, St. Louis, MO, USA) and 10 min incubation at 37 °C. We previously verified that, once TMH1-PRF was bound to target membranes, the effect of DTT on its native disulphide bonds did not change the pore forming functionality.[13] Mobile TMH1-PRF assemblies were fixed by addition of glutaraldehyde 8% EM grade (TAAB Laboratories, Aldermaston, UK) to a final concentration of 0.04% v/v, and 10 min incubation at room temperature. Fixed TMH1-PRF assemblies were removed from their DOPS membrane substrate by chelating calcium with EGTA. To this end, the samples were washed 6 times with 80 µL of buffer containing 150 mM NaCl, 20 mM HEPES, and 4 mM EGTA, pH 7.4. EGTA was immediately removed from the sample afterwards by 6 further washes with 80 µL buffer containing 150 mM NaCl, 5 mM Ca^{2+}, and 20 mM HEPES, pH 7.4.

AFM imaging and analysis

AFM images were recorded by either force–distance curve-based imaging (Peak-Force Tapping) on a MultiMode 8 system (Bruker, Santa Barbara, CA, USA) or photothermal excitation (blueDrive) on a Cypher ES AFM (Oxford Instruments, Abingdon, UK). The imaging conditions with commercial MSNL cantilevers (Bruker) for PeakForce Tapping are outlined in ref. 14. In brief, PeakForce Tapping was performed at 2 kHz and a maximum tip–sample separation of 5–20 nm. Images were recorded at 0.75 Hz scan speed and tip–sample interaction forces between 50 and 100 pN on an E-Scanner (Bruker, Santa Barbara, CA, USA) with temperature control. For blueDrive, we used BL-AC40TS probes (Olympus, Tokyo, Japan). The UV laser for photothermal excitation was focussed onto the cantilever base. The laser was tuned to the resonance frequency of the cantilever in liquid (*ca.* 25 kHz) and the amplitude was adjusted to 1 V. Imaging was performed at an amplitude setpoint of *ca.* 750 mV and 1 Hz scan speed. All samples were imaged at 37 °C to retain thermotropic properties, or at room temperature.

Raw AFM images were background subtracted with reference to the lipid surface, masking perforin and applying second-order flattening. Height values of perforin prepores/pores indicated in the manuscript are given with ±1 nm confidence intervals, with the uncertainty due to scanner calibration and possible sample deformation caused by the probe–sample interaction. The same colour/height scale was applied to all images (except for the insets in Fig. 2A and 4A as specified in their captions), spanning 25 nm and 9 nm below and 16 nm above the membrane surface (set to 0 nm). The colour scale is only depicted once, in Fig. 2. Values for perforin coverage were estimated either by the area above a height threshold located 6–8 nm above the membrane surface and adjusted to counteract tip broadening effects; or, when sufficient images at a higher pixel resolution were available, by tracing pore shapes with 3dmod 4.9.4 (BL3DEMC & Regents of the University of Colorado[52]). Perforin coverages obtained by both methods are normalized with respect to a 100% DOPC reference and given as values between 0 and 1. One-way analysis of variance (ANOVA) with Dunnett's post-hoc analysis was performed in R-3.6.3 using the multcomp package.[53]

Perforin binding to lipid strips

Membrane lipid strips (Echelon Biosciences, Salt Lake City, UT, USA) were incubated in 4 mL of blocking buffer containing 3% w/w BSA (Roche Diagnostics GmbH, Mannheim, Germany), 150 mM NaCl, and 20 mM HEPES, pH 7.4 for 1 h at room temperature. 2 μg mL^{-1} TMH1-GFP-PRF was added to a lipid strip in 4 mL of blocking buffer supplemented with 2 mM CaCl$_2$, pH 7.4. The use of the GFP fusion construct allowed readout of the lipid strips without the need for antibody labelling. To assess calcium-independent (non-specific) perforin binding, 2 μg mL^{-1} TMH1-GFP-PRF in 4 mL of blocking buffer was added to a lipid strip. After 1 h of incubation at room temperature, the lipid strips were washed three times with 4 mL of blocking buffer (with or without adding 2 mM CaCl$_2$ to match the initial incubation). GFP fluorescence (of wet lipid strips) as a measure of perforin binding was recorded on an iBright 1500 western blot imaging system (Thermo Fisher Scientific, Waltham, MA, USA). The strips were stored in the blocking buffer for *ca.* 96 hours at 4 °C and imaged again.

Results and discussion

Effect of lipid order on perforin binding and pore formation

To visualize the binding of perforin to different phase domains of different levels of lipid order, we used a disulphide-locked mutant, TMH1-PRF, that can bind to and assemble on, but not insert into the target membrane.[13] Its pore forming functionality was fully restored after reducing the disulphide bond with the reducing agent dithiothreitol (DTT). This mutant has an advantage over the wild-type protein in that membrane binding and pore insertion can be uncoupled and studied as two separate events.[13,14]

Using TMH1-PRF, we first verified the earlier observation[11] that perforin does not bind to lipids that are below their gel-transition temperature, *i.e.*, in the gel or

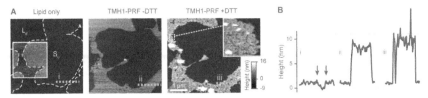

Fig. 2 Prepore-locked TMH1-PRF preferentially binding to L_d domains in a phase separated L_d/S_o membrane. (A) AFM images of a supported lipid bilayer composed of DOPC/DPPC mixed in a 1 : 7 molar ratio. The first panel ('Lipid only') shows the empty membrane, where the colour contrast has been enhanced in the inset (colour scale: 4 nm) to better visualize the lipid phase separation. The phase boundaries are outlined for the whole image by dashed white lines. Addition of TMH1-PRF ('TMH1-PRF − DTT') leads to the formation of a diffuse plateau limited to L_d domains. Similar plateaus were earlier interpreted as mobile prepore assemblies.[13] After addition of DTT ('TMH1-PRF + DTT'), the mobile assemblies insert into the membrane, and a dense layer of arc- and ring-shaped pores is formed (see inset), still confined to L_d domains. Size of the inset, 150 nm. (B) Height profiles extracted along the dashed coloured lines in (A). The profiles depict the 0.5–1 nm height change at the phase boundaries (i) (the boundaries are highlighted by arrows), and the *ca.* 7–11 nm tall prepore and pore layers (ii and iii). Dashed grey lines indicate the height of the L_d membrane (0 nm) and the height of a perforin monomer (11 nm (ref. 7)). Note that perforin features can appear compressed due to tip–sample interaction forces. The data was recorded at 37 °C.

S_o phase. This could be best articulated by visualizing the binding of perforin on phase separated bilayers that contained both L_d and S_o domains. To this end, we mixed high T_m DPPC and low T_m DOPC and verified the phase separation by AFM, with the S_o domains appearing *ca.* 1 nm higher than the L_d domains (see Fig. 2). Upon exposure to TMH1-PRF, the L_d domains showed extensive protein coverage by the appearance of diffuse plateaus with a height close to 10 nm above the membrane. As in previous work, we interpret these plateaus as membrane-bound but not inserted and, hence, highly mobile perforin prepores.[13] After addition of DTT, the TMH1-PRF transitioned from the prepore to the membrane-inserted pore state, while remaining localized within L_d domains.

Besides bilayer-based experimental substrates, membrane strips blotted with different types of lipids have been used to characterize the lipid specificity of several pore forming proteins (*e.g.* ref. 29 and 54–56). We find that commercially available strips fail to detect perforin binding to, for example, PC (Fig. 3), in agreement with previous reports using lipid strips;†[55] this is in apparent contradiction to the scientific literature spanning from the 1980s[10] to today.[14] This contradiction can be simply explained by noting that the blotted (phospho-) lipids have a high T_m and were not in an L_d state at the physiological/experimental temperature, and are thus likely to cause erroneous readings when lipid order is a factor of importance for protein binding (such as for perforin). Moreover,

---

† Some of the observed perforin–lipid binding was different from previously published results. Without investigating this further, we point out that we used a recombinant mouse perforin mutant (*vs.* native human perforin elsewhere) and detected fluorescence directly (*vs.* primary/secondary antibody detection elsewhere). Furthermore, we noted some perforin binding that exclusively occurred in the absence of Ca^{2+} and disappeared after prolonged incubation in the washing buffer. The reasons for this are unknown to us.

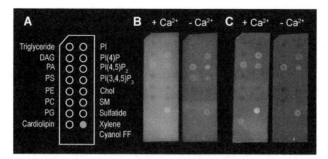

Fig. 3 Binding of TMH1-GFP-PRF to lipid strips in the presence and absence of 2 mM Ca^{2+}, visualized by fluorescence imaging.† (A) Schematic layout of the lipid strips, with spots of lipids blotted where indicated (xylene cyanol FF is a non-lipid control). (B) Detection of TMH1-GFP-PRF binding to lipid strips in the presence and absence of Ca^{2+} ('+Ca^{2+}' and '−Ca^{2+}' respectively) immediately after washing. (C) The same lipid strips as in (B) after 4 days at 4 °C in blocking buffer.

lipids that bound perforin on these lipid strips (Fig. 3B and C, '+Ca^{2+}' vs. '−Ca^{2+}') did so independently of the calcium that is required for perforin binding to target membranes; such interactions may therefore be artefacts due to defects in the lipid covered membrane surfaces.

As reported previously[14] and reiterated here for completion and for comparison with Fig. 2, we observe the same preference of perforin for L_d domains when adding TMH1-PRF to phase separated lipid membranes that contain both L_d and L_o domains,[14] which is also in agreement with previous results using WT-PRF.[13,14] As shown in Fig. 4, perforin again preferentially binds to and forms pores in L_d domains, albeit that some rare examples of perforin binding may be observed on L_o domains. This binding dependence on

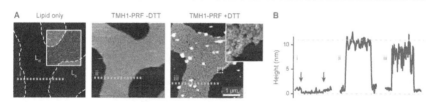

Fig. 4 TMH1-PRF preferentially binding to L_d domains in a phase separated L_d/L_o membrane, analogous to Fig. 2. (A) AFM images of an approximately equimolar DOPC/egg SM/cholesterol supported lipid bilayer. 'Lipid only' shows the empty membrane with the phase boundaries between L_d and L_o domains highlighted by the inset (with a 4 nm colour scale) and dashed white lines. TMH1-PRF exclusively binds L_d domains ('TMH1-PRF − DTT') and remains (mostly) confined there after addition of DTT ('TMH1-PRF + DTT') and the formation of transmembrane pores (see inset). Size of the inset, 150 nm. (B) Height profiles extracted along the dashed coloured lines in A. The profiles depict the 0.5–1 nm height change at the phase boundaries (i) (the boundaries are highlighted by arrows), and the ca. 7–11 nm tall prepore and pore layers (ii and iii). Dashed grey lines indicate the height of the L_d membrane (0 nm) and the height of a perforin monomer (11 nm (ref. 7)). Note that perforin features can appear compressed due to tip–sample interaction forces. The data was recorded at 37 °C. Figure reproduced from ref. 14, under a Creative Commons Attribution 4.0 International License (CC BY 4.0).

lipid order is also in agreement with the reduction of WT-PRF pore formation on highly ordered egg sphingomyelin membranes compared with L_d 18 : 1 sphingomyelin membranes.[14]

In the experiments reported above, lipid order was varied by using lipids with identical headgroups but different hydrophobic tails. In addition, membrane order can be dependent on divalent ions that intercalate with lipid headgroups, modulating intermolecular attractions.[57] In most of our AFM work on model membranes, we used up to 25 mM Mg^{2+} in our buffers to stabilize the supported lipid bilayers on the negatively charged mica substrate. This concentration is about one order of magnitude higher than blood levels.[58] To test how the presence of Mg^{2+} affects perforin function, we designed experiments in which we washed samples to remove Mg^{2+} from the buffer before adding perforin (WT-PRF, at 37 °C as usual) onto model membranes in either the L_d (pure DOPC), L_o (DOPC/cholesterol or egg SM/cholesterol, both 47/53 molar ratio), or S_o (pure egg SM) state.[37]

By comparison with previously published data acquired in the presence of Mg^{2+}, we found the differences between pore formation at high and low levels of Mg^{2+} to be mostly insignificant (see Fig. 5). However, for low levels of Mg^{2+}, a significant but small increase in pore formation was found on the L_o membranes consisting of DOPC/cholesterol and egg SM/cholesterol. We did not note any phase separation in any of these lipid substrates at either level of Mg^{2+}, suggesting that perforin binding was uniformly affected, if at all, in all samples. In summary, the suggested increase in membrane order due to Mg^{2+} may be present and affect perforin pore formation in membranes of intermediate order, but this effect is small compared to the effects on lipid order due to high amounts of cholesterol or introduction of gel-phase lipids as reported above.

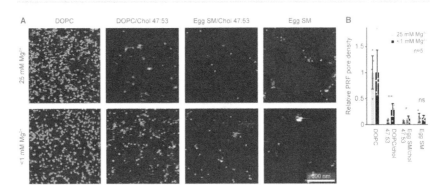

Fig. 5 Perforin lipid specificity as a function of Mg^{2+} concentration. (A) Representative AFM images of WT-PRF pores incubated on magnesium-depleted (<1 mM Mg^{2+}, see Experimental) membranes of different lipid compositions and at 25 mM Mg^{2+}, as indicated. The data was recorded at 37 °C. (B) Average perforin pore formation on different lipid mixtures, at Mg^{2+} concentrations of 25 mM and <1 mM, normalized to the number of pores on DOPC-only membranes. Here, perforin was incubated for 2 min instead of 5 min (see Experimental) to match the experimental conditions of the two datasets. Error bars represent standard deviations. The statistical significance was assessed using ANOVA with Dunnett's post-hoc analysis, where 'ns' is not significant, $*p < 0.05$, $**p < 0.01$. The data for 25 mM Mg^{2+} are reproduced from ref. 14.

Effect of lipid charge on perforin pore formation

By the here described variations in perforin binding with lipid order, we can explain the reduced binding of perforin to CTLs that has been shown in earlier work.[14] However, when incubated with higher concentrations of recombinant perforin, CTLs were found to still resist perforin pore formation in spite of binding amounts of perforin that were lytic to target cells, which we attributed to the presence of PS in the outer leaflets of the lymphocyte membranes.[14]

Perforin can bind to PS-rich membranes, but pore formation is decreased: instead of pores, perforin aggregates into dysfunctional plaques.[14] PS lipids have a net negative charge at physiological pH, and we previously hypothesized that this negative charge is the underlying cause of perforin dysfunction. We therefore tested the effect of the negatively charged DOPG and cholesterol sulphate on perforin pore formation. As predicted, the decrease in perforin pore formation

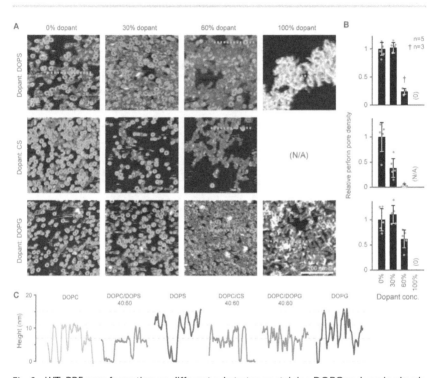

Fig. 6 WT-PRF pore formation on different substrates containing DOPC and varying levels of either DOPS, DOPG, or cholesterol sulphate (CS). (A) Representative AFM images of perforin pores and aggregations on the different substrates. For pure CS, no bilayer could be formed. All data was recorded at room temperature. (B) Quantification of pore formation (mean ± SD) in the samples shown in (A), relative to the 0% dopant/100% DOPC reference. (C) Height profiles extracted along the dashed lines in (A); the 'DOPC' reference profile was extracted from the first tile in (A). The different profiles show the membrane level adjusted to 0 nm and the height of perforin pores (ca. 11 nm) and aggregates (ca. 7 nm at 60% dopant levels, up to 15 nm at 100% dopant levels), as highlighted by horizontal dashed lines. The colour tone of the profiles is darker compared to the 'DOPC' reference, corresponding to the level of negative charge present in the membrane substrates. Panels (A) and (B) are reproduced from ref. 14, under a Creative Commons Attribution 4.0 International License (CC BY 4.0).

was proportional to the levels of negatively charged lipids in the membranes and, possibly, further affected by an ordering effect induced by cholesterol sulphate (Fig. 6 A and B). This leads us to conclude that it is a generic negative surface potential, rather than specific lipid headgroups, that prevent perforin pore formation here.

On the membranes with higher negative charge, the decrease in WT-PRF pore formation was accompanied by an increase in the presence of plaques of protein aggregates. Perhaps the most striking feature of these plaques is their height. Perforin aggregations appear at *ca.* 7 nm in height when PC is doped with 30–

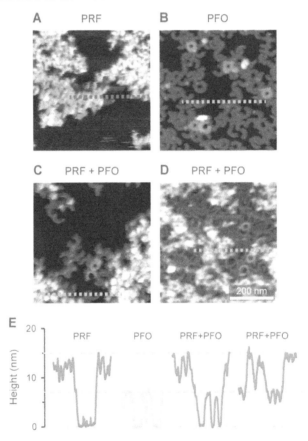

Fig. 7   AFM images of WT-PRF and PFO on 70 : 30 mol% DOPS/cholesterol bilayers for height referencing. (A) WT-PRF forms protein plaques on the lipid bilayer. (B) Cholesterol dependent PFO pore formation, visible as arc- and ring-shaped assemblies. A small number of the assemblies appeared higher, probably due to incomplete membrane insertion: PFO collapses from *ca.* 10 nm to 7 nm height upon membrane insertion.[45]. (C and D) When the DOPS/cholesterol bilayers were incubated first with WT-PRF and next with PFO, perforin plaques were observed adjacent to PFO pores. We here show two samples incubated with different amounts of PFO, *ca.* 150 nM in (C) and *ca.* 450 nM in (D). Consequently, the membrane surface is still visible in (C), while in (D) the PFO pores cover most of the remaining membrane. (E) Height profiles extracted along the dashed lines in (A)–(D). Horizontal lines at 0, 7, and 15 nm highlight the membrane surface and the heights of PFO pores and perforin plaques, respectively. All AFM data were recorded at room temperature.

60 mol% negatively charged lipids; of note, the actual proportion of negatively charged lipids in the outer layer of supported lipid bilayers is likely to be reduced by at least one half due to leaflet asymmetry in negatively charged supported lipid bilayers.[49] However, on pure DOPS and DOPG membranes, the aggregations appear as plaques with a height of (up to) 15 nm above the membrane surface (Fig. 6C). This height is to be compared with the *ca.* 11 nm height of membrane-bound perforin prior to and after membrane insertion.[13,14]

Intriguingly, the observed 15 nm height above the membrane agrees with the full height of perforin pores including the hairpins that span the membrane.[7] This suggests that the protein has initiated the transition from its prepore to pore state, yet while unfurling these hairpins, it has failed to insert into the membrane. Given this possible interpretation, we sought to first further verify the height measurements of perforin plaques on pure PS membranes, by including the cholesterol dependent cytolysin (CDC) perfringolysin O (PFO) as a height ruler in our experiments. Like other CDCs, PFO forms pores that protrude *ca.* 7 nm above cholesterol-containing membranes.[45,46,59,60] To this end, we prepared 70 : 30 mol% DOPS/cholesterol membranes, on which perforin behaved similarly as on pure DOPS membranes (Fig. 7A) but which – by the inclusion of cholesterol – allowed CDC binding and pore formation too (Fig. 7B). By first incubating these membranes with perforin and next with the CDC PFO, we observed PFO pores in addition to perforin plaques (Fig. 7C and D), with the perforin plaques being approximately double the height above the membrane as the PFO pores, which were taken as a height reference of *ca.* 7 nm (Fig. 7E). This fully confirms the extraordinary height of perforin on DOPS-only bilayers.

The preconditions and structural changes necessary to form such perforin plaques are unknown. In our earlier studies with the non-functional TMH1-PRF on pure DOPS bilayers, it emerged that in the membrane-binding and early assembly stage, *i.e.*, before pore insertion, the behaviour of perforin is similar to that observed on DOPC bilayers:[13,14] TMH1-PRF on DOPS and DOPC (i) showed a similar distribution of subunits per assembly (quantification in ref. 14), (ii) was *ca.* 10 nm in height above the membrane (similar to the height of an upstanding perforin molecule), and (iii) diffused freely and could be removed from the membrane surface by chelation of Ca^{2+} (demonstrated by the removal of oligomers after chelating calcium from the buffer, see Fig. 8A). After adding DTT and thus unlocking TMH1-PRF, the short oligomers clustered together and increased their height to *ca.* 15 nm (Fig. 8B). Taken together, this supports the interpretation that the formation of plaques on PS is linked to the unfurling of the protein as it attempts – unsuccessfully – to insert into the membrane.

To further investigate how this behaviour depends on electrostatic interactions, we varied the concentration of divalent ions in solution, thus changing the screening of surface charges. Firstly, when 5 mM Ca^{2+} and an excess concentration of Mg^{2+} (25 mM) were present in the buffer, perforin (WT-PRF) would not bind to or form plaques on a DOPS membrane, even at higher perforin concentrations (Fig. 9A and B). Secondly, in the absence of Mg^{2+}, the appearance of the plaques was dependent on the Ca^{2+} concentration in the buffer: at higher concentrations of Ca^{2+}, there was a decrease in the spread of plaques over the membrane surface (Fig. 9C and D). The dependence of perforin membrane binding on the concentration of divalent cations further confirms that the observed behaviour on DOPS is mediated by electrostatic interactions. These two

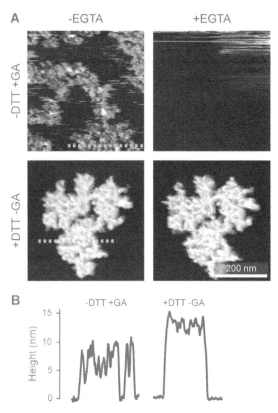

Fig. 8 Membrane binding to DOPS membranes prior to plaque formation, assessed with TMH1-PRF. (A) AFM images of locked and unlocked TMH1-PRF on DOPS membranes. To visualize initially mobile TMH1-PRF features (not shown here) in the AFM images, we fixed the protein by addition of glutaraldehyde (GA). The patches of crosslinked perforin ('−DTT + GA', '−EGTA') are removed by washing the samples with the calcium-chelating agent EGTA ('−DTT + GA', '+EGTA'). If TMH1-PRF is unlocked by addition of DTT, perforin plaques are formed similarly to WT-PRF ('+DTT − GA', '−EGTA'). The plaques are not visibly affected by washing with EGTA ('+DTT − GA', '+EGTA'). (B) Height profiles of cross-linked ('−DTT + GA') and unlocked ('+DTT − GA') TMH1-PRF extracted from the panels in A, as indicated by dashed lines. All images were recorded at room temperature. The panels in (A) are reproduced from ref. 14, under a Creative Commons Attribution 4.0 International License (CC BY 4.0).

observations are different from what we observe on DOPC membranes, where a similar increase in Ca^{2+} concentration produced no effect on the formation of arc- and ring-shaped pores.

For functional perforin, the initial membrane binding occurs through its C2 domain. By mutating this domain in D429A-PRF,[5] we could also test whether the initial perforin binding depends on lipid composition: the mutation completely abrogated D429-PRF binding to DOPC bilayers, but on DOPS membranes the mutant still formed plaques with heights of mostly ca. 7 nm and up to 15 nm (Fig. 10), roughly consistent with the plaques formed by WT-PRF (Fig. 6).

Taken together, the experimental data indicate that negatively charged membranes disrupt perforin function due to electrostatic interactions, and that

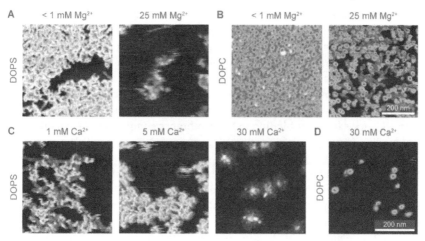

Fig. 9 Interaction of WT-PRF with DOPS membranes at various levels of Ca^{2+} and Mg^{2+}. (A) WT-PRF on DOPS membranes at low (<1 mM) and high (25 mM) Mg^{2+} levels in the buffer. Perforin only forms the *ca.* 15 nm high plaques at low Mg^{2+} levels, although some protein binds at a high Mg^{2+} level. Here, we used 400 nM WT-PRF (instead of the 150 nM used for other experiments, see Experimental) to test the effect at high perforin concentrations. (B) Analogous experiment to (A) on DOPC instead of DOPS membranes, as a control. At both low and high Mg^{2+} concentrations, arc- and ring-shaped perforin pores are visible.‡ (C) Perforin plaques formed on DOPS membranes at Ca^{2+} concentrations of 1, 5, and 30 mM Ca^{2+} (with no Mg^{2+}). For larger Ca^{2+} concentrations, the plaques appear less dispersed. (D) For comparison, arc- and ring-shaped pores formed on DOPC at 30 mM Ca^{2+}. All images were recorded at room temperature.

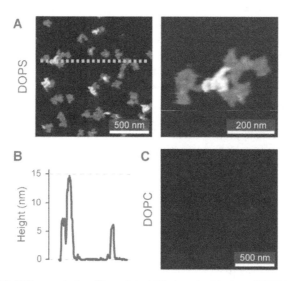

Fig. 10 D429A-PRF, a perforin with mutated C2 domain, binds to DOPS but not DOPC membranes. (A) D429A-PRF forming plaques on a DOPS membrane. (B) Height profile extracted along the dashed line in (A). (C) A DOPC membrane incubated with the same amount of D429A-PRF as in (A) does not show perforin binding. AFM data was recorded at room temperature.

the disruptions are manifested at the stage of membrane insertion. The data in Fig. 8 and 9 show that the interaction of perforin with PS membranes depends on calcium, suggesting the involvement of calcium binding sites within the C2 domain.[9] As illustrated by locked TMH1-PRF in Fig. 8, perforin initially binds to PS membranes in an upright orientation, further indicating that the C2 domain is facing the membrane surface (an alternative, but somewhat far-fetched hypothesis is that the protein is bound upside down). In contrast, the C2 domain mutant perforin shows different binding to PS membranes compared to PC membranes, *i.e.*, binding is completely absent on PC membranes. To reconcile the seemingly contradictory data, we can imagine two scenarios: (I) disruption of perforin function is caused by small differences in the binding geometry between the C2 domain and charged lipid substrates,§ leading to protein misfolding and plaque formation; (II) only a (possibly small and imperceptible in AFM images) fraction of perforin is required to bind DOPS in a different fashion, possibly independent of the C2 domain, and this fraction disables otherwise correctly bound perforin when it attempts to insert into the membrane. Of note, a similar disruptive effect was observed when functional WT-PRF was co-incubated with excess non-functional TMH1-PRF,[13] although in that case, the pore forming functionality could be fully restored by subsequent addition of DTT (unlocking the disulphide lock in TMH1-PRF).

Effect of membrane tension on perforin pore formation

Besides lipid order and charge,[14] another physical membrane property that may modulate perforin pore formation is membrane tension, which has been suggested to enhance perforin function in the immune synapse.[61] To some extent, such effects can be tested in supported lipid bilayers by the inclusion of curvature-inducing lipids. For example, phosphatidylethanolamine (PE) is a zwitterionic lipid with no net charge and a relatively small headgroup compared with the width of its hydrophobic tail. This causes PE to favour curved membrane arrangements, consistent with its prevalence in the inner leaflet of the eukaryotic plasma membrane and implying interfacial tension when forced to arrange in planar membranes; indeed, bilayers containing only (unsaturated) PE lipids do not form under physiological conditions.[62] PE can be synthesized from PS by decarboxylation and is co-located with PS in the inner plasma membrane leaflet;[63,64] their externalization is regulated by the same transporters.[65]

To test how the addition of PE affects pore formation by perforin, we doped a DOPC bilayer with up to 60 mol% DOPE, and exposed the resulting membranes to WT-PRF. As shown in Fig. 11A and B, the addition of DOPE had no significant

---

§ D429A-PRF is distinct from WT in its dual effect on the C2 domain: it is unable to bind two out of five C2 domain Ca^{2+} ions,[9] and it also fails to undergo the conformational change required for the reorientation of W427 and Y430 residues – a critical Ca^{2+}-dependent step required for perforin binding to a membrane.[8] The fact that D429A binds to PS, but not to PC, suggests that perforin binding to PS occurs through a non-canonical mechanism that is independent of the hydrophobic interactions of W427, Y430, Y486 and W488 with the membrane.[8]

‡ We note that in the images depicted here, the overall perforin coverage at high Mg^{2+} concentration appears lower (at 400 nM WT-PRF concentration). It is not clear if this difference is significant; to date, we do not have sufficient AFM data and repeats of this experiment to rigorously quantify the number of perforin pores at high *versus* low Mg^{2+} concentrations.

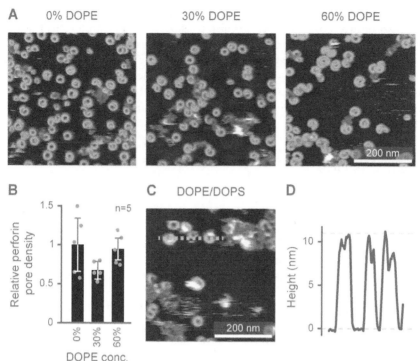

Fig. 11 Effect of DOPE on WT-PRF pore formation. (A) AFM images of perforin pore formation on membranes containing DOPC and increasing amounts of DOPE. (B) Quantification of the pore formation normalized to the 0% DOPE/100% DOPC reference. Error bars represent standard deviations. (C) AFM image of perforin on an 80 : 20 DOPS/ DOPE membrane, showing at least partial restoration of pore formation. (D) Height profile extracted along the dashed line in (C). AFM images were recorded at room temperature.

influence on the pore forming capacity of perforin *per se*. We also performed an alternative experiment in which we tested whether addition of 80 mol% DOPE would restore pore forming capacity in a DOPS host bilayer. A qualitative assessment of the AFM images indeed indicated that the inclusion of PE caused some restoration of pore formation on DOPS membranes (Fig. 11C). This

Fig. 12 Schematic illustrations of perforin (blue) interacting with lipid substrates of different packing and charge. (A) On zwitterionic (net neutral) and disordered membranes, perforin oligomerizes into membrane spanning pores. (B) On more ordered membranes, such as liquid-ordered membranes that contain cholesterol (yellow), perforin cannot bind. (C) On membranes containing lipids with negatively charged headgroups (pink), perforin is sequestered into plaques. Membrane lesions are not formed, and the protein orientation on the surface is not known (symbolized by the question mark).

preliminary result could be explained by presuming that perforin directly binds to PE lipid headgroups. However, it is also possible that PE, with its small head-group, provides no direct perforin binding site and, instead, introduces membrane defects that expose areas otherwise buried underneath the membrane surface or generally acts as a spacer between DOPS molecules. As a result, perforin might access and bind the PS headgroup differently, thus (partially) restoring its functionality. Future experiments will need to determine if perforin can bind PE directly.

Conclusions

As discussed in this paper, the physical properties of membranes play essential roles in determining their sensitivity to perforin pore formation. This applies to the lipid order and packing, which reduce perforin binding to the membrane,[11,14] and to the lipid charge, which causes perforin to be trapped in dysfunctional aggregates,[14] as illustrated in Fig. 12. We briefly discussed membrane tension as a possible factor, which has been reported to enhance perforin function in the immune synapse.[61]

Compared with previous results, we have here (i) demonstrated the power of AFM and model membranes in investigating the lipid specificity of pore forming proteins and of perforin in particular; (ii) used AFM to demonstrate how membrane order in gel-phase lipids completely prevents perforin binding, as previously observed for liquid-ordered domains;[14] (iii) demonstrated that this lipid specificity for liquid-disordered membranes is robust against varia-tions in divalent ion concentration (Mg^{2+}, and Ca^{2+} above the threshold needed to facilitate perforin binding to the membrane); (iv) verified the extraordinary height (ca. 15 nm above the membrane) of dysfunctional perforin aggregates observed on negatively charged membranes; (v) confirmed the electrostatic nature of how such membranes disable perforin; and (vi) showed that perforin pore formation is relatively insensitive to interfacial membrane tension, although it may play a role in restoring perforin functionality on PS-rich membranes.

To assess the physiological relevance of these findings, they need to be compared with cell-based assays, e.g., possible correlations of perforin lysis with lipid order in target cell membranes,[12] which confirm that reduction of lipid order sensitizes CTL membranes to perforin and that non-lytic perforin is co-localized with externalized (non-apoptotic) PS on CTLs.[14]

Finally, it is noted that related pore forming proteins have been reported to show specificity for particular lipids, e.g., the membrane attack complex[30] and gasdermin[31] show specificity for negatively charged lipids, whereas bacterial CDCs prefer cholesterol-rich and hence liquid-ordered domains in phase sepa-rated membranes.[46] These observations indicate a wide range of biomedically relevant processes in which the physical properties of membranes may be determinants of the function of pore forming proteins.

Conflicts of interest

There are no conflicts to declare.

Acknowledgements

We thank Richard Thorogate, Elena Taran, and Tian Zheng for technical support and access to AFM facilities; Peter Andrew and Rana Lonnen for providing PFO; and Sandra Verschoor and Annette Ciccone for expression and purification of perforin and perforin mutants. This work has been funded by an NHMRC Project Grant (1128587), an NHMRC Fellowship (1059126), the BBSRC (BB/J005932/1, BB/J006254/1 and BB/N015487/1); the EPSRC (EP/M028100/1); the Sackler Foundation; and a Swiss National Science Foundation Grant (p2skp3_187634).

Notes and references

1 I. Voskoboinik, J. C. Whisstock and J. A. Trapani, *Nat. Rev. Immunol.*, 2015, **15**, 388–400.

2 P. Golstein and G. M. Griffiths, *Nat. Rev. Immunol.*, 2018, **18**, 527–535.

3 J. A. Lopez, M. R. Jenkins, J. A. Rudd-schmidt, A. J. Brennan, J. C. Danne, S. I. Mannering, J. A. Trapani and I. Voskoboinik, *J. Immunol.*, 2013, **191**, 2328–2334.

4 M. R. Jenkins, J. A. Rudd-Schmidt, J. A. Lopez, K. M. Ramsbottom, S. I. Mannering, D. M. Andrews, I. Voskoboinik and J. A. Trapani, *J. Exp. Med.*, 2015, **212**, 307–317.

5 I. Voskoboinik, M. C. Thia, J. Fletcher, A. Ciccone, K. Browne, M. J. Smyth and J. A. Trapani, *J. Biol. Chem.*, 2005, **280**, 8426–8434.

6 R. U. Moreno, J. Gil, C. Rodriguez-Sainz, E. Cela, V. LaFay, B. Oloizia, A. B. Herr, J. Sumegi, M. B. Jordan and K. A. Risma, *Blood*, 2009, **113**, 338–346.

7 R. H. P. Law, N. Lukoyanova, I. Voskoboinik, T. T. Caradoc-Davies, K. Baran, M. A. Dunstone, M. E. D'Angelo, E. V. Orlova, F. Coulibaly, S. Verschoor, K. a Browne, A. Ciccone, M. J. Kuiper, P. I. Bird, J. A. Trapani, H. R. Saibil and J. C. Whisstock, *Nature*, 2010, **468**, 447–451.

8 D. A. K. Traore, A. J. Brennan, R. H. P. Law, C. Dogovski, M. A. Perugini, N. Lukoyanova, E. W. W. Leung, R. S. Norton, J. A. Lopez, K. A. Browne, H. Yagita, G. J. Lloyd, A. Ciccone, S. Verschoor, J. A. Trapani, J. C. Whisstock and I. Voskoboinik, *Biochem. J.*, 2013, **456**, 323–335.

9 H. Yagi, P. J. Conroy, E. W. W. Leung, R. H. P. Law, J. A. Trapani, I. Voskoboinik, J. C. Whisstock and R. S. Norton, *J. Biol. Chem.*, 2015, **290**, 25213–25226.

10 J. Tschopp, S. Schäfer, D. Masson, M. C. Peitsch and C. Heusser, *Nature*, 1989, **337**, 272–274.

11 R. Antia, R. A. Schlegel and P. Williamson, *Immunol. Lett.*, 1992, **32**, 153–157.

12 D. M. Ojcius, S. Jiang, P. M. Persechini, J. Storch and J. D. Young, *Mol. Immunol.*, 1990, **27**, 839–845.

13 C. Leung, A. W. Hodel, A. J. Brennan, N. Lukoyanova, S. Tran, C. M. House, S. C. Kondos, J. C. Whisstock, M. A. Dunstone, J. A. Trapani, I. Voskoboinik, H. R. Saibil and B. W. Hoogenboom, *Nat. Nanotechnol.*, 2017, **12**, 467–473.

14 J. A. Rudd-Schmidt, A. W. Hodel, T. Noori, J. A. Lopez, H. J. Cho, S. Verschoor, A. Ciccone, J. A. Trapani, B. W. Hoogenboom and I. Voskoboinik, *Nat. Commun.*, 2019, **10**, 1–13.

15 N. Rojko and G. Anderluh, *Acc. Chem. Res.*, 2015, **48**, 3073–3079.

16 N. V. Dudkina, B. A. Spicer, C. F. Reboul, P. J. Conroy, N. Lukoyanova, H. Elmlund, R. H. P. Law, S. M. Ekkel, S. C. Kondos, R. J. A. Goode, G. Ramm, J. C. Whisstock, H. R. Saibil and M. A. Dunstone, *Nat. Commun.*, 2016, **7**, 10588.

17 N. Lukoyanova, B. W. Hoogenboom and H. R. Saibil, *J. Cell Sci.*, 2016, **129**, 2125–2133.

18 R. J. C. Gilbert, M. D. Serra, C. J. Froelich, M. I. Wallace and G. Anderluh, *Trends Biochem. Sci.*, 2014, **39**, 510–516.

19 C. F. Reboul, J. C. Whisstock and M. A. Dunstone, *Biochim. Biophys. Acta, Biomembr.*, 2016, **1858**, 475–486.

20 P. Drücker, I. Iacovache, S. Bachler, B. Zuber, E. B. Babiychuk, P. S. Dittrich and A. Draeger, *Biomater. Sci.*, 2019, **7**, 3693–3705.

21 B. B. Johnson, P. C. Moe, D. Wang, K. Rossi, B. L. Trigatti and A. P. Heuck, *Biochemistry*, 2012, **51**, 3373–3382.

22 J. J. Flanagan, R. K. Tweten, A. E. Johnson and A. P. Heuck, *Biochemistry*, 2009, **48**, 3977–3987.

23 L. D. Nelson, A. E. Johnson and E. London, *J. Biol. Chem.*, 2008, **283**, 4632–4642.

24 J. P. Sáenz, D. Grosser, A. S. Bradley, T. J. Lagny, O. Lavrynenko, M. Broda and K. Simons, *Proc. Natl. Acad. Sci. U. S. A.*, 2015, **112**, 11971–11976.

25 E. J. Dufourc, *J. Chem. Biol.*, 2008, **1**, 63–77.

26 T. Tomita, K. Noguchi, H. Mimuro, F. Ukaji, K. Ito, N. Sugawara-Tomita and Y. Hashimoto, *J. Biol. Chem.*, 2004, **279**, 26975–26982.

27 K. Ota, A. Leonardi, M. Mikelj, M. Skočaj, T. Wohlschlager, M. Künzler, M. Aebi, M. Narat, I. Križaj, G. Anderluh, K. Sepčić and P. Maček, *Biochimie*, 2013, **95**, 1855–1864.

28 N. Lukoyanova, S. C. Kondos, I. Farabella, R. H. P. Law, C. F. Reboul, T. T. Caradoc-Davies, B. A. Spicer, O. Kleifeld, D. A. K. Traore, S. M. Ekkel, I. Voskoboinik, J. A. Trapani, T. Hatfaludi, K. Oliver, E. M. Hotze, R. K. Tweten, J. C. Whisstock, M. Topf, H. R. Saibil and M. A. Dunstone, *PLoS Biol.*, 2015, **13**, e1002049.

29 M. Novak, T. Krpan, A. Panevska, L. K. Shewell, C. J. Day, M. P. Jennings, G. Guella and K. Sepčić, *Biochim. Biophys. Acta, Biomembr.*, 2020, 183307.

30 E. S. Parsons, G. J. Stanley, A. L. B. Pyne, A. W. Hodel, A. P. Nievergelt, A. Menny, A. R. Yon, A. Rowley, R. P. Richter, G. E. Fantner, D. Bubeck and B. W. Hoogenboom, *Nat. Commun.*, 2019, **10**, 1–10.

31 P. Broz, P. Pelegrín and F. Shao, *Nat. Rev. Immunol.*, 2020, **20**, 143–157.

32 R. N. Lewis, B. D. Sykes and R. N. McElhaney, *Biochemistry*, 1988, **27**, 880–887.

33 K. Baran, M. Dunstone, J. Chia, A. Ciccone, K. A. Browne, C. J. P. Clarke, N. Lukoyanova, H. Saibil, J. C. Whisstock, I. Voskoboinik and J. A. Trapani, *Immunity*, 2009, **30**, 684–695.

34 J. A. Lopez, A. J. Brennan, J. C. Whisstock, I. Voskoboinik and J. A. Trapani, *Trends Immunol.*, 2012, **33**, 406–412.

35 S. S. Metkar, M. Marchioretto, V. Antonini, L. Lunelli, B. Wang, R. J. C. Gilbert, G. Anderluh, R. Roth, M. Pooga, J. Pardo, J. E. Heuser, M. D. Serra and C. J. Froelich, *Cell Death Differ.*, 2015, **22**, 74–85.

36 R. N. Lewis, N. M. Nanette Mak and R. N. McElhaney, *Biochemistry*, 1987, **26**, 6118–6126.

37 S. L. Veatch and S. L. Keller, *Phys. Rev. Lett.*, 2005, **94**, 3–6.

38 R. F. M. De Almeida, A. Fedorov and M. Prieto, *Biophys. J.*, 2003, **85**, 2406–2416.

39 K. Furuya and T. Mitsui, *J. Phys. Soc. Jpn.*, 1979, **46**, 611–616.

40 R. A. Demel, F. Paltauf and H. Hauser, *Biochemistry*, 1987, **26**, 8659–8665.

41 P. W. M. Van Dijck, *Biochim. Biophys. Acta, Biomembr.*, 1979, **555**, 89–101.

42 E. B. Smaal, K. Nicolay, J. G. Mandersloot, J. de Gier and B. de Kruijff, *Biochim. Biophys. Acta, Biomembr.*, 1987, **897**, 453–466.

43 N. Yilmaz and T. Kobayashi, *Biochim. Biophys. Acta, Biomembr.*, 2016, **1858**, 500–511.

44 A. W. Hodel, C. Leung, N. V. Dudkina, H. R. Saibil and B. W. Hoogenboom, *Curr. Opin. Struct. Biol.*, 2016, **39**, 8–15.

45 D. M. Czajkowsky, E. M. Hotze, Z. Shao and R. K. Tweten, *EMBO J.*, 2004, **23**, 3206–3215.

46 C. Leung, N. V Dudkina, N. Lukoyanova, A. W. Hodel, I. Farabella, A. P. Pandurangan, N. Jahan, M. Pires Damaso, D. Osmanović, C. F. Reboul, M. A. Dunstone, P. W. Andrew, R. Lonnen, M. Topf, H. R. Saibil and B. W. Hoogenboom, *eLife*, 2014, **3**, e04247.

47 S. D. Connell and D. A. Smith, *Mol. Membr. Biol.*, 2006, **23**, 17–28.

48 V. R. Sutton, N. J. Waterhouse, K. Baran, K. Browne, I. Voskoboinik and J. A. Trapani, *Methods*, 2008, **44**, 241–249.

49 R. P. Richter, N. Maury and A. R. Brisson, *Langmuir*, 2005, **21**, 299–304.

50 M. J. Hope, M. B. Bally, G. Webb and P. R. Cullis, *Biochim. Biophys. Acta, Biomembr.*, 1985, **812**, 55–65.

51 C. P. S. Tilcock, *Chem. Phys. Lipids*, 1986, **40**, 109–125.

52 J. R. Kremer, D. N. Mastronarde and J. R. McIntosh, *J. Struct. Biol.*, 1996, **116**, 71–76.

53 T. Hothorn, F. Bretz and P. Westfall, *Biom. J.*, 2008, **50**, 346–363.

54 A. J. Farrand, S. LaChapelle, E. M. Hotze, A. E. Johnson and R. K. Tweten, *Proc. Natl. Acad. Sci. U. S. A.*, 2010, **107**, 4341–4346.

55 X. Liu, Z. Zhang, J. Ruan, Y. Pan, V. G. Magupalli, H. Wu and J. Lieberman, *Nature*, 2016, **535**, 153–158.

56 S. S. Pang, C. Bayly-Jones, M. Radjainia, B. A. Spicer, R. H. P. Law, A. W. Hodel, E. S. Parsons, S. M. Ekkel, P. J. Conroy, G. Ramm, H. Venugopal, P. I. Bird, B. W. Hoogenboom, I. Voskoboinik, Y. Gambin, E. Sierecki, M. A. Dunstone and J. C. Whisstock, *Nat. Commun.*, 2019, **10**, 1–9.

57 G. Pabst, A. Hodzic, J. Štrancar, S. Danner, M. Rappolt and P. Laggner, *Biophys. J.*, 2007, **93**, 2688–2696.

58 R. J. Elin, *DM, Dis.-Mon.*, 1988, **34**, 166–218.

59 S. J. Tilley, E. V. Orlova, R. J. C. Gilbert, P. W. Andrew and H. R. Saibil, *Cell*, 2005, **121**, 247–256.

60 K. van Pee, E. Mulvihill, D. J. Müller and Ö. Yildiz, *Nano Lett.*, 2016, **16**, 7915–7924.

61 R. Basu, B. M. Whitlock, J. Husson, A. Le Floc'h, W. Jin, A. Oyler-Yaniv, F. Dotiwala, G. Giannone, C. Hivroz, N. Biais, J. Lieberman, L. C. Kam and M. Huse, *Cell*, 2016, **165**, 100–110.

62 V. A. Frolov, A. V. Shnyrova and J. Zimmerberg, *Cold Spring Harbor Perspect. Biol.*, 2011, **3**, a004747.

63 G. van Meer, D. R. Voelker and G. W. Feigenson, *Nat. Rev. Mol. Cell Biol.*, 2008, **9**, 112–124.

64 P. A. Leventis and S. Grinstein, *Annu. Rev. Biophys.*, 2010, **39**, 407–427.

65 J. H. Stafford and P. E. Thorpe, *Neoplasia*, 2011, **13**, 299–308.

Faraday Discussions

Theoretical and experimental studies of complex peptide–membrane systems: general discussion

Mibel Aguilar, Kareem Al Nahas, Francisco Barrera, Patricia Bassereau, Margarida Bastos, ⬛ Paul Beales, Burkhard Bechinger, Boyan Bonev, Izabella Brand, Amitabha Chattopadhyay, Ronald J. Clarke, ⬛ William DeGrado, Evelyne Deplazes, Ana J. Garcia Saez, Bart Hoogenboom, Reidar Lund, Paula Milán Rodríguez, Paul O'Shea, Georg Pabst, Sreetama Pal, Aurélien Roux, John Sanderson, Enrico Federico Semeraro, ⬛ Durba Sengupta, David P. Siegel, ⬛ Leonie van 't Hag, Aishwarya Vijayakumar and Larisa Zoranić

DOI: 10.1039/d1fd90066h

Mibel Aguilar opened a general discussion of the paper by Ronald J. Clarke: Did you try other negatively charged lipids to determine if the electrostatic interaction was specific for PS? Also, what was the P : L ratio for the QCM and CD experiments? Would you have expected the helical content to be higher than you observed?

Ronald J. Clarke replied: Yes, in fact we did investigate other negatively charged lipids. In the course of our studies we found that a simple but very effective method to screen for electrostatic interactions is to measure the static light scattering of suspensions of lipid vesicles or membrane fragments containing the Na^+,K^+-ATPase. If an interaction occurs, aggregation and flocculation of the colloidal system occurs, resulting initially in a large increase in light scattering.[1] At relatively high concentrations of polyamino acids, the light scattering increase is followed by a slower decay in scattering, probably due to permeabilisation and disruption of the lipid membrane. For these measurements we repurposed a fluorimeter in our laboratory and measured the scattering at 90° to the incident light at 826 nm, corresponding to a high intensity line of the instrument's xenon arc lamp. To determine whether the peptide interaction with the membrane is specific to phosphatidylserine, we carried out light scattering measurements using vesicles composed of 85 mol% dioleoylphosphatidylcholine (DOPC) and 15 mol% of either dioleoylphosphatidylserine (DOPS), dioleoylphosphatidylglycerol (DOPG) or dioleoylphosphatidic acid (DOPA). Vesicles containing DOPS, DOPG or DOPA all showed similar large increases in light scattering after the addition of poly-L-lysine, indicating that the electrostatic interaction of lysine with the membrane is not specific to PS.[2] Lysine also interacts with PG and PA. In contrast, no increase in light scattering was observed when poly-L-lysine was added to uncharged vesicles composed of 100% DOPC or 85% DOPC and 15% dioleoylphosphatidylethanolamine (DOPE). Using the same technique we also found that the addition of poly-L-arginine to Na^+,K^+-ATPase-containing membrane fragments caused slightly larger increases in light scattering than those observed with poly-L-lysine, indicating that membrane interaction is not specific to lysine. Experiments using a negatively charged polyamino acid, poly-L-glutamic acid, showed no increase in light scattering. The interaction, therefore, appears to be nonspecific, simply requiring a negatively charged membrane surface and a positively charged amino acid side chain.

In the case of the QCM-D experiments, a precise peptide : lipid molar ratio (P : L) is difficult to estimate because the lipid is bound to the quartz crystal surface, whereas the peptide is delivered to the surface of the crystal within a continuous flow system. Thus, the lipid is situated on a two-dimensional surface, while the peptide is in a three-dimensional solution, so that they are strictly speaking, not comparable. Nevertheless, the concentration of lipid used in forming the bilayer was 3 mM and the concentrations of peptide used to study their interaction with the membrane were 10 μM for the Na^+,K^+-ATPase peptide and 9 μM for the H^+,K^+-ATPase peptide. Of the 3 mM lipid used in the flow system, a significant fraction of the lipid molecules would have passed through the system without binding to the crystal surface. When the peptides were added to the flow system, a very slow rate of 50 μl min^{-1} was used to allow sufficient time for the peptides to bind to the membrane.

In the case of the CD measurements the lipid concentration was 500 μM and the peptide concentrations were 100 μM for the Na^+,K^+-ATPase peptide and 90 μM for the H^+,K^+-ATPase peptide. However, prior to recording the CD spectrum extravesicular peptide was removed from the system by centrifugation, so that only intravesicular peptide was measured. The reason for this was to maximise the probability of membrane interaction and to avoid a large background of non-membrane-bound extravesicular peptide. Nevertheless, there is still probably a significant amount of unbound intravesicular peptide present, which could account for the fact that the measured helicities of the peptides do not reach 100%. It is worth bearing in mind that in the native protein systems, the peptides would be permanently anchored close to the membrane surface by the transmembrane domains of both ATPases. Therefore, the likelihood of membrane interaction would be much higher than in the studies we have reported here using free peptides.

1 A. Gorman, K. R. Hossain, F. Cornelius and R. J. Clarke, Penetration of phospholipid membranes by poly-L-lysine depends on cholesterol and phospholipid composition, *Biochim. Biophys. Acta, Biomembr.*, 2020, **1862**, 183128.
2 K. Nguyen, A. Garcia, M.-A. Sani, D. Diaz, V. Dubey, D. Clayton, G. Dal Poggetto, F. Cornelius, R. J. Payne, F. Separovic, H. Khandelia and R. J. Clarke, Interaction of N-terminal peptide analogues of the Na^+,K^+-ATPase with membranes, *Biochim. Biophys. Acta, Biomembr.*, 2018, **1860**, 1282–1291.

Leonie van 't Hag asked: Why do you think there is a small interaction with DOPC/cholesterol bilayers in the QCM experiment whereas there was none with pure DOPC? Did you also try DOPC/DOPS without cholesterol? This would be a nice addition to see if there was also a difference in that case.

Ronald J. Clarke replied: The small interaction with DOPC/cholesterol bilayers could perhaps be explained by the cholesterol-induced increase in membrane dipole potential.[1] Because the polarity of the dipole potential is positive in the membrane interior and negative on the membrane surface, cholesterol would be expected to make the membrane surface slightly more negatively charged, which could attract more peptide. However, the effect of cholesterol on peptide binding is quite small and much smaller than the effect observed after including phosphatidylserine, with a net negative charge. We haven't looked at peptide binding to DOPC/DOPS membranes in the absence of cholesterol. I agree that this could be useful to see what effect the omission of cholesterol makes. The native membranes with which the peptide would interact, however, contain a high mole percentage of cholesterol of 40%. Therefore, we have included cholesterol at the same level so as to reproduce the native situation as closely as possible.

1 T. Starke-Peterkovic, N. Turner, M. F. Vitha, M. P. Waller, D. E. Hibbs and R. J. Clarke, Cholesterol effect on the dipole potential of lipid membranes, *Biophys. J.*, 2006, **90**, 4060–4070.

Leonie van 't Hag commented further: I was wondering if you tried circular dichroism spectroscopy with vesicles in the presence of salt, and if that showed enhanced peptide folding? I guess the experiment was done in the absence of salt due to the light absorption of chloride in the far-UV region. This experiment works very well, however, when one uses fluoride salts. We have for example done this for CD measurements with peptides in cubosomes.[1]

1 L. van't Hag, X. Li, T. G. Meikle, S. V. Hoffmann, N. C. Jones, J. S. Pedersen, A. M. Hawley, S. L. Gras, C. E. Conn and C. J. Drummond, How peptide molecular structure and charge influence the nanostructure of lipid bicontinuous cubic mesophases: model synthetic WALP peptides provide insights, *Langmuir*, 2016, **32**, 6882–6894.

Ronald J. Clarke replied: Thank you for your good suggestion. At this stage we have only carried out measurements in 10 mM Tris buffer, because of its low absorbance above 190 nm[1] and in the absence of salt in order to maximise the strength of attraction of the peptides for the membrane. A problem we have is that the CD signals are quite small, because we are only measuring peptide encapsulated within the vesicles. Free non-encapsulated peptide was removed by centrifugation and resuspension of the pellet in order to maximise the probability of detecting membrane-bound peptide. In the native system the N-terminal peptide is attached to the rest of the protein, *i.e.*, the Na^+,K^+- or H^+,K^+-ATPase, and can, therefore, never escape the vicinity of the membrane. In the future we intend to synthesise peptides with hydrophobic anchors at the C-termini to more closely approximate the native situation. This should allow us to measure circular dichroism at higher peptide concentrations and obtain an improved signal-to-noise ratio. Once we've achieved this we could then more easily study the effect of the addition of salts, such as NaF.

1 S. M. Kelly and N. C. Price, The use of circular dichroism in the investigation of protein structure and function, *Curr. Protein Pept. Sci.*, 2000, **1**, 349–384.

Larisa Zoranić remarked: In the article, there is a reference: "the proteins of hyperthermophilic organisms preferentially contain lysine residues (*i.e.*, over arginine), because of their greater number of rotameric conformations, which leads to an enhanced thermostability *via* an entropic stabilization mechanism". Could you say more about the preferences of lysine over arginine (residues of the N-termini) in the case of ion pumps? Is this preference also connected to the proposed electrostatically based detection mechanism for the changes in lipid membrane composition?

Ronald J. Clarke answered: This is a very interesting question. As in the case of hyperthermophilic organisms, the lysine residues of the N-terminal peptide of the Na^+,K^+- and H^+,K^+-ATPases could provide an entropic stabilisation when the peptide is not bound to the membrane. Berezovsky *et al.*[1] attributed the preference of hyperthermophilic bacteria for lysine over arginine to a higher number of accessible rotamers in lysine. This could partly be due to the fact that the lysine side chain has one more tetrahedral carbon than the arginine side chain. To estimate the possible order of magnitude of the entropic stabilisation due to the N-terminal peptide of the ATPases, we performed a prediction of the *Homo sapiens* peptides' secondary structures using the QUARK server[2] and inputted the resulting pdb files into the PLOPS server[3] to obtain a value of the entropy change for the complete unfolding of the peptides. The calculations yielded values of $-T\Delta S$ of -234 kJ mol^{-1} for the Na^+,K^+-ATPase and -322 kJ mol^{-1} for the H^+,K^+-ATPase.[4] These values are much greater than the free energy of ATP hydrolysis under physiological conditions of -55 kJ mol^{-1}.[5] Although it is unlikely that the N-termini undergo complete unfolding, and the entropy changes could be partially compensated by changes in the entropy of solvating water molecules, the calculations indicate that entropic stabilisation of the N-terminal peptides could make a significant contribution to the proteins' energetics and kinetics, *i.e.*, by modifying the activation energy barriers of partial reactions of the ion pumping cycles.

When lysines of the N-termini interact electrostatically with the surrounding membrane, they produce an enthalpic stabilisation, which could also modify activation energy barriers for ion pumping. In this case lysine may perhaps be preferred to arginine, because, according to molecular dynamics simulations,[6] arginine penetrates further into the membrane than lysine and, therefore, may bind so strongly that conformational transitions of the ATPases involving the release of the N-terminal peptide from the membrane, and which may be required for ion pumping, might be inhibited. The deeper penetration of an arginine residue into a lipid membrane than a lysine residue, can also be rationalised simply by considering the structures of the two amino acid residues at neutral pH. Both have positive charges, but in the case of lysine the charge is localized on an ammonium group, whereas in the case of arginine the charge is delocalized across a guanidinium group. Thus, the charge density and hence the Born energy within a membrane would be lower in the case of arginine than lysine, allowing greater penetration by arginine.

1 I. N. Berezovsky, W. W. Chen, P. J. Choi and E. I. Shakhnovich, Entropic stabilization of proteins and its proteomic consequences, *PLoS Comput. Biol.*, 2005, **1**, e47.
2 D. Xu and Y. Zhang, *Ab initio* protein structure assembly using continuous structure fragments and optimized knowledge-based force field, *Proteins*, 2012, **80**, 1715–1735.

3 M. C. Baxa, E. J. Haddadian, J. M. Jumper, K. F. Freed and T. R. Sosnick, Loss of conformational entropy in protein folding calculated using realistic ensembles and its implications for NMR-based calculations, *Proc. Natl. Acad. Sci. U. S. A.*, 2014, **11**, 15396–15401.
4 K. R. Hossain, X. Li, T. Zhang, S. Paula, F. Cornelius and R. J. Clarke, Polarity of the ATP
binding site of the Na$^+$,K$^+$-ATPase, gastric H$^+$,K$^+$-ATPase and sarcoplasmic reticulum Ca^{2+}-
ATPase, *Biochim. Biophys. Acta, Biomembr.*, 2020, **1862**, 183138.
5 R. J. Clarke, M. Catauro, H. H. Rasmussen, and H.-J. Apell, Quantitative calculation of the
role of the Na$^+$,K$^+$-ATPase in thermogenesis, *Biochim. Biophys. Acta, Bioenerg.*, 2013, **1827**,
1205–1212.
6 L. B. Li, I. Vorobyov and T. W. Allen, The different interactions of lysine and arginine side
chains with lipid membranes, *J. Phys. Chem. B*, 2013, **117**, 11906–11920.

Paul O'Shea commented: Regarding your use of 150 mM KCl to subdue the
electrostatic interactions; if Cl$^-$ gets in the way of CD measurements and the
other suggestion of using fluoride doesn't work (*i.e.* exchanging chloride for
something else), I am sure you know that 150 mM monovalent K$^+$ chloride is
about the same as a few mM divalent cations (*e.g.* Ca^{2+}) in electrostatic effects on
membranes, so perhaps this could be used in CD?

Ronald J. Clarke answered: Yes, it is true that using CaCl$_2$ instead of NaCl or
KCl one could achieve a higher ionic strength with a lower concentration of salt.
However, in the case of Ca^{2+}, any effects seen could not simply be attributed to
a nonspecific charge screening effect. Ca^{2+} reduces the membrane dipole
potential of phosphatidylcholine membranes, probably by binding to its phosphate group.[1] It has also long been known that Ca^{2+} and Mg^{2+} bind to the
headgroup of phosphatidylserine.[2,3] Therefore, assuming we do see changes in
membrane binding of our peptides after adding CaCl$_2$, it's likely that these
changes would be due to Ca^{2+} binding to the lipid headgroups rather than simply
charge screening. This said, it would actually be very interesting to study the effect
of Ca^{2+} ions because of their known role in cell signalling.

1 R. J. Clarke and C. Luepfert, Influence of anions and cations on the dipole potential of
phosphatidylcholine vesicles: a basis for the Hofmeister effect, *Biophys. J.*, 1999, **76**, 2614–
2624.
2 H. Hauser, D. Chapman and R. M. C. Dawson, Physical studies of phospholipids: XI Ca^{2+}
binding to monolayers of phosphatidylserine and phosphatidylinositol, *Biochim. Biophys.
Acta, Biomembr.*, 1969, **183**, 320–333.
3 C. Newton, W. Pangborn, S. Nir and D. Papahadjopoulos, Specificity of Ca^{2+} and Mg^{2+}
binding to phosphatidylserine vesicles and resultant phase changes of bilayer membrane
structure, *Biochim. Biophys. Acta, Biomembr.*, 1978, **506**, 281–287.

Paul O'Shea continued: Ron, the satellite peptide on the ATPase is very
interesting, particularly the involvement of phosphorylation. Such a dramatic
localised change of the charge density could lead to lateral movement of the
ATPase (see *e.g.* ref. 1) as well as changes of the activity. I wonder therefore, does it
have any bearing on the sorting of the ATPase in the membrane? There is
precedence for electrostatic localisation of ATPases, as many years ago Jim Barber
at Imperial College promoted the idea of chloroplast ATPase sorting and thylakoid stacking due to changes of the surface electrostatics.[2] I was intrigued by your
story of the peptide flipping in or out based on phosphorylation and electrostatics. Does this have any basis on the sorting of the ATPases?

1 J. L. Richens, J. S. Lane, J. P. Bramble and P. O'Shea, The electrical interplay between proteins and lipids in membranes, *Biochim. Biophys. Acta, Biomembr.*, 2015, **1848**, 1828–1836.

2 J. Barber, Influence of surface charges on thylakoid structure and function, *Annu. Rev. Plant Physiol.*, 1982, **33**, 261–295.

Ronald J. Clarke responded: The question of the quaternary structure of the Na^+,K^+-ATPase is a very interesting one, which has been the topic of debate for many years. It is known that the minimal functional unit of the Na^+,K^+-ATPase consists of two subunits; the catalytic alpha subunit with ten transmembrane helices and the beta subunit with a single transmembrane helix, which is necessary for its trafficking to the plasma membrane and is probably involved in the pump's regulation. In addition, the Na^+,K^+-ATPase can be further modified by another single transmembrane FXYD protein, termed the gamma subunit in the kidney enzyme.[1] In 2007, with the publication of the first crystal structure of the Na^+,K^+-ATPase, it became clear that the ion transport pathway of the enzyme is located within the alpha subunit.[2] Therefore, further oligomerisation is not necessary to create the ion pathway. Nevertheless, there is a substantial amount of data indicating that the Na^+,K^+-ATPase is able to aggregate within the membrane to form $\alpha_2\beta_2$ dimers or $\alpha_4\beta_4$ tetramers.[3] The exact role of this aggregation of the protein is still uncertain, but kinetic evidence obtained *via* the stopped-flow technique indicates that the degree of aggregation depends on the concentration of the substrate ATP and the ionic strength of the surrounding solution, and that the aggregation into a dimer or higher oligomer decreases the rate of ion pumping but increases the ATP binding affinity.[4–7] The dependence of the state of aggregation of the Na^+,K^+-ATPase on the ionic strength of the surrounding solution is a strong indication that the protein dimers or oligomers are stabilized by an electrostatic interaction. Considering the data which we have shown in our paper, as well as in previously published work,[8] indicating an electrostatic interaction between the alpha subunit's N-terminus with the surrounding membrane, it appears likely that the N-terminus could be involved in determining the state of aggregation of the protein within the membrane, as you suggest.

1 M. V. Clausen, F. Hilbers and H. Poulsen, The structure and function of the Na^+,K^+-ATPase isoforms in health and disease, *Front. Physiol.*, 2017, **8**, 371.

2 J. P. Morth, B. P. Pedersen, M. S. Toustrup-Jensen, T. L. M. Sørensen, J. Petersen, J. P. Andersen, B. Vilsen and P. Nissen, Crystal structure of the sodium-potassium pump, *Nature*, 2007, **459**, 446–450.

3 R. J. Clarke and X. Fan, Pumping ions, *Clin. Exp. Pharmacol. Physiol.*, 2011, **38**, 726–33.

4 C. Lüpfert, E. Grell, V. Pintschovius, H.-J. Apell, F. Cornelius and R. J. Clarke, Rate limitation of the Na^+,K^+-ATPase pump cycle, *Biophys. J.*, 2001, **81**, 2069–2081.

5 R. J. Clarke and D. J. Kane, Two gears of pumping, *Biophys. J.*, 2007, **93**, 4187–4196.

6 R. J. Clarke, Mechanism of allosteric effects of ATP on the kinetics of P-type ATPases, *Eur. Biophys. J.*, 2009, **39**, 3–17.

7 Q. Jiang, A. Garcia, M. Han, F. Cornelius, H.-J. Apell, H. Khandelia and R. J. Clarke, Electrostatic stabilization plays a central role in autoinhibitory regulation of the Na^+,K^+-ATPase, *Biophys. J.*, 2017, **112**, 288–299.

8 K. Nguyen, A. Garcia, M.-A. Sani, D. Diaz, V. Dubey, D. Clayton, G. Dal Poggetto, F. Cornelius, R. J. Payne, F. Separovic, H. Khandelia and R. J. Clarke, Interaction of N-terminal peptide analogues of the Na^+,K^+-ATPase with membranes, *Biochim. Biophys. Acta, Biomembr.*, 2018, **1860**, 1282–1291.

Paul O'Shea continued: Does the peptide perhaps talk to cholesterol rich domains when it becomes incorporated in the membrane, and so may have a role in targeting or localisation of the ATPase? Have you seen any domain formation as a result of the peptide interaction or localisation of the ATPases or even 'polymerisation' of the ATPases?

Ronald J. Clarke answered: Cholesterol certainly does have a strong effect on Na^+,K^+-ATPase activity, which increases significantly with increasing cholesterol content of the membrane up to the physiological level of around 40 mol% of the total lipid.[1] Cholesterol favours conversion of the enzyme into its E1 conformational state, the state to which Na^+ binds. The conformational transition into the E1 state is also necessary to allow phosphorylation of the Na^+,K^+-ATPase by ATP and transfer of Na^+ across the membrane. There is clear evidence from studies looking at the pattern of proteolytic digestion of the protein in its E1 and E2 (K^+ binding) conformations that the N-terminus does undergo significant movement during the E2-to-E1 transition.[2] Therefore, it is indeed possible that there could be some interplay between cholesterol and the protein's N-terminus involved in localising the Na^+,K^+-ATPase to cholesterol-rich domains. It would be interesting to perform some experiments imaging the distribution of the Na^+,K^+-ATPase to see if one could detect any localisation in lipid rafts or liquid-ordered domains.

1 A. Garcia, B. Lev, K. R. Hossain, A. Gorman, D. Diaz, T. H. N. Pham, F. Cornelius, T. W. Allen and R. J. Clarke, Cholesterol depletion inhibits Na^+,K^+-ATPase activity in a near-native membrane environment, *J. Biol. Chem.*, 2019, **194**, 5966–5969.
2 P. L. Jorgensen and J. H. Collins, Tryptic and chymotryptic cleavage sites in sequence of α-subunit of $(Na^+ + K^+)$-ATPase from outer medulla of mammalian kidney, *Biochim. Biophys. Acta*, 1986, **860**, 570–576.

Patricia Bassereau asked: If the pumping activity of the ATP-ase depends on peptide binding, how are the protein activity and the phosphorylation synchronized?

Ronald J. Clarke responded: The proteins' ion pumping activity is not synchronized with the phosphorylation of the conserved serine on the extramembranous N-terminus. P-type ATPase activity is driven by the hydrolysis of ATP, which causes an autophosphorylation (*i.e.*, not catalysed by another enzyme) of a conserved aspartic acid residue on the proteins' α-subunit. Although the entire α-subunit is of course a single polypeptide chain, this aspartic acid residue is not located within the proteins' N-terminal tail which we are investigating. Autophosphorylation of the aspartic acid residue is a catalytic phoshorylation required for ion pumping activity, whereas phosphorylation of the serine of the N-terminal tail by protein kinase C is a regulatory phosphorylation, which alters the ion pumping rate but isn't necessary to allow pumping to occur.

Margarida Bastos said: I am not sure about the relevance of deepness of insertion between Lys and Arg you referred to when answering our colleague, Larisa Zoranić , as although Arg might go deeper into membranes, the CPPs (cell penetrating peptides), rich in Arg, enter the cells, even with cargo, without destroying the membranes as AMPs do. This might be one of the reasons why Arg is the most common positively charged amino acid in CPPs.

Ronald J. Clarke replied: Yes, I agree that arginine is preferable to lysine in cell penetrating peptides to promote transport across the membrane. In the case of the lysine-rich N-terminal peptide of the Na^+,K^+-ATPase which I spoke about, we believe that the peptide moves on and off the membrane during the protein's ion pumping cycle. If the peptide was arginine-rich instead of lysine-rich, the peptide may be so strongly bound to the membrane that it acts as a permanent anchor rather than switching between on and off states. This interpretation is supported by molecular dynamics simulations comparing the arginine and lysine side-chains,[1] which showed enhanced binding of arginine, disrupting and permeabilizing the membrane.

1 L. Li, I. Vorobyov and T. W. Allen, The different interactions of lysine and arginine side chains with lipid membranes, *J. Phys. Chem. B*, 2013, **117**, 11906–11920.

Burkhard Bechinger added: Previously H-bonding interactions between Arg and lipid phosphates have been suggested that pull the lipid head groups into the membrane interior.[1] A difference between lysine and arginine is that the positive charge of the latter is 'smeared out' over a larger part of the molecule when compared to lysine, making the energy of placing an ion in a hydrophobic environment less unfavourable.[2] Furthermore, the arginine with three H-bond acceptors on the side chain should be able to associate with a larger hydration sphere which eases membrane insertion of the charge.

1 M. Tang, A. J. Waring and M. Hong, Phosphate-mediated arginine insertion into lipid membranes and pore formation by a cationic membrane peptide from solid-state NMR, *J. Appl. Chem. Sci.*, 2007, **129**, 11438–11446.
2 J. N. Israelachvili, S. Marcelja and R. G. Horn, Physical principles of membrane organization, *Q. Rev. Biophys.*, 1980, **13**, 121–200.

Sreetama Pal opened a general discussion of the paper by Paul Beales: It might be interesting to explore the role of calcium in these systems. Since you have phosphatidylserine in the system, the presence of calcium could induce isothermal phase transitions. The resulting spatial heterogeneity might provide additional control on the action of ESCRT.

Paul Beales responded: In these experiments we start with phase separated vesicles using a lipid composition close to the middle of the liquid–liquid coexistence region of the phase diagram. We already have divalent Mg^{2+} ions in solution for the ATP hydrolysis.

In future work, an interesting next step could be to make single phase systems close to the phase coexistence boundary and investigate how environment responsive phase separation couples with ESCRT activity. This wouldn't necessarily need calcium ions. ESCRT-II has been shown previously to trigger phase separation in supported lipid bilayers in the work by the group of James Hurley.[1] It is likely that this is a multivalent electrostatic clustering effect as many multivalent lipid clustering interactions have been shown to be able to trigger phase separation in the literature.

1 E. Boura, V. Ivanov, L.-A. Carlson, K. Mizuuchi and J. H. Hurley, *J. Biol. Chem.*, 2012, **287**, 28144–28151.

Patricia Bassereau asked: The L_o phase is more difficult to bend than L_d. So why would the L_o domain be recruited? The Snf7 assembly alone induces a local flat membrane shape but, bends it as soon as other ESCRT-III proteins are recruited, which should in principle recruit the more flexible L_d membrane.

Paul Beales responded: The curvature of the buds is still relatively small as their radius of curvature is still in the μm range. Therefore the energy cost of bending is not likely to be severe. The energy cost of bending will also be compensated by the reduction in line tension at the domain edge. The high curvature region of the buds is in the neck, where we still see evidence of the importance of the L_d domains in these regions.

Patricia Bassereau remarked: How would your mechanism work on asymmetric membranes, such as the plasma membrane?

Paul Beales replied: Great question. This would be a really interesting experiment to do but isn't something we have looked at. Asymmetric GUVs are not a standard system and so present challenges. If I was to speculate, bilayer domains are still likely to occur in the plasma membrane since this has been observed in asymmetric model membranes. Therefore the mechanisms we observe in these model membranes may well be relevant *in vivo* too.

Patricia Bassereau asked: You propose that the L_o domain is nucleated or recruited under the ESCRTs filaments. But, when the bud forms, the ESCRTs are expected to be at the neck where you also find the L_d membrane if I understand, and the L_o domain is more underneath but in the bud. Can you explain more about this?

Paul Beales responded: The model we present to explain our observations proposes that the L_o phase may be recruited beneath the growing ESCRT complex due to an entropic effect where suppressed membrane thermal undulations would favour the more rigid phase (analogous to a similar effect reported previously by Gordon and Deserno *et al.*[1]). Once the membrane begins to bud, either the L_o domain is kinetically trapped in this position, or the reduction in line tension by constriction of the neck becomes the energetic factor that favours the L_o domain localisation in the bud. This is currently our best hypothesis using plausible biophysical mechanisms. However, we do not present this as a proven mechanism and we are open to alternative models.

1 V. D. Gordon, M. Deserno, C. M. J. Andrew, S. U. Egelhaaf and W. C. K. Poon, Adhesion promotes phase separation in mixed-lipid membranes, *EPL*, 2008, **84**, 48003.

Durba Sengupta asked: Could the ESCRT cause L_o/L_d local "remixing" by modulating the line tension?

Paul Beales responded: On the contrary, ESCRT-II has been shown to induce phase separation on surface-supported membranes.[1] Furthermore, we do not see any evidence for phase remixing in our experiments – the timescale for which would be expected to be slow enough to be observable. As these membranes are in a L_o–L_d region of the phase diagram, the bulk free energy of each phase is favourable in equilibrium. It would need a very large line tension to disfavour the phase separation and based on other reports, where there is a high line tension between membrane domains, membrane bending and vesicle fission has been reported. Interestingly, ESCRTs do drive membrane fission, so I would not discount that modulation of line tension could play a role in this function.

1 E. Boura, V. Ivanov, L.-A. Carlson, K. Mizuuchi and J. H. Hurley, Endosomal sorting complex required for transport (ESCRT) complexes induce phase-separated micro-domains in supported lipid bilayers, *J. Biol. Chem.*, 2012, **287**, 28144–28151.

Amitabha Chattopadhyay said: Although asymmetric vesicles (liposomes) are difficult to prepare, Erwin London (Stony Brook University) has been able to generate asymmetric liposomes of different sizes using cyclodextrin-mediated exchange.

Paul Beales responded: I agree that asymmetric vesicles have been prepared but I feel this is a long way from a robust and standardised approach for the field. Knowing/controlling the precise compositions of the two leaflets of the membrane and characterising the lifetime of the asymmetry, which will dissipate over time due to lipid flip-flop, are examples of challenges in working with asymmetric vesicles. I agree that this is an important future direction for membrane biophysics research, but not something we have investigated at this stage.

Patricia Bassereau commented: A comment on phase-separation induced by ESCRT: it is a well-known effect that many multi-valent proteins that bind to lipids change the phase diagram and facilitate phase-separation (entropic effect). This is the case for toxins that bind to glycolipids, such as the Shiga toxin (15 Gb3) or Cholera toxin (5 GM1), or actin filament binding to PiP2 lipids.

Paul Beales answered: I agree. This has been seen in many systems as a general biophysical phenomenon where extraneous matter binds to the membrane and clusters specific lipids.

Patricia Bassereau also asked: If I understand you correctly, having L_d lipids in the "neck" can reduce the overall bending energy, but it will be higher that a uniform L_d bud? Maybe this is not an equilibrium process?

David P. Siegel answered: Let k_o and k_d be the bending elastic moduli of bilayers in the ordered and disordered phase, respectively. Let κ_o and κ_d be the Gaussian curvature elastic moduli of bilayers in those two phases, respectively. The Gaussian curvature of a separate GUV is 4π. In the limit of large vesicle size, introducing a "neck" connecting the GUV to the flat parent membrane makes a Gaussian curvature contribution of -4π. Let's call the GUV connected to the flat membrane a "bud." The curvature energy of a disordered phase GUV relative to the original flat bilayer $= 8\pi k_d + 4\pi\kappa_d$; that of a non-fissioned L_d-phase bud with an L_d phase "neck" is $8\pi k_d$, and that of a bud mostly in the L_o phase but with an L_d phase "neck" is $= 8\pi k_o + 4\pi(\kappa_o - \kappa_d)$. The difference in curvature energy between the ordered phase bud with the L_d phase "neck" and the bud completely in the L_d phase is $8\pi(k_o - k_d) + 4\pi(\kappa_o - \kappa_d)$. In theory, for $(\kappa_d - \kappa_o) > 2(k_o - k_d)$, the curvature energy of the bud with the L_d neck and L_o phase body is lower than the bud completely in the L_d phase. However, as far as I am aware, the values of the Gaussian curvature moduli of bilayers in these phases are unknown.

Paul Beales added: This effect can be seen in the classic 2003 *Nature* paper from Baumgart,[1] Hess and Webb where they see the phase boundary is not in the neck of budding GUVs in L_o–L_d (which would minimise line tension) but the neck is L_d and the phase boundary sits just outside this due to the contribution of bending energy. Furthermore, by having the phase boundary near the neck of the bud, this lowers the overall line tension between coexisting phases. The reduction in line tension also needs to be considered alongside the curvature energies of the bulk phases.

1 T. Baumgart, S. Hess and W. Webb, Imaging coexisting fluid domains in biomembrane models coupling curvature and line tension, *Nature*, 2003, **425**, 821–824, DOI: 10.1038/nature02013.

David P. Siegel addressed Paul Beales and Patricia Bassereau: The curvature energy of a vesicle produced by budding is determined by the values of both the bilayer bending modulus and the bilayer Gaussian curvature modulus. This is also true of the curvature energy of the intermediate in budding, in which the vesicle is still connected to the parent membrane by a "neck." The Gaussian curvature elastic energy contribution of the "neck" is considerable, and different in sign than that of the final budded vesicle. The values of the Gaussian curvature modulus for the L_d and L_o phases of this composition are unknown, and they are difficult to measure. Depending on the values of these Gaussian curvature moduli with respect to the bending moduli, it is conceivable that a vesicle which is primarily in the L_o phase, but which has a neck region in the L_d phase, could have a total curvature energy which is lower than a bud completely in the L_o phase, and conceivably comparable to a bud totally in the L_d phase. Therefore Professor Beales' finding is not necessarily inconsistent with a simple membrane curvature energy model, even before we consider the effects of line tension.

Paul Beales responded: It would be great to be able to give full consideration of the Gaussian curvature moduli but these are elusive and have only been measured experimentally in a limited number of special cases.

Patricia Bassereau asked: The ESCRT complexes do not remain flat on the GUV if partners of Snf7 are present, and also Vps4. It is expected to induce the neck formation. Are the L_o domains moving away from the ESCRT filaments, but remain trapped in the forming bud?

Paul Beales replied: What we can say from these experiments is that L_o domains preferentially located in the ESCRT-induced buds and that PS is required in the L_d phase, which recruits the ESCRT complex. It also appears that the L_d phase locates to the necks of budding vesicles. We hypothesise in the paper that the suppression of entropic thermal undulations under the assembling ESCRT complex may energetically favour the recruitment of the L_o phase. The budding and neck formation also minimises the line tension between phases. However, I think you'd agree that this is an interesting and unexpected observation, where further work will be required to fully pin down the precise mechanistic explanation.

Paul Beales added, in response to a comment from Aurélien Roux regarding VPS4: There still appears to be debate on this topic in the literature and we do not perform specific experiments that probe the role of Vps4. What is your opinion on the role of Vps4 given your experience with experiments in this area and the possible difference between *in vivo* and *in vitro*?

Aurélien Roux replied: I know that people are debating about Vps4 being a sole depolymerization factor. In my opinion, there is numerous evidence that Vps4 does not solely promote disassembly. First, overexpression of the dominant negative ATPase dead mutant also blocks the fission reaction *in vivo*. It is true for HIV viruses, cytokinetic abscission, and ILV formation. These is quite strong evidence that Vps4 is needed for fission. And we found the same in our *in vitro* reconstituted assay.

Paul Beales responded: Certainly we have seen that without the Vps4, we can get some ILV formation activity. We don't have the resolution in confocal microscopy studies to see what is happening with individual ESCRT components on the membrane. But we do then see a second round of ILV formation events with the addition of Vps4 and ATP.[1] So there appears to be some recycling of the ESCRT activity.

1 A. Booth, C. J. Marklew, B. Ciani and P. A. Beales, *In vitro* membrane remodeling by ESCRT is regulated by negative feedback from membrane tension, *iScience*, 2019, **15**, 173-184.

Paula Milán Rodríguez opened a general discussion of the paper by Reidar Lund: To what extent are the results (degree of insertion in the membrane) sensitive to the membrane lipid composition and P/L ratio? Have you tried other compositions and other ratios?

Reidar Lund replied: The experiments were performed on a variety of P : L ratios (reported in the paper) and we generally find that the interaction is peptide-dependent. For most peptides there is mostly an increase in amount accumulated, but for LL-37 we see a clear concentration dependent interaction pattern where the peptide seems to presumably bundle and insert deeper into the membrane with higher concentrations. We have previously reported data on DMPC/DMPG vesicles containing 2.5 mol% pegylated vesicles. We also varied the fraction of PG, and thereby the overall charge of the membrane, in the case of indolicidin. Surprisingly, we did not see significant changes in peptide insertion in this case.[1] We plan to study the effect of varying the amount of negatively charged lipids on the membrane interaction of other peptides besides indolicidin in future work.

1 J. E. Nielsen, V. A. Bjørnestad and R. Lund, Resolving the structural interactions between antimicrobial peptides and lipid membranes using small-angle scattering methods: the case of indolicidin, *Soft Matter*, 2018, **14**, 8750–8763.

Paula Milán Rodríguez asked: What are the other polar heads and aliphatic tails present in the *E. coli* membranes?

Reidar Lund answered: The lipid composition of *E. coli* has been widely studied, and found to depend on, amongst other things, the growth temperature of the bacteria. With growth at 37 °C the phospholipid headgroup composition of the inner membrane of *E. coli* is reported to be 75±1% phosphatidylethanol-amine (PE), 19±1% phosphatidylglycerol (PG) and 6±2% diphosphatidylglycerol (DPG). While the acyl chain composition is 3.7% 14 : 0, 0.3% 14 : 1c7, 0.4% 15 : 0, 42.8% 16 : 0, 31.9% 16 : 1c9, 1.6% 18 : 0, 14.7% of 18 : 1c11, and 4.6% uniden-tified. This equals 46.9% unsaturated and 50.3% saturated acyl chains (not including the unidentified).[1]

1 M. Sven, A. Ann-Sofie, R. Leif and L. Göran, Wild-type *Escherichia coli* cells regulate the membrane lipid composition in a "window" between gel and non-lamellar structures, *J. Biol. Chem.*, 1996, **271**, 6801–6809.

Georg Pabst said: Your contribution seems to really rely on your specific need to have the unilamellar vesicles. This led you to the use of PEGylated lipids and studies of DMPE/DMPG in the gel phase. As a consequence, you compare gel phase data of DMPE/DMPG to fluid phase data of DMPC/DMPG (*i.e.* completely different states of membrane structure). How can you make conclusions on the activity of your studied peptides in the physiologically relevant state of membranes (which is fluid)? Instead, you could use other and physiologically more relevant lipid mixtures of *e.g.* POPE/POPG (possibly even add cardiolipin) and dose the peptide concentration in a way that does not lead to morphological or topological changes of your vesicles. This is fully compatible with scattering experiments on vesicles, including SANS, to enhance contrast (see *e.g.* reports from our group, *e. g.* ref. 1). Alternatively, you could use the peptide induced topological/morphological change as the readout for peptide activity.

1 M. Pachler, I. Kabelka, M.-S. Appavou, K. Lohner, R. Vácha and G. Pabst, Magainin 2 and PGLa in bacterial membrane mimics I: Peptide-peptide and lipid-peptide interactions, *Biophys. J.*, 2019, **117**, 1858–1869, DOI: 10.1016/j.bpj.2019.10.022.

Reidar Lund responded: First of all it should be noted that accurate determi-nation of the peptide position within the membrane, unilamellar vesicles are needed. Having said that, we agree with the comment that fluid lipid membranes are desirable and we plan more experiments using unsaturated lipids in the future. Our intention in the TR-SANS experiment was to increase the temperature to above the melting point of the lipids. However, the stability of the current system we were working with did not allow us to do this. We do have experiments using other lipids, including cardiolipin, planned, but these have so far been postponed due to Covid-19 and neutron facility shut-downs.

Enrico Federico Semeraro commented: The scattering data analysis of the mixed lipid/AMP systems presents some gaps that the authors should address.

Specifically, (i) the presence of AMPs only on the outer bilayer leaflet, and (ii) the lack of critical comments on the uniqueness of the solution and on the confidence degree of the adjustable parameters.

(i) The presented results show that indolicidin, LL-37 and lacticin Q reside exclusively in the outer membrane leaflet, and aurein 1.2 penetrates deeper into the hydrophobic core. However, samples were measured waiting for equilibration (an aspect that should be specified), and it is unlikely that the spatial distribution of peptides is not equilibrated (at least partially) among both leaflets. Firstly, as examples, Ulmschneider reported the translocation of PGLa;[1] the laboratory of Vácha and colleagues reported extensively on the free energy barrier to cross lipid bilayers [see, *e.g.*, ref. 2 and 3]. Secondly, in order to justify the lack of effects induced by colistin, the authors write: "intracellular targets like peptide binding to ribosomes indicating passage through the outer and inner membrane has been presented." In this respect, there is evidence for intracellular targets for indolicin, LL-37 and an analog of aurein (to make the list short, see the review[4]). Thus, according to the authors themselves, peptide translocation and redistribution on both membrane leaflet is a required physical condition. Thirdly, the authors observed a detergent-like effect that cause micellization. Given the degree of perturbation of this process (membrane remodeling), it is physically extremely unlikely for the peptides to have an absolutely inaccessible inner leaflet.

(ii) The analysis of SAXS data that aims to retrieve the volume probability distribution of each component of a lipid membrane is not trivial and requires several constrains even in simple systems.[5] Often, contrast variation SANS measurements are used to reduce the degree of variability of the adjustable parameters. The addition of peptides, modeled as Gaussian volume distribution regardless of their size, can easily lead to overparameterization issues. These bias the uniqueness of the results, which then strongly depend on the initial assumptions and constrains of the applied model. Firstly, in the presence of peptides, the observed lift-off of the low-q minimum of the scattering patterns was addressed to peptide-induced transbilayer asymmetry. Unfortunately, asymmetry is not the only cause of such an empirical observation. Also changes in contrast profile, or volume of the unit cell (which depends on the ratio of partitioned peptides per lipids) can affect the scattering pattern in that region (just to name a few). As an example, the scattering analysis of the coexisting quasi-interdigitated phase has been reported,[6] where high-q data were fitted properly. Secondly, the precision of the structural details (in the nanometre length-scale) has to be reflected in the quality of the fitting in the high-q range (including $q > 0.1$ A A^{-1}, Fig. 3 in ref. 6), because it is the q-range that contains most of the information about such details. It appears, instead, that the authors only focused on the low-q region. Thirdly, the degree of confidence of the adjustable parameters should be included given the significant number of parameters. Without that, the interpretation of the results remains speculative. To conclude, considering all these points, I would appreciate from the authors critical comments on both data analysis and interpretation of results.

1 J. P. Ulmschneider, Charged antimicrobial peptides can translocate across membranes without forming channel-like pores, *Biophys. J.*, 2017, **113**, 73–81.
2 I. Kabelka and R. Vácha, Optimal Hydrophobicity and Reorientation of Amphiphilic Peptides Translocating through Membrane, *Biophys. J.*, 2018, **115**, 1045–1054.

3 I. Kabelka, R. Brožek and R. Vácha, Selecting collective variables and free-energy methods for peptide translocation across membranes, *J. Chem. Inf. Model.*, 2021, **61**, 819–830.
4 C.-F. Le, C.-M. Fang and S. D. Sekaran, Intracellular targeting mechanisms by antimicrobial peptides, *Antimicrob. Agents Chemother.*, 2017, **61**, e02340-16.
5 N. Kučerka, J. F. Nagle, J. N. Sachs, S. E. Feller, J. Pencer, A. Jackson and J. Katsaras, Lipid bilayer structure determined by the simultaneous analysis of neutron and X-ray scattering data, *Biophys. J.*, 2008, **95**, 2356–2367.
6 E. Sevcsik, G. Pabst, A. Jilek and K. Lohner, How lipids influence the mode of action of membrane-active peptides, *Biochim. Biophys. Acta, Biomembr.*, 2007, **1768**, 2586–2595.

Reidar Lund responded: This is an interesting question. However, we do not quite see why symmetry must be imposed *a priori*. First of all, difference in curvature of the inner/outer leaflet may change the interaction and lead to different lipid distribution, and consequently charge distribution. Secondly, the peptides were added to the outside of the vesicle (trying to crudely mimic the situation when applied to bacteria). There might thus be significant barriers for translocation of the peptide. Since we measured the mixtures over time in hours, and no change was detected, this means that we are in a stable, although possibly, metastable, quasi- equilibrium situation. We measured the time evolution and we did not detect any change at all. In fact, trial experiments using stopped-flow SAXS showed that, for example, indolicidin inserted on time scales down to a few milliseconds without detectable changes in the bilayer structure or peptide distribution. In ref. 1 we show how differences in asymmetric *versus* symmetric insertion of peptides can be resolved from SAXS data. Regarding the solubilization we do agree that resolving the peptide location in the remaining liposomes imposes a challenge, as the number of free parameters increases significantly. Furthermore, the scattering contribution from the micelles interferes with the first minimum which we show is essential to resolve the peptide insertion. This provides less confidence for more potent peptides at large P : L values which is why we aimed to start at the smallest P : L values that are detectable. We have also seen that even for rather weakly interacting peptides, such as indolicidin, solubilization indeed also occurs, but at very high P : L values not directly relevant biologically. However interestingly, such high P : L values of 1 : 1 are typical values for surface techniques such as QCM and reflectometry.

We are well aware that the fit analysis using the detailed model is complicated and needs constraints. However, for the neat bilayers there are well established data both from experiments and simulations published by, among others, Kučerka *et al.* We have applied all possible constraints in our analysis to minimize the free parameters. In fact, only the thickness is varied within soft constrains based on literature values by Kučerka *et al.*[4] For the peptide we introduce only two more – the width and main position within the bilayer. We try first to only vary the peptide-related parameters which usually works fine at low P : L ratios. At higher values, depending on the peptide, the thickness of the bilayer might change; and at even higher rations we observe some solubilization. We agree that at significant amounts of solubilization the conclusions become less clear. Concerning asymmetry, we are fully aware that there are several factors that influence not only the position but also the shape of the first minimum. However, we use low P : L values, below the solubilization limit, and vary the fraction of peptides to detect such changes. As discussed, the model allows us to explicitly let the density distribution of the peptide vary freely while properly calculating the electron density distribution throughout the bilayer. See *e.g.* ref. 2 for detailed discussions, and ref. 3 for comparison with neutron reflectometry results. Previously reported work sets to be under the constraint of either needing to restrict the peptide to the head or tail, and symmetric distribution on both leaflets simultaneously. Concerning the negative remark about the fit quality at high q, we do not see that there are large deviations considering that we use a complete fit model for the whole q range. Please note that electron density distribution should give consistent results both at high and intermediate q. We do not believe in cutting the SAXS q range to improve the fit, we rather use data over as large q-range as possible doing our experiments using high resolution synchrotron SAXS instruments, and attempting to analyse complete data sets.

1 J. E. Nielsen, V. A. Bjørnestad, V. Pipich, H. Jenssen and R. Lund, Beyond structural models for the mode of action: How natural antimicrobial peptides affect lipid transport, *J. Colloid Interface Sci.*, 2021, **582**, 793–802.
2 J. E. Nielsen, V. A. Bjørnestad and R. Lund, Resolving the structural interactions between antimicrobial peptides and lipid membranes using small-angle scattering methods: the case of indolicidin, *Soft Matter*, 2018, **14**, 8750–8763.
3 J. E. Nielsen, T. K. Lind, A. Lone, Y. Gerelli, P. R. Hansen, H. Jenssen, M. Cárdenas, R. Lund, A biophysical study of the interactions between the antimicrobial peptide indolicidin and lipid model systems, *Biochim. Biophys. Acta, Biomembr.*, **1861**(7), 2019, 1355–1364.
4. N. Kučerka, J. F. Nagle, J. N. Sachs, S. E. Feller, J. Pencer, A. Jackson and J. Katsaras, Lipid Bilayer Structure Determined by the Simultaneous Analysis of Neutron and X-Ray Scattering Data, *Biophys. J.*, 2008, **95**, 2356–2367.

Margarida Bastos stated: You state in the abstract that "However, model systems based on PE-lipids (phosphatidylethanolamine) are more prone to destabilization upon addition of peptides, with formation of multilamellar structures and morphological changes. These properties of PE-vesicles lead to less conclusive results regarding peptide effect on structure and dynamics of the membrane." Nevertheless, this lipid is present in most bacterial membranes, mainly in the inner leaflet, so it is very important to use it, as has been discussed in the literature and as you also comment. I believe that this 'instability' is very important, and in fact is responsible for many AMP effects – AMPs are known to destroy membranes in various ways, not only by pore formation, they can form cubic structures, stacked bilayers, *etc.* So, lipids that have a tendency to form inverted structures, and that 'induce curvature' are very important for AMP action.

Aurélien Roux said: This is a very fair statement, and I guess the different modes of membrane destabilization will depend on the exact structure of the inverted lipid, and of its density. I guess one can draw phase diagrams of destabilization. I would be happy to see those diagrams.

Reidar Lund added: This is indeed a very good point. We specifically wanted to avoid multilamellar and non-lamellar phases in order to investigate the peptide organization at/in the membrane- and specifically, whether well-defined pores or channels can be detected. Also, we aimed at measuring the lipid dynamics in as "clean conditions as possible". We will look more into other mechanisms in the near future.

Boyan Bonev responded: Yes, but largely in model systems. With few exceptions, cytoskeleton/peptidoglycan-stabilised lipid membranes, as found in the bacterial envelope, do not buckle or undergo phase conversion to non-bilayer phases.

Kareem Al Nahas asked: Would the observed differences between PC and PE remain for other phospholipids with hydrophobic tails longer than DMPC/DMPE (*e.g.* DOPE/DOPC or DPPC/DPPE)?

Reidar Lund replied: Good question, we have not studied this yet. We plan to use POPE rather than DMPE in the future, due to the lower phase transition temperature. Through these experiments we will be able to also compare with previous collected data on POPC liposomes.

John Sanderson queried: What are the typical errors in measurement when you model the thickness? For example you state "The estimated overall thickness of the DMPE/DMPG membrane based on model fits was found to be 45 Å".

Reidar Lund replied: The thickness of the bilayer is constrained to literature values by Lee *et al.*,[1] Pan *et al.*[2] and Kučerka *et al.*[3]

1 T.-H. Lee, C. Heng, M. J. Swann, J. D. Gehman, F. Separovic and M.-I. Aguilar, Real-time quantitative analysis of lipid disordering by aurein 1.2 during membrane adsorption, destabilisation and lysis, *Biochim. Biophys. Acta, Biomembr.*, 2010, **1798**, 1977–1986.
2 J. Pan, F. A. Heberle, S. Tristram-Nagle, M. Szymanski ,M. Koepfinger, J. Katsaras, N. Kučerka, Molecular structures of fluid phase phosphatidylglycerol bilayers as determined by small angle neutron and X-ray scattering, *Biochim. Biophys. Acta, Biomembr.*, 2012, **1818**, 2135–2148.
3 N. Kučerka, B. van Oosten, J. Pan, F. A. Heberle, T. A. Harroun and J. Katsaras, Molecular structures of fluid phosphatidylethanolamine bilayers obtained from simulation-to-experiment comparisons and experimental scattering density profiles, *J. Phys. Chem. B*, 2015, **119**, 1947–1956.

John Sanderson asked: What is the timescale for the detergent effects of aurein 1.2 at high P : L?

Reidar Lund replied: We freshly mixed the solution just before the synchrotron SAXS experiment, so the solubilization we observe happens within a few minutes. To resolve the actual time scale of the solubilization we hope to utilize stopped-flow SAXS in future experiments.

Mibel Aguilar said: Aurein is a potent membrane lytic peptide at quite low concentrations. What concentrations did you use and did you see any evidence of bilayer disruption?

Reidar Lund responded: We think that the most relevant parameter in this case is the peptide–lipid (P : L) ratio rather than the concentration of peptide, as concentration can be misleading when comparing, for example, MIC values which are at a given bacterial cell amount and therefore not an independent concentration. In this work, we varied the P : L ratios 1 : 20, 1 : 50 and 1 : 100 which correspond to approximately 0.27–0.6 mg ml^{-1} aurein 1.2. At 1 : 20 we did see significant solubilisation (∼48% micelles) of the liposomes in the case of aurein 1.2 (significantly more than for the other peptides included in this study as seen from Table S1 of the ESI in our paper). We also observe some solubilisation at 1 : 50 P : L ratio (∼5% micelles). This indeed supports aurein 1.2 being a potent membrane lytic peptide.

Evelyne Deplazes remarked: Do you have any insights into why colistin does not work?

Reidar Lund answered: We need to investigate this in more detail before concluding. Please also see my answer to the question from Georg Pabst below, addressed to Evelyne Deplazes and I.

Georg Pabst commented: Your time-resolved SANS data is a convolution of several processes (lipid transfer through the aqueous phase with and without peptide, peptide-mediated lipid flip-flop,…). How can you conclude on lipid flip-flop only? A speeding up of SANS contrast changes might be also due to peptide-mediated transfer between vesicles (e.g. via micellar-like aggregates). Have you tried writing down rate equations and seeing if they give reasonable results?

Reidar Lund responded: Indeed, we probe both exchange between vesicles and flip-flop. Both may be relevant to the AMP mechanism. This gives rise to two rate constants as reported by Nakano et al.[1] (and several follow-up papers), and also by us in a similar system. In the present work, we also observe two rate constants which become faster with the addition of peptides. In ref. 2 we show similar data for DMPC/DMPG liposomes, and is in this case we were able to extract quantitative values for the rate constants, as well as activation energies. In the present case, temperature variation was not straightforward since the structural stability was impaired at temperatures higher than about 37 °C. However, for exchange processes that occur in parallel either freely or via peptide complexation, we would expect the rate laws to give effective rate constants (∼ sum of the two rate constants). Different exchange rates would thus be very challenging to detect using the current contrast design and set-up.

1 M. Nakano, M. Fukuda, T. Kudo, H. Endo and T. Handa, Determination of interbilayer and transbilayer lipid transfers by time-resolved small-angle neutron scattering, *Phys. Rev. Lett.*, 2007, **98**, 238101.
2 J. E. Nielsen, V. A. Bjørnestad, V. Pipich, H. Jenssen and R. Lund, Beyond structural models for the mode of action: How natural antimicrobial peptides affect lipid transport, *J. Colloid Interface Sci.*, 2021, **582**, 793–802.

Izabella Brand remarked: In the introduction you wrote: "… AMPs in some way or another mainly target the cytoplasmic membrane of the microorganisms". Do you know something about the transfer of these AMPs through the OM? Do they disrupt the structure of the OM?

Reidar Lund answered: This is an important point, We have not investigated this in detail and are not attempted to mimic the OM. However these are interesting questions that might be answered e.g. by floating bilayer mimics of OM. See e.g., ref. 1.

1 L. A. Clifton, S. A. Holt, A. V. Hughes, E. L. Daulton, W. Arunmanee, F. Heinrich, S. Khalid, D. Jefferies, T. R. Charlton, J. R. P. Webster, C. J. Kinane and J. H. Lakey, An accurate *in vitro* model of the *E. coli* envelope, *Angew. Chem.*, 2015, **127**, 12120–12123.

Georg Pabst addressed Reidar Lund and Evelyne Deplazes: I am reflecting on the question above from Evelyne Deplazes to Reidar Lund on colistin:

Thermodynamic considerations of partitioning of amphiphilic compounds (*e.g.* AMPs) in membranes show that at the high lipid concentrations (> few mM) typically used for SAXS experiments all peptides can be considered to be bound (*i.e.*, the unbound fraction is negligibly small) – see the work by Steve White[1] or Heiko Heerklotz.[2] This means all colistin can be considered membrane bound (no if's and but's). The fact that no changes on the SAXS patterns are observed then strongly implies that the technique is simply not sensitive to the presence of colistin in the studied *q*-range (maybe because there is no change in overall membrane thickness). Possibly, some signatures are observed in the WAXS range (where you should have a signal because of measuring in the gel phase). Note also that raw inspection of SAXS data can be misleading if one looks only for changes such as shifts or lift-offs of the form factor minima. Compositional modelling may tell. In any case, the bottom-line of my comment is: not seeing changes in SAXS does not necessarily imply that there are no interactions.

1 S. H. White and W. C. Wimley, *Biochim. Biophys. Acta*, 1998, **1376**, 339–352.
2 H. Heerklotz and J. Seelig, *Eur. Biophys. J.*, 2007, **36**, 305–314.

Reidar Lund answered: Please first note that the conclusion of an apparent lack of interaction is not only relying on results from SAXS. We have also done TR-SANS (no change in lipid dynamics) and DSC (no significant change in lipid packing).[1] We are fully aware that a raw inspection of the SAXS curve is misleading. For example, there is a common misconception that changes in the minimum at intermediate Q towards higher Q, automatically can be interpreted as "membrane thinning". We showed in previous work that this is not necessarily the case as even subtle changes in the electron density distribution/contrast due to peptide insertion will affect this minimum. For SAXS we are in fact very sensitive to any change in electron density as the lipid component (head and tail) have a large difference and opposite sign of the contrast towards water. This means that the scattering cross-term between tail and head region is negative and the resulting structure is almost contrast matched on average. Technically speaking:

$$V_{(tails)}\Delta\rho_{(tails)} \approx V_{(head)}\Delta\rho_{(head)}$$

In fact, for certain lipid vesicles such as POPC, the vesicles are almost perfectly matched at low Q. Our system is thus very sensitive to insertion as well as the position within the membrane because the peptide will affect the electron density of the tails and head region quite differently. In summary, since the contrast for colistin is very similar to other AMPs, and the fact that other techniques do not show any effects of colistin, we conclude that for this lipid mixture, as well as DMPC/DMPG previously measured, colistin sulfate does not seem to insert at all. This might be different for other lipid mixtures. However, why colistin does not insert in a similar manner as the other peptides needs to be further investigated.

We also remark that colistin is in fact a mixture of compounds, and in this study we used commercially available colistin from Sigma-Aldrich.

1 J. E. Nielsen, V. A. Bjørnestad, V. Pipich, H. Jenssen and R. Lund, Beyond structural models for the mode of action: How natural antimicrobial peptides affect lipid transport, *J. Colloid Interface Sci.*, 2021, **582**, 793–802.

Evelyne Deplazes added: Thanks Dr Lund. It is very helpful information to know that colistin can be considered bound (no if's and but's) and that the effect might not be changes in thickness but changes in other properties not sensitive to SAXS measurements.

Izabella Brand said: You mentioned in the answer to Dr Pabst that you want to increase temperature to 44 °C, to observe the phase transition in DMPE vesicles. At temperature above 40 °C some proteins/peptides become denatured. Will such temperature increase affect the structure of your peptides?

Reidar Lund replied: Good point, we will validate this in the case of all of the helical peptides studied before doing experiments above 40 °C. While for indo-licidin we have previously shown that the structure does not change at temperatures above 40 °C.[1]

1 J. E. Nielsen, V. A. Bjørnestad and R. Lund, Resolving the structural interactions between antimicrobial peptides and lipid membranes using small-angle scattering methods: the case of indolicidin, *Soft Matter*, 2018, **14**, 8750–8763.

Burkhard Bechinger remarked: In the paper you state that aurein can penetrate deeply into the membrane because of the low net charge. However there are many charges on this 13-residue peptide: 2K, E, D and the potentially charged N-terminus. The energies to insert all these charges into the hydrophobic interior are expected to be very unfavourable. How do you think this happens? Do you assume the pKs shift due to salt bridges or that dipoles formation is enough to neutralize the charges?

Reidar Lund replied: This is true, there are several charged side chains in aurein 1.2. Insertion of these charged units into the membrane could be viewed as rather unfavorable. However, it is well known that most AMPs are unstructured in buffered environments, and transform into more stable confined secondary structures upon interaction with lipid membranes. The folded structure of aurein 1.2 when interacting with micelles is described as a short well defined helix (PDB.file 1VM5).[1] In this graphic illustration it is rather clear that the charge residues are positioned in such a way that one has to assume that salt bridges are formed, thus neutralizing the overall charge and enabling integration into the lipid bilayer.

1 Wang, Guangshun, Y. Li and X. Li, Correlation of three-dimensional structures with the antibacterial activity of a group of peptides designed based on a nontoxic bacterial membrane Anchor, *J. Biol. Chem.*, 2005, **280**, 5803–5811.

Burkhard Bechinger commented: Could you explain in more detail what you mean by 'partial charge of indole ring'.

Reidar Lund replied: Tryptophan bears an indole ring which is well known for its ability to dock into the interfacial regions of different biological membranes and is well described, and their depth of interaction is stabilized by strong NOEs between the NH proton in the indole and the choline methyl and alkyl chain proton in a DPC system,[1] enabled as a result of mild charge distribution in the indole.[2] Specifically, in relation to peptide membrane interactions, the indole ring has been described to align on the membrane surface, stabilizing the peptide in the membrane.[3] Furthermore, it is not only the number of indoles, but also their sequential positioning in the peptide structure, which determines the local hydrophobic face, which eventually determines the peptide internalization effect.[4]

1 J. Koehler, E. S. Sulistijo, M. Sakakura, H. J. Kim, C. D. Ellis and Charles R. Sanders, Lysophospholipid micelles sustain the stability and catalytic activity of diacylglycerol kinase in the absence of lipids, *Biochemistry*, 2010, **49**, 7089–7099.
2 Norman and Kristen, *APS Southeastern Section Meeting Abstracts*, 2005, vol. 72.
3 W. Hu, K. C. Lee and T. A. Cross, Tryptophans in membrane proteins: indole ring orientations and functional implications in the gramicidin channel, *Biochemistry*, 1993, **32**, 7035–7047.
4 M.-L. Jobina, M. Blanchet, S. Henry, S. Chaignepain, C. Manigand, S. Castano, S. Lecomte, F. Burlina, S. Sagan and I. D. Alves, *Biochim. Biophys. Acta, Biomembr.*, 2015, **1848**, 593–602.

David P. Siegel opened the discussion of the paper by Durba Sengupta: If I understand you correctly, you define leaflet curvature as a deformation towards or away from the bilayer midplane. Did you measure a displacement of the mono-layer as a whole, or the membrane as a whole? Or is the displacement merely of the lipid–water interface? The radii of curvature are quite large (order 60 nm) compared to the thickness of the bilayer (Fig. 6 in your article). Moreover, the curvature fluctuations occur over an area roughly 5–10 nm wide; making the fluctuations rather local. Therefore, it seems to me that the variations in curvature around cav-1 could actually correspond to local variations in leaflet thickness. For

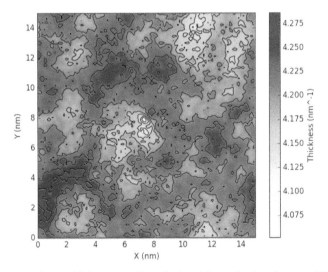

Fig. 1 The membrane thickness profile calculated from the head-group (*P*) distance between the two leaflets binding over the membrane plane.

purposes of comparing the curvature energy of different regions of a monolayer, as in caveolae and the surrounding planar membrane, there is an important distinction between variation in monolayer thickness and variation in the principal radii of curvature of a leaflet. At least according to continuum elastic theory, the curvature energy is determined by values of the bending modulus, the spontaneous curvature and the local curvature. In contrast, the energy required to vary the thickness of the leaflet depends on a different, chain stretching elastic constant. You seem to propose that the deformations which cav-1 induces in your simulations may play a role in stabilizing the 50 nm-scale membrane curvature observed in caveolae. I do not immediately see how variations in local leaflet thickness rationalize that. Obviously, a much higher concentration of the protein in the membrane may have that effect, but I am unsure that the present observations indicate that. I apologize if I have simply misunderstood your definitions.

Durba Sengupta replied: The sign of the curvature (positive or negative) is defined based on the mid-plane. The membrane thickness doesn't appear to change substantially (see Fig. 1 in this discussion).

Paul O'Shea asked: Can you comment on what your model would need to reconcile it with what we would see in a biological system?

Durba Sengupta replied: The model predictions such as cholesterol clustering, PS lipid interactions have already been validated by experimental approaches.[1] However, full-length protein structural models, protein oligomerization, and further post-translational modifications would be important steps in the future.

1 Y. Zhou, N. Ariotti, J. Rae, H. Liang, V. Tillu, S. Tee, M. Bastiani, A. T. Bademosi, B. M. Collins, F. A. Meunier and J. F. Hancock, Caveolin-1 and cavin1 act synergistically to generate a unique lipid environment in caveolae, *J. Cell Biol.*, 2021, **220**, e202005138.

Paul O'Shea remarked: Durba: your simulations imply the radii of the invaginations is a few nm, but in cells the radii is of the order of 80–100 nm – do you think you are looking at the same mechanism, *i.e.* is there perhaps a temporal feature that is missing or a time-domain not covered by your model?

Or in other words: What would your model need to reconcile with what we would see in a real biological system? This may help inform your model.

Durba Sengupta responded: Indeed fluctuations of these large bilayer models and instantaneous local curvature events need to be accounted for. We have averaged over multiple simulation sets and over time and don't observe further changes in the time evolution of curvature or lipid clustering. A single caveolin peptide construct would perhaps not stabilize the radii of 80–100 nm, but the models report on the local membrane remodeling induced by a single caveolin monomer.

Paul O'Shea further asked: Durba: This is interesting but I wonder, can you run the model for longer to perhaps see something that is stabilising; you seem to be out of equilibrium still?

Durba Sengupta replied: We did test the convergence of lipid clustering and curvature in time-blocks and don't observe large effects. However, we will take these comments on board and check multi microsecond time evolution as well.

Patricia Bassereau said: Caveolin monomers do not really exist. They are generally found as oligomers only. Since you show there is an interplay between curvature and the lipid composition, if the curvature created by a larger assembly of caveolins is not so strong, what would be your conclusions?

Durba Sengupta answered: Indeed caveolin oligomerization is a very critical aspect of caveolae. As of now, the model doesn't include the terminal domains required for oligomerization. The monomer simulations provide insight into the molecular mechanism of specific lipid–protein interactions and the intrinsic curvature induced. I hope that in the future we will be able to include oligomerization in our models.

Izabella Brand asked: Could you comment on the effect of the palmitoyl chain insertion in caveolin-1 into the membrane? Does this insertion have an effect on the membrane curvature?

Durba Sengupta replied: The lipid chain insertion appears to occur after the initial protein–membrane interaction. We have previously compared the protein-only construct and the lipidated-protein and don't discern differences in membrane curvature. However, further work is required for these complex membranes.

Aishwarya Vijayakumar queried: Does caveolin undergo post-translational modifications? How would it affect membrane curvature?

Durba Sengupta responded: The most significant post-translational modification is the addition of palmitoyl lipid tails. The construct we have considered has a single lipid tail modification (the full length protein has 3). In previous work, we compared the effects of the peptide with and without the lipid tail in simple model membranes. Although the membrane binding is modulated, we were unable to discern any curvature effects.

Aishwarya Vijayakumar asked: If we consider proteins that are instrinsically disordered, they do undergo post translational modifications that impact oligomerisation. In the case of caveolin, how would the oligomerisation be impacted?

Durba Sengupta answered: I would like to clarify that we have considered a protein construct that comprises of the main membrane interacting domains. The intrinsically disordered terminal domains have not been considered. Indeed specific phosphorylation sites have been identified in these domains, and careful work is needed to analyze their role in oligomerization.

Mibel Aguilar opened a general discussion of the paper by Bart Hoogenboom: I understand that cholesterol is important for some perforins. Does cholesterol play a role in TMH1-PRF function, and have you looked at the effect of different cholesterol concentrations?

Bart Hoogenboom replied: Yes, we have looked at the role of cholesterol. We find that increased cholesterol content makes the membrane more ordered and therefore more resistant to perforin. Cholesterol is not required for perforin pore formation. Perforin pore forms absolutely fine in PC-only membranes. These observations are consistent between WT and TMH1-mutant perforin.

Cholesterol is required, however, for bacterial cholesterol dependent cytolysins (CDC), which are pore forming proteins of the same superfamily as perforin. Many of these CDCs use cholesterol as a binding target on the membrane.

Mibel Aguilar asked: Could you elaborate on the rationale for using cholesterol sulfate?

Bart Hoogenboom responded: We had found that PS scavenged perforin in a dysfunctional state and wanted to verify if that was due to the negative charge in the PS head-group. To test this, we investigated other negatively charged membranes. We used PG as well as cholesterol sulfate, as these result in a negative membrane surface charge, yet are chemically different from PS. The effects on perforin assembly were consistent between PS, PG, and cholesterol sulfate containing membranes, allowing us to attribute the "scavenging effect" to the negative surface charge.

Mibel Aguilar added: Have you done any QNM analysis on these materials to see if there are different properties in terms of the modulus and the adhesion?

Bart Hoogenboom responded: We have not, mainly because the mechanical properties of such supported lipid bilayers have been well documented in the literature, with ordered lipid domains showing enhanced stiffness.

Mibel Aguilar further commented: When you form these beautiful pores, where do you think the lipid that formed the initial membrane bilayers has gone? Is there lipid associated with these protein pores? Are you using a solution that causes the lipid to be washed away?

Bart Hoogenboom replied: There are different possible mechanisms by which the lipid may be removed. In earlier work on related cholesterol-dependent cytolysis,[1] we showed that lipids are essentially ejected, once protein assemblies cut into the membrane like cookie cutters. This can be explained as the inner walls of the pore-forming beta-barrel are hydrophilic, such that lipid patches inside the pore become unstable. This also emerged as a potential mechanism in more recent MD simulations.[2]

For perforin however, this is more subtle, since perforin's pore forming mechanism is more like a can-opener sliding into the membrane around the pore edge. We have not observed any signature of lipid-ejection for perforin, and speculate that there is a receding lipid edge as the pore size increases.

1 C. Leung, N. V. Dudkina, N. Lukoyanova, A. W. Hodel, I. Farabella, A. P. Pandurangan, N. Jahan, M. P. Damaso, D. Osmanović, C. F. Reboul, M. A. Dunstone, P. W. Andrew, R. Lonnen, M. Topf, H. R. Saibil and B. W. Hoogenboom, Stepwise visualization of membrane pore formation by suilysin, a bacterial cholesterol-dependent cytolysin, *eLife*, 2014, 3, e04247.
2 M. Vögele, R. M. Bhaskara, E. Mulvihill, K. van Pee, Ö. Yildiz, W. Kühlbrandt, D. J. Müller and G. Hummer, *Proc. Natl. Acad. Sci. U. S. A.*, 2019, **116**, 13352–13357, DOI:10.1073/pnas.1904304116.

Paul O'Shea said: Bart: This is a tricky problem; the need for Ca^{2+} at 1 mM as well as an ambient electrolyte concentration to alter the electrostatic interactions. Does it imply one set of interactions is specific *i.e.* for Ca^{2+}?

Bart Hoogenboom responded: Perforin specifically needs Ca^{2+} to bind to the membrane. This is not a purely electrostatic effect, as *e.g.* Mg^{2+} cannot assume this role.

Ana J. Garcia Saez asked: Did you see an effect of line tension at the domain borders?

Bart Hoogenboom responded: We did not observe significantly enhanced pore formation at boundaries between lipid domains. We did observe enhanced pore formation at the edges of lipid patches when we had incomplete lipid coverage of our substrates, but that may be of little physiological relevance.

Patricia Bassereau commented: You propose that a gel-phase forms in the lymphocyte membrane to resist the insertion of perforin in the contact zone with the target cell. How is the gel-phase triggered by the lymphocyte? How can the cell locally switch its membrane to an ordered organization? This should perturb its capacity for exocytosis in the contact zone. Is this a problem?

Bart Hoogenboom answered: That is a really intriguing question, but needs a bit of nuance.

Firstly, the lipid phases we discuss are not in the gel-phase, but in a liquid-ordered phase. In our experience, gel-phase membranes are generally quite robust against membrane-targeting peptides and proteins, but are of little physiological relevance.

Secondly, we find that lymphocytes are overall protected against perforin, also in experiments where they are exposed to recombinant perforin (so without synapse formation). This overall protection is reduced when we reduce the overall lipid order of the plasma membrane, using keta-cholesterol.

But then we would indeed expect this protection mechanism to be further enhanced at the immune synapse, as there is extensive experimental evidence that the lymphocyte membrane shows (further) enhanced order at the synapse. It coincides with the clearance in the actin cortex at the synapse, and *in vitro* work suggests that interactions between actin and the membrane can aid in pinning particular membrane domains.

Yet I don't think we fully understand how this local enhancement of lipid order might work.

Francisco Barrera asked: Some cancer cells have increased plasma membrane fluidity. Would this circumstance protect them from perforin's attack? Could they be more sensitive? This could be a nuanced situation because cancer cells can also expose PS, which might be expected to have the opposite effect.

Bart Hoogenboom replied: Since we observe that increased membrane packing/order (and thus reduced fluidity, I presume) protects against perforin, I would expect increased membrane fluidity to make such cells more sensitive to perforin attack. Indeed the exposure of PS could protect them, resulting in the opposite effect. Without actual experiments on such cells, we can speculate about the directions this may take, but it is hard to predict which effects will dominate.

Bart Hoogenboom opened a general discussion of the paper by Mibel Aguilar: How quickly do bacteria adapt their membrane lipid composition in response to changes in the environment?

Mibel Aguilar answered: Lipid production in bacteria is very finely tuned by the environment. Bacteria respond to various environmental stressors and conditions and can alter lipid synthesis at the genetic level. With the emergence of lipidomics, the complex changes are now becoming more evident, at least for bacteria, and the metabolic and genetic changes in response to environmental stresses is very quick.

The changes occur relatively quickly. For example, as indicated by Dr Bonev above, the conversion of acyl chain double bonds to cyclopropyl lipids is almost complete during the mid-phase of growth. With on-going improvements in lipidomic analysis, we will be able to track this more easily in future.

Amitabha Chattopadhyay addressed Bart Hoogenboom and Mibel Aguilar: How will bacterial growth rate affect your observations?

Mibel Aguilar replied: As discussed later in responses to Paula Milán Rodríguez, Paul O'Shea, Patricia Bassereau and Burkhard Bechinger during Session 4 (DOI: 10.1039/d1fd90068d), there are many cell properties that we need to monitor given that the lipid composition changed so significantly with the growth phase. We harvest the cells at a specific OD, so at least in the stationary phase, the culture is in a somewhat stable growth phase.

Conflicts of interest

William F. DeGrado is an advisor to Innovation Pharmaceuticals, and there are no other conflicts to declare.

PAPER

Peptide lipidation in lysophospholipid micelles and lysophospholipid-enriched membranes†

Vian S. Ismail,‡[a] Hannah M. Britt, §[a] Jackie A. Mosely [b]
and John M. Sanderson [*a]

Received 17th April 2021, Accepted 16th July 2021

DOI: 10.1039/d1fd00030f

Acyl transfer from lipids to membrane-associated peptides is a well-documented process, leading to the generation of a lipidated peptide and a lysolipid. In this article, we demonstrate that acyl transfer from lysophosphatidylcholines (lysoPCs) to the peptide melittin also occurs, both in micelles of pure lysolipid and in lipid/lysolipid mixtures. In the case of bilayers containing lysolipids, acyl transfer from the lysolipid is marginally favoured over transfer from the lipid. In pure bilayers of saturated lipids, the introduction of even small amounts of lysolipid appears to significantly increase the reactivity towards lipidation.

1. Introduction

It is now well established that acyl transfer can potentially occur from ester-based lipids to any molecules that are able to interact with lipid membranes and possess suitable acceptor groups.[1] Examples of molecules known to undergo this process, termed intrinsic lipidation, include low molecular weight organics[2,3] and peptides.[4-7] Suitable acceptor groups include amino groups and alcohols. When the donor is a lipid, a lysolipid byproduct is formed alongside the lipidated acceptor, leading to the formation of mixed lipid/lysolipid systems (Fig. 1a). Many of these acyl transfer experiments have been conducted *in vitro* in liposomes composed of synthetic lipids, for which the origin of the acyl group is unambiguous. Early studies on intrinsic lipidation were conducted using melittin, obtained synthetically to avoid complications from the presence of

[a]Chemistry Department, Durham University, Durham, DH1 3LE, UK. E-mail: j.m.sanderson@durham.ac.uk

[b]National Horizons Centre, School of Health & Life Sciences, Teesside University, Darlington, DL1 1HG, UK. E-mail: j.mosely@tees.ac.uk

† Electronic supplementary information (ESI) available. See DOI: 10.1039/d1fd00030f

‡ Current address: Chemistry Department, Soran University, Kurdistan Region, Iraq, vian.esmaeil@soran.edu.iq.

§ Current address: Institute of Structural & Molecular Biology, University College London, Gower Street, London, WC1E 6BT, UK, h.britt@ucl.ac.uk.

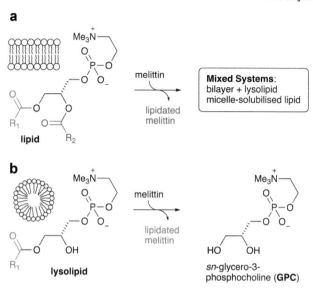

a

b

lysolipid

sn-glycero-3-phosphocholine (**GPC**)

c   H-GIGAVLKVLTTGLPALISWIKRKRQQ-NH$_2$

Fig. 1 Summary of the reactivity of melittin with lipids: (a) reaction with ester-linked diacyl lipids leads to the generation of lipidated melittin and a lysolipid byproduct; (b) reaction with lysolipids leads to the formation of lipidated melittin and the corresponding deacylated glycerol derivative GPC; (c) sequence of melittin, with the main lipidation residues highlighted in bold and the minor sites underlined.

phospholipase PLA$_2$ found in melittin extracted from bee venom. It was found that melittin undergoes ready intrinsic lipidation in membranes composed of unsaturated phosphatidylcholine (PC) based lipids. The reaction was generally favoured at the N-terminal amino group of the peptide, but significant lipidation of Lys23 was also noted, alongside minor amounts of lipidation of the side chains of Lys21, Lys7 and Ser18 (Fig. 1).[6,7] Lipidation activity was generally influenced by the presence of other lipid components such as PE, PS or PG, but the manner of the influence was very composition dependent.[5,6] More recent work has further demonstrated the complexity of this process, as cholesterol incorporation into PC membranes both significantly increases lipidation activity, despite decreasing the affinity of the peptide for the membrane, and changes the relative reactivity at different sites on the peptide. Moreover, membranes composed of saturated PC lipids are found to be unreactive in the absence of cholesterol.[4]

In a number of these experiments, there appears to be generally little selectivity between transfer from the *sn*-1 and *sn*-2 acyl groups of the lipids, especially at longer timescales after peptide/lipid mixing. In the initial periods after mixing, however, *sn*-1/*sn*-2 selectivity is sometimes apparent in liquid chromatography-mass spectrometry (LC-MS) analyses, which may reflect minor differences in product ionisability near the detection limit, or alternatively may reflect a significant difference in reactivity between lipids and lysolipids. Small amounts of lysolipids may be pre-existing as a consequence of hydrolysis, or formed in the membrane as the by-products of lipidation. In the latter case, as lysolipids are potential acyl group donors, any intrinsic preference for reaction at the sn-1 and sn-2 positions of the lipid may be masked if the lysolipid is more reactive than the lipid.

It is therefore of fundamental importance to determine the level of reactivity of lysolipids towards membrane-associated peptides. Given the sensitivity of intrinsic lipidation to membrane composition, the generation of lysolipids *in situ* is also expected to have a significant impact on the extent and selectivity of lipidation. This article describes measurements of melittin lipidation undertaken over a range of surfactant compositions, ranging from pure lysolipid, through lysolipid/lipid mixtures, to pure lipids.

2.   Methods

Materials

Synthetic melittin (≥97% by HPLC) and bee venom melittin (>85% by HPLC) were obtained from Sigma-Aldrich, Dorset, UK. A stock solution of the peptide at an approximate concentration of 1 mg ml^{-1} was prepared in water and used fresh. The concentration of peptide was more accurately determined before use by absorbance measurements (ε at 250 nm = 5500 M^{-1} cm^{-1}).[7,8] Authentic samples of N^1- and K^{23}-oleoyl and palmitoyl melittin were obtained from Almac Group (Craigavon, UK) at >95% purity by HPLC.

1,2-Dioleoyl-*sn*-glycero-3-phosphocholine (DOPC) and 1,2-dipalmitoyl-*sn*-glycero-3-phosphocholine (DPPC) were obtained from Sigma-Aldrich, Dorset, UK. 1-Oleoyl-2-hydroxy-*sn*-glycero-3-phosphocholine (OPC) and 1-palmitoyl-2-hydroxy-*sn*-glycero-3-phosphocholine (PPC) were obtained from Avanti Polar Lipids, USA. Water was purified using a Milli-Q Direct Q system from Millipore (Millipore (UK) Ltd) to give a resistivity of ≥18 MΩ cm^{-2}. Stock dispersions of lysoPCs in water were prepared at a concentration of 1.3 mM in aqueous buffer (90 mM NaCl, 10 mM sodium phosphate, pH 7.4). Other solvents and reagents were obtained from Fisher Scientific, UK.

Liposome preparation

Liposomes were prepared by concentrating a solution of lipid (1 mg) in chloroform (100 μl) *in vacuo* to form a thin film. The lipid film was hydrated with 1 ml of aqueous medium containing 90 mM NaCl buffered with 10 mM sodium phosphate at pH 7.4 and mixed thoroughly before being subjected to five freeze thaw cycles and extruded 10 times through a 100 nm laser-etched polycarbonate membrane (Whatman) at 50 °C under a stream of nitrogen using a thermobarrel extruder (Northern Lipids, Burnaby, Canada).

Lipidation experiments

Standard lipidation conditions employed a buffer of 90 mM NaCl buffered with 10 mM sodium phosphate at pH 7.4 at 37 °C. Mixtures of lysolipid and lipid were prepared by adding a dispersion of the lysolipid in buffer to an extruded dispersion of unilamellar liposomes, followed by agitation. Melittin was then added to the mixture to give a final peptide concentration of 50 μm and a lipid concentration of 0.65 mM, and the sample was agitated. Samples for analysis were diluted by a factor of 5 in water. The times given in the text refer to the time post-mixing after melittin addition.

Liquid chromatography-mass spectrometry (LC-MS)

Separations were performed using an Xbridge C18 column (3.5 μm particle size, 2.1 mm internal diameter, 100 mm length; Waters UK, Manchester, UK). Typical analyses used a 5 μl sample injection at a flow rate of 200 μl min^{-1} with a gradient of mobile phases A (0.1% (v/v) formic acid (FA) in water), B (0.1% (v/v) FA in acetonitrile) and C (0.1% (v/v) FA in methanol) as follows: 95% A : 5% B to 5% A : 95% B over 10 min, followed by 5 min at 100% B and 15 min at 100% C, plus 5 min re-equilibration at 95% A : 5% B.

Positive ion mass spectra were recorded over m/z 200–2000 using a LTQFT mass spectrometer equipped with a 7 T magnet (ThermoFisher Corp., Bremen, Germany). Electrospray ionisation (ESI) was used to generate positive ions using the following conditions: source voltage 4.0 kV, capillary voltage 30.0 V, capillary temperature 350 °C and tube lens 100.0 V. The auxiliary gas flow and sweep gas flow were set at 5.0 arbitrary units and the nitrogen sheath gas flow at 15.0 arbitrary units. Collision-induced dissociation experiments were performed entirely within the linear ion trap with a fixed isolation window of 4 m/z, using helium as a collision gas and an optimized normalized collision energy level of 25%. Spectra were analysed using XcaliburQualBrowser version 2.0.7 (Thermo Fisher Scientific Inc.) and processed using the embedded program Qual Browser.

Peptide CMC measurements

A peptide stock solution of approximately 1 mg ml^{-1} in water was added sequentially to a solution of Rhodamine 6G (0.08 μM, 45 μl) in water in a 40 μl ultra-micro quartz cuvette to give peptide concentrations in the range 1×10^{-6} m to 1×10^{-4} M. Following each addition, fluorescence emission was measured in the range 500 nm to 600 nm, with an excitation wavelength of 480 nm.[9,10] Emission intensities at 550 nm were corrected for dilution.

3.   Melittin reactivity with lysolipids

Initial experiments were conducted by mixing synthetic melittin with lysophosphatidylcholine (lysoPC) micelles. Experiments were conducted using a lysoPC with either a saturated acyl chain (palmitoyl-sn-glycero-3-phosphocholine, PPC), an unsaturated acyl chain (oleoyl-sn-glycero-3-phosphocholine, OPC) or a 1 : 1 mixture of the two. These glycerol-based lysolipids all exist as a mixture of 1- and 2-acyl isomers in equilibrium, with the 1-acyl being predominant. In all cases (Fig. 2), lipidated melittin products were readily detectable after 24 h and were the predominant melittin species present after 168 h.

Two features of the lipidation in the lysoPC micelles were particularly striking. First, there was a shift in selectivity towards lipidation on the Lys23 side chain in OPC micelles (Fig. 2b: ii > i), rather than the N-terminal amino group preference found for PPC (Fig. 2d: vii > viii) and mixed OPC/PPC micelles (Fig. 2f: I > ii and vii > iii), and observed previously with lipids.[4-6] Second, the levels of melittin bearing

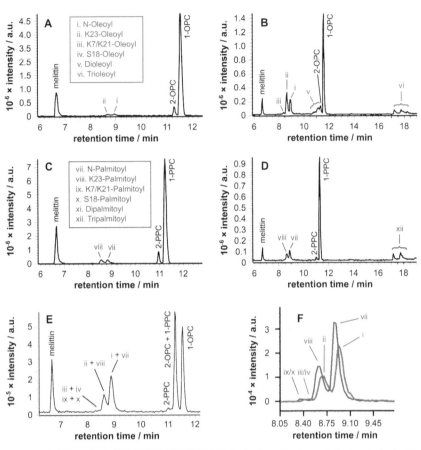

Fig. 2 LC-MS analysis (ESI-FTICR-MS (LTQFT), C_{18}) of synthetic melittin incubated with lysolipid micelles at 37 °C in buffer (10 mM $NaHCO_3$/90 mM NaCl) at pH 7.4. (a) 1/2-Oleoyl-sn-glycero-3-phosphocholine (OPC), 24 h; (b) OPC, 168 h; (c) 1/2-palmitoyl-sn-glycero-3-phosphocholine (PPC), 24 h; (d) PPC, 168 h; (e) 1 : 1 PPC/OPC, 168 h; (f) extracted ion chromatograms (EICs) from spectrum (e) for m/z 772.00 and 778.51, corresponding to palmitoyl and oleoyl melittin respectively ($z = 4$). MS data for the lipidated products are in Fig. S1 to S4 and Tables S1 to S4.†

multiple acyl groups was significant, particularly in OPC micelles, in which melittin lipidated with both two and three acyl groups was detected. Triple-acylated melittin was also observed in PPC micelles, although curiously the double palmitoylated species was absent, which could be accounted for by the faster addition of the third acyl group in PPC micelles compared to the addition of the second.

4.   Melittin reactivity in lysolipid/lipid mixtures

In order to examine qualitatively the relative reactivity of PC-based lysolipids and lipids with melittin, a series of experiments was conducted using a range of systems composed of lipids, lysolipids, and mixtures of the two. The experiments were undertaken using lipids with the same acyl group at the sn-1 and sn-

2 positions and a lysoPC with a different acyl chain in order to permit transfer from the lysolipid and the lipid to be distinguished. For the 1 : 1 mixtures of lysolipid and lipid, melittin analogues modified with single, double, and triple acyl modifications were observed after extended time periods (Fig. 3). An interesting feature of the product profiles for the double acylated products is the observation of similar product ratios regardless of whether they are formed in PPC/DOPC or OPC/DPPC mixtures. For example, the product ions for melittin modified with oleoyl + palmitoyl and 2× oleoyl or 2× palmitoyl are of similar relative abundance in both mixtures (Fig. S2†). The same observations apply for the triple acylated melittin (Fig. S3 and S4†). These observations suggest that there is little selectivity between lysolipids and lipids when it comes to the addition of an extra acyl group to a melittin molecule that is already lipidated.

Tandem mass spectrometry analysis (Fig. S5 to S19 and Tables S5 to S32†) revealed that the predominant double-acylated product arose from lipidation at the both the N-terminus and at K23, in line with the relative abundance of the single acyl modifications to these sites. The second most abundant double modification was to the side chains of K23 and K7. Modification of the K7 side chain was usually of a relatively low abundance, suggesting that the presence of

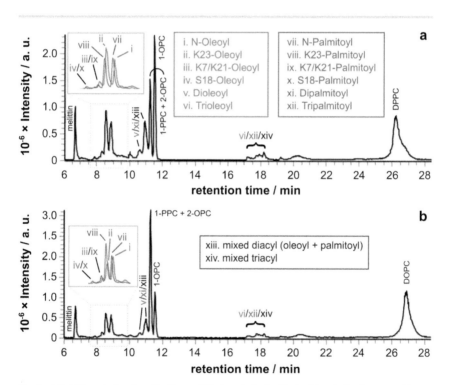

Fig. 3 LC-MS analysis of synthetic melittin incubated with 1 : 1 (molar) mixtures of lyso-lipid and lipid at 37 °C in buffer (10 mM NaHCO₃/90 mM NaCl) at pH 7.4. (a) OPC/1,2-dipalmitoyl-sn-glycero-3-phosphocholine (DPPC); (b) PPC/1,2-dioleoyl-sn-glycero-3-phosphocholine (DOPC). The insets show EICs for m/z 772.00 and 778.51, corresponding to palmitoyl and oleoyl melittin respectively (z = 4). MS data for the lipidated products are in Fig. S1 to S4 and Tables S1 to S4.†

a single fatty acyl modification changes the reactivity at other sites, most likely as a consequence of a change in the interfacial positioning of the peptide. Two products were also tentatively identified as arising from lipidation at the N-terminus and the side chain of either R22 or R24. These materials had the retention time characteristics of the double-acylated materials, but intact molecular ions were challenging to observe. Acyl modifications to arginine side chains would be expected to be highly labile and fragment easily in the source. Modification at other nearby sites, K21 and K23, could be ruled out as these products were identified elsewhere in the chromatogram.

A more extended series of lysolipid/lipid mixtures was examined. The series as a whole covered lipid bilayers, bilayers containing lysolipids, lysolipid micelles containing lipids, mixed micelle/lipid systems and micelles (Fig. 4).

For mixtures of OPC with DPPC (Fig. 4a–e), oleoyl transfer was always favoured to the side chain of Lys23 of the peptide (peak ii), regardless of whether the mixture was a micelle or bilayer containing detergent. The relative abundance of palmitoylated products arising from transfer to the N-terminal amino group (peak

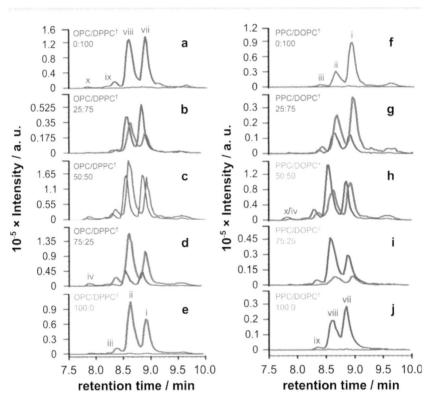

Fig. 4 EICs for m/z 772.00 and 778.51, corresponding to palmitoyl and oleoyl melittin respectively ($z = 4$), from LC-MS analyses of synthetic melittin incubated for 168 h with mixtures of the lysolipid and lipid at 37 °C in buffer (10 mM NaHCO$_3$/90 mM NaCl) at pH 7.4. (a to e) DPPC + OPC; (f to j) DOPC + PPC. The initial molar ratio of lysolipid to lipid is indicated on each trace. Traces are labeled according to their expected phase behavior on the basis of literature precedents:[11-14] bilayer ± lysolipid (†); mixed bilayer/detergent (*); or micelles (‡). Peak annotations are identified in Fig. 2 and 3. The corresponding data after 48 h are given in Fig. S20.†

vii) or the side chain of Lys23 (peak viii) was always around 50/50. In previous work, melittin was found to be unreactive in DPPC membranes.[4] However, this was not the case in the work reported here (Fig. 4a). In seeking to account for this difference it was found that small levels of OPC were detectable in these DPPC samples. Blank LC-MS runs acquired immediately before the DPPC samples were analysed did not reveal any detectable OPC, so the likely source of this material in the DPPC experiments is presumably a small amount of carry over from previous work in the same apparatus. On the basis of this finding, we hypothesise that the presence of even small quantities (<1 mol%) of lysolipid in a bilayer with saturated acyl chains is sufficient to perturb structure and packing to the extent that lipidation becomes feasible. This hypothesis is in line with previous work that has demonstrated the significant instability of DPPC membranes containing 1 mol% OPC.[11] If this bilayer disruption can be proven to extend to lower concentrations of lysolipid, it will have significant ramifications for bilayer stability in scenarios where membrane damage can occur, as even small amounts of hydrolysis or oxidation may facilitate other kinds of reactivity. For mixtures of PPC with DOPC (Fig. 4f–j), oleoyl transfer from the lipid to the N-terminal amino group of the peptide was always preferred (peak i), in line with previous experiments.[4–7] Palmitoyl transfer, however, exhibited more complex behavior. In systems containing micelles and lipids (Fig. 4h, i), transfer from the lysolipid to the side chain of Lys23 was always preferred (peak viii). In pure lysolipid micelles however (Fig. 4j), reaction at the N-terminal amino group of melittin was favoured (peak vii). In bilayers containing lysolipid (Fig. 4g), products formed by transfer from the lysolipid to Lys23 and the N-terminal amino group were of similar relative abundance.

The data in Fig. 4 show only the regions of the analyses corresponding to melittin modified by a single acyl group at a single time point. Given the low selectivity observed for the subsequent transfer of acyl groups to a melittin molecule that has already been lipidated (Fig. 3), the product distributions in Fig. 4 are likely to reflect the selectivity for the addition of the first acyl group, with the caveat that any initial deviations in the overall reaction rate immediately following peptide addition to the membranes will not be captured. In each sample in Fig. 4, the relative proportion of melittin lipidated by the transfer of a single acyl group from the lysolipid or the lipid is broadly in line with the molar ratio of these lipids, consistent with each (diacyl) lipid reacting only once. However, comparison of the data for both 1 : 1 samples (Fig. 4c and h) suggests that lysolipid reactivity with melittin is marginally higher than lipid reactivity. This is also supported by the comparison of each 25 : 75 mixture with its corresponding 75 : 25 mixture, such as Fig. 4g and 4i, in which the proportion of product formed by acyl transfer from the lysolipid is higher in the 25 : 75 mixture than transfer from the lipid in the 72 : 25 mixture.

Overall, the pattern of product formation is indicative of a marginally higher reactivity of melittin towards lysolipids than lipids. Product distributions, in terms of the relative levels of acyl transfer to different residues in the melittin sequence, are much more sensitive to bilayer composition. This sensitivity to bilayer composition most likely reflects corresponding differences in the interfacial orientation and location of melittin in each lipid mixture.

5. Reactivity of bee venom melittin in PC/PS mixtures

Melittin extracted from bee venom (BVM) is usually obtained in the presence of phospholipase A2 (PLA$_2$) as a contaminant. For most applications, the level of PLA$_2$ activity can be minimised by the use of a Ca^{2+} chelator such as ethylene glycol-bis(β-aminoethyl ether)-N,N,N',N'-tetraacetic acid (EGTA). For the work reported here, however, it was desirable to examine whether PLA$_2$ activity, leading to the formation of lysolipids, was able to influence the lipidation process. Therefore, BVM was added to the liposomes in the absence of any metal chelator. In the case of the pure PC membranes, the major product, by a considerable margin, resulted from the transfer of the sn-1 chain from the lysolipid form by PLA$_2$ activity: oleoyl in the case of 1-oleoyl-2-palmitoyl-sn-glycero-3-phosphocholine (OPPC) and palmitoyl for 1-palmitoyl-2-oleoyl-sn-glycero-3-phosphocholine (POPC). For membranes composed of mixtures of PC with phosphatidylserine (PS) lipids (Fig. 5), the major lipidation product arose, as expected, by the transfer of the sn-1 acyl chain of the PC component of the mixture.

In a similar manner, the level of stearoyl transfer from SLPS was greater than that of linoleoyl transfer, particularly in the case of the OPPC/SLPS mixture, reflecting the reactivity of bee venom PLA$_2$ with SLPS.[15]

These data are consistent with significant reactivity towards the lysolipids formed by PLA$_2$ activity, assuming that the rate of lipase-catalysed hydrolysis is faster than that of lipidation, which is a reasonable assumption based on literature precedents.[16,17]

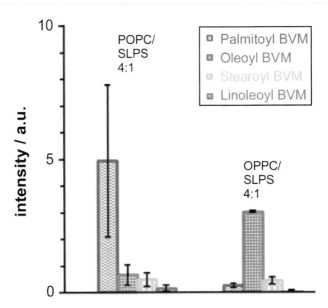

Fig. 5 Peak areas for singly lipidated melittin products following LC-MS analysis of mixtures of bee venom melittin (BVM) with POPC/1-stearoyl-2-linoleoyl-sn-glycero-3-phosphoserine (SLPS; 4 : 1) and OPPC/SLPS (4 : 1) liposomes incubated for 24 h at 37 °C, pH 7.4. Errors are plotted as the SEM of the normalised peak area ($n = 2$). Normalisation of BVM single acylation was performed relative to the non-acylated BVM in the trace.

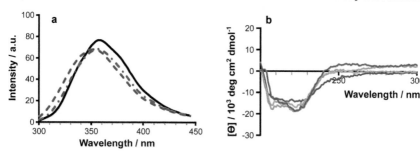

Fig. 6 The effects of lipidation on melittin structure. (a) Fluorescence emission (λ_{ex} 280 nm) for melittin (solid black line), N^1-palmitoyl melittin (dotted and dashed red line) and N^1-oleoyl melittin (dashed blue line). (b) CD spectra of N^1-palmitoyl melittin (red); K^{23}-palmitoyl melittin (orange); N^1-oleoyl melittin (blue); and K^{23}-oleoyl melittin (green). Authentic samples of peptides acylated at the N-terminus and K23 side chain were obtained commercially.

6.   The effects of lipidation on peptide structure

Melittin is known to be unfolded in solution as a monomer, forming a folded helix in the membrane and a tetramer at high concentrations.[18–22] Lipidation has the potential to disrupt the balance between these structural forms, or to push the peptide towards additional conformational states. Authentic samples of lipidated melittin exhibited fluorescence and CD spectral profiles typical of a folded helical form, with blue shifted Trp emission (Fig. 6a) and negative ellipticity minima at approximately 210 nm and 225 nm (Fig. 6a).

Modifications at both the N-terminus and side chain of Lys23 were able to induce folding of the peptide in solution. Fluorescence measurements of solutions of acylated melittin prepared in the presence of Rhodamine 6G[23] indicated that the acylated peptides are able to form micelles, with critical micelle concentrations of ≤10 μM (Fig. S22†). Micelle formation may be a factor in driving the adoption of the peptide secondary structure.

7.   Conclusions

As a whole, the data demonstrate that melittin is readily lipidated by acyl transfer from lysophosphatidylcholines (lysoPCs) in all of the systems studied: micelles, PC bilayers containing lysolipid, and mixed lipid/micelle dispersions. The reactivity of the peptide with lysoPCs in these mixtures is marginally higher than that with lipids and leads to the modification of melittin lipidated at a number of sites, predominantly the N-terminus and the side chain of Lys23. In mixtures of lysoPCs without lipid, there is some selectivity according to the nature of the acyl group of the lysolipid, with palmitoyl transfer being favoured over oleoyl. Although the same range of products is formed by reaction with either lipids or lysolipids, the ratio of lipidation at the N-terminus and Lys23 is sensitive to the composition of the mixture, favouring the N-terminus in systems rich in DOPC and Lys23 in most other mixtures apart from 100% PPC. These changes in the product profile are likely to be a consequence of the composition of the heterogeneous phase influencing both the preferred depth and orientation of melittin binding in the interface and the positioning of lysolipid and lipid acyl groups.

Melittin is known to associate favourably with regions of high positive curvature,[24] a tendency that may be associated with the ability of the peptide to form pores. As membrane remodelling and pore formation are both expected to occur faster than lipidation, it is therefore possible that the changes in the product profile result from lipid segregation and aggregation. Such phenomena were not probed here, however.

In general terms, how the interplay between peptide structure and membrane composition influences lipidation, both in selectivity and overall reactivity, are matters for further discussion.

For membranes composed of saturated lipids such as DPPC, for which the lipidation activity of melittin is normally low, the presence of even a small amount of lysolipid appears to be sufficient to promote reactivity. The mechanism by which this occurs is unclear, but it is likely to result from a disruption to lipid packing that would, for example, enable better water penetration into the interface. The presence of interfacial water has the potential to facilitate the formation of products from intermediates in the acylation reaction.[6,25,26] Should this be the case, other scenarios that can facilitate water penetration into the membrane, such oxidative damage, should also produce increased rates of acyl transfer reactions from lipids to suitable acceptors. The role that physical and chemical stresses play in promoting intrinsic lipidation is worthy of further consideration.

For melittin, the addition of a fatty acyl group at either of the two main sites of modification by transfer from lipids is sufficient to promote the adoption of a secondary structure in the peptide and drive micellisation. At low concentrations of acylated peptide, it is likely that acylation both increases the affinity of the peptide for the heterogeneous phase and modifies peptide orientation in the interface. These changes in affinity and orientation may account for the extensive formation of peptides modified with multiple acyl groups in lysolipid micelles. At higher peptide concentrations in all systems, the formation of peptide-containing micelles is likely. It is notable that the formation of micelles following the addition of melittin to lipid membranes has been documented.[27,28] A key issue for debate is whether lipidation is able to steer the folding of peptides into particular conformations with adverse biological activity. In this regard, it has been proposed that by driving secondary structure formation and tethering peptides to the membrane, lipidation offers a route for facilitating the nucleation of amyloid fibrils.[1]

Author contributions

JMS and JAM conceived of and supervised the project. VSE performed the lipidation work with melittin. VSE and JMS analysed the data. HMB performed the CMC work with authentic samples of acylated melittin and analysed the data. The manuscript was written by JMS, with contributions from all authors.

Conflicts of interest

There are no conflicts of interest to declare.

Acknowledgements

The authors thank the Ministry of Higher Education and Scientific Research (grant number 1391) of the Kurdistan Regional Government for support to VSE, and the Engineering and Physical Sciences Research Council (EP/M506321/1) for support to HMB.

References

1 J. M. Sanderson, *BioEssays*, 2020, **42**, e1900147.
2 H. M. Britt, A. S. Prakash, S. Appleby, J. A. Mosely and J. M. Sanderson, *Sci. Adv.*, 2020, **6**, eaaz8598.
3 H. Britt, C. A. García-Herrero, P. W. Denny, J. A. Mosely and J. M. Sanderson, *Chem. Sci.*, 2019, **10**, 674–680.
4 H. M. Britt, J. A. Mosely and J. M. Sanderson, *Phys. Chem. Chem. Phys.*, 2019, **21**, 631–640.
5 R. H. Dods, B. Bechinger, J. A. Mosely and J. M. Sanderson, *J. Mol. Biol.*, 2013, **425**, 4379–4387.
6 R. H. Dods, J. A. Mosely and J. M. Sanderson, *Org. Biomol. Chem.*, 2012, **10**, 5371–5378.
7 C. J. Pridmore, J. A. Mosely, A. Rodger and J. M. Sanderson, *Chem. Commun.*, 2011, **47**, 1422–1424.
8 C. N. Pace, F. Vajdos, L. Fee, G. Grimsley and T. Gray, *Protein Sci.*, 1995, **4**, 2411–2423.
9 P. P. Bonsen, G. H. de Haas, W. A. Pieterson and L. L. van Deenen, *Biochim. Biophys. Acta, Lipids Lipid Metab.*, 1972, **270**, 364–382.
10 P. Becher, *J. Phys. Chem.*, 1962, **66**, 374–375.
11 C. J. Van Echteld, B. de Kruijff, J. G. Mandersloot and J. De Gier, *Biochim. Biophys. Acta, Biomembr.*, 1981, **649**, 211–220.
12 J. R. Henriksen, T. L. Andresen, L. N. Feldborg, L. Duelund and J. H. Ipsen, *Biophys. J.*, 2010, **98**, 2199–2205.
13 D. V. Zhelev, *Biophys. J.*, 1998, **75**, 321–330.
14 D. Needham and D. V. Zhelev, *Ann. Biomed. Eng.*, 1995, **23**, 287–298.
15 E. A. Dennis, J. Cao, Y. H. Hsu, V. Magrioti and G. Kokotos, *Chem. Rev.*, 2011, **111**, 6130–6185.
16 L. Feng, K. Manabe, J. C. Shope, S. Widmer, D. B. DeWald and G. D. Prestwich, *Chem. Biol.*, 2002, **9**, 795–803.
17 F. Ghomashchi, Y. Lin, M. S. Hixon, B. Z. Yu, R. Annand, M. K. Jain and M. H. Gelb, *Biochemistry*, 1998, **37**, 6697–6710.
18 I. Constantinescu and M. Lafleur, *Biochim. Biophys. Acta, Biomembr.*, 2004, **1667**, 26–37.
19 C. E. Dempsey, *Biochim. Biophys. Acta, Rev. Biomembr.*, 1990, **1031**, 143–161.
20 W. Wilcox and D. Eisenberg, *Protein Sci.*, 1992, **1**, 641–653.
21 P. Yuan, P. J. Fisher, F. G. Prendergast and M. D. Kemple, *Biophys. J.*, 1996, **70**, 2223–2238.
22 J. F. Faucon, J. Dufourcq and C. Lussan, *FEBS Lett.*, 1979, **102**, 187–190.
23 H. S. Courtney, W. A. Simpson and E. H. Beachey, *Infect. Immun.*, 1986, **51**, 414–418.

24 P. Wessman, M. Morin, K. Reijmar and K. Edwards, *J. Colloid Interface Sci.*, 2010, **346**, 127–135.

25 W. Jencks, *Protein Sci.*, 1994, **3**, 2459–2464.

26 A. C. Satterthwait and W. P. Jencks, *J. Am. Chem. Soc.*, 1974, **96**, 7018–7031.

27 M. Monette, M. R. Van Calsteren and M. Lafleur, *Biochim. Biophys. Acta, Biomembr.*, 1993, **1149**, 319–328.

28 T. Pott and E.-J. Dufourc, *Biophys. J.*, 1995, **68**, 965–977.

Faraday Discussions

ROYAL SOCIETY OF CHEMISTRY

PAPER

Modulation of a host's cell membrane nano-environment by mycobacterial glycolipids: involvement of PI(4,5)P$_2$ signaling lipid?†

Manjari Mishra and Shobhna Kapoor ⓘ *

Received 3rd May 2020, Accepted 30th June 2020

DOI: 10.1039/d0fd00051e

Virulence-associated glycolipids from *Mycobacterium tuberculosis* (*Mtb*) act as effector molecules during infection—in addition to proteins. Upon insertion, they alter the host cell's membrane properties modifying the host's functions to aid *Mtb* survival and disease course. Here we combine tether force experiments and microscopy to reveal previously unknown insights on the potential involvement of the phosphatidylinositol 4,5-bisphosphate (PI(4,5)P$_2$) lipid in the *Mtb* lipid–host interaction landscape. Our data shows that *Mtb* lipids, having different structural and chemical make-up, distinctly alter a host's PI(4,5)P$_2$ membrane abundance/organization and PI(4,5)P$_2$-actin colocalization, thus impacting the plasma membrane–cytoskeletal adhesion forces. Combined with our previous findings that underscore the role of exogenous *Mtb* lipids in remodeling host plasma membrane organization and mechanics, this work builds upon a lipid-centric view of tubercular infections. Dynamically changing a host's plasma membrane lipid content – in response to virulent lipids – might represent a so far unexplored mechanism invoked by *Mtb* to modulate the host cell's adhesive properties to escape immune surveillance. These findings will deepen our collective understanding of the functional role of *Mtb* lipids in hijacking the host cell processes amenable to pharmacological inhibition.

Introduction

Cell membranes orchestrate essential cellular processes such as signaling, immune responses, differentiation, and molecular recognition.[1] The plasma membrane (PM), functions by compartmentalizing specific lipids and proteins in fluid-membrane bounded structures. One such type of structure that is best described as "lipid rafts", are enriched in cholesterol, glycosphingolipids, sphingomylein and raftophilic proteins. Fluctuations in local PM lipid

Department of Chemistry, Indian Institute of Technology Bombay, Mumbai 400076, India. E-mail: shobhnakapoor@chem.iitb.ac.in

† Electronic supplementary information (ESI) available. See DOI: 10.1039/d0fd00051e

composition leads to lipid raft domain formation that acts as a major signaling hotspot, regulating cellular processes.[2,3] Membrane physical properties such as order, fluidity, and elasticity, regulate membrane lateral organization and control lipid/protein diffusion,[4,5] lipid/protein localization[6,7] and lipid–protein interactions.[7,8] The aforementioned membrane properties have recently been in the spotlight of membrane biologists, due to their marked alteration in cancer, inflammation, and pathogen invasion.

PM–cytoskeleton adhesive interaction is one of the central lipid–protein interactions that drive fundamental processes, including endocytosis, exocytosis, signaling, formation and retraction of lamellipodia/filopodia.[9,10] Though membrane adhesion to external substrates has been extensively studied, the intracellular adhesion between the PM and cytoskeleton, and its modulation under diseased conditions, is far less explored. This is in part due to the complex interaction between the ever-expanding list of contributing components – lipids, membrane receptors and cytoskeleton-associated regulatory proteins such as, talin, WASP, gesolin, cofilin, and small Rho GTPases.[11,12] PM mechanics also substantially influence the mechanical association of membranes with the underlying cytoskeleton.[13,14] The major lipid players that regulate membrane–cytoskeletal binding directly or indirectly are cholesterol, ceramide, and sphingolipids.[15] For instance, cholesterol alteration increases or decreases PM–cytoskeletal interactions in a cell-type specific manner by modifying the local lipid composition.[16,17] Another prime regulator of membrane–cytoskeleton association,[18] and consequently of cell mechanical properties,[19] is phosphatidylinositol 4,5-bisphosphate ($PI(4,5)P_2$) lipid, by virtue of its direct association to many cytoskeletal proteins. Further, $PI(4,5)P_2$ alters the activity/localization of small GTPases that regulate actin assembly. $PI(4,5)P_2$ forms clusters/domains in PM; and membrane-localized PI kinases, phosphatases and other lipids control its steady-state membrane distribution.[20] This drives the fine-tuning of cellular adhesion domains, eventually regulating the membrane–cytoskeletal interactome. For instance, cholesterol-induced alteration of membrane organization modulates the local abundance of ($PI(4,5)P_2$) within lipid domains, subsequently leading to down-regulated activity of the $PI(4,5)P_2$-binding protein, gesolin[7].

As changes in PM phospholipid lipid number and repertoire modifies the membrane–cytoskeleton interactions, they are tapped into by pathogens for their survival, uptake and infection.[21] For example, the effector released from *S. flexneri* hydrolyses $PI(4,5)P_2$ to $PI(4)P$ decreasing the tethering force of PM–$PI(4,5)P_2$ binding and cytoskeletal-anchoring proteins. This consequently causes the extension of filopodia and membrane blebbing, promoting pathogen uptake.[22] Similarly, the VPA0450 effector from *V. parahaemolyticus* contains a catalytic motif from inositol polyphosphate 5-phosphatase, which allows it to hydrolyze $PI(4,5)P_2$. This causes the disruption of actin dynamics and detachment of cortical cytoskeleton from the PM.[23] In contrast, enteropathogenic *E. coli* (EPEC) using its effector – an intimin receptor – triggers host receptor clustering to activate signaling cascades that consequently regulate actin assembly at the host cell PM.[24]

Mycobacterium tuberculosis (*Mtb*) also influences a number of membrane-dependent processes involved in host–pathogen interplay through its various chemically diverse lipids.[25,26] Insertion of exogenous *Mtb* lipids alters the host cell's PM lateral membrane organization and associated biophysical properties.[27-31] Further, *Mtb* lipids selectively modify the actin network underneath the PM by redistributing actin structures such as filaments, patches and puncta.[28] The knowledge of molecular mechanisms and key players involved in *Mtb* lipid-mediated remodeling of the host's PM and actin cytoskeleton remains limited though. Here we explore the effects of various *Mtb* lipids on the host's cell membrane–cytoskeleton interactions using atomic force tether measurements and confocal microscopy. The findings suggest a plausible functional involvement of host membrane lipid PI(4,5)P$_2$, which *hitherto* remained unknown. The results show that *Mtb* lipids modify PM–cytoskeleton adhesion forces, PI(4,5)P$_2$ localization, clustering and co-localization with actin, dictated by the chemical nature of the *Mtb* lipid. These provide support to the paradigm that *Mtb* can temporally modulate a host's cell functions, that depend on the PM and cytoskeletal network, by synthesizing and employing distinct virulent lipids at various stages of infection.

Together, these results exemplify the role of specific lipids from pathogenic mycobacteria to fine-tune the host's lipid membrane structure and its association with the cytoskeleton by interfering with the abundance and localization of the prominent signaling lipid, PI(4,5)P$_2$. This could confer regulatory control on the downstream membrane-associated host's signaling by virtue of modified PM–cytoskeleton organization, structure and function. These findings enhance our understanding of far-less explored pathogenic lipid effectors in tubercular infections and the associated host factors mediating their effect. This work could inspire the development of novel lipid-based therapeutic approaches against *Mtb*.

Experimental materials and methods

Cell lines, mycobacterial lipids, plasmid and reagents

Roswell Park Memorial Institute medium (RPMI-1640), 20× Antibiotic-Antimycotic Solution, 0.25% trypsin-EDTA, Fetal Bovine Serum (FBS) HyClone (USA), Laurdan, sodium carbonate, 1× Dulbecco's phosphate buffer saline (D-PBS), DMSO, paraformaldehyde, PMA (phorbol 12-myristate-13-acetate), DAPI, Phalloidin–F-actin dye (Sigma), PLCδ1PH–GFP plasmid was a kind gift from Prof. Tamas Balla. THP-1 cells were a kind gift from Prof. Sarika Mehra (Chemical Engineering, IIT Bombay). THP-1 monocytic cell lines were maintained as per our previous published reports.[28,36] Cell lines were routinely checked for the absence of mycoplasma contamination and experiments were performed using the cells with a passage no of 18–20. Purified mycobacterium lipids from H37Rv strain were obtained from BEI Resources, NIH; purified sulfolipid-1 (NR-14845), phthiocerol dimycocerosate (NR-20328), trehalose dimycolate (NR-14844), lipoarabinomannan (NR-14848) and lipoarabinomannan (NR-14849) from *Mycobacterium smegmatis*, strain mc2155. Monoclonal anti-*Mycobacterium* LAM, clone CS-40 antibody (BEI Resources, USA) and CF 488A, goat anti-mouse IgG (H + L) (Biotium) secondary fluorescent antibody (Biotium).

Cell maintenance

THP-1 cells were maintained in RPMI-1640 medium supplemented with 10% heat-inactivated FBS and penicillin (100 U mL^{-1}), streptomycin (100 mg mL^{-1}), and gentamicin (20 µg mL^{-1}) at 37 °C in humidified air containing 5% CO$_2$. For the experiments, THP-1 cells were differentiated using 20 nM PMA, 72 h. Non-adherent cells were removed by washing with RPMI without FBS. Cells were routinely tested for Mycoplasma and found free of contamination.

Preparation of mycobacterial lipid suspensions

Lipid suspensions were reconstituted and prepared by gentle hydration and freeze thaw method reported previously.[28,36] All lipids were prepared at a concentration of 1 mg mL^{-1} and used at a working concentration of 10 μg mL^{-1}.

Confocal microscopy

Plasma membrane staining with N-Rh-DHPE and NBD-DPPE-loaded mycobacterium lipid suspensions. Cells were seeded and differentiated in confocal dishes as described above. Fluorescent *Mtb* lipid suspensions (10 μg mL^{-1}) containing N-RhDHPE (2 μg mL^{-1}) or NBD-DPPE (5 μg mL^{-1}) were prepared using the hydration method and added to cells. Images were recorded after 1 h and 4 h using a laser scanning confocal microscope (Carl Zeiss 780), 63×/1.4 oil objective.

Localization of lipoarabinomanan (LAM-v) in THP-1 macrophage membranes. THP-1 macrophages were fixed with PBS/4% (v/v) paraformaldehyde for 15 min at room temperature (RT) and further permeabilized with 0.1% Triton-X 100 in 1 × PBS for 10 min. After each step, slides were washed 3 times for 3 min with 1 × PBS. Cells were blocked with 1% BSA in 1 × PBS for 2 h at RT and washed 5 times for 2 min. Primary monoclonal anti-*Mycobacterium* LAM, clone CS-40 antibody raised in mouse cells, was diluted in 1% BSA and allowed to incubate (1 : 100–1 : 50) for 2 h at 4 °C. Further, cells were washed 5 times using 1 × PBS at RT. Thereafter, cells were incubated with CF 488A, goat anti-mouse IgG (H + L) secondary fluorescent antibody (1 : 100) diluted in 1% BSA for 50 min at RT. Slides/coverslips were washed again and finally mounted. Samples were stored at 4 °C until imaging was performed with the laser scanning confocal microscope (Carl Zeiss 780), 63×/1.4 oil objective.

Laurdan two-photon generalized polarization (GP) Imaging

Laurdan GP imaging experiments were performed and quantified as per our previously published reports.[28,36] Briefly, THP-1 cells were seeded in glass bottom dishes and differentiated using 20 nM PMA for 72 h. Cells were treated with *Mtb* lipid suspension at the indicated concentrations and time, at 37 °C and 5% CO$_2$ followed by washing with phosphate buffer saline (PBS) and the addition of 5 μM Laurdan in serum free media for 30 min, followed by washing. Spectral imaging was performed on a Zeiss LSM 780 confocal microscope (Carl Zeiss, Germany). Laurdan was excited at 780 nm using a multi-photon (titanium sapphire) laser (coherent radiation, CA) was applied, and images were reordered from 420 nm to 600 nm with 10 nm steps. The images were analyzed using a reported plug-in compatible with ImageJ as briefly described below. Images were restricted using an intensity threshold, as explained in detail in ref. 28. GP was then calculated at each pixel using the intensities from the images collected at 440 and 490 using eqn (1), and pseudo colored GP maps were generated.

$$GP = \frac{I_{440} - GI_{490}}{I_{440} + GI_{490}} \tag{1}$$

Calibration images were collected to calculate G factor using 5 µM Laurdan solution in DMSO and eqn (2):

$$G = \frac{GP_{ref} + GP_{ref}GP_{mes} - GP_{mes} - 1}{GP_{mes} + GP_{ref}GP_{mes} - GP_{ref} - 1} \tag{2}$$

where, GP_{ref} is the reported reference GP value (0.207) of the Laurdan dye, and GP_{mes} is the GP value of the dye in pure DMSO measured with the same experimental set-up used for the experiments with cells.

Transfection of PLCδ1PH–GFP reporter in THP-1 macrophages

Differentiated THP-1 macrophages were grown on glass coverslips and transfected using Lipofectamine (Life Technology) with PLCδ1PH–GFP plasmids. After 8–10 h of transfection, cells were replenished with fresh serum media and left overnight. Cells were checked under the fluorescence microscope for GFP expression.

Phalloidin–TRITC F-actin imaging

PLCδ1PH–GFP transfected THP-1 macrophages were incubated with various mycobacterial lipids (PDIM, TDM, LAM$_A$ and LAM$_V$) for 4 h at a concentration of 10 µg mL^{-1}. Thereafter, cells were washed and fixed with 4% paraformaldehyde for 10 min at RT and washed with 1 × PBS 3 times. Then they were permeabilized using 0.1% Triton-X-100 in 1 × PBS followed by blocking in 1% BSA for 1.5 h and rinsed with PBS/PBST. Phalloidin–TRITC (Invitrogen) was used to stain F-actin (2 µg mL^{-1}) for 30 min in the dark, then it was rinsed thoroughly with 1 × PBS and slides mounted with mounting media. Images were taken on a laser scanning confocal microscope (Carl Zeiss 780) using a 63×/1.4NA objective.

Image and data analysis for quantitative localization of the PLCδPH domain in plasma membrane

All post-acquisition image analyses for F-actin and PLCδ1PH–GFP reporter were performed using Zen 3.0 software. Fluorescence intensity of both the channels were quantified by drawing the line across the cells using the line profile module in Zen 3.0 module. To evaluate the changes in plasma membrane PI(4,5)P$_2$ in cells treated with *Mtb* lipids and their role in virulence, we tracked the relative PLCδ1PH–GFP plasma membrane and cytosolic intensity. Transfected cells were acquired with good expression of PLCδ1PH–GFP reporter and pixel intensity histograms were created through a selected line across the cells, intensity was not saturated at highest level. The right side panel of Fig. 4 shows the ratio, which helps us to assess the localization of PI(4,5)P$_2$ in the membrane after *Mtb* lipid treatment. More than one line was drawn on each cell, to maintain the reproducibility of intensities. A minimum of 30 cells was analyzed per condition per three independent experiments, which was obtained manually using Blue edition Zen software (Zeiss, Jena, Germany). Further, the data were analyzed and plotted using Microsoft Excel 2007 (Redmond, WA) and Graph Pad Prism 5.0. Significance level was assessed by One way-ANOVA test using Graph pad Prism software version 5.0 (San Diego, CA).

Integrin immunofluorescence

THP-1 macrophages were fixed with PBS/4% (v/v) paraformaldehyde for 15 min at room temperature (RT) and further permeabilized with 0.1% Triton-X 100 in 1 ×

PBS for 10 min. After each step, slides were washed three times for 3 min with 1 × PBS. Cells were blocked with 1% BSA in 1 × PBS for 2 h at RT and washed five times for 2 min. Primary monoclonal integrin β-1 antibody (CD29) (Thermo Fischer Scientific; 16-0291-85) raised in mouse was diluted in 1% BSA and allowed to incubate (1 : 100–1 : 50) for 2 h at 4 °C. Further, cells were washed for five times using 1 × PBS at RT. Thereafter, cells were incubated with Alexa flour 546, goat anti-mouse IgG (H + L) secondary fluorescent antibody (1 : 100) diluted in 1% BSA for 50 min at RT. Slides/coverslips were washed again and finally mounted. Samples were stored at 4 °C until imaging was performed with laser scanning confocal microscope (Carl Zeiss 780), 63×/1.4 oil objective.

Laurdan fluorescence spectroscopy

In a 6-well plate, 3×10^5 THP-1 macrophages were treated with PDIM, TDM, SL-1, LAM$_A$ and LAM$_V$ at a concentration of 10 µg mL^{-1} respectively, for 4 h. The cells were washed with 1 × PBS, trypsinized, centrifuged for 5 min at 1500 rpm and suspended in 1 mL of 1 × PBS. Laurdan stock was dissolved and prepared in DMSO and was added for the corresponding fluorescence measurements such that the final concentration of the dye in 1 mL of the cell suspension was 5 µM after which the cells were incubated at RT for 20 min, centrifuged at 1500 rpm for 10 min and re-suspended in 1 mL of 1 × PBS. Laurdan G. P. was measured using the temperature-controlled Varian Cary fluorescence spectrophotometer. Laurdan generalized polarization (GP, ex: 350 nm) was calculated, using the eqn (3).

$$GP = \frac{I_{440} - I_{490}}{I_{440} + I_{490}} \tag{3}$$

where I_{440} and I_{490} refer to the intensities at 440 nm and 490 nm, which are characteristic for an ordered (gel) lipid phase and fluid liquid-crystalline phase, respectively. The GP values range from +1 to −1.

Atomic force spectroscopy

Force spectroscopy experiments were conducted in contact mode. MFP-3D atomic force microscope (Asylum Research, Santa Barbara, CA, USA) with silicon nitride cantilevers (Oxford Instruments), nominal resonance frequency and a spring constant of 22 kHz and 0.16 N m^{-1}, were used. Cantilever calibration was performed to obtain a spring constant of 0.10–0.19 N m using a thermal noise system. The load applied was 60 nN, and the velocity of the cantilever was set at 6 µm s^{-1}. PDIM, TDM, SL-1, LAM$_A$ and LAM$_V$ were incubated with THP-1 macrophages for 2 h at a concentration of 10 µg mL^{-1}, respectively. The force curves were recorded in the region of the cell area, which was between the cell periphery and the nucleus. At least 100 cells were used for the spectroscopy in each set of independent experiments. For the analysis of membrane tethers (tether force, tether number and length), a reported custom-made MATLAB program was used to analyze the data.[37] Tethers were observed in a force $vs.$ distance curve as well-defined plateaus of constant force, where the stepwise force shift corresponds to the extension and rupture of the individual tethers. Membrane rupture events followed by a force plateau with a distance of more than 2 µm were considered as tethers in the analysis. The tether forces (F) were then used to calculate cytoskeleton–membrane adhesion energy (γ) as per previous reports,[14,32,33] and eqn (4) and eqn (5), which show that the tether force is proportional to the square root of the cytoskeleton–membrane adhesion energy using first approximation (eqn (4));

$$\gamma = \frac{F^2}{8\pi^2 B} \tag{4}$$

where, B is the membrane bending modulus, which can further be calculated using eqn (5)

$$F = \frac{2\pi B}{R} \tag{5}$$

where, R is the tether radius. In our study we have estimated the value for B from a previous study detailing the mechanical properties of macrophage cells, and a value of 8.0×10^{-19} J was implemented in our analysis.[14]

Results and discussion

Insertion of virulence-associated glycolipids impacts the host's cell membrane hydration and order

Mtb possesses three distinct classes of virulence-associated glycolipids: trehalose-containing lipids including trehalosedimycolate (TDM), lipomannan (LM), and sulfoglycolipid-1 (SL-1); phosphatidylinositol-derived lipoglycans *e.g.*, lipoarabinomannan (LAM); and mycocerosate-containing lipids such as phthiocerol dimycocerosate (PDIM), see Fig. 1. The surface exposed location of these glycolipids poises them for effective interaction with the host cell's PM. Successful insertion of exogenously added *Mtb* lipids within the host cell PM (Fig. S1 and S2†) modifies host membrane properties in a precise fashion. In this regard, by mapping lipid hydration and membrane order, we have recently shown that natural *Mtb* lipids remodel the host cell plasma membrane surface of THP-1 macrophages in a spatio-temporal fashion.[28] A ratio-metric parameter called generalized polarization (GP) was used to access these cell membrane properties using confocal microscopy and specific z-sections. High positive GP values stand for rigid and ordered membrane surfaces, and low negative GP values specify increased membrane fluidity and subsequently decreased order. Psuedo-coloured GP images for control cells displayed heterogeneous lipid hydration/order distribution with distinctly high GP domains on the cell surface (Fig. 2A, colored red); consistent with previous reports on this cell type.[34] These high GP domains reflect ordered lipid domains (mostly aggregated) or lipid rafts on the THP-1 cell surface. GP histograms demonstrated a bimodal distribution centered at distinct GP values ($C1 = -0.06$, $C2 = 0.46$). Comparison of the GP values with earlier values acquired on cellular and model membranes indicate the presence of co-existing ordered and fluid regions in THP-1 lipid membranes, *i.e.*, heterogeneous membrane organization.[34] *Mtb* SL-1 led to a decrease in the host cell membrane order by increasing the abundance (surface coverage of fluid domains increases to 70.5% from 36%) and fluidity (decrease of GP value ($C2$) from -0.06 to $-0.1/-0.53$) of fluid membrane regions (Fig. 2). Concomitantly a reduction in ordered cell membrane regions was observed (surface coverage of this region dropped from 54% to approx. 15%).

Marked differences within the *Mtb* lipid structures were reflected in significantly different host cell membrane hydration patterns. PDIM showed no effect

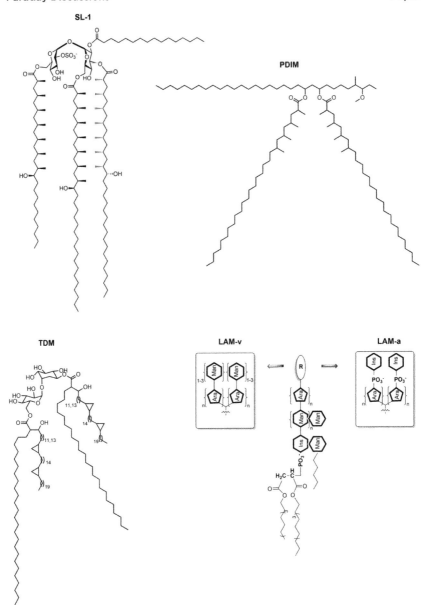

Fig. 1 Schematic representation of the virulence-associated and avirulent mycobacterial glycolipids. Phthiocerol dimycocerosate (PDIM), trehalose dimycolate (TDM), sulfogly-colpid-1 (SL-1) and virulent/avirulent lipoarabinomannan (LAM$_V$/LAM$_A$).

on the cell surface organization and order, and the abundance of fluid/ordered regions (Fig. 2) was similar to untreated control cells. On the other hand, virulent LAM (LAM$_V$) increased the cell membrane order, *i.e.*, the $C1$ GP increased to -0.02 and $C2$ GP to 0.49, unlike SL-1 (Fig. 2). It also increased the surface coverage of ordered membrane regions by approx. 10%. These findings are rationalized based on the structural make-up of *Mtb* lipid acyl chains and head groups. The

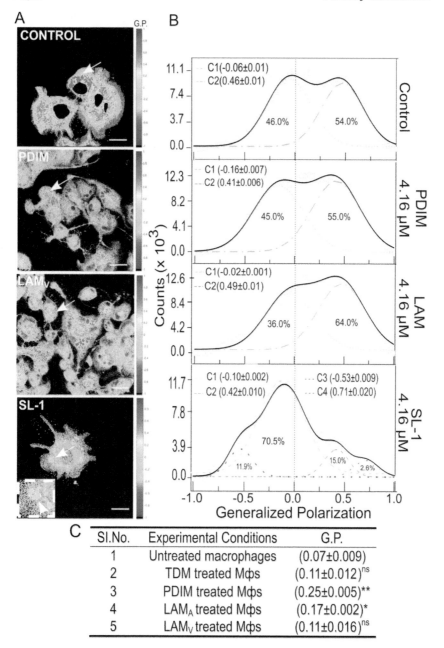

Fig. 2 Virulence-associated *Mtb* lipids regulate the host's cell membrane order and hydration in a structure-dependent fashion. (A) Pseudo-colored GP images of THP-1 macrophages in the absence and presence of the various *Mtb* lipids, treated for 4 h at a given concentration of 10 μg mL^{-1}. Inset for SL-1 shows distinct domains induced by SL-1 treatment. (B) GP distribution associated with the stack of GP images ($n = 90$, $N = 3$) fitted to Gaussian distributions. Graph shows peak centroids (C1–C4), which represent the GP values of all Laurdan-stained pixels for each respective population. The surface coverage of each Gaussian population (area under the curve) is indicated as shown. Scale bar: 10 μm, 40× water objective. (C) Quantitative measure of global cellular GP of THP-1 macrophages in the presence of, and without, the indicated *Mtb* lipids. *Fluorescence spectroscopy. Data are mean ± SEM from three independent experiments. ns: not significant. **$P < 0.01$.

shorter lipid anchors in LAM facilitate high conformational packing with similar carbon chain length host lipids leading to less water penetration at the lipid head group and hence the high GP or low hydration.

Another striking observation was the nature of lateral organization induced by SL-1 relative to LAM_V or PDIM. SL-1 treated THP-1 cell membranes demonstrated a multimodal GP distribution and distinct domains with outer ring-like structures of relative higher GP, indicating selective lipid phase separation.[28] On the contrary, with LAM_V, the ordered membrane regions were uniformly distributed around the cell boundary compared to control cells; wherein they were clustered together. These results suggest higher perturbation of ordered membrane regions on the THP-1 cell surfaces by *Mtb* virulent lipids. Moreover, the ordered lipid regions affected by *Mtb* lipids have been shown to be cholesterol enriched.[28] Consistently, previous findings have shown that pathogens, including *Mtb*, markedly influence cholesterol-enriched lipid domains to modulate a host's cells in their favor.[35–37] In this regard, very little information is available on the pathogenic factors that induce this phenotype and our results (using purified bacterial glycolipids) point towards the role of exposed *Mtb* lipids in altering the host membrane organization with modulation of ordered lipid domains.

Laurdan spectroscopy measurements on THP-1 cells treated with various *Mtb* lipids (Fig. 2C) corroborate the above results. PDIM demonstrated contrasting results by showing a significant increase of spectroscopic GP values. This could be due to dye internalization and labeling of internal membranes. Thus, we consider these changes as global membrane changes in lipid packing. Nonetheless, some important insights could be obtained. Comparison of virulent and avirulent LAM spectroscopy data indicate that branched mannose capping in the virulent lipid possibly acts as a handle to maintain an optimum level of water hydration at the lipid interface, *i.e.*, the GP value is closer to the control cells. This could be due to molecular interactions with either host lipids or receptors or both, and at present the contribution of these factors remain unknown. Next, TDM treated cells displayed no significant effect on the cell membrane hydration. This is in contrast to previous reports that has shown that TDM (from *N. asteroids*, C48) inhibits vesicle fusion by reducing the membrane fluidity in natural and host mimetic model membranes.[38] We assign this discrepancy to the carbon chain length differences in TDM from different bacterial origin. The conformationally constrained and highly hydrophobic long mycolic chains (C_{80}–C_{90}) might mitigate water penetration at the lipid head group interface, thus leading to no change in the host cell's membrane GP.

Collectively, these results suggest that structurally unrelated *Mtb* lipids distinctly remodel the host cell's membrane with prominent effects on the ordered lipid regions *via* mechanisms that remain under explored. In an independent study, we delved into the molecular mechanism behind these observations, and revealed that the distinct conformational states sampled by SL-1 acyl chains play a significant role in the host cell's membrane interactions.[39] In addition to acyl chain conformations, the chemical nature of the lipid head group – rendered by studies with synthesized SL-1 analogs – further fine-tunes the lipid–lipid interactions during host–pathogen crosstalk. Thus, the varied chemical nature of lipid acyl chains and the head group within *Mtb* lipids plays a crucial role in modulating the host's lipid membrane organization and related properties, necessitating further exploration. In this regard, we are currently exploring the conformational landscape of *Mtb* lipid chains using molecular dynamics simulations, as well as gaining insights into the nature of the host's PM domain remodeling induced by *Mtb* lipids.

Modulated membrane nanomechanics alters membrane–cytoskeleton interactions in infected host's cells

Cell membrane stiffness regulates membrane lateral organization and distribution of lipids/protein in the bilayer plane which is critical for physiological processes such as migration, differentiation, adhesion and signaling.[40] Response to external stimuli, including pathogens, means membrane stiffness can be an early biophysical disease marker. We previously investigated the host membrane stiffness (elastic modulus) upon insertion of exogenous *Mtb* lipids (ref. 28) using atomic force microscopy (AFM). The results revealed that the cortical stiffness of THP-1 macrophages increases in the presence of virulent lipids PDIM and TDM; underlined by the tight packing of the long saturated acyl (C_{34}–C_{40}) and mycolic acid (MA, C_{60}–C_{80}) chains within PDIM and TDM, respectively, with the host PM lipids (Fig. 3). Moreover, the MA chains in TDM exist in a W-shape or folded extended configuration, with high inherent packing efficiency, and are governed by the

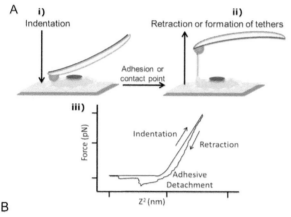

S.No	Conditions	Tether Force (pN)	Adhesion Energy $\kappa(10^{-19}$ J)
1	Untreated macrophages	76.4±1.2	20.4±1.2
2	TDM treated Mɸs	(69.2±1.1)***	16.4±1.1
3	PDIM treated Mɸs	(77.6±1.2)ⁿˢ	20.6±1.2
4	LAM_A treated Mɸs	(62.4±1.3)***	13.3±1.3
5	LAM_V treated Mɸs	(67.8±1.2)***	15.7±1.2
6	SL-1 treated Mɸs	74.2±2.1**(85.7±3.1)	18.8±2.1(25.1±3.1)

Fig. 3 *Mtb* lipids distinctly modulate the membrane–cytoskeletal adhesion energy. (A) Schematic representation of cell surface (i) indentation using AFM tip, (ii) tether formation upon retraction of AFM tip, which defines the term "adhesion energy", (iii) graphical representation of force *vs.* deflection, which determines the tether force and adhesion energy of the plasma membrane during detachment. (B) Quantitative measurement of tether force and adhesion energy of THP-1 macrophages under the indicated conditions. The values for THP-1 macrophages treated with SL-1 and the appropriate control (in parenthesis) were taken from a previously published report.[28] Data are mean ± SEM from three independent experiments ns: not significant; (***P < 0.001, **P < 0.01, two-tailed Student's *t*-test).

conformational constraints imposed by the cyclopropane rings. In contrast, LAM and SL-1 significantly reduced the host PM cortical stiffness (*i.e.*, softer cell membranes) due to the CH_3-bracnched acyl chains within SL-1 and the sugar appendages in LAM, packing inefficiently with the host lipids. We also observed that the terminal mannose capping, that differentiates virulent LAM from its avirulent counterpart, attenuated the cortical softening of the THP-1 cell membrane, adding to the functional differences between the LAM variants from different mycobacterial strains. For instance, while virulent LAM reorganizes membrane domains to disrupt membrane-associated signaling and subsequently reduces phagosome–lysosome fusion and apoptosis, in contrast, avirulent LAM counteracts the activation of the host's apoptotic response.[31,41] Our data indicates that LAM and SL-1 inserted into THP-1 membranes, are more elastic; and TDM or PDIM-inserted membranes are less elastic compared to the control cell membranes.

Collectively, our previous work demonstrates that different *Mtb* lipids distinctly impact the host's membrane mechanics invoking the fine features of their chemical structures that regulates acyl chain packing, van der Waals, electrostatic, and head group interactions with host lipids and/or membrane receptors. Various factors could contribute to these observations, such as membrane stiffness, in-plane bilayer tension, lipid composition changes and membrane–cytoskeletal interactions; the specific contributions of each remains to be investigated. In this work, we shed some light on the membrane–cytoskeletal interactions in the presence of *Mtb* lipids, and the identity of the host's involved molecular players.

Modulation of membrane–cytoskeleton interactions is linked with the regulation of cell functions such as exocytosis, endocytosis and migration, all of which are intimately exploited by pathogens to escape the host's immune surveillance.[21] Here, we explore the same during *Mtb* lipid–host membrane interactions. One measure of membrane–cytoskeletal interactions is the adhesion of membrane to the underlying cytoskeleton, wherein the two adhere with some optimal adhesion energy – the membrane adhesion energy.[42] Measurement of this energy requires that the membrane be separated from the cytoskeleton in some form. To this view, two approaches have been used and involve the formation of membrane blebs or tethers; among them the latter is preferred.[10] Membrane tethers are thin nanotubes that are pivotal regulators of cellular communication, adhesion and immune regulation and act as readouts of membrane stiffness, tension, bending, and interaction with the cytoskeleton.[43] Decreased plasma membrane–cytoskeletal adhesion leads to decreased tether force and depends on a number of factors including lipid composition, cytoskeleton density/structure and abundance and localization of cytoskeleton-associated proteins/lipids.[9] Findings have shown that tethers lack an actin cytoskeleton due to their small radius, and hence tether force could be a reliable measure of energy required to move the membrane from PM into tethers;[42] although some reports have documented the presence of actin monomers in tethers.[44] In this work, we explore the correlation of tether force with membrane–cytoskeleton adhesion energy. Tether force was measured by drawing tethers using an AFM tip (Fig. 3A), wherein it appears as a region of constant force during tether extension (AFM tip retraction) in a force-extension curve. Membrane adhesion energies were then calculated using previously reported studies. According to the theories of tether extraction and energy minimization,[14,32,33] tether force (F) is proportional to the square root of the cytoskeleton–membrane adhesion energy (γ) (using eqn (4) and (5)).

Mtb lipids SL-1, LAM$_A$, LAM$_V$ and TDM treated THP-1 cells displayed a significant reduction in the tether forces (9.4–18%), indicating that these tethers require less force to be extended compared to the untreated control membrane tethers[28] (Fig. 3B). Consequently, reduced adhesion energy was obtained in all treated conditions (albeit with varying amounts; LAM$_A$ (34%) < SL-1 (27%) < LAM$_V$ (23%) < TDM (19.6%)), except for PDIM (Fig. 3B). Reduced adhesion energy implies that collective binding interactions between the PM and cytoskeleton that influences adhesion are down regulated or re-distributed. This most likely involves alteration in the local abundance, or localization of the cytoskeleton anchoring/binding lipids and proteins within the membrane plane, which is impacted distinctly by the various *Mtb* lipids. Interaction of *Mtb* lipids with their bona fide receptors on the host's cell surface as well as interactions with the host's lipids in an *Mtb* lipid-dependent manner is implicated by these observations. In lieu of this, cholesterol depletion has shown to weaken PM–cytoskeleton interactions,[16] and given the re-distribution/reduction of cholesterol enriched domains (high GP raft-like domains in Laurdan GP imaging) by *Mtb* lipids (specifically by SL-1 and LAM in our case), we speculate a plausible role of membrane cholesterol in regulating the adhesion energy in THP-1 cells in response to *Mtb* glycolipids. Though at this point, the role of additional host lipids and proteins cannot be ignored. For PDIM, absence of any change in the adhesion energy indicates that it does not markedly alter the interactions between PM and the underlying actin, and hence possibly harnesses a mutually exclusive mechanism of action compared to other mycobacterial lipids.

As pathogens regulate membrane–cytoskeleton interactions to augment their uptake and enhance intracellular survival (*vida supra*), modulation of membrane adhesion energy represents a plausible mode of action of *Mtb* lipids during host interaction. Collectively, the tether force and adhesion energy analysis shows that the insertion of exogenously added *Mtb* lipids into the THP-1 membrane modulates lateral membrane organization, ordered lipid domains, and consequently PM–cytoskeletal binding interactions in a structure-dictated fashion. Together with Laurdan GP measurements, the above findings strongly support the hypothesis that diverse *Mtb* lipids modify the host cell membrane in a selective fashion pointing towards a temporally regulated attack on the host cell, driven by distinct *Mtb* lipids synthesized during various stages of infection.

Alteration of membrane–cytoskeletal interactions: participation of signaling lipid PI(4,5)P$_2$?

Key determinants of membrane–cytoskeletal interactions or membrane adhesion energy are adhesion molecules such as integrins, cathedrins, specific PM lipids and their respective association to the underlying cytoskeleton. Studies of membrane adhesion have revealed the role of membrane cholesterol in influencing membrane–cytoskeletal interactions and have brought to the forefront the signaling lipid, phosphatidylinositol-4,5-bisphosphate (PI(4,5)P$_2$), in regulating adhesion.[9] This is further underscored by observations that PI(4,5)P$_2$ is a master regulator of PM adhesion energy due to a number of factors. First it binds and regulates activity of many cytoskeletal proteins such as gesolin, talin, vinculin,

MARCKS, WASP, Arp2/3 and cofilin *etc.* Second, it acts as a precursor for second messengers such as inositol phosphate (IP), diacylglycerol (DAG) and Ca^{2+}, which together with $PI(4,5)P_2$ form a feedback network to fine-tune the cytoskeletal structure and organization.[10] Thus, local changes in $PI(4,5)P_2$ levels or localization can regulate membrane–cytoskeletal structure by altering the interactions between $PI(4,5)P_2$ and actin-anchoring/remodeling proteins. Subsequently, these modulated binding interactions regulate the local adhesion energy. Given our tether force measurements that implicate modulated adhesion energies, as well as previous observations that show *Mtb* lipids induce active re-structuring of cortical actin morphology, we sought out to explore changes, if any, in $PI(4,5)P_2$ lipids within THP-1 cell membrane upon treatment with exogenous *Mtb* lipids.

We used PLCδ1PH–GFP reporter to monitor $PI(4,5)P_2$ intracellular distribution in THP-1 macrophages, since this probe is known to selectively bind $PI(4,5)P_2$ lipids and is widely used for deciphering $PI(4,5)P_2$ dynamics *in cellulo*.[45] First, in contrast to reports that show uniform distribution of $PI(4,5)P_2$ along the plasma membrane, we observed a punctate/patchy appearance in the THP-1 cell membrane (Fig. 4A). We attribute this to the $PI(4,5)P_2$ localization pattern specific to THP-1 cells, as cell-type specific $PI(4,5)P_2$ staining has already been documented.[46] For example, in endothelial cells, $PI(4,5)P_2$ shows nuclear localization, homogenous distribution throughout the cell, as well as punctate formation along the cell membrane. The majority of $PI(4,5)P_2$ is found in PM; but some pools are associated with internal membranes such as in Golgi, ER and some vesicular structures.[47] The punctate-like appearance of $PI(4,5)P_2$ in THP-1 in Fig. 4A could also signal towards specific lipid domains that sequester $PI(4,5)P_2$.[20] Further, lipid micro domains such as lipid rafts and caveole are known to attract acidic lipids such as $PI(4,5)P_2$, thus the presence of high GP domains on the THP-1 cell surface could materialize in the form of highly fluorescent $PI(4,5)P_2$-enriched lipid domains (Fig. 4A).

Upon exogenous addition of physiologically relevant concentrations of *Mtb* lipids to THP-1 membranes, distinct $PI(4,5)P_2$ localization patterns were observed (Fig. 4A). *Mtb* SL-1 treated THP-1 cell membrane demonstrated an enhanced and uniform $PI(4,5)P_2$ localization along the PM with some intermittent patches. TDM and PDIM, on the other hand, showed uniform $PI(4,5)P_2$ fluorescence throughout the cell body, and most likely correspond to marked cytosolic translocation/ distribution of $PI(4,5)P_2$. LAM_V exhibited a pattern similar to SL-1, though the abundance of $PI(4,5)P_2$ patches was enhanced. Avirulent LAM showed definitive punctate distribution throughout the cell membrane with a qualitatively higher number of punctate/cell compared to the control. This could be, in part, attributed to modification of cholesterol enriched lipid domains by *Mtb* lipids. Cholesterol alteration has been previously linked with marked changes in $PI(4,5)P_2$ localization, wherein a patchy localization pattern was converted to a more uniform distribution across the cell by loss of lipid rafts/caveole.[46] As cellular levels of $PI(4,5)P_2$ are altered upon pathogen attack,[22,23] the modification of host signaling lipid $PI(4,5)P_2$ by *Mtb* lipids presented here might present an articulated strategy used by mycobacteria to induce signaling alterations in host cells during infection.

To further quantify these differences, fluorescence intensity was recorded along the line-intensity profile as shown in Fig. 4A and the PM *versus* cytosolic $PI(4,5)P_2$ intensity ratios were calculated (Fig. 4C). *Mtb* SL-1 and LAM_V

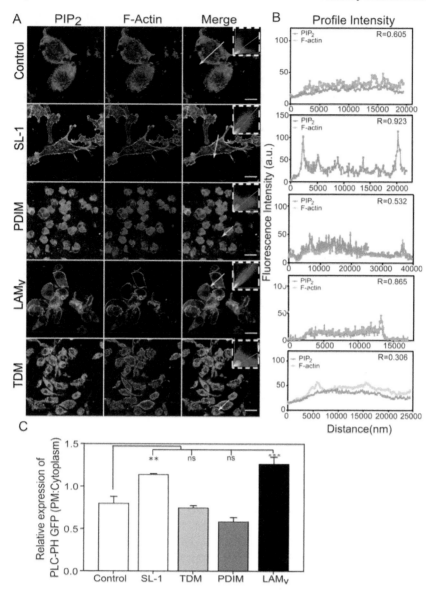

Fig. 4 Distinct cellular distribution and translocation responses of PLCδ1PH–GFP reporter from plasma membrane to cytoplasm in THP-1 macrophages: (A) THP-1 macrophages were transfected with PLCδ1PH–GFP. Cells were treated with mycobacterial lipids with the given concentration (10 µg mL^{-1}) for 4 h. Representative confocal images were acquired after staining with Phalloidin–F-actin. The Pearson correlation coefficient was calculated using Colocalization test Plugin in ImageJ, correlation scatter plots are shown in the white dotted box, in the top right corner of the merged image. Scale bar: 10 µm, 63× oil objective. (B) Average intensity histograms were calculated in each case; and a line profile was taken across the section of the cells as depicted above with a white arrow. (C) Average PM and cytoplasmic fluorescence intensity values were computed for calculating the PM/Cyto ratios: relative translocation or cellular distribution of PLCδ1PH–GFP from PM to cytoplasm. Data are mean ± SEM from three independent experiments (***P value < 0.001, **P value < 0.001, one way-ANOVA test).

significantly increased the PM : cytoplasmic PI(4,5)P$_2$ ratio, indicating higher PI(4,5)P$_2$ PM abundance; though the localization pattern in each of these cases were different (for PDIM and TDM, the ratio should be taken cautiously due to a diffuse labeling of the cell). Taken together, Fig. 4A shows the unique translocation responses of PI(4,5)P$_2$ upon treatment with various *Mtb* lipids suggesting that *Mtb* lipids distinctly alter the PI(4,5)P$_2$ membrane localization and/or abundance. The increase in PM PI(4,5)P$_2$ levels as seen with SL-1 and LAM$_V$ may represent a positive feedback mechanism, which can up regulate signaling processes, since a number of signaling proteins (containing PH domains) and cortical cytoskeletal structures are attached to the PM by binding to PI(4,5)P$_2$. Further, cytosolic PI(4,5)P$_2$ localization observed in the presence of TDM and PDIM imply modulation of mutually exclusive PI(4,5)P$_2$-dependent host processes compared with SL-1 and LAM, thus requiring different PI(4,5)P$_2$ localization patterns within the host cell. Furthermore, given the remodeling of cholesterol lipid domains by *Mtb* lipids (SL-1 and LAM), participation of membrane cholesterol on PI(4,5)P$_2$ distribution (as shown by other reports) is also implicated by our studies and would require further exploration.

The translocation response of PLCPH–GFP PI(4,5)P$_2$ reporter (*i.e.* loss of PM signal) has also been attributed to increasing levels of IP3 following PI(4,5)P$_2$ hydrolysis *via* PLC and cannot be ruled out in the present work. Nonetheless, to the best of our knowledge, this is the first report that demonstrates a link between pathogenic *Mtb* factors and PI(4,5)P$_2$ lipid re-distribution in host cells; the full biological implication of these observations is unknown at present. Currently, we are performing molecular dynamics simulations to obtain insights into the PIP(4,5)$_2$ lipid domain organization within the PM inner leaflet in the presence of *Mtb* lipids.

Motivated by the observations of cellular PI(4,5)P$_2$ within host cells, we next investigated the co-localization of PI(4,5)P$_2$ and actin (reflective of membrane–cytoskeletal interactions) in the presence of exogenous *Mtb* lipids. We found strong co-localization of PI(4,5)P$_2$ and actin in SL-1 and LAM$_V$ treated cells with high Pearson's coefficients compared to the control membranes (Fig. 4B). In PDIM treated THP-1 cells, PI(4,5)P$_2$–actin co-localization remained unchanged and for TDM a reduced co-localization was obtained. This suggests that SL-1 and LAM induced changes in how PI(4,5)P$_2$ percolates the actin network, but for TDM- and PDIM-treated cells, decoupling of the two takes place. In addition, reduced plasma membrane PI(4,5)P$_2$ in TDM and PDIM treated cells would consequently reduce anchorage points for actin in the plasma membrane, and hence less co-localization. The opposite is expected for SL-1 and LAM treated cells. Finally, alteration in membrane–cytoskeletal attributes coupled with modulated PI(4,5)P$_2$ lipid domains may lead to differential modulation of distinct host signaling cascades by SL-1 and LAM compared with TDM and PDIM. Indeed, we have observed differential cytokine signaling by these lipids;[39] though the link between these observations requires further testing. At present, involvement of additional host proteins mediating PI(4,5)P$_2$–actin interactions, or lack thereof, in the presence of *Mtb* lipids remains limited and mandates suitable proteomic investigations. Nonetheless, our observations provide exciting starting points to explore host's PI(4,5)P$_2$ interactome in response to *Mtb* infection for therapeutic targeting and gaining an exhaustive understanding of the lipid centric host–pathogen landscape.

Next, we monitored integrin receptors in response to *Mtb* lipid insertion. Integrins are transmembrane (TM) receptors that mediate cell–cell and cell–matrix interactions during cell adhesion and migration.[48] It has been reported that changes in the local plasma membrane lipid composition, specifically sphingolipids, attune clustering and dynamics of integrin receptors due to increased decoupling to the actin cytoskeleton, thus impacting a host cell's adhesion.[15] This was proposed as an efficient avenue exploited by pathogens during host interaction, for instance measles virus induces activation of sphingomyelinase (SMase) and decreases SM levels by driving its conversion to ceramide in the plasma membrane. This displaces cholesterol from lipid rafts and forms large ceramide-enriched domains, which in turn influence the function of PM receptors and consequently modifies the host's signaling cascades in favor of the pathogen.[49]

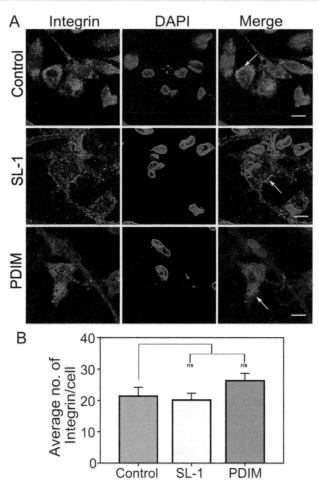

Fig. 5 Cellular distribution of plasma membranes β1-integrin receptor is not altered by *Mtb* lipids. (A) THP-1 macrophages were treated with the indicated mycobacterial lipids (10 μg mL^{-1}) for 4 h. Representative confocal images were acquired after staining with β-1 integrin antibody. Scale bar: 10 μm, 63× oil objective. (B) Quantification of β-1 integrin puncta/spots using Image J. data are mean ± SEM from three independent experiments (one way-ANOVA test; ns: non-significant).

As the TM regions of integrins are in direct contact with the PM; modulation of PM lipids such as $PI(4,5)P_2$ and actin localization/abundance can impact their location, diffusion and subsequently their function. This prompted us to investigate any effect of *Mtb* lipids on β1-integrin in host cells. We did not observe any change in β1-integrin signals in SL-1 and PDIM treated THP-1 plasma membranes compared to the control (Fig. 5). Unfortunately, due to the recent COVID-19 outbreak, this work was halted and no other *Mtb* lipids could be tested. Thus, we cautiously assign these as preliminary results only, awaiting further investigation. Nonetheless, no significant effect was observed on host β1-integrin by *Mtb* SL-1 and PDIM, indicating the potential non-involvement of integrins in *Mtb* lipid-mediated modulation of a host's cell membrane–cytoskeletal adhesion.

Conclusions

Tubercular infections stand apart among other infectious diseases due to the involvement of a complex lipid repertoire synthesized by *mycobacterium tuberculosis*. These act as virulence factors in addition to the highly explored mycobacterial proteins. Recently, host cell membrane insertion and modification of cellular immune processes has been substantiated as a well-accepted mode of action of virulent mycobacterial lipids. This mechanism is further supported by the highly hydrophobic nature of *Mtb* lipids predisposing them to insertion into the host cell. There are three main classes of virulence-associated *Mtb* glycolipids that are structurally diverse both in their acyl chain (harbouring long saturated mycolic chains, C_{80}–C_{90} or CH_3 branches) and the head group regions (containing mannose or arabinose sugar moieties). However, the understanding of how these structural variations differentially impact a host's cell functions is rather limited.

Our previous work has revealed uncharacterized mechanisms by which *Mtb* utilizes its atypical lipids endowed with unprecedented conformational properties to modulate the host's cell membrane structure and function, as well as modulating the actin cytoskeleton. Further, it was demonstrated that *Mtb* virulence-associated lipids modulate host autophagy and cytokine signaling pathways distinctly, dictated by the lipid chemical structure. The host's PM orchestrates various signaling cascades by compartmentalizing signaling lipids and proteins into specific domains, and has been evolutionarily hijacked by various pathogens, including *Mtb*, to enhance their survival and infection. This emphasizes the urgency to investigate the host's lipids/lipid domains and their plasma membrane distribution that are affected by *Mtb* lipid insertion, which remains understudied at present. To this end, gaining insight into the host (lipid) factors impacted by *Mtb* lipids during infection has the potential to not only improve our understanding of tubercular infections but also inspire development of suitable anti-Tb therapeutics.

Here, using tether force measurements along with microscopy and cell biology we demonstrate *Mtb*-lipid selective perturbation of a host's plasma membrane–cytoskeleton adhesion forces. Delving deeper into the plasma membrane–cytoskeletal landscape upon *Mtb* lipid interaction, we observed significant alteration in the cellular distribution of a key host signaling lipid implicated in membrane–cytoskeleton association – phosphatidylinositol 4,5-bisphosphate, $PI(4,5)P_2$. Further, we demonstrate modulated $PI(4,5)P_2$-actin co-localization patterns, indicating mutually exclusive mechanisms of action of various virulent *Mtb* glycolipids on membrane–cytoskeleton interactions. In summary, our results for the first time, provide a previously unknown, preliminary insight into the role of a host's lipid $PI(4,5)P_2$ in mediating the effect of *Mtb* lipids. These findings enable the generation of a framework wherein pertinent questions emerge such as; how does the host lipid $PI(4,5)P_2$ communicate with cytoskeletal machinery upon *Mtb* lipid insertion? Does the $PI(4,5)P_2$ interactome in the host cell correlate with distinct *Mtb* glycolipids? Further, it is tempting to speculate that *Mtb* modulates its cell wall lipid synthesis at distinct stages of infection within the host, and links it with the $PI(4,5)P_2$ re-distribution to temporally alter a host's signaling, to be favourable for itself.

Collectively, the structural differences in the *Mtb* lipid acyl chains and head group serve as a pathogenic handle to fine-tune lipid–lipid interactions during host contact implicating the host's lipid $PI(4,5)P_2$. These eventually modulate membrane–cytoskeleton interactions and subsequently coordinate host cellular functions in favor of the pathogen during infection.

Funding sources

This work was supported by the Seed Grant, IRCC, IIT Bombay, DBT Ramalingaswami fellowship, DST-Inspire Faculty Award and SERB, DST (Grant EMR/2016/005414).

Conflicts of interest

The authors declare that they have no conflict of interest.

Acknowledgements

We acknowledge the AFM, SAIF-MS and confocal facilities at IIT Bombay. We thank Prof. Tamas Balla for sharing the PLCδ1PH–GFP plasmid. We acknowledge BEI Resources, NIAID, NIH for purified sulfolipid-1 (NR-14845), phthiocerol dimycocerosate (NR-20328), trehalose dimycolate (NR-14844), lipoarabinomannan (NR-14848) from *Mycobacterium tuberculosis*, strain H37Rv and lipoarabinomannan (NR-14849) from *Mycobacterium smegmatis*, strain MC^2155. Monoclonal anti-mycobacterium LAM, clone CS-40 antibody (BEI Resources, USA). SK acknowledges her newborn baby girl, Maya Kapoor Narayan, for allowing her to finish the manuscript on time.

References

1 H. I. Ingolfsson, *et al.*, Lipid organization of the plasma membrane, *J. Am. Chem. Soc.*, 2014, **136**(41), 14554–14559.

2 D. Lingwood and K. Simons, Lipid rafts as a membrane-organizing principle, *Science*, 2010, **327**(5961), 46–50.

3 K. Jacobson, O. G. Mouritsen and R. G. Anderson, Lipid rafts: at a crossroad between cell biology and physics, *Nat. Cell Biol.*, 2007, **9**(1), 7–14.

4 R. Sheng, *et al.*, Lipids Regulate Lck Protein Activity through Their Interactions with the Lck Src Homology 2 Domain, *J. Biol. Chem.*, 2016, **291**(34), 17639–17650.

5 D. M. Owen, *et al.*, Sub-resolution lipid domains exist in the plasma membrane and regulate protein diffusion and distribution, *Nat. Commun.*, 2012, **3**, 1256.

6 P. V. Escriba, M. Sastre and J. A. Garcia-Sevilla, Disruption of cellular signaling pathways by daunomycin through destabilization of nonlamellar membrane structures, *Proc. Natl. Acad. Sci. U. S. A.*, 1995, **92**(16), 7595–7599.

7 Y. H. Wang, R. Bucki and P. A. Janmey, Cholesterol-Dependent Phase-Demixing in Lipid Bilayers as a Switch for the Activity of the Phosphoinositide-Binding Cytoskeletal Protein Gelsolin, *Biochemistry*, 2016, **55**(24), 3361–3369.

8 S. Kapoor, *et al.*, Pressure modulation of Ras-membrane interactions and intervesicle transfer, *J. Am. Chem. Soc.*, 2013, **135**(16), 6149–6156.

9 D. Raucher, *et al.*, Phosphatidylinositol 4,5-bisphosphate functions as a second messenger that regulates cytoskeleton-plasma membrane adhesion, *Cell*, 2000, **100**(2), 221–228.

10 M. P. Sheetz, J. E. Sable and H. G. Dobereiner, Continuous membrane-cytoskeleton adhesion requires continuous accommodation to lipid and cytoskeleton dynamics, *Annu. Rev. Biophys. Biomol. Struct.*, 2006, **35**, 417–434.

11 V. DesMarais, *et al.*, Cofilin takes the lead, *J. Cell Sci.*, 2005, **118**(Pt 1), 19–26.

12 S. McLaughlin, *et al.*, PIP(2) and proteins: interactions, organization, and information flow, *Annu. Rev. Biophys. Biomol. Struct.*, 2002, **31**, 151–175.

13 A. Diz-Munoz, D. A. Fletcher and O. D. Weiner, Use the force: membrane tension as an organizer of cell shape and motility, *Trends Cell Biol.*, 2013, **23**(2), 47–53.

14 B. Pontes, *et al.*, Membrane elastic properties and cell function, *PLoS One*, 2013, **8**(7), e67708.

15 C. Eich, *et al.*, Changes in membrane sphingolipid composition modulate dynamics and adhesion of integrin nanoclusters, *Sci. Rep.*, 2016, **6**, 20693.

16 J. Kwik, *et al.*, Membrane cholesterol, lateral mobility, and the phosphatidylinositol 4,5-bisphosphate-dependent organization of cell actin, *Proc. Natl. Acad. Sci. U. S. A.*, 2003, **100**(24), 13964–13969.

17 M. Sun, *et al.*, The effect of cellular cholesterol on membrane-cytoskeleton adhesion, *J. Cell Sci.*, 2007, **120**(Pt 13), 2223–2231.

18 A. S. Sechi and J. Wehland, The actin cytoskeleton and plasma membrane connection: PtdIns(4,5)P(2) influences cytoskeletal protein activity at the plasma membrane, *J. Cell Sci.*, 2000, **113**(Pt 21), 3685–3695.

19 P. De Camilli, *et al.*, Phosphoinositides as regulators in membrane traffic, *Science*, 1996, **271**(5255), 1533–1539.

20 L. Chierico, *et al.*, Live cell imaging of membrane/cytoskeleton interactions and membrane topology, *Sci. Rep.*, 2014, **4**, 6056.

21 H. Ham, A. Sreelatha and K. Orth, Manipulation of host membranes by bacterial effectors, *Nat. Rev. Microbiol.*, 2011, **9**(9), 635–646.

22 K. Niebuhr, *et al.*, Conversion of PtdIns(4,5)P(2) into PtdIns(5)P by the S. flexneri effector IpgD reorganizes host cell morphology, *EMBO J.*, 2002, **21**(19), 5069–5078.

23 C. A. Broberg, *et al.*, A Vibrio effector protein is an inositol phosphatase and disrupts host cell membrane integrity, *Science*, 2010, **329**(5999), 1660–1662.

24 R. D. Hayward, *et al.*, Exploiting pathogenic Escherichia coli to model transmembrane receptor signalling, *Nat. Rev. Microbiol.*, 2006, **4**(5), 358–370.

25 O. Neyrolles and C. Guilhot, Recent advances in deciphering the contribution of Mycobacterium tuberculosis lipids to pathogenesis, *Tuberculosis*, 2011, **91**(3), 187–195.

26 M. Jackson, The mycobacterial cell envelope-lipids, *Cold Spring Harbor Perspect. Med.*, 2014, **4**(10), a021105.

27 J. Quigley, *et al.*, The Cell Wall Lipid PDIM Contributes to Phagosomal Escape and Host Cell Exit of Mycobacterium tuberculosis, *mBio*, 2017, **8**(2), e00148-17.

28 M. Mishra, *et al.*, Dynamic Remodeling of the Host Cell Membrane by Virulent Mycobacterial Sulfoglycolipid-1, *Sci. Rep.*, 2019, **9**(1), 12844.

29 C. Astarie-Dequeker, *et al.*, Phthiocerol dimycocerosates of M. tuberculosis participate in macrophage invasion by inducing changes in the organization of plasma membrane lipids, *PLoS Pathog.*, 2009, **5**(2), e1000289.

30 H. Nakayama, *et al.*, Lipoarabinomannan binding to lactosylceramide in lipid rafts is essential for the phagocytosis of mycobacteria by human neutrophils, *Sci. Signaling*, 2016, **9**(449), ra101.

31 A. Welin, *et al.*, Incorporation of Mycobacterium tuberculosis lipoarabinomannan into macrophage membrane rafts is a prerequisite for the phagosomal maturation block, *Infect. Immun.*, 2008, **76**(7), 2882–2887.

32 I. Derenyi, F. Julicher and J. Prost, Formation and interaction of membrane tubes, *Phys. Rev. Lett.*, 2002, **88**(23), 238101.

33 T. R. Powers, G. Huber and R. E. Goldstein, Fluid-membrane tethers: minimal surfaces and elastic boundary layers, *Phys. Rev. E: Stat., Nonlinear, Soft Matter Phys.*, 2002, **65**(4 Pt 1), 041901.

34 K. Gaus, *et al.*, Visualizing lipid structure and raft domains in living cells with two-photon microscopy, *Proc. Natl. Acad. Sci. U. S. A.*, 2003, **100**(26), 15554–15559.

35 G. Viswanathan, *et al.*, Macrophage sphingolipids are essential for the entry of mycobacteria, *Chem. Phys. Lipids*, 2018, **213**, 25–31.

36 G. A. Kumar, M. Jafurulla and A. Chattopadhyay, The membrane as the gatekeeper of infection: cholesterol in host-pathogen interaction, *Chem. Phys. Lipids*, 2016, **199**, 179–185.

37 G. A. Kumar, *et al.*, Statin-induced chronic cholesterol depletion inhibits Leishmania donovani infection: relevance of optimum host membrane cholesterol, *Biochim. Biophys. Acta*, 2016, **1858**(9), 2088–2096.

38 B. J. Spargo, *et al.*, Cord factor (alpha,alpha-trehalose 6,6′-dimycolate) inhibits fusion between phospholipid vesicles, *Proc. Natl. Acad. Sci. U. S. A.*, 1991, **88**(3), 737–740.

39 R. Dadhich, *et al.*, A Virulence-Associated Glycolipid with Distinct Conformational Attributes: Impact on Lateral Organization of Host Plasma Membrane, Autophagy, and Signaling, *ACS Chem. Biol.*, 2020, **15**(3), 740–750.

40 G. Batta, *et al.*, Alterations in the properties of the cell membrane due to glycosphingolipid accumulation in a model of Gaucher disease, *Sci. Rep.*, 2018, **8**(1), 157.

41 R. A. Fratti, *et al.*, Mycobacterium tuberculosis glycosylated phosphatidylinositol causes phagosome maturation arrest, *Proc. Natl. Acad. Sci. U. S. A.*, 2003, **100**(9), 5437–5442.

42 R. M. Hochmuth and W. D. Marcus, Membrane tethers formed from blood cells with available area and determination of their adhesion energy, *Biophys. J.*, 2002, **82**(6), 2964–2969.

43 M. Sun, *et al.*, Multiple membrane tethers probed by atomic force microscopy, *Biophys. J.*, 2005, **89**(6), 4320–4329.

44 D. V. Zhelev and R. M. Hochmuth, Mechanically stimulated cytoskeleton rearrangement and cortical contraction in human neutrophils, *Biophys. J.*, 1995, **68**(5), 2004–2014.

45 T. Balla and P. Varnai, Visualization of cellular phosphoinositide pools with GFP-fused protein-domains, *Curr. Protoc. Cell Biol.*, 2009, Ch. 24, Unit 24.4.

46 Z. Hong, *et al.*, How cholesterol regulates endothelial biomechanics, *Front. Physiol.*, 2012, **3**, 426.

47 S. A. Watt, *et al.*, Subcellular localization of phosphatidylinositol 4,5-bisphosphate using the pleckstrin homology domain of phospholipase C delta1, *Biochem. J.*, 2002, **363**(Pt 3), 657–666.

48 T. A. Springer and M. L. Dustin, Integrin inside-out signaling and the immunological synapse, *Curr. Opin. Cell Biol.*, 2012, **24**(1), 107–115.

49 E. Avota, E. Gulbins and S. Schneider-Schaulies, DC-SIGN mediated sphingomyelinase-activation and ceramide generation is essential for enhancement of viral uptake in dendritic cells, *PLoS Pathog.*, 2011, 7(2), e1001290.

Faraday Discussions

PAPER

Interactions of polymyxin B with lipopolysaccharide-containing membranes

Alice Goode, (ID) Vivien Yeh (ID) and Boyan B. Bonev (ID)*

Received 1st June 2021, Accepted 23rd July 2021

DOI: 10.1039/d1fd00036e

Bacterial resistance to antibiotics constantly remodels the battlefront between infections and antibiotic therapy. Polymyxin B, a cationic peptide with an anti-Gram-negative spectrum of activity is re-entering use as a last resort measure and as an adjuvant. We use fluorescence dequenching to investigate the role of the rough chemotype bacterial lipopolysaccharide from *E. coli* BL21 as a molecular facilitator of membrane disruption by LPS. The minimal polymyxin B/lipid ratio required for leakage onset increased from 5.9×10^{-4} to 1.9×10^{-7} in the presence of rLPS. We confirm polymyxin B activity against *E. coli* BL21 by the agar diffusion method and determined a MIC of $291 \ \mu g \ ml^{-1}$. Changes in lipid membrane stability and dynamics in response to polymyxin and the role of LPS are investigated by ^{31}P NMR and high resolution ^{31}P MAS NMR relaxation is used to monitor selective molecular interactions between polymyxin B and rLPS within bilayer lipid membranes. We observe a strong facilitating effect from rLPS on the membrane lytic properties of polymyxin B and a specific, pyrophosphate-mediated process of molecular recognition of LPS by polymyxin B.

Introduction

Bacterial infections remain the primary cause of morbidity and mortality in humans despite the success of antibiotic chemotherapy. Bacteria have natural defences against xenobiotics but also the ability to adapt as a population under sustained antibiotic pressure that selects favourable phenotypes. In addition, such adaptations can be transferred within and between bacterial populations, which leads to a growing prevalence of bacterial phenotypes that do not respond well or are resistant to antibiotics. Particularly challenging are Gram-negative bacteria, which have an outer membrane (OM) that hinders antibiotic accessibility to target sites and reduces bacterial susceptibility to antibiotics. One approach to the management of such infections is the use of adjuvants in combination with conventional antibiotics.

The definitive feature of Gram-negative bacteria is the presence of an outer membrane that envelopes the entire bacterial cell and forms a periplasmic

School of Life Sciences, University of Nottingham, Queen's Medical Centre, Nottingham, NG7 2UH, UK. E-mail: boyan.bonev@nottingham.ac.uk

compartment containing bacterial peptidoglycan and a number of specialised enzymes, often with roles in the post-translational modification of proteins. The bacterial OM is an asymmetric bilayer with an inner leaflet made largely of phospholipids common to the inner membrane (IM), while the outer leaflet of the OM is almost entirely made of lipopolysaccharide (LPS). The OM also contains a significant fraction of membrane proteins with a unique, beta-barrel fold.

LPS lines the bacterial exterior where it performs a number of important functions, including cell recognition and xenobiotic defence. While the molecular structure of LPS has unique features that identify the bacterial species, its overall architecture commonly contains three parts – lipid A, the core oligosaccharide and O-antigen (Fig. 1). Lipid A has a largely conserved structure, which comprises a N-acetylated disaccharide, commonly derivatised at positions 2,3,2′ and 3′ with six or seven 14-carbon saturated chains and often phosphorylated or pyrophosphory-lated in positions 1 and 4′. Also conserved, the polysaccharide core attached at 6′ on lipid A contains unique 3-deoxy-D-*manno*-oct-2-ulosonic acid (Kdo) and heptose (Hep) monosaccharides that can be phosphorylated, pyrophosphorylated or

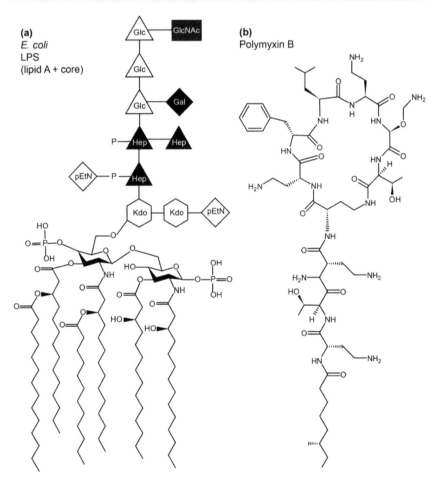

Fig. 1 Schematic structure of LPS from *E. coli* (a) and of polymyxin B (b).

derivatised with phosphoethanolamine (pEtN). A variable length, species-specific O-antigen, consisting of up to 40–50 saccharide repeats, extends beyond the core and serves as the characteristic recognition motif that identifies the bacterial species. LPS that has all three components is of the smooth chemotype (sLPS), while LPS lacking the O-antigen is of the rough chemotype (rLPS). Commonly studied *Escherichia coli* strains BL21 and K12 have truncated rLPS (Fig. 1). Phosphorylation and pyrophosphorylation in LPS are essential for establishing and maintaining OM stability *via* divalent cation-mediated (chiefly Ca^{2+} or Mg^{2+}) LPS/LPS interactions. Using solid state NMR, we have shown recently that such phosphorylation and pyrophosphorylation is extensive in the core region but sub-stoichiometric within lipid A, revealing the pivotal role of the core in OM stability.[1]

The conserved structure of LPS has driven co-evolution of antibacterial proteins,[2] peptides[3] and other xenobiotics that rely on molecular LPS recognition to disrupt bacterial OMs and exercise antibacterial functions. Polymyxins and colistins are a family of cyclic polycationic peptides derived from progenitors naturally produced by *Bacillus polymyxa*.[4] They were in clinical use until the 1980s, when their use declined due to relatively high nephrotoxicity.[5] Due to the rise in antibiotic resistance, despite their nephrotoxicity, polymyxins find use as last defence drugs against carbapenem-resistant Gram-negative infections[6] and can be used as adjuvants in antibiotic cocktails.[7]

Polymyxins disrupt both the outer and inner bacterial membranes.[8] Crossing the OM is sensitive to variation in the LPS chemotype, suggesting an LPS-dependent receptor-specific mechanism, while crossing the IM is sensitive to lipid composition. Besides the disruption of both membranes, cellular targets have been suggested to include DNA and ribosomes where polymyxin induces DNA clotting and ribosomal condensation.[9,10] The mechanism of antimicrobial action and bacterial adaptations and resistance have been reviewed recently.[11]

Modifications of LPS, specifically the pEtN-ation of lipid A, reduce susceptibility to polymyxins through the reduction of the LPS negative charge (for a review see ref. 12). Polymyxin resistant strains of *P. aeruginosa* have been shown to produce less phospholipid,[13] most likely through a pressure selection rather than as an adaptive response in protecting the bacterial IM. A study of polymyxin mediated glucose release from lipid vesicles has shown reduced release from phosphatidylcholine and methyl phosphatidyl ethanolamine compared to phosphatidyl ethanolamine membranes.[14]

In this report, we investigate the role of rLPS as a molecular receptor for polymyxin in lipid membranes using fluorescence dequenching and we monitor its antimicrobial activity. We investigate specific molecular interactions between polymyxin and LPS using longitudinal relaxation ^{31}P magic angle spinning (MAS) NMR. Relative changes in fast molecular dynamics are compared to the membrane phospholipid. We monitor membrane stability and collective dynamics by following changes in the ^{31}P chemical shift anisotropy (CSA) using wideline ^{31}P solid state NMR.

Experimental

Phospholipids, 1,2-dioleoyl-*sn*-glycero-3-phosphocholine (DOPC) and 1,2-dimyristoyl-*sn*-glycero-3-phosphocholine (DMPC) (Avanti Polar Lipids Inc. Alabaster, Alabama, USA) were used as purchased at >98% purity.

Phenol extraction of rLPS

Rough chemotype lipopolysaccharide (rLPS) was purified from *E. coli* (BL21) as previously described.[15,16] Briefly, 3 g of the *E. coli* (BL21) bacterial pellet was resuspended in deionised water and heated to 68 °C. To this, 90% phenol (w/v) was added dropwise and stirred at 68 °C, cooled on ice for 15 min and centrifuged at 1000*g* for 45 min at 10 °C. The aqueous phase was collected and a further 10 ml of deionised water at 68 °C was added to the remaining sample, which was cooled and centrifuged and the aqueous phase was collected and dialysed to remove any remaining phenol, then freeze dried. The crude product was resuspended in 0.1 M Tris pH 7.0 and 175 mM NaCl, 0.05% w/v sodium azide at 10 mg ml^{-1} and underwent a further purification process, as described previously.[16] Briefly, the sample was treated with DNase and RNase (both 50 µg per 10 mg of crude LPS) at 37 °C for 30 min. This was followed by treatment with proteinase K (50 µg per 10 mg of crude LPS) at 55 °C for 3 h and then at room temperature for 16 h with fresh proteinase K added. The proteinase K treatments were repeated and LPS was precipitated with 4 volumes of methanol at 18 °C for 2 h. The precipitate was recovered by centrifugation at 6000*g* for 15 min at 4 °C and freeze-dried.

LPS-containing membranes

LPS was suspended in distilled water and incubated at 56 °C for 15 min, vortexed for 2 min and then cooled to 4 °C. This was repeated thrice. Samples were incubated with either DOPC (Avanti) (1 : 1 w/w ratio of DOPC : LPS) or DMPC (Avanti) (2 : 1 ratio of DMPC : LPS w/w) prepared as small unilamellar vesicles by sonication, as previously described.[16] The sample was then freeze dried.

Polymyxin B preparation

20 mg of polymyxin B (Fisher bioreagents) was resuspended in 1 ml HPLC water and then centrifuged for 10 min at 3000*g* to remove precipitates. This was then loaded onto a C18 RP-HPLC column. The elution of polymyxin B was with a linear gradient from 3.5% to 70% acetonitrile with 0.1% v/v TFA over 30 min, measured using a UV-vis detector at 220 nm. The acetonitrile was then removed and a small amount was sent to ESI MS (Bruker MicroTOF) to confirm the presence of polymyxin B. Polymyxin B was resuspended at 50 mg ml^{-1} in HPLC water and then lyophilised LPS in DMPC was either hydrated in HPLC water or polymyxin B in HPLC water (50 mg ml^{-1}). Polymyxin B was added at a 1 : 1 molar ratio of LPS : polymyxin B. When the samples were completely hydrated and mixed with a glass stirring rod, they were freeze–thawed 5 times and loaded into 4 mm MAS NMR rotors.

Solid state NMR

Solid state NMR experiments were carried out on a Varian 400 MHz VNMRS widebore spectrometer equipped with a 4 mm T4 MAS NMR probe. The temperature was regulated with balanced heated/vortex tube-cooled gas flow.[17] The phosphorus-31 MAS NMR were referenced externally to H_3PO_4 at 0 ppm at a frequency of 161.82 MHz.

The results of the phosphorous-31 static wideline NMR experiments were recorded at 28 °C, above the transition temperature of DMPC, using a Hahn echo sequence with 100 kHz $\pi/2$ and π pulses separated by 12 μs intervals and pre-acquisition delays. Spectra were recorded with 20 ms acquisition time with a recycle delay of 5 s, under SPINAL-64 heteronuclear decoupling.[18]

The results of the high resolution ^{31}P experiments were acquired at 5 kHz MAS frequency at either 4 °C or 28 °C. Inversion recovery was used to investigate the ^{31}P longitudinal relaxation, with delay times varying between 10 ms and 1.5 s between initial π pulses and $\pi/2$ reading pulse. Spectra were recorded with 50 ms acquisition time under SPINAL64 decoupling[18] with a recycle delay of 9 s to exceed five-fold [31]P T_1 values in membranes. Relaxation times T_1 were obtained assuming a single exponential relaxation mechanism by fitting

$$M(\tau) = M(0)\left(1 - 2\exp\left(\frac{\tau}{T_1}\right)\right)$$

using Excel (Microsoft). All spectra were processed and analysed using ACDLabs 2015.

Dye release fluorescence studies

Carboxyfluorescein (CF) fluorescence dequenching studies were performed as previously described.[15] Briefly, polymyxin B was suspended in 100 mM NaCl and 10 mM Hepes (pH 7.4) and equilibrated overnight at 4 °C. DOPC films or DOPC : rLPS (1 : 1 w/w) were hydrated in 1 ml 5(6)-carboxyfluorescein (Acros organics) buffer (50 mM CF, 50 mM NaCl, 10 mM Hepes pH 7.4) for 1 h. The solution then underwent 5 cycles of freeze–thawing until the lipid films were fully suspended. The resulting suspension was extruded 11 times through a 100 nm polycarbonate filter with an Avanti extruder (Avanti Polar Lipids). CF-loaded vesicles were separated from non-encapsulated CF using a PD-10 column (GE Healthcare) equilibrated with 100 mM NaCl in 10 mM Hepes, pH 7.4, and used within 24 h.

Polymyxin B-induced CF release was monitored using the fluorescence increase (excitation 490 nm, emission 515 nm, 400 V) over 300 s, at which the time intensity changes with time were within 1%. CF-loaded large unilamellar vesicles (LUVs) in buffer (100 mM NaCl, 10 mM Hepes, pH 7.4) were equilibrated to achieve steady background fluorescence. Polymyxin B was added after 60 s (with a final concentration range of 0.025–50 μg ml^{-1}). After equilibrium (120 s), residual liposomes were dispersed with Triton X-100 (Fluka BioChemika). For each polymyxin B concentration, experiments were repeated in triplicate. CF leakage was expressed as a fraction of CF release upon Triton X-100 addition, normalised to background fluorescence:

$$\% \text{ leakage} = (F_{pol} - F_0)/(F_{T_x} - F_0) \times 100$$

where F_0 is the baseline fluorescence recorded before the addition of polymyxin B, F_{pol} is the steady state fluorescence after adding polymyxin B and F_{T_x} is the maximum fluorescence release after the addition of Triton X-100 to destroy any remaining vesicles.

Antimicrobial susceptibility testing

Bacterial susceptibility to polymyxin was assessed by the agar diffusion method to determine the minimal inhibitory concentration (MIC), as described previously.[19] Briefly, 2×10^7 colony forming units (CFU) of BL21 *E. coli* inocula were spread onto solid agar Luria–Bertani (LB) plates, 3 mm holes were punched with a sterile cork borer and the wells were filled with 9 µl of varying concentrations of poly-myxin B between 5 and 0.625 mg ml^{-1}. Plates were incubated at 37 °C for 16 h and the polymyxin diffusion distances measured from the size of bacterial growth inhibition zones less the well diameter. Plates were individually analysed and three biological repeats of three technical plates (nine repeats per point) were taken to determine the final MIC. The MIC was determined using the webtool http://www.agardiffusion.com.

Results & discussion

Role of LPS in membrane disruption by polymyxin

To investigate the role LPS plays as a receptor for polymyxin, we used a dye release assay, in which CF-loaded large unilamellar vesicles (LUV) of DOPC without or with *E. coli* rLPS were incubated with increasing amounts of polymyxin B. The choice of phosphatidyl choline (PC) ensured membrane stability and lack of non-specific interactions between polymyxin B and the lipid headgroups. While in Gram-negative bacterial membranes phosphatidyl ethanolamine (PE) is present at 70–80% molar fraction, unsupported bilayers favour negative curvature and show higher instability in the presence of polymyxin than PC.[14]

CF-loaded LUVs of DOPC alone or of DOPC containing 1 : 1 w : w rLPS were incubated with polymyxin B at molar ratios of polymyxin to DOPC between 0.003 and 3 (Fig. 2). The molecular weight of rLPS was estimated at roughly 2400 ± 100 Da from the ^{31}P MAS NMR phosphate (P) to pyrophosphate (PP) intensities, as previously described.[3] To estimate the impact of LPS on membrane stability, we extrapolated the linear fits from the CF release and used the zero intercept to obtain the minimal polymyxin/DOPC ratios required for inducing lysis. Leakage onset in pure DOPC liposomes was observed at a polymyxin/DOPC ratio of 5.9×10^{-4}, while the presence of rLPS reduced this to 1.9×10^{-7}.

Polymyxin binding to membranes and changes in lipid dynamics

We used wideline ^{31}P solid state NMR to monitor changes in the molecular organisation and slow collective dynamics in DMPC membranes that result from the addition of polymyxin, the presence of rLPS or both (Fig. 3). The ^{31}P NMR wideline spectrum from hydrated DMPC in the liquid crystalline phase at 28 °C is reflective of the fast axial motions of phospholipid molecules with spherically symmetric angular orientations. The spectroscopic features are characterised by axially symmetric effective chemical shift anisotropy, which results in a charac-teristic powder or Pake spectral intensity distribution.[20] The observed ^{31}P effective CSA from the DMPC multilamellar vesicle (MLV) suspensions is approximately 45 ppm (Fig. 3), which is consistent with reported values.[21,22] The addition of polymyxin at a 1 : 1 molar ratio does not disrupt the membrane structure but markedly increases the lipid disorder. The spectral features remain reflective of

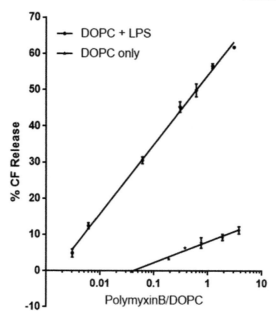

Fig. 2   Polymyxin-dependent CF release from membranes.

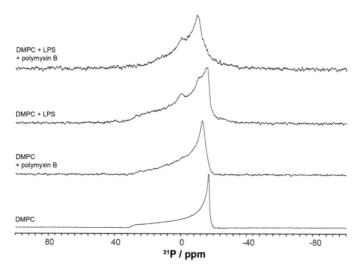

Fig. 3   Phosphorus-31 wideline NMR spectra of the membranes, recorded at 28 °C, above the DMPC main transition temperature.

a powder distribution but at a much reduced effective CSA of 38.3 ppm (Fig. 3), clearly revealing the incorporation of polymyxin B into the DMPC bilayers.

The ^{31}P wideline NMR spectrum from DMPC/LPS 1 : 1 w/w reveals the self-assembly of a stable bilayer membrane, reflected in the CSA-dominated powder distribution with a slightly reduced width of 43.6 ppm (Fig. 3). In addition to the DMPC-dominated symmetric CSA wideline, we observe a much more mobile environment at −7.2 ppm, reflective of the increased mobility in the lipid A region, as well as a broad isotropic resonance at 3.1 ppm. The latter arises from the LPS phosphates and pyrophosphates in the outer core.[2,3,23] In contrast to the phospholipid phosphates, due to the high flexibility between lipid A and the LPS core, the phosphate and pyrophosphate CSA collapses completely and we observe isotropic resonances superimposed onto the DMPC wideline spectrum. Such high phosphate and pyrophosphate mobilities are also observed in the membrane embedded polyisoprenoid cell wall intermediates lipid II, lipid I and undecaprenyl mono- and pyrophosphate.[24,25]

The addition of polymyxin B at a 1 : 1 ratio to the DMPC/rLPS bilayers also preserved the powder distribution and the underlying mixed phospholipid bilayers. Akin to DMPC alone, the incorporation of polymyxin B into the mixed DMPC/rLPS membranes further increases lipid disorder compared with that in pure DMPC, which is observed as a greater reduction in the effective CSA to approximately 30 ppm.

Selective targeting of LPS by polymyxin

High resolution ^{31}P MAS NMR spectroscopy permits the independent and quantitative observation of individual membrane components and the selective effects of polymyxin addition. Due to pyrophosphorylation, membrane rLPS can be followed by a well resolved resonance at −11 ppm (Fig. 4). The single sharp resonance of DMPC observed at −0.97 ppm is broadened slightly upon the addition of polymyxin B. The presence of LPS in the DMPC membranes leads to a significant increase in the isotropic line width, which is countered slightly by the addition of polymyxin B (Fig. 4). Despite this line broadening, the pyrophosphate resonance remains clearly resolved from the compound intensity of DMPC and LPS monophosphates. A contribution from the 3.1 ppm LPS phosphate is seen as a shoulder on the main monophosphate resonance.

Longitudinal nuclear relaxation is a sensitive reporter of changes in fast, GHz molecular motions, such as the axial rotation in membrane lipids that leads to CSA and dipolar coupling modulation. We use inversion recovery ^{31}P MAS NMR to

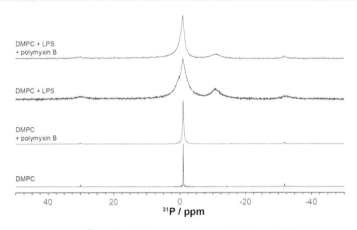

Fig. 4 High resolution ^{31}P MAS NMR spectra from DMPC or DMPC/LPS membranes without or with polymyxin, acquired at 28 °C and 5 kHz MAS.

determine the relaxation behaviour of the lipid and LPS ^{31}P nuclear systems and to explore specific molecular interactions between polymyxin B and rLPS within the DMPC membranes. Phosphorus-31 within phosphates or pyrophosphates is a particularly appropriate nuclear reporter, as the pure O-linking severely reduces coupling to protons and obviates the need of decoupling during the long relaxation intervals.

Longitudinal ^{31}P MAS NMR relaxation times T_1 were determined at 28 °C and at 4 °C (Fig. 5, Table 1). At 28 °C both DMPC and rLPS are in the fast motion regime and we observe a reduction in the rLPS pyrophosphate T_1 from 130 to 90 ms, while the DMPC-dominated monophosphate T_1 remained almost unaffected at 240 and 230 ms, respectively. The selective reduction in the pyrophosphate relaxation time reflects reduced mobility and the formation of an LPS/polymyxin membrane complex. By contrast, the DMPC mobility remained unchanged, which is consistent with the previously reported lack of molecular interactions.[14]

The relaxation time T_1 from LPS pyrophosphates remained unchanged at 130 ms when the temperature was lowered from 28 °C to 4 °C, suggesting that the system is crossing a T_1 minimum and enters slow motion at the lower temperature. The molecular motions are in the slow regime within DMPC in the gel phase and the reduction in the phosphate T_1 upon the addition of polymyxin B reflects the increase in membrane disorder, consistent with the wideline and high resolution ^{31}P MAS observations. The LPS pyrophosphate relaxation time increased at 4 °C from 130 to 160 ms, which is reflective of motional restrictions in a slow motion system and of molecular complex formation, specifically between membrane rLPS and polymyxin B.

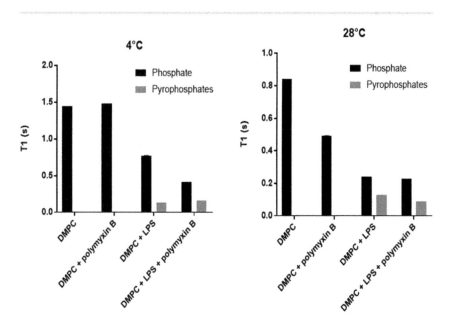

Fig. 5 Phosphorus-31 MAS NMR longitudinal relaxation times T_1 for DMPC membranes without or with LPS or polymyxin. Phosphate T_1 values are shown in black and pyrophosphate T_1 values from LPS are in grey.

Table 1 Phosphorus-31 MAS NMR longitudinal relaxation times T_1 at 4 °C and 28 °C from DMPC membranes without or with LPS before or after polymyxin B (PMB) addition

	T_1 [s] at 4 °C		T_1 [s] at 28 °C	
	P	PP	P	PP
DMPC	1.44		0.84	
DMPC/PMB	1.48		0.49	
DMPC/LPS	0.77	0.13	0.24	0.13
DMPC/LPS/PMB	0.42	0.16	0.23	0.09

Biological activity of polymyxin against *E. coli* BL21

To confirm the efficacy of polymyxin against our test strain of *E. coli* BL21, we carried out susceptibility assays using the agar diffusion method.[19] The results were obtained from three technical and three biological repeats and the minimal inhibitory concentration (MIC) was calculated, using the free MIC web calculator at http://www.agardiffusion.com, to be 291 µg ml^{-1} using the linear zone size model. The average regression coefficient for the linear d-model was $R^2 = 0.974$, while for the d^2 model $R^2 = 0.964$. This closer fit to the linear model reflects some loss of polymyxin during diffusion within the agar, either through degradation or through interactions with the agarose matrix.[19]

Polymyxin B is an anti-Gram-negative peptide antimicrobial with a complex mechanism of action that involves disruption and crossing both the bacterial OM and IM. OM translocation is conditional on the availability of LPS and specific modifications, such as pEtN-ation, reduce the ability of polymyxin B to engage bacterial targets. Dye release studies in this study show that the presence of LPS in lipid membranes significantly enhances the ability of polymyxin B to destabilise and permeabilise bilayer membranes without the formation of stable non-bilayer products. While polymyxin B binds to zwitterionic PC membranes, causing a reduction in the lipid orientational order, membrane leakage only occurs at high polymyxin B to lipid ratios, most likely through charge repulsion-induced local membrane curvature. This model aligns with the reported role of membrane charge as a destabilising factor in the membrane response to polymyxin B, with better chain packing countering this effect.[26]

The presence of LPS significantly enhances the phospholipid motional freedom following the addition of polymyxin B. In contrast to the pure lipid case, a different mechanism comes to the fore, in which pyrophosphate recognition by polymyxin B leads to the assembly of lytic binary complexes that more efficiently disrupt the LPS-containing membranes. This model is consistent with reported bacterial adaptations in pEtN-derivatised LPS, which reduce bacterial susceptibility to polymyxins.[12,13] The OM translocation of polymyxin B relies on hijacking natural LPS pyrophosphate groups, where the polycationic peptide challenges membrane integrity through competition for the divalent cation binding sites which are responsible for maintaining OM integrity. Capping LPS mono- and pyrophosphates in polymyxin resistant strains results in a reduction of the LPS negative charge, as well as restricting the access to pyrophosphates as docking motifs for polymyxin B.

Conclusions

As last resort antimicrobials such as polymyxin B are re-entering the battle against bacterial resistance to antibiotics, novel molecular targets and antimicrobial mechanisms play an increasingly important role. In this work, we explore the role of bacterial rLPS as a molecular receptor for polymyxin B and used solid state NMR to show the specific pyrophosphate-mediated molecular interaction mechanism. We contrast and quantify non-specific membrane disruption by polymyxin B to a pyrophosphate-mediated mechanism relying on the formation of LPS/polymyxin B membrane complexes. Biologically active polymyxin B is significantly more lytic in the presence of LPS in the target membranes and solid state ^{31}P MAS NMR relaxation revealed LPS pyrophosphates to be the specific loci of polymyxin B/LPS recognition.

Conflicts of interest

The authors have no conflicts of interest to declare.

Acknowledgements

The authors would like to thank Georgina Taylor for her technical assistance with the leakage assays and susceptibility testing. We acknowledge funding from the Biotechnology and Biological Sciences Research Council grant BB/N010426/1 to B. B. B. that supports V. Y. and from the Medical Research Council grant MR/N010477/1 to B. B. B. that supports A. G.

References

1 F. Ciesielski, D. C. Griffin, M. Rittig, I. Moriyón and B. B. Bonev, Interactions of lipopolysaccharide with lipid membranes, raft models—a solid state NMR study, *Biochim. Biophys. Acta, Biomembr.*, 2013, **1828**(8), 1731–1742.

2 C. L. Johnson, H. Ridley, R. Marchetti, A. Silipo, D. C. Griffin, L. Crawford, *et al.*, The antibacterial toxin colicin N binds to the inner core of lipopolysaccharide and close to its translocator protein, *Mol. Microbiol.*, 2014, **92**(3), 440–452.

3 A. B. M. Lanne, A. Goode, C. Prattley, D. Kumari, M. R. Drasbek, P. Williams, *et al.*, Molecular recognition of lipopolysaccharide by the lantibiotic nisin, *Biochim. Biophys. Acta, Biomembr.*, 2019, **1861**(1), 83–92.

4 G. C. Ainsworth, A. M. Brown and G. Brownlee, Aerosporin, an antibiotic produced by Bacillus aerosporus Greer, *Nature*, 1947, **160**(4060), 263.

5 M. A. K. Azad, B. A. Finnin, A. Poudyal, K. Davis, J. H. Li, P. A. Hill, *et al.*, Polymyxin B Induces Apoptosis in Kidney Proximal Tubular Cells, *Antimicrob. Agents Chemother.*, 2013, **57**(9), 4329–4335.

6 M. D. F. Vattimo, M. Watanabe, C. D. da Fonseca, L. Neiva, E. A. Pessoa and F. T. Borges, Polymyxin B Nephrotoxicity: From Organ to Cell Damage, *PLoS One*, 2016, **11**(8), e0161057.

7 M. Vaara, Polymyxin Derivatives that Sensitize Gram-Negative Bacteria to Other Antibiotics, *Molecules*, 2019, **24**(2), 249.

8 Z. Z. Deris, J. D. Swarbrick, K. D. Roberts, M. A. K. Azad, J. Akter, A. S. Horne, *et al.*, Probing the Penetration of Antimicrobial Polymyxin Lipopeptides into Gram-Negative Bacteria, *Bioconjugate Chem.*, 2014, **25**(4), 750–760.

9 P. R. G. Schindler and M. Teuber, Action of Polymyxin B on Bacterial Membranes: Morphological Changes in the Cytoplasm and in the Outer Membrane of Salmonella typhimurium and *Escherichia coli* B, *Antimicrob. Agents Chemother.*, 1975, **8**(1), 95–104.

10 A. M. L. Scavuzzi, L. C. Alves, D. L. Veras, F. A. Brayner and A. C. S. Lopes, Ultrastructural changes caused by polymyxin B and meropenem in multiresistant Klebsiella pneumoniae carrying bla(KPC-2) gene, *J. Med. Microbiol.*, 2016, **65**, 1370–1377.

11 M. J. Trimble, P. Mlynarcik, M. Kolar and R. E. W. Hancock, Polymyxin: Alternative Mechanisms of Action and Resistance, *Cold Spring Harbor Perspect. Med.*, 2016, **6**(10), a025288.

12 B. D. Needham and M. S. Trent, Fortifying the barrier: the impact of lipid A remodelling on bacterial pathogenesis, *Nat. Rev. Microbiol.*, 2013, **11**(7), 467–481.

13 M. L. Han, Y. Zhu, D. J. Creek, Y. W. Lin, D. Anderson, H. H. Shen, *et al.*, Alterations of Metabolic and Lipid Profiles in Polymyxin-Resistant *Pseudomonas aeruginosa*, *Antimicrob. Agents Chemother.*, 2018, **62**(6), e02656-17.

14 C. C. Hsuchen and D. S. Feingold, Mechanism of polymyxin B action and selectivity toward biologic membranes, *Biochemistry*, 1973, **12**(11), 2105–2111.

15 A. B. Lanne, A. Goode, C. Prattley, D. Kumari, M. R. Drasbek, P. Williams, *et al.*, Molecular recognition of lipopolysaccharide by the lantibiotic nisin, *Biochim. Biophys. Acta, Biomembr.*, 2019, **1861**(1), 83–92.

16 F. Ciesielski, D. C. Griffin, M. Rittig, I. Moriyón and B. B. Bonev, Interactions of lipopolysaccharide with lipid membranes, raft models—a solid state NMR study, *Biochim. Biophys. Acta, Biomembr.*, 2013, **1828**(8), 1731–1742.

17 F. Ciesielski, D. C. Griffin, M. Rittig and B. B. Bonev, High-resolution J-coupled 13C MAS NMR spectroscopy of lipid membranes, *Chem. Phys. Lipids*, 2009, **161**(2), 77–85.

18 B. M. Fung, A. K. Khitrin and K. Ermolaev, An improved broadband decoupling sequence for liquid crystals and solids, *J. Magn. Reson.*, 2000, **142**(1), 97–101.

19 B. Bonev, J. Hooper and J. Parisot, Principles of assessing bacterial susceptibility to antibiotics using the agar diffusion method, *J. Antimicrob. Chemother.*, 2008, **61**(6), 1295–1301.

20 G. E. Pake, Nuclear resonance absorption in hydrated crystals – fine structure of the proton line, *J. Chem. Phys.*, 1948, **16**(4), 327–336.

21 J. Seelig, P-31 nuclear magnetic resonance and headgroup structure of phospholipids in membranes, *Biochim. Biophys. Acta, Rev. Biomembr.*, 1978, **515**(2), 105–140.

22 V. Yeh, A. Goode, G. Eastham, R. P. Rambo, K. Inoue, J. Doutch, *et al.*, Membrane Stability in the Presence of Methacrylate Esters, *Langmuir*, 2020, **36**(33), 9649–9657.

23 F. Ciesielski, D. C. Griffin, J. Loraine, M. Rittig, J. Delves-Broughton and B. B. Bonev, Recognition of membrane sterols by polyene antifungals amphotericin B and natamycin, a 13C MAS NMR study, *Front. Cell Dev. Biol.*, 2016, **4**, 57.

24 J. Parisot, S. Carey, E. Breukink, W. C. Chan, A. Narbad and B. Bonev, Molecular mechanism of target recognition by subtilin, a class I lanthionine antibiotic, *Antimicrob. Agents Chemother.*, 2008, **52**(2), 612–618.

25 B. B. Bonev, E. Breukink, E. Swiezewska, B. De Kruijff and A. Watts, Targeting extracellular pyrophosphates underpins the high selectivity of nisin, *FASEB J.*, 2004, **18**(15), 1862–1869.

26 A. Khondkerg, A. K. Dhaliwal, S. Saem, A. Mahmood, C. Fradin, J. Moran-Mirabal, *et al.*, Membrane charge and lipid packing determine polymyxin-induced membrane damage, *Commun. Biol.*, 2019, **2**(1), 67.

Faraday Discussions

ROYAL SOCIETY
OF CHEMISTRY

PAPER

Membrane electrostatics sensed by tryptophan anchors in hydrophobic model peptides depends on non-aromatic interfacial amino acids: implications in hydrophobic mismatch†

Sreetama Pal, [iD] *abc Roger E. Koeppe, II [iD] d
and Amitabha Chattopadhyay [iD] *ac

Received 18th May 2020, Accepted 11th September 2020
DOI: 10.1039/d0fd00065e

WALPs are synthetic α-helical membrane-spanning peptides that constitute a well-studied system for exploring hydrophobic mismatch. These peptides represent a simplified consensus motif for transmembrane domains of intrinsic membrane proteins due to their hydrophobic core of alternating leucine and alanine flanked by membrane-anchoring aromatic tryptophan residues. Although the modulation of mismatch responses in WALPs by tryptophan anchors has been reported earlier, there have been limited attempts to utilize the intrinsic tryptophan fluorescence of this class of peptides in mismatch sensors. We have previously shown, utilizing the red edge excitation shift (REES) approach, that interfacial WALP tryptophan residues in fluid phase bilayers experience a dynamically constrained membrane microenvironment. Interestingly, emerging reports suggest the involvement of non-aromatic interfacially localized residues in modulating local structure and dynamics in WALP analogs. In this backdrop, we have explored the effect of interfacial amino acids, such as lysine (in KWALPs) and glycine (in GWALPs), on the tryptophan microenvironment of WALP analogs in zwitterionic and negatively charged membranes. We show that interfacial tryptophans in KWALP and GWALP experience a more restricted microenvironment, as reflected in the substantial increase in magnitude of REES and apparent rotational correlation time, relative to those in WALP in zwitterionic membranes. Interestingly, in

^a*CSIR-Centre for Cellular and Molecular Biology, Hyderabad 500 007, India. E-mail: sreetama@ccmb.res.in; amit@ccmb.res.in*

^b*CSIR-Indian Institute of Chemical Technology, Hyderabad 500 007, India*

^c*Academy of Scientific and Innovative Research, Ghaziabad 201 002, India*

^d*Department of Chemistry and Biochemistry, University of Arkansas, AR 72701, USA*

† Electronic supplementary information (ESI) available: Section S1, preparation of unilamellar vesicles; Section S2, steady state fluorescence measurements; Section S3, time-resolved fluorescence measurements; Fig. S1, representative fluorescence emission spectra of WALP tryptophans showing the shift in emission maximum upon red edge excitation; Fig. S2, representative time-resolved fluorescence intensity decay profile of WALP tryptophans. See DOI: 10.1039/d0fd00065e

contrast to WALP, the tryptophan anchors in KWALP and GWALP appear insensitive to the presence of negatively charged lipids in the membrane. These results reveal a subtle interplay between non-aromatic flanking residues in transmembrane helices and negatively charged lipids at the membrane interface, which could modulate the membrane microenvironment experienced by interfacially localized tryptophan residues. Since interfacial tryptophans are known to influence mismatch responses in WALPs, our results highlight the possibility of utilizing the fluorescence signatures of tryptophans in membrane proteins or model peptides such as WALP as markers for assessing protein responses to hydrophobic mismatch. More importantly, these results constitute one of the first reports on the influence of lipid headgroup charge in fine-tuning hydrophobic mismatch in membrane bilayers, thereby enriching the existing framework of hydrophobic mismatch.

1   Introduction

Membrane-mediated regulation of cellular responses could arise due to the interaction of membrane proteins with specific lipids, or the influence of global bilayer properties on the conformational dynamics and partitioning of membrane proteins, or a combination of both.[1] The modulation of membrane protein function by the membrane bilayer has its genesis, in part, in the strong coupling of the protein hydrophobic core to the lipid membrane hydrophobic thickness.[2,3] This hydrophobic coupling, termed the hydrophobic matching principle,[4] is implicated in membrane protein sorting, localization, stability, topology and function.[5–8] Unraveling the molecular basis and consequences of hydrophobic (mis)match at the protein–lipid interface requires a systematic exploration of the organization and dynamics of transmembrane domains of membrane proteins in the context of the global properties of their immediate microenvironment.

The WALP class of peptides (acetyl-GWW(LA)$_n$LWWA-ethanolamide), synthesized based on gramicidin as a template, constitutes the most well-studied model system for hydrophobic mismatch. WALPs consist of a helical hydrophobic core, constructed from alternating alanine and leucine residues, capped at each end by aromatic amino acids, such as tryptophan (see Fig. 1a).[9] The terminal residues are typically unwound from the core helix.[10,11] These peptides, therefore, represent a consensus motif for transmembrane domains of intrinsic membrane proteins due to the leucine and alanine residues, which are characterized by a high propensity for adopting α-helical conformations. This model, albeit simplistic, is appropriate since transmembrane segments of integral membrane proteins consist of a primarily hydrophobic α-helical core, flanked at each end by aromatic and charged residues.[12–14] In addition, the number of alanine–leucine repeats in this class of peptides can be used as a handle to tune the hydrophobic length and subsequent response of these peptides to a wide spectrum of (positive or negative) hydrophobic mismatch.

WALPs have been shown to respond to hydrophobic mismatch conditions by inducing non-bilayer structures in model membranes[9,15] and *E. coli* membrane mimics.[16] The stabilization of non-lamellar membrane phases in the presence of WALPs has been ascribed to the flanking tryptophan residues, which may override the commonly observed peptide response to mismatch conditions.[17] This suppression of peptide responses to mismatch conditions by interfacial

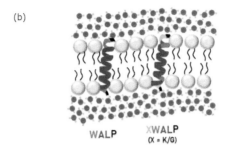

(a) WALP23 Ac-GW²W³(LA)₈LW²¹W²²A-NHCH₂CH₂OH
     KWALP23 Ac-GKALW³(LA)₆LW¹⁹LAKA-NHCH₂CH₂OH
     GWALP23 Ac-GGALW³(LA)₆LW¹⁹LAGA-NHCH₂CH₂OH

Fig. 1 A schematic showing (a) the amino acid sequence of the model transmembrane peptide WALP23 and its XWALP23 analogs, KWALP23 and GWALP23 (Ac represents an acetyl group), and (b) the membrane topology of these analogs. WALP analogs are a class of synthetic peptides that consist of a hydrophobic core of alternating alanine and leucine residues (shown in purple), capped on each side by aromatic amino acids such as tryptophan (marked in pink). This characteristic sequence ensures that WALP peptides represent a consensus sequence for α-helical transmembrane domains of membrane proteins. The XWALP (X = K/G) analogs represent variations of the prototypical WALP sequence, each with only one tryptophan anchor near each terminal and the other tryptophan on both termini replaced with lysine or glycine (shown in green). The hydrophobic length of the peptide can be tuned by changing the number of alanine–leucine repeats, thereby making these peptides excellent systems for exploring the signatures and consequences of hydrophobic mismatch in membranes. The phospholipid headgroups are depicted as gray beads and fatty acyl chains are shown in black. Water molecules are depicted in shades of blue. See the text for more details.

tryptophan residues could be attributed to the preference of this amino acid for the membrane interface[14,18–20] and possibly originates from its participation in hydrogen bonding with lipid carbonyls and interfacial water molecules.[21] Due to the design and architecture of these peptides, including the interfacial anchoring, their responses to hydrophobic mismatch are manifested predominantly as changes in tilt angles[22–24] and helix rotations or helical distortions,[25] and are dictated by the nature of the interfacial residues.

Despite studies highlighting the contribution of tryptophan anchors in modulating the membrane interaction of the WALP class of peptides, there have been very few attempts to explore this phenomenon utilizing the intrinsic fluorescence of these tryptophan residues. This issue assumes significance since interfacially localized tryptophan residues are indispensable in supporting membrane protein function and are known to define the protein hydrophobic length by acting as membrane anchors.[14] In this context, fluorescence readouts of interfacial tryptophans have emerged as sensitive markers for membrane protein organization, dynamics and function.[26–28] The restricted microenvironment experienced by tryptophan residues localized at the membrane interface has been shown to correlate with the function of several membrane-interacting peptides, including the ion channel peptide gramicidin[29,30] and the lytic peptide melittin.[31] In addition, the hydrophobic mismatch experienced by gramicidin in membrane bilayers has been shown to manifest as changes in motional restrictions in the tryptophan microenvironment.[32]

We have previously shown, utilizing red edge excitation shift (REES), that interfacial WALP tryptophans experience a dynamically constrained membrane microenvironment.[33] Interestingly, emerging evidence suggests the involvement of interfacially localized non-aromatic amino acids in modulating local structure and dynamics in WALP analogs.[11,24] With this backdrop, we probed whether the distinct interfacial electrostatics imposed by zwitterionic and negatively charged membranes is sensed differentially by tryptophan residues in WALP analogs due to the presence of non-aromatic flanking residues such as lysine (in KWALP: acetyl-GKALW(LA)$_{n-2}$LWLAKA-ethanolamide) or glycine (in GWALP: acetyl-GGALW(LA)$_{n-2}$LWLAGA-ethanolamide). Our results suggest that non-aromatic flanking residues at the membrane interface indeed may contribute to membrane protein organization and function by modulating the extent to which tryptophan residues can sense their immediate microenvironment. These results highlight the possible involvement of the membrane interface in modulating hydrophobic (mis)match responses and could enrich the existing framework of hydrophobic mismatch. These observations assume significance in the overall backdrop of the well-documented role of tryptophan residues in shaping membrane protein structure and function,[14,29–31,34] and in the specific context of tryptophan-mediated protein response to hydrophobic mismatch in membranes.[19]

2 Experimental

2.1 Materials

1-Palmitoyl-2-oleoyl-*sn*-glycero-3-phosphocholine (POPC) and 1-palmitoyl-2-oleoyl-*sn*-glycero-3-phosphoglycerol (POPG) lipids were obtained from Avanti Polar Lipids (Alabaster, AL). WALP23, KWALP23 and GWALP23 peptides were synthesized as described previously.[9,24] Lipids were checked for purity by thin layer chromatography and the phospholipid concentration was determined by a phosphate assay, as described earlier.[33] The peptide concentration was calculated based on tryptophan abundance, using a molar extinction coefficient (ε) of 22 400 M^{-1} cm^{-1} for WALP and 11 200 M^{-1} cm^{-1} for KWALP or GWALP at 280 nm.[35] Chemicals of the highest available purity, UV spectroscopy grade solvents, and water purified through a Millipore (Bedford, MA) Milli-Q system were used for all experiments.

2.2 Preparation of unilamellar vesicles

Unilamellar vesicles (ULVs) of POPC or POPC/POPG (70/30, mol/mol) containing 2 mol% of the peptide (WALP23 or KWALP23 or GWALP23) suspended in 10 mM sodium phosphate and 150 mM sodium chloride (pH 7.2 buffer) were used for all experiments. The total amounts of lipid and of each peptide were kept constant at 550 and 11 nmol, respectively, which translates to a lipid/peptide ratio of 50 (mol/mol). ULVs were prepared by sonication using a Sonics Vibra-Cell VCX 500 sonifier (Sonics & Materials Inc, Newtown, CT) fitted with a titanium microtip, as described previously.[33] The protocol for ULV preparation is described in the ESI.† The final concentrations of lipid and peptide in the ULVs were 550 and 11 µM, respectively. Background samples (without the peptide) were prepared in a similar manner. All experiments were carried out with at least three sets of samples at room temperature ($\sim$23 °C).

2.3 Steady state fluorescence measurements

Steady state fluorescence data were acquired on a Fluorolog-3 Model FL3-22 spectrofluorometer (Horiba Jobin Yvon, Edison, NJ) using semi-micro quartz cuvettes. Slit widths of 2 and 4 nm were used for excitation and emission, respectively. In each case, the reported emission maxima were identical to (or within ± 1 nm of) the values reported. Fluorescence anisotropy data were acquired using the Glan–Thompson polarization accessory in the same setup, with the same slit width parameters. Details of the acquisition parameters are outlined in the ESI.†

2.4 Time-resolved fluorescence measurements

Time-resolved fluorescence intensity decays were recorded using a Delta-D TCSPC setup (Horiba Jobin Yvon IBH, Glasgow, UK) in the time-correlated single photon counting (TCSPC) mode, as described earlier.[36] A pulsed light-emitting diode (DD-290), with a typical pulse width of 0.8 ns, was used as an excitation source. The LED profile (reflecting the instrument response function (IRF)) was measured using colloidal silica (Ludox) as a scatterer. Data were collected, stored and analyzed using the in-built plugins in the EzTime software version 3.2.2.4 (Horiba Scientific, Edison, NJ). The fluorescence intensity decay curves were deconvoluted with the IRF and analyzed as a sum of exponential terms to yield the pre-exponential factors and the corresponding fluorescence decay times. A fit was considered acceptable when plots of the weighted residuals and their autocorrelation functions exhibited random deviation about zero, with a χ^2 value of not more than 1.2. The intensity-averaged mean fluorescence lifetime ($\langle \tau \rangle$) was calculated from the values extracted for these decay parameters. The data reported is representative of at least three independent measurements. Further details of the data acquisition parameters and analysis are described in the ESI.†

2.5 Data analysis and plotting

Student's two-tailed unpaired t-test, performed using GraphPad Prism software version 4.0 (San Diego, CA), was used to check statistical significance of the data. Microcal Origin version 8.0 (OriginLab, Northampton, MA) was used for all other data analysis and plotting.

3 Results

WALPs are prototypical α-helical transmembrane peptides that represent a consensus sequence for transmembrane domains of integral membrane proteins due to the presence of leucine–alanine repeats flanked by membrane-anchoring tryptophan residues at each end (Fig. 1a). These tryptophan residues ensure a transmembrane orientation of the peptide due to the stability associated with their interfacial localization,[14,19] even in mismatch conditions.[17] In this work, we have explored the organization and dynamics of interfacial tryptophan residues in WALP23 and its analogs (KWALP23 and GWALP23) in zwitterionic POPC and negatively charged POPC/POPG (70/30, mol/mol) ULVs utilizing the wavelength-selective fluorescence toolbox.[26,27,37] KWALP and GWALP represent variations of the prototypical WALP sequence, each with only one tryptophan anchor near each terminal, since the other tryptophan on both termini has been replaced with lysine or glycine.[24] We used POPC and POPC/POPG bilayers as model systems because the hydrophobic length of these bilayers in the fluid phase[38,39] is similar to the hydrophobic length of the transmembrane helix in WALP analogs (see Fig. 1b). The peptides utilized for the present work are known to adopt an α-helical conformation in a membrane milieu.[24,33]

Wavelength-selective fluorescence (which includes REES) represents a sensitive toolbox to assess the spectroscopic signature of polar fluorophores in motionally restricted microenvironments, such as membranes and proteins.[26,27,37,40] The phenomenon of REES is *only* observed if solvent relaxation in the immediate microenvironment of the fluorophore occurs at timescales slower than or comparable to fluorescence timescales. This leads to a temporal overlap of solvation dynamics with fluorophore dynamics in the excited state and allows us to utilize REES to explore the organization and dynamics of fluorophores at the membrane interface. Operationally, REES is defined as the magnitude of red shift in the fluorescence emission maximum due to a concomitant shift in the excitation wavelength toward the red edge. The magnitude of REES is proportional to the extent of restriction imposed on solvation dynamics in the vicinity of the fluorophore.

The emission spectrum of WALP in POPC membranes displayed a characteristic maximum at 334 nm when excited at 280 nm (Fig. S1 in the ESI†), in agreement with our previous results.[33] As shown in Fig. 2a, the emission maximum of WALP remained unchanged in POPC/POPG membranes. The shift in fluorescence emission maxima of the tryptophan residues of WALP analogs in POPC and POPC/POPG membranes with increasing excitation wavelength is also shown in Fig. 2a. For WALP in POPC membranes, the emission maximum exhibited a shift from 334 to 340 nm as the excitation wavelength was increased from 280 to 310 nm. This corresponds to REES of 6 nm (see Fig. 2b) and indicates that the interfacial WALP tryptophans in POPC membranes experience a microenvironment characterized by constrained solvation dynamics, consistent with our previous results.[33] Interestingly, the presence of negatively charged POPG lipids resulted in REES of 4 nm (*i.e.*, 334 to 338 nm, see Fig. 2a) for WALP, implying a reduction in the constrained dynamics experienced by WALP tryptophans. A similar effect of interfacial membrane electrostatics on tryptophan organization and dynamics using REES has been previously reported.[31]

The corresponding REES data for WALP analogs KWALP and GWALP are shown in Fig. 2a and b. The emission maximum of KWALP in POPC was observed at 331 nm upon excitation at 280 nm. This was shifted to 340 nm upon red edge excitation at 310 nm, resulting in REES of 9 nm (Fig. 2b). The magnitude of REES for KWALP was therefore higher relative to that of WALP in POPC membranes, implying increased motional restriction in the fluorophore microenvironment. Interestingly, the observed emission maximum (331 nm) and REES (9 nm) of KWALP in negatively charged POPC/POPG membranes turned out to be the same as those observed in zwitterionic POPC membranes. In other words, the fluorescence spectral signatures of KWALP appear to lack sensitivity to lipid composition and membrane interfacial electrostatics. In the case of GWALP, the

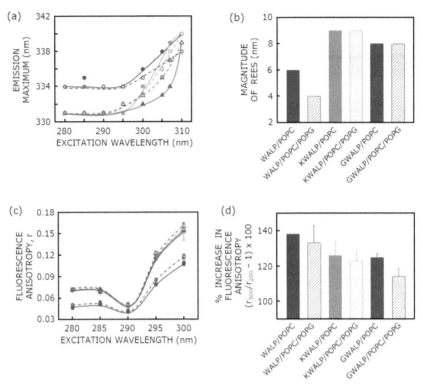

Fig. 2 Dependence of fluorescence emission maximum and anisotropy on excitation wavelength for WALP analogs in zwitterionic POPC and negatively charged POPC/POPG (70/30, mol/mol) membranes. (a) Red shift in the emission maxima of WALP (●, blue), KWALP (■, green) and GWALP (▲, maroon) in POPC (filled symbols) and POPC/POPG (open symbols) membranes upon increasing the excitation wavelength. The data shown represent at least three independent measurements. Panel (b) shows the corresponding magnitude of REES, operationally defined as the shift in fluorescence emission maximum upon changing the excitation wavelength from 280 to 310 nm, for each WALP analog in POPC (filled bars) and POPC/POPG (hatched bars) membranes. (c) Increase in fluorescence anisotropy of tryptophan residues in WALP analogs with increasing excitation wavelength. The emission wavelength was kept constant at the emission maximum of each peptide observed upon excitation at 280 nm, i.e., 334 nm for WALP and 331 nm for KWALP/GWALP. Panel (d) shows the corresponding magnitude of increase in fluorescence anisotropy, calculated as the percentage increase in anisotropy at 300 nm relative to that at 280 nm. Data represent means ± SE for at least three independent measurements. The color-coding for panels (c) and (d) is the same as in panels (a) and (b), respectively. The lines joining the data points in panels (a) and (c) represent viewing guides. The concentration of each WALP analog was 11 µM and the lipid/peptide ratio was 50 (mol/mol) in all cases. See Experimental for more details.

emission maximum in POPC membranes was found to be at 331 nm when excitation was carried out at 280 nm. Upon increasing the excitation wavelength to 310 nm, the emission maximum was shifted to 339 nm, giving rise to REES of 8 nm in POPC membranes (Fig. 2b). The motional constraint of GWALP tryptophans in POPC membranes, therefore, appears to be more than that experienced by WALP tryptophans. Even in this case, the observed emission maximum (331 nm) and REES (8 nm) in negatively charged POPC/POPG membranes turned out to be identical to those observed for GWALP in zwitterionic POPC membranes. Taken together, the influence of negatively charged lipids on tryptophan dynamics appears to be attenuated in KWALP and GWALP analogs, thereby suggesting that these flanking residues could be involved in modulating the immediate microenvironment of tryptophan residues at the membrane interface.

Fig. 2c shows that tryptophan fluorescence anisotropy, calculated using eqn (S1),† exhibits an increase with increasing excitation wavelength in WALP analogs

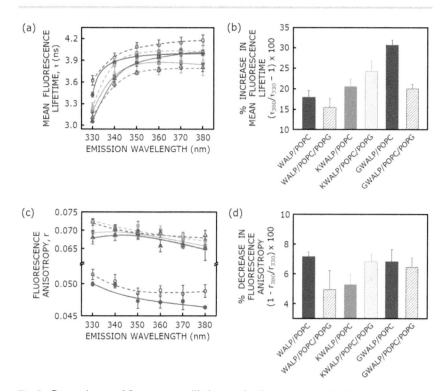

Fig. 3   Dependence of fluorescence lifetime and anisotropy on emission wavelength for WALP analogs in zwitterionic POPC and negatively charged POPC/POPG (70/30, mol/mol) membranes. (a) Effect of increasing emission wavelength on the intensity-averaged mean fluorescence lifetime of WALP (●, blue), KWALP (■, green) and GWALP (▲, maroon) in POPC (filled symbols) and POPC/POPG (open symbols) membranes. The excitation wavelength was 297 nm in all cases. Mean fluorescence lifetimes were calculated using eqn (S3).† Panel (b) shows the corresponding magnitude of increase in mean fluorescence lifetime, calculated as the percentage increase in lifetime at 380 nm relative to that at 330 nm, for each WALP analog in POPC (filled bars) and POPC/POPG (hatched bars) membranes. (c) Reduction in fluorescence anisotropy of tryptophan residues in WALP analogs with increasing emission wavelength. The excitation wavelength was kept constant at 280 nm. Panel (d) shows the corresponding magnitude of decrease in fluorescence anisotropy, calculated as the percentage decrease in anisotropy at 380 nm relative to that at 330 nm. Data represent means ± SE for at least three independent measurements. The color-coding for panels (c) and (d) is the same as in panels (a) and (b), respectively. The lines joining the data points in panels (a) and (c) are viewing guides. The concentration of each WALP analog was 11 μM and the lipid/peptide ratio was 50 (mol/mol) in all cases. See Experimental for more details.

in zwitterionic and negatively charged membranes. The increase in fluorescence anisotropy (calculated as the percentage increase in anisotropy at 300 nm relative to that at 280 nm) is plotted in Fig. 2d. Fluorescence anisotropy was found to display a substantial change with increasing excitation wavelength, with a pronounced enhancement toward the red edge. Data for excitation wavelengths beyond 300 nm are not shown due to poor signal-to-noise ratio. Strong dipolar interactions of solvent-relaxed fluorophores (photoselected by excitation at the red edge) with the surrounding solvent molecules[37] would impose restrictions on tryptophan rotational dynamics, which is manifested as an increase in fluorescence anisotropy. The initial dip in fluorescence anisotropy at ~290 nm shown in Fig. 2c is a spectroscopic signature of tryptophan residues in a restricted environment[37] and originates from complex tryptophan photophysics.[27,37,41]

Fig. 3a shows the dependence of mean fluorescence lifetime on emission wavelength for tryptophan residues in WALP analogs. A representative fluorescence decay profile of WALP tryptophans in POPC membranes is included as Fig. S2 in the ESI.† The intensity-averaged mean fluorescence lifetimes at different emission wavelengths were calculated using eqn (S3)† and are shown in Tables 1–3. The increase in mean fluorescence lifetime (calculated as the percentage increase in fluorescence lifetime at 380 nm relative to that at 330 nm) of interfacial tryptophan residues in POPC and POPC/POPG membranes is shown in Fig. 3b. The increase in mean fluorescence lifetime with increasing emission wavelength could be explained as follows.[37,41] Emission at shorter wavelengths photoselects unrelaxed fluorophores with a dual mode of relaxation. This subpopulation relaxes *via* both fluorescence emission at the given excitation

Table 1 Representative mean intensity-averaged fluorescence lifetimes of WALP tryptophan residues in POPC and POPC/POPG membranes[a]

Emission wavelength (nm)	α_1	τ_1 (ns)	α_2	τ_2 (ns)	α_3	τ_3 (ns)	$\langle \tau \rangle^b$ (ns)	χ^2
POPC membranes								
330	0.17	1.73	0.12	5.00	0.71	0.19	3.43	1.16
334[c]	0.21	1.91	0.16	5.03	0.63	0.24	3.57	1.03
340	0.44	2.07	0.29	5.35	0.27	0.87	3.85	1.10
350	0.41	2.31	0.33	5.26	0.26	0.81	3.97	1.04
360	0.45	2.11	0.34	5.23	0.21	0.59	3.99	1.10
370	0.42	2.42	0.32	5.39	0.26	0.80	4.04	1.16
380	0.44	2.29	0.31	5.39	0.25	0.63	4.02	1.03
POPC/POPG membranes								
330	0.27	2.00	0.20	5.04	0.53	0.33	3.61	1.12
334[c]	0.34	1.99	0.23	5.12	0.43	0.33	3.72	1.06
340	0.35	2.19	0.30	5.21	0.35	0.35	4.02	1.15
350	0.42	2.56	0.32	5.35	0.26	0.87	4.02	1.09
360	0.44	2.39	0.34	5.33	0.22	0.73	4.06	1.11
370	0.44	2.72	0.32	5.47	0.24	1.00	4.10	1.10
380	0.42	3.01	0.29	5.67	0.29	1.24	4.15	1.07

[a] The excitation wavelength was 297 nm. A total of 10 000 photons were collected at the peak channel for robust data acquisition. The concentration of WALP was 11 µM and the lipid/peptide ratio was 50 (mol/mol). See Experimental for more details. [b] Calculated using eqn (S3) (see ESI). [c] Emission maximum.

Table 2 Representative mean intensity-averaged fluorescence lifetimes of KWALP tryptophan residues in POPC and POPC/POPG membranes[a]

Emission wavelength (nm)	α_1	τ_1 (ns)	α_2	τ_2 (ns)	α_3	τ_3 (ns)	$\langle\tau\rangle^b$ (ns)	χ^2
POPC membranes								
330	0.09	1.75	0.09	4.76	0.82	0.18	3.19	1.06
331[c]	0.05	1.80	0.05	4.86	0.90	0.09	3.26	1.04
340	0.31	2.04	0.25	5.04	0.44	0.36	3.75	1.10
350	0.38	2.13	0.36	5.03	0.26	0.71	3.91	1.08
360	0.41	2.08	0.31	5.13	0.28	0.42	3.90	1.10
370	0.39	2.11	0.37	5.06	0.24	0.89	3.92	1.09
380	0.39	2.55	0.29	5.35	0.32	1.05	3.88	1.07
POPC/POPG membranes								
330	0.08	1.84	0.07	4.78	0.85	0.11	3.27	1.07
331[c]	0.13	1.90	0.11	4.80	0.76	0.16	3.37	1.16
340	0.21	2.08	0.28	4.94	0.51	0.26	3.98	1.12
350	0.42	2.02	0.38	4.91	0.20	0.74	3.84	1.05
360	0.31	2.34	0.35	5.18	0.34	0.29	4.22	1.08
370	0.31	2.30	0.35	5.11	0.34	0.27	4.17	1.11
380	0.45	2.61	0.34	5.46	0.21	0.54	4.22	1.10

[a] The excitation wavelength was 297 nm. A total of 10 000 photons were collected at the peak channel for robust data acquisition. The concentration of KWALP was 11 μM and the lipid/peptide ratio was 50 (mol/mol). See Experimental for more details. [b] Calculated using eqn (S3) (see ESI). [c] Emission maximum.

wavelength and decay to longer wavelengths, although the latter remains unobserved due to the emission wavelength window chosen for data acquisition. These rapidly relaxing fluorophores photoselected at shorter emission wavelengths contribute to the shorter fluorescence lifetimes. In addition, fluorophore populations emitting at the red edge are subjected to greater extents of solvent relaxation (since they reside longer in the excited state), and therefore exhibit higher fluorescence lifetime.

Fig. 3c shows the reduction in tryptophan fluorescence anisotropy in WALP analogs with increasing emission wavelength. The decrease in fluorescence anisotropy (calculated as the percentage decrease in anisotropy at 380 nm relative to that at 330 nm) in zwitterionic POPC and negatively charged POPC/POPG membranes is shown in Fig. 3d. Longer-lived fluorophores are photoselected at longer emission wavelengths. These fluorophores are characterized by lower values of fluorescence anisotropy at the red edge due to greater tumbling (rotation) in the excited state and subsequent depolarization. An additional factor complicating the wavelength-dependence of fluorescence anisotropy in WALP is the possibility of homo-fluorescence resonance energy transfer (homo-FRET) due to the pairwise distribution of tryptophan residues at each end of WALP. This is due to a lack of orientational correlation between the photoselected donor and the subsequently excited acceptor involved in homo-FRET, which leads to fluorescence depolarization.[42,43] The lower values of tryptophan fluorescence anisotropy observed in WALP, relative to KWALP and GWALP peptides (Fig. 2c and 3c), lend credence to this possibility.

Table 3 Representative mean intensity-averaged fluorescence lifetimes of GWALP tryptophan residues in POPC and POPC/POPG membranes[a]

Emission wavelength (nm)	α_1	τ_1 (ns)	α_2	τ_2 (ns)	α_3	τ_3 (ns)	$\langle\tau\rangle^b$ (ns)	χ^2
POPC membranes								
330	0.10	1.75	0.07	4.77	0.83	0.14	3.06	1.06
331[c]	0.15	1.83	0.10	4.91	0.75	0.20	3.21	1.12
340	0.32	2.08	0.23	5.10	0.45	0.37	3.70	1.07
350	0.43	2.36	0.30	5.28	0.27	0.78	3.89	1.11
360	0.43	2.43	0.30	5.37	0.27	0.87	3.94	1.13
370	0.43	2.27	0.32	5.31	0.25	0.77	3.97	1.09
380	0.41	2.54	0.29	5.51	0.30	0.97	4.00	1.07
POPC/POPG membranes								
330	0.13	1.89	0.08	5.01	0.79	0.21	3.09	1.09
331[c]	0.18	1.79	0.12	4.77	0.70	0.21	3.20	1.05
340	0.41	2.07	0.25	5.05	0.34	0.61	3.56	1.05
350	0.47	2.05	0.27	5.27	0.26	0.69	3.74	1.05
360	0.31	2.93	0.22	5.63	0.47	1.31	3.78	1.13
370	0.35	2.64	0.26	5.39	0.39	1.17	3.78	1.05
380	0.48	2.10	0.27	5.32	0.25	0.70	3.77	1.04

[a] The excitation wavelength was 297 nm. A total of 10 000 photons were collected at the peak channel for robust data acquisition. The concentration of GWALP was 11 µM and the lipid/peptide ratio was 50 (mol/mol). See Experimental for more details. [b] Calculated using eqn (S3) (see ESI). [c] Emission maximum.

Since fluorescence anisotropy and lifetime represent sensitive parameters for monitoring the dynamics and environment of excited state fluorophores, an essential prerequisite for interpreting trends in fluorescence anisotropy is to ensure negligible interference from fluorescence lifetime induced artifacts. For this, apparent rotational correlation times were calculated using Perrin's equation:[44]

$$\tau_c = \frac{r\langle\tau\rangle}{r_o - r} \tag{1}$$

where r_o is the fundamental anisotropy of tryptophan residues, r corresponds to the fluorescence anisotropy (from Fig. 2c), and $\langle\tau\rangle$ is the mean fluorescence lifetime (from Tables 1–3). Fluorescence anisotropy and lifetime values obtained with excitation at 295 and 297 nm respectively, and emission wavelengths set to the corresponding emission maximum (i.e., 334 nm for WALP and 331 nm for KWALP/GWALP) were used. The value of r_o was taken to be 0.16 for tryptophan residues.[45]

Fig. 4 shows the apparent (average) rotational correlation time of tryptophan residues in WALP analogs. In physical terms, rotational correlation time reflects the average time taken for a fluorophore (tryptophan, in this case) to rotate about its axis and is a reporter of order in the fluorophore microenvironment.[44] The rotational correlation time of WALP tryptophans in negatively charged POPC/POPG membranes was found to be significantly higher relative to that in zwitterionic POPC membranes. This indicates that interfacial WALP tryptophan residues in negatively charged membranes experience a more structured

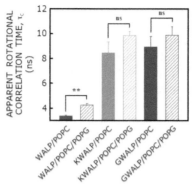

Fig. 4 Effect of the membrane microenvironment and non-aromatic flanking residues on the rotational dynamics of tryptophan in WALP analogs in zwitterionic POPC and negatively charged POPC/POPG (70/30, mol/mol) membranes. Apparent rotational correlation times of interfacial tryptophan residues in WALP (blue), KWALP (green) and GWALP (maroon) peptides embedded in POPC (filled bars) and POPC/POPG (hatched bars) membranes are shown. Apparent rotational correlation times were calculated using eqn (1) from fluorescence anisotropy values acquired with an excitation wavelength of 295 nm (data from Fig. 2c) and mean fluorescence lifetimes recorded at an excitation wavelength of 297 nm (representative data shown in Tables 1–3). The emission wavelength for both fluorescence anisotropy and lifetime readouts was kept constant at the emission maximum of each peptide analog observed upon excitation at 280 nm, *i.e.*, 334 nm for WALP and 331 nm for KWALP/GWALP. Data represent means ± SE for at least three independent measurements (** corresponds to a significant ($p < 0.01$) difference in the apparent rotational correlation time of WALP tryptophans in POPC/POPG membranes relative to that in POPC membranes and ns denotes a statistically insignificant change). See Experimental and text for more details.

microenvironment, probably due to a higher fraction of ordered water molecules present at the membrane interface.[46,47] Interestingly, Fig. 4 shows that the rotational correlation time of tryptophan residues in KWALP/GWALP peptides in POPC membranes was significantly higher ($p < 0.01$, not shown) relative to WALP tryptophan residues. Therefore, the presence of non-aromatic interfacial flanking residues, such as lysine in KWALP and glycine in GWALP, appears to result in slower rotational dynamics of interfacial tryptophan residues. Importantly, the rotational correlation times of KWALP and GWALP tryptophan residues remained invariant in the presence of negatively charged POPG lipids, thereby hinting at a subtle interplay of negatively charged lipids and non-aromatic flanking residues in modulating the microenvironment of interfacial tryptophan residues. These interpretations are reinforced by the trends observed in Fig. 2a and b.

4   Discussion

In this work, we have utilized the wavelength-selective fluorescence approach to explore the organization and dynamics of membrane-anchoring tryptophan residues in WALP and its analogs, KWALP and GWALP, in zwitterionic POPC and negatively charged POPC/POPG membranes. Our results reveal that tryptophans in all the WALP analogs exhibit REES, indicating that these aromatic residues are localized at or near the membrane interface and their microenvironment is characterized by restricted solvation dynamics. The presence of negatively charged lipids leads to a loss in motional restriction experienced by WALP tryptophans, as shown by a decrease in the magnitude of REES (see Fig. 2b). This change could be attributed to increased solvent exposure at the negatively charged membrane interface, as has been reported for other membrane-interacting peptides.[31] Nevertheless, the apparent rotational correlation time for WALP tryptophans exhibited an increase in the presence of negatively charged lipids. This is probably due to the presence of water molecules held in a more ordered collective network at the membrane interface by negatively charged lipids.[46,47] The apparently opposing trends in REES (Fig. 2a) and apparent rotational correlation times (Fig. 4) for WALP in negatively charged membranes relative to zwitterionic ones could originate from subtle differences in fluorophore properties represented by these observables. The wavelength-selective fluorescence approach reports predominantly on solvation dynamics in the fluorophore microenvironment, whereas apparent rotational correlation times reflect the restriction imposed on rotational dynamics of the fluorophore.

The restriction in solvation and rotational dynamics experienced by KWALP and GWALP tryptophan residues is significantly higher than that experienced by those in WALP (see Fig. 2b and 4). This restriction could be due to the presence of non-aromatic flanking residues such as lysine in KWALPs since lysine side chains are known to interact with the tryptophan indole ring *via* cation–π interactions.[48] The high motional restriction experienced by GWALP tryptophans could be a different effect, perhaps due to the greater tilt angle of GWALP relative to WALP (see Fig. 1b),[49] which would translate to deeper interfacial localization of GWALP tryptophan residues. Interestingly, tryptophan residues in the KWALP and GWALP analogs appear to be insensitive to the lipid headgroup charge in negatively charged membranes, both with respect to solvation dynamics and apparent rotational correlation times (see Fig. 2b and 4). For KWALPs, cation–π–anion interactions[50] between lysine, tryptophan and negatively charged POPG could oppose the expected reduction in motional restriction of the tryptophan microenvironment. This would result in dampening the influence of negatively charged lipids in tryptophan solvation and rotational dynamics, giving rise to invariant REES and apparent rotational correlation time for KWALP in zwitterionic and negatively charged membranes. In fact, electrostatic interactions between membrane lipids and proteins have been recently reported to complement the contribution of interfacial tryptophans toward stabilization of optimal (and functionally relevant) anchoring of transmembrane helices.[51] Taken together, these insights collectively point to the presence of novel and largely unexplored regulatory mechanisms at the helix–lipid interface, where a subtle interplay between non-aromatic flanking residues in transmembrane helices and negatively charged lipids could tune the microenvironment of interfacially localized tryptophan residues.

Membrane proteins (and peptides) are known to respond to hydrophobic mismatch in their immediate microenvironment by changing helical tilt angles,[6,8] which would correspond to changes in the effective length of the hydrophobic core. In the case of WALP analogs, the extent of helical tilt is coupled to the position of the tryptophan residues.[25,49] The fluorescence signatures of WALP tryptophans reported in this work could therefore be utilized as an indirect marker for helical tilt of transmembrane helices. In addition, judicious introduction or repositioning of tryptophan residues in the transmembrane domains of integral membrane proteins (represented by WALPs here) could induce not only differential tilt angles but also changes in dynamics and inter-helical packing of the transmembrane helix bundle.[52] This has important consequences for the modulation of membrane protein function by physico-chemical aspects (including mismatch conditions) of the lipid microenviron-ment, since optimal helical tilt angles have been correlated to membrane protein stability and activity.[53–57] In particular, lysine could emerge as a key determinant in these local structural changes because these residues,[14] along with negatively charged lipids,[58] are known to be preferentially localized at the inner leaflet in biological membranes.

Taken together, our results suggest that non-aromatic residues, such as lysine and glycine, at the membrane interface could influence membrane protein stability, structure and function by modulating the extent to which tryptophan

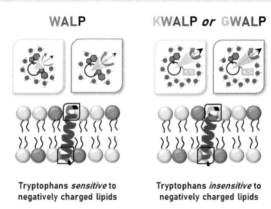

Fig. 5 Differential sensing of membrane interfacial electrostatics by tryptophan residues in WALP analogs due to the presence of non-aromatic flanking residues. Tryptophan residues are shown in pink, lysine/glycine in green, and leucine and alanine in purple. Phospholipids are depicted with gray (zwitterionic POPC) and brown (negatively charged POPG) headgroups, and black acyl chains. Schematic representations of the tryptophan microenvironment in WALP analogs in the presence of POPC (gray box) and POPG (brown box) are shown at the top. Dynamically constrained water molecules at the membrane interface are represented in shades of blue and reflect the extent of motional restriction experienced by the tryptophan residues. Fluorophore rotational dynamics is depicted with shaded cones and curly arrows, with the cones color coded to match the corresponding lipid headgroup. The left panel shows the decreased solvation dynamics (represented by the lower number of slow relaxing interfacial water molecules) and increased apparent rotational correlation time (represented by the increased size of the shaded cone and the increased thickness of the pink curly arrow) characteristic of WALP tryptophans in the presence of POPG, relative to POPC. The right panel shows the significant difference in the tryptophan microenvironment in KWALP and GWALP relative to that in WALP in POPC membranes. The tryptophan residues in KWALP and GWALP appear insensitive to the presence of negatively charged lipids (see text and Fig. 2b and 4). These insights collec-tively point to the presence of novel and largely unexplored regulatory mechanisms at the helix–lipid interface, where a subtle interplay between non-aromatic flanking residues in transmembrane helices and negatively charged lipids could tune the microenvironment of interfacially localized tryptophan residues. These results highlight the possible involve-ment of membrane interfacial electrostatics in fine-tuning hydrophobic mismatch responses. See the text for more details.

residues can sense their immediate microenvironment (Fig. 5). Since tryptophan interfacial localization has been shown to affect hydrophobic mismatch responses in WALP peptides,[17] the involvement of non-aromatic flanking residues in modulating the tryptophan microenvironment could be expected to have important consequences for mismatch responses of transmembrane helical domains in membrane proteins. The current concept of hydrophobic mismatch is based primarily on the length of fatty acyl chains of membrane lipids, although the presence of cholesterol can also contribute.[8] Lipid acyl chains and cholesterol are both responsible for the mismatch between the hydrophobic core of the membrane and transmembrane domains of membrane peptides and proteins.[8] Our present results show that the charge of the lipid headgroup could be an additional player in modulating hydrophobic mismatch in membranes. To the best of our knowledge, these results constitute one of the early reports on the involvement of membrane interfacial electrostatics in modulating hydrophobic mismatch responses. These observations assume significance in light of the crucial role of interfacial tryptophan residues in the organization, dynamics, topology and stability of several membrane-interacting peptides.[14,29–31,34] We envision that insights from our present results would enrich the prevalent framework of hydrophobic mismatch formalism and enhance our overall understanding of the importance of tryptophan localization, organization and dynamics in supporting membrane protein structure and function.

Conflicts of interest

There are no conflicts of interest to declare.

Acknowledgements

This work was supported by a SERB Distinguished Fellowship grant (Department of Science and Technology, Govt. of India) to A. C. and core support from CSIR-Centre for Cellular and Molecular Biology. R. E. K. acknowledges support from MCB grant 1713242 from the United States National Science Foundation. S. P. thanks the University Grants Commission for the award of a Senior Research Fellowship. A. C. is a Distinguished Visiting Professor at the Indian Institute of Technology Bombay (Mumbai), an Adjunct Professor at Tata Institute of Fundamental Research (Mumbai) and the Indian Institute of Science Education and Research (Kolkata), and an Honorary Professor at the Jawaharlal Nehru Centre for Advanced Scientific Research (Bengaluru). We thank G. Aditya Kumar for useful input in making some of the figures and members of the Chattopadhyay laboratory for comments and discussions.

References

1 M. Jafurulla, G. A. Kumar, B. D. Rao and A. Chattopadhyay, *Adv. Exp. Med. Biol.*, 2019, **1115**, 21.

2 S. H. White and W. C. Wimley, *Biochim. Biophys. Acta, Rev. Biomembr.*, 1998, **1376**, 339.

3 O. S. Andersen and R. E. Koeppe II, *Annu. Rev. Biophys. Biomol. Struct.*, 2007, **36**, 107.

4 O. G. Mouritsen and M. Bloom, *Biophys. J.*, 1984, **46**, 141.

5 M. S. Bretscher and S. Munro, *Science*, 1993, **261**, 1280.

6 J. A. Killian, *Biochim. Biophys. Acta, Rev. Biomembr.*, 1998, **1376**, 401.

7 F. Dumas, M. C. Lebrun and J.-F. Tocanne, *FEBS Lett.*, 1999, **458**, 271.

8 B. D. Rao, S. Shrivastava and A. Chattopadhyay, in *Membrane Organization and Dynamics*, ed. A. Chattopadhyay, Springer, Heidelberg, 2017, vol. 16, pp. 375–387.

9 J. A. Killian, I. Salemink, M. R. R. de Planque, G. Lindblom, R. E. Koeppe II and D. V. Greathouse, *Biochemistry*, 1996, **35**, 1037.

10 A. Mortazavi, V. Rajagopalan, K. A. Sparks, D. V. Greathouse and R. E. Koeppe II, *ChemBioChem*, 2016, **17**, 462.

11 F. Afrose, M. J. McKay, A. Mortazavi, V. S. Kumar, D. V. Greathouse and R. E. Koeppe II, *Biochemistry*, 2019, **58**, 633.

12 I. T. Arkin and A. T. Brunger, *Biochim. Biophys. Acta, Protein Struct. Mol. Enzymol.*, 1998, **1429**, 113.

13 L. Adamian, V. Nanda, W. F. DeGrado and J. Liang, *Proteins: Struct., Funct., Bioinf.*, 2005, **59**, 496.

14 D. A. Kelkar and A. Chattopadhyay, *J. Biosci.*, 2006, **31**, 297.

15 P. C. A. van der Wel, T. Pott, S. Morein, D. V. Greathouse, R. E. Koeppe II and J. A. Killian, *Biochemistry*, 2000, **39**, 3124.

16 S. Morein, R. E. Koeppe II, G. Lindblom, B. de Kruijff and J. A. Killian, *Biophys. J.*, 2000, **78**, 2475.

17 M. R. R. de Planque, B. B. Bonev, J. A. A. Demmers, D. V. Greathouse, R. E. Koeppe II, F. Separovic, A. Watts and J. A. Killian, *Biochemistry*, 2003, **42**, 5341.

18 J. A. Killian and G. von Heijne, *Trends Biochem. Sci.*, 2000, **25**, 429.

19 A. J. de Jesus and T. W. Allen, *Biochim. Biophys. Acta, Biomembr.*, 2013, **1828**, 864.

20 R. E. Koeppe II, *J. Gen. Physiol.*, 2007, **130**, 223.

21 J. A. Ippolito, R. S. Alexander and D. W. Christianson, *J. Mol. Biol.*, 1990, **215**, 457.

22 E. Strandberg, S. Özdirekcan, D. T. S. Rijkers, P. C. A. van der Wel, R. E. Koeppe II, R. M. J. Liskamp and J. A. Killian, *Biophys. J.*, 2004, **86**, 3709.

23 S. Özdirekcan, D. T. S. Rijkers, R. M. J. Liskamp and J. A. Killian, *Biochemistry*, 2005, **44**, 1004.

24 V. V. Vostrikov, A. E. Daily, D. V. Greathouse and R. E. Koeppe II, *J. Biol. Chem.*, 2010, **285**, 31723.

25 A. E. Daily, D. V. Greathouse, P. C. A. van der Wel and R. E. Koeppe II, *Biophys. J.*, 2008, **94**, 480.

26 S. Haldar, A. Chaudhuri and A. Chattopadhyay, *J. Phys. Chem. B*, 2011, **115**, 5693.

27 A. Chattopadhyay and S. Haldar, *Acc. Chem. Res.*, 2014, **47**, 12.

28 D. A. M. Catici, H. E. Amos, Y. Yang, J. M. H. van den Elsen and C. R. Pudney, *FEBS J.*, 2016, **283**, 2272.

29 S. S. Rawat, D. A. Kelkar and A. Chattopadhyay, *Biophys. J.*, 2004, **87**, 831.

30 A. Chattopadhyay, S. S. Rawat, D. V. Greathouse, D. A. Kelkar and R. E. Koeppe II, *Biophys. J.*, 2008, **95**, 166.

31 A. K. Ghosh, R. Rukmini and A. Chattopadhyay, *Biochemistry*, 1997, **36**, 14291.

32 D. A. Kelkar and A. Chattopadhyay, *Biochim. Biophys. Acta, Biomembr.*, 2007, **1768**, 2011.

33 S. Pal, R. E. Koeppe II and A. Chattopadhyay, *J. Fluoresc.*, 2018, **28**, 1317.

34 A. Chaudhuri, S. Haldar, H. Sun, R. E. Koeppe II and A. Chattopadhyay, *Biochim. Biophys. Acta, Biomembr.*, 2014, **1838**, 419.

35 N. J. Gleason, V. V. Vostrikov, D. V. Greathouse, C. V. Grant, S. J. Opella and R. E. Koeppe II, *Biochemistry*, 2012, **51**, 2044.

36 S. Pal, R. Aute, P. Sarkar, S. Bose, M. V. Deshmukh and A. Chattopadhyay, *Biophys. Chem.*, 2018, **240**, 34.

37 S. Mukherjee and A. Chattopadhyay, *J. Fluoresc.*, 1995, **5**, 237.

38 N. Kučerka, M.-P. Nieh and J. Katsaras, *Biochim. Biophys. Acta, Biomembr.*, 2011, **1808**, 2761.

39 J. Pan, F. A. Heberle, S. Tristam-Nagle, M. Szymanski, M. Koepfinger, J. Katsaras and N. Kučerka, *Biochim. Biophys. Acta, Biomembr.*, 2012, **1818**, 2135.

40 A. P. Demchenko, *Methods Enzymol.*, 2008, **450**, 59.

41 S. Mukherjee and A. Chattopadhyay, *Biochemistry*, 1994, **33**, 5089.

42 F. T. S. Chan, C. F. Kaminski and G. S. K. Schierle, *ChemPhysChem*, 2011, **12**, 500.

43 S. Ganguly, A. H. A. Clayton and A. Chattopadhyay, *Biophys. J.*, 2011, **100**, 361.

44 J. R. Lakowicz, *Principles of Fluorescence Spectroscopy*, Springer, New York, 3rd edn, 2006.

45 M. R. Eftink, L. A. Selvidge, P. R. Callis and A. A. Rehms, *J. Phys. Chem.*, 1990, **94**, 3469.

46 L. Janosi and A. A. Gorfe, *J. Chem. Theory Comput.*, 2010, **6**, 3267.

47 S. Pal, N. Samanta, D. Das Mahanta, R. K. Mitra and A. Chattopadhyay, *J. Phys. Chem. B*, 2018, **122**, 5066.

48 D. A. Dougherty, *Acc. Chem. Res.*, 2013, **46**, 885.

49 V. V. Vostrikov, C. V. Grant, A. E. Daily, S. J. Opella and R. E. Koeppe II, *J. Am. Chem. Soc.*, 2008, **130**, 12584.

50 D. Kim, E. C. Lee, K. S. Kim and P. Tarakeshwar, *J. Phys. Chem. A*, 2007, **111**, 7980.

51 A. J. Situ, S.-M. Kang, B. B. Frey, W. An, C. Kim and T. S. Ulmer, *J. Phys. Chem. B*, 2018, **122**, 1185.

52 M. J. McKay, A. N. Martfeld, A. A. De Angelis, S. J. Opella, D. V. Greathouse and R. E. Koeppe II, *Biophys. J.*, 2018, **114**, 2617.

53 J. Le Coutre, L. R. Narasimhan, C. K. N. Patel and H. R. Kaback, *Proc. Natl. Acad. Sci. U. S. A.*, 1997, **94**, 10167.

54 C.-S. Chiang, L. Shirinian and S. Sukharev, *Biochemistry*, 2005, **44**, 12589.

55 J. E. Donald, Y. Zhang, G. Fiorin, V. Carnevale, D. R. Slochower, F. Gai, M. L. Klein and W. F. DeGrado, *Proc. Natl. Acad. Sci. U. S. A.*, 2011, **108**, 3958.

56 F. Hilbers, W. Kopec, T. J. Isaksen, T. H. Holm, K. Lykke-Hartmann, P. Nissen, H. Khandelia and H. Poulsen, *Sci. Rep.*, 2016, **6**, 20442.

57 Z. Ren, P. X. Ren, R. Balusu and X. Yang, *Sci. Rep.*, 2016, **6**, 34129.

58 G. van Meer and A. I. P. M. de Kroon, *J. Cell Sci.*, 2011, **124**, 5.

Faraday Discussions

ROYAL SOCIETY
OF CHEMISTRY

PAPER

Common principles of surface deformation in biology

Aurélien Roux [iD] *ab

Received 9th July 2021, Accepted 14th July 2021

DOI: 10.1039/d1fd00040c

Living organisms, whether they are cells or multicellular organisms, are separated from their environment by an interface. For example, cells are delimited by lipid bilayers while embryos or individuals are delimited by epithelia, ectoderms or epiderms. These biological interfaces, while being different in nature and composition, and at very different scales, share common properties: they are surfaces, their thickness being very small compared to their size. They are materials of chemical composition or cell type that is unique and different from the core of the material they envelop. They are visco-elastic sheets, meaning that components can flow in the plane of the surface. The shape of cells and of embryos is inherently dictated by the shape of their envelope, and because these interfaces have common properties, we explore in this commentary article the different mechanisms that remodel these different biological surfaces, and their common principles.

Introduction

All living units are separated from their environment by a surface, which is unique in that it materializes as a sheet of visco-elastic material. For example, cells are delimited by lipid membranes; and multi-cellular organisms, at least the vast majority of them, by a cell monolayer that protects them from the environment – epiderms, ectoderms, epithelia, or endothelia are common examples. While being at very different scales, and very different in their nature, these envelopes share common properties and have functions essential to life: first, they are diffusion barriers, with selective permeability – while plasma membranes are impermeable to most water solutes because of their hydrophobic core, they are still permeable to partially hydrophobic molecules, and transmembrane transporters and channels control molecules through them. Cell monolayers that protect multicellular organisms are known to be strong osmotic barriers, even reducing water efflux to avoid air-living organisms from drying out. Second, they are visco-elastic surfaces, which means that they can deform by changing their

Department of Biochemistry, University of Geneva, CH-1211 Geneva, Switzerland. E-mail: aurelien.roux@unige.ch

National Center of Competence in Research Chemical Biology, University of Geneva, CH-1211 Geneva, Switzerland

area and curvature elastically over short time scales, and viscously over longer times. This is essential because all living units have the property of changing their shape and size – cells grow and divide during cycles, and during early development many organs are formed by deforming the ectoderm of the embryo. Also, during the life of multicellular organisms, growth of new parts in plants, or muscle-driven movements in animals, will require their envelope to smoothly adapt to the new shape. Yet, this shape is given by the visco-elastic surface, which conserves the living units integrity.

The common visco-elastic nature of living surfaces makes us wonder, do these surfaces share common mechanisms of shape acquisition? Of course, while comparing very different biological structures in order to find common properties, one should not forget about their differences: lipid membranes are composed of amphiphilic molecules which spontaneously assemble into bilayers of three to five nanometers; while cell monolayers are usually composed of polarized cells spreading from one side to the other over tens of microns. Thus, while lipid bilayers are basically non-active materials, cell monolayers are intrinsically active – they can proliferate, generate internal forces, and composing units (cells) can actively change their shape with time. On the other hand, lipids in bilayers form non-polarized structures, at least if the compositions of the two leaflets are the same, they are subject to Brownian motion, and cannot multiply themselves. Some of these intrinsic properties can however be changed in the cellular context, where the activity of proteins can create asymmetry in the leaflet composition, or change the number of lipids, or their chemical nature. Another crucial difference between lipid bilayers and cell monolayers is the time scale at which they acquire their shape: while lipid bilayers can change shape in seconds, or even faster, cell monolayers require at least several hours if not days for the same transformation. Also, the characteristic visco-elastic time of lipid bilayers and cell monolayers – the threshold time of deformation application below which the surface responds elastically, and above which it responds viscously – is very different, a few tens of milliseconds for lipid membranes compared to a few tens of minutes for cell monolayers.

In this review, while comparing examples taken from the literature, I discuss principles by which living surfaces deform. While detailed mechanisms are clearly different, their mechanistic principles are usually similar.

A – Curvature of surfaces is coupled to the asymmetric shapes of the basic components

Because living surfaces have a given thickness, the spontaneous (at equilibrium, or minimal energy) shape of these surfaces is usually coupled to the shapes of their components. Numerous examples in the literature couple the curvature of lipid membranes to lipid shapes, and cell monolayer shapes to cell shapes. However, the specific fluid properties of the surface can make this coupling weaker.

1   Lipid *vs.* cell shapes and spontaneous curvature of lipid membranes *vs.* epithelia

Phospholipids, the main components of lipid membranes have two chemically different parts: the hydrophilic headgroup and the hydrophobic acyl chains. The steric hindrance of the head, and the length and number of acyls can modify the

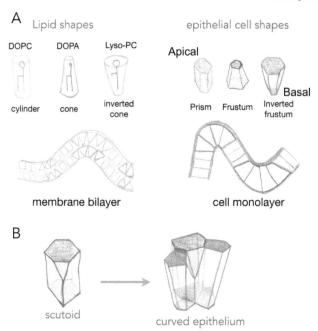

Fig. 1 Shapes of lipids and epithelial cells that adapt to curved lipid bilayers and cell monolayers. (A) different types of lipids (DOPC, dioleyl-phosphatidylcholine; DOPA, dio-leyl-phosphatidic acid; lyso-PC, lyso-phosphatidylcholine), have different shapes that fit specific curvatures of the bilayer. Epithelial cells, which are polarized having an apical and a basal side, have also variable shapes that are adapted to different curvatures of the cell monolayer. (B) A scutoid shape and how it adapts locally epithelial cells to curve.

overall shape of the lipids from being cylindrical to being conical or an inverted shape (see Fig. 1). These shapes, of course, from very simple geometrical arguments, are more adapted to different curvatures: cylinders like phosphatidyl-choline will adapt better to flat bilayers, while inverted cone shapes like phosphatidic acid will adapt better to concave leaflets, and conical lysolipids to convex leaflets. In the 70s, Isra-elachvili *et al.* detailed how the spontaneous curvature of lipid self-assembled structures is coupled to the shape of the lipids.[2] The spontaneous curvature of lipids calculated from their chemical structure and molecular properties is a rather predictive parameter of the formation of micelles, inverted micelles or bilayers.

While the situation in mixed lipid composition bilayers with low curvature may be quite different (see discussion point 1.2 in the article[2]), the same reasoning, coupling cell shape to curvature of the epithelium for example, has been made for cell monolayers (Fig. 1). The most common example is the one of intestinal villi, where cells in crypts have a smaller apical area, and cells at the tip of villis have a larger apical area. The exact area difference between apical and basal sides of the cells not only depends on the curvature, but also on the thickness of the cells. This is best represented in the recent example where epithelial cells were grown on wavy substrates, and where the thickness of the cells dramatically changed from concave to convex structures.[3] As it will be discussed further in active process curving cell monolayers (see Section 2), these observations support that several shape transitions with varying apical-basal area difference and cell thickness, are required to generate the proper shape of organs.[4]

An interesting case is the one of anisotropic curvatures, such as in cylinders: in one direction (along the cylinder axis) the curvature is null, while being constant and high in the perpendicular axis. Many epithelial or endothelial structures have such anisotropic curvatures, at least in branched conductive networks such as lungs or secretory organs. In these situations, the easiest way to accommodate cell shape with curvature is to create anisotropic cell shapes, which are more conical in the direction of highest curvature in the plane of the cell monolayer. But recently, several reports show that the anisotropy may arise in the number of neighbors for cells (see Fig. 1B). As cells will extend more on the basal side in the direction of curvature, and much less in the apical side, the number of neighbors may change to accommodate the anisotropic curvature. These cell shapes, equivalent to a T1 transition in the normal direction, are called scutoids, and have been shown to represent a fairly high percentage of cell shapes in tubular organ structures.[5] However, they do not represent a minimal energy shape, and their dynamics (whether they are stable shapes or transitory shapes between two or more stable shapes) is not known.

Interestingly, lipids spontaneously form different types of micelles, with isotropic shapes (spheres) or anisotropic shapes (tubules). At least, the anisotropy of curvature should also be linked in this case to an anisotropy in the specific spontaneous curvature of lipids. In this case a certain degree of order, or persistent orientation of the lipids should allow to propagate the anisotropy over large distances. But the link between anisotropy in the chemical structure, order and the final curvature anisotropy in the lipid assembly, is not well characterized to my knowledge.

2   Brownian motion and diffusion counteract lipid shape coupling to the curvature

In self-assembled structures of lipids such as micelles, the curvature of the structure is very high, and close to the specific spontaneous curvature of the lipid that composed it. However, in cellular membranes, the overall structure is a bilayer with curvature more than ten times the size of the lipids (1.5 nm), and the composition mixed hundreds of different types of lipids. In this context, whether the shape of lipids inserted in the bilayer is a strong determinant of the spontaneous curvature of the membrane is still debated. Moreover, the bilayer nature of the lipid membrane means the contribution of spontaneous curvature of the lipids in one leaflet is compensated by the ones in the opposite leaflet, as long as the two leaflets have the same composition. A counter effect could be linked to lipid diffusion, as lipids favorable to convex curvature may diffuse and accumulate in the convex leaflet, while ones preferring concave curvature may accumulate in the opposite leaflet, in the same place. Overall, the curvature of the membrane would be zero, but the local concentration of curvature sensitive lipids may enhance curvature changes locally.

However, most probably, Brownian motion and diffusion strongly interfere with these effects. A simple calculation (see Box 1) shows that the energy gain of localizing a single lipid molecule to an area that matches its curvature better, is much lower than kT, meaning that this curvature sorting of single lipid molecules is overruled by Brownian motion.

Box 1: Estimate of energy gain for curvature sorting of lipids

One can estimate the gain of energy ΔF upon curvature-dependent lipid sorting, in the extreme case of a difference of bending rigidity $\Delta\kappa$ between the soft and stiff phase of $50k_BT$, for a membrane curvature C of 1/20 nm and an area per lipid $a \approx 0.5$ nm^2:

$$\Delta F = \frac{1}{2}\Delta\kappa C^2 \times a = \frac{1}{2}50k_BT \times \frac{1}{(20\,\text{nm})^2} \times 0.5\,\text{nm}^2$$

$$\Delta F = 0.5 \times 50 \times k_BT \times \frac{1}{4 \times 400} = \frac{1}{64}k_BT$$

Due to the very small size of lipid molecules, the gain of energy per lipid is then much smaller than k_BT, which means that at room temperature, entropic motion counteracts lipid sorting by curvature. Increasing a through lipid nanoclustering will increase the energy gain, thereby making lipid nanoclusters more sensitive to curvature then single lipids. Adapted from ref. 1.

As a matter of fact, curvature sorting of lipids strongly depends on the capacity of lipids to form more viscous domains, and to be close to a phase separation.[6,7]

All these counteracting processes that reduce the coupling of lipid shapes to curvature in lipid bilayers are, by nature, absent in cell monolayers. Thus, the coupling between cell shape and curvature is much stronger in a cell monolayer, and physiological curvatures of cell monolayers are close to the thickness of cells in the tissue (microns to tens of microns). Moreover, and contrary to lipids, cells can actively change their shape, dynamically changing the shapes of cell monolayers. But, as we will discuss in the following, proteins involved in lipid membrane remodeling have evolved to use similar principles for deforming membranes than cells use to induce curvature of cell monolayers.

B – Dynamic change of surface shape by shape changes in the basic components

While the shape and curvature of surfaces may correlate with the shape of their constituents, active changes of a constituent's shape can drive deformation of living surfaces. In the following, I try to find similar principles to examples of mechanisms found at different scales and on different biological systems.

1 Apical constriction and head group compaction

The most well understood principle of cell monolayer deformation is called apical constriction. In a polarized epithelium, in which cells have a contractile acto-

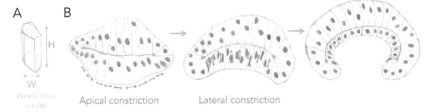

Fig. 2 Changes of cell aspect ratio enhance invagination. (A) Definition of cell aspect ratio. (B) Sequence of apical constriction followed by cell aspect ratio change (lateral constriction) helps a tissue to reverse its curvature (adapted from ref. 4).

myosin ring on the apical side, activation of contractility in a group of cells results in their apical area reduction, forcing the cells to adopt a conical shape bending the epithelium (Fig. 2). This mechanism was first proposed a century ago, by pioneer studies of gastrulation of sea anemones.[8,9] A large pool of data has confirmed this mechanism as essential for triggering epithelium folding in many morphogenetic events.[4] It is also one of the most beautiful examples in which single cell mechanics and geometries can directly explain the shape of simple tissues, like the sea anemone embryo. There are, however, several theoretical descriptions that can generate the same shapes as the ones experimentally observed.[10]

Interestingly, parallels can be drawn between apical constriction, and some lipid membrane remodeling mechanisms driven by proteins. The most obvious one is the mechanism by which the Shiga toxin forms its own membrane carrier to enter cells by toxin-induced endocytosis.[11,12] In this mechanism, the pentameric beta domain of the Shiga toxin binds 15 Gb3 molecules, a ganglioside, by their headgroups. The binding induces a reduction of the specific lipid headgroup volume, condensing the 15 lipids on the side to which the toxin binds. This condensation leads to membrane curvature, by transforming cylindrical lipids into inverted cone-shaped lipids, allowing the formation of membrane buds that accumulate the toxin. Not surprisingly this mechanism was found in other lipid binding molecules, such as lectins,[13] and requires a high affinity for the lipids, and high valency in order to cluster enough lipids together to compensate for entropic dissipation.

Importantly, in lipid membranes, the reverse mechanism of apical constriction, which consists of area expansion to induce curvature in the opposite direction, is a very common mechanism used by membrane traffic proteins. For example, amphipathic helices found in coat proteins can insert in between lipid headgroups to generate a cone-shaped insertion.[14] When clustered together by polymerization of the coat, the membrane will deform locally. It is important to notice that in this mechanism, there are two effects causing curvature that are at play. The first one is the wedge-like effect, pushing the headgroups away from each other, and forcing conical insertions locally. The second one is to create an area difference between the 2 leaflets, by increasing the number of constituents in one of the two leaflets.[15] While the first effect is local and can be very strong (high curvature induced for a very low number of proteins binding to the membrane), the second effect is global, and requires a substantial increase of the number of lipids/proteins in one leaflet to obtain low curvatures.[16] Thus, the wedge-like effect is often considered to be the dominant effect in curvature induced by a protein inserting into the membrane.

All the mechanisms that cause curvature by generating an area difference between the two sides of the surface strongly depend on the thickness of the surface, as the thicker they are, the less area difference is required for creating the same curvature. Thus, mechanisms that can change the thickness of the surface, locally or globally, may help curvature acquisition.

2   Mechanisms of surface thickness changes associated with curvature

Cell monolayers display a large variety of thickness, usually linked to the cell aspect ratio of specific cell lines (ratio between height and width of cells, see Fig. 2A). Cell shortening during apical constriction further promotes invagination.[4] Importantly, changes in lateral tension of cells may also cause invagination, by locally changing the aspect ratio of cells, and if cells remain attached to a flat surface.[17] Thus, cells found in monolayers have multiple ways to change their shape locally or globally in order to induce a specific curvature of the entire tissue, all of which are based on localizing acto-myosin contractility in cells, that can redistribute in cells in order to change tissue shape with time. One of the best examples of this is the growth of the fish eye cup, which requires reorganization of the actin cytoskeleton in the neuroepithelium, to change the aspect ratio of cells while they proliferate, in order to keep the overall general proportions of the eye.[18]

Similar principles of thickness change in lipid bilayers seem *a priori* excluded. Furthermore, as discussed above, since typical radii of curvatures in lipid bilayers are usually more than ten times larger than their thickness, changing the thickness may not strongly affect curvatures induced by other means. However, two processes may strongly interfere with membrane thickness and susceptibility of the membrane to curving. First, phase transition, as lipids in the liquid-disordered state are less packed than in ordered phases, and thus membrane phase transitions to more ordered states increase their thickness, but also their bending rigidity as rigidity increases with order. Another important aspect is the lack of saturation in lipid acyl chains. Besides strongly affecting the order, reducing melting temperatures dramatically, it also makes the membrane more deformable, as lipid acyl chains with poly-unsaturations can form conformations that change thickness necessary for bending.[19] All the dynamic processes of curvature generation discussed above result from internal and local stresses within the surface. But can external, or global stresses shape living surfaces?

C – Global and external forces involved in surface shaping

Surfaces can be deformed by applying global or local forces externally to the surface. In the following I review known mechanisms of surface deformation by external forces.

1   Global: buckling of surface through growth under confinement

Growth can generate internal stresses, in particular when growth is confined. For surfaces, growth means area expansion, which can lead to folding when surfaces grow in confined volumes. In this situation, the surface is buckling, forming local folds from global compressive stresses. This is a very general and simple, force driven mechanism of deformation, which relies on the elastic properties of surfaces.

Buckling of surfaces was proposed as a folding mechanism of cell monolayers a century ago, by the same pioneers that discovered apical constriction.[8,20] Indeed, many of the shapes observed in developing cell monolayers are reproduced by theory and simulations of buckling surfaces.[10] However, it turns out to be difficult to show that biological shapes are indeed generated by buckling, as other mechanisms such as apical constriction can generate similar shapes, and measuring global forces within epithelia remains technically challenging. My team recently used an artificial system (see Fig. 3A), in which epithelial cells growing in elastic hollow spheres formed a cell monolayer that spontaneously buckled while growing. Combining theory and global pressure measurements (obtained from the elastic deformations of the spherical shell), we could show that the shapes obtained were generated from buckling.[21]

Is buckling a mechanism by which lipid membranes can be deformed? Theoretically, rapid changes of membrane area by large fluxes of lipids can induce curvature instabilities.[22] Another buckling instability was proposed for the deformation of lipid membranes by growing circular cytoskeletal filaments,[23] in order to explain the beautiful membrane protrusions emerging from the center of filamentous spirals formed by ESCRT-III assemblies in cells.[24,25] ESCRT-III polymers are thought to be the most evolutionary ancient membrane remodeling machinery, as they are the only ones present in archaea, and the only ones to work

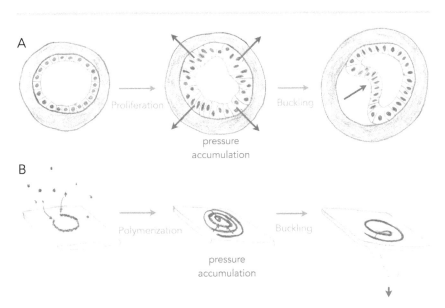

Fig. 3   Buckling of an epithelium and of a lipid membrane by growth-generated pressure accumulation. (A) Proliferation of epithelial cells under spherical confinement (elastic, hollow shell) generates pressure within the epithelium that causes inward buckling (from ref. 21). (B) Elongation of a curved ESCRT-III filament into a spiral accumulates elastic pressure, causing its buckling into a tubular shape, deforming the membrane beneath (from ref. 27).

on virtually all cell membrane organelles.[26] In the ESCRT-III buckling mechanism (see Fig. 3B), filaments of ESCRT-III protein initially grow with a preferred radius of curvature of about 35 nm on the surface of the membrane.[27] When about to reach the closed ring state, it can continue growing further into a flat spiral if the tips do not anneal. While growing as a spiral, the ESCRT-III filament accumulates elastic energy because filament turns inside the spiral are at a lower radius than their preferred one, and filament turns outside have a larger radius than their preferred one. In this situation, filament elastic stress can be released by buckling, transforming a flat spiral into a helical filament, in which most turns are at their preferred radius of curvature. Buckling of the ESCRT-III spiral will be in competition with the deformation of the membrane, but conditions in which buckling instability can deform the membrane constitute a fairly large area of the parameter space.[23] But most mechanical properties of the ESCRT-III filaments are compatible with this mechanism, even though the trigger for buckling may come from the addition of secondary subunits.[28–30]

So what is common to buckling an epithelium and buckling of the membrane by the ECSRT-III? First, growth is the source of compressive stresses in both cases: proliferation of cells in one case, addition of subunits to the polymer in the other case. Second, the confinement permits accumulation of the stress, and is provided by the elastic shell or the structure of the spirals, making a sort of auto-confinement in the ESCRT-III case. And finally, the shape of protrusion, only dictated by the mechanics of surfaces, are very similar. In physiological cases of cell tissue morphogenesis, confinement may be generated by fast growing domains of cells within slowly growing tissues.[31]

2   Local: local forces applied by external elements

Surface deformations can be generated by local forces applied by external elements. This situation is often seen in membrane traffic, in which protein coats, made of polymerized multi-protein complexes that have a given shape, force the membrane to adopt this shape. This mechanism is called scaffolding, and the best known examples are spherical coats in the form of clathrin and COPs.[32] But many other coats, from dynamin-like proteins to BAR domain proteins tubulate membranes by making helical cylindrical polymers.

Scaffolding is rather unknown to epithelium folding during morphogenesis. One putative example though, is the formation of the eye cup, the precursor of the eye lens, which is pinched by a growing and folding neuroepithelium that will become the retina. Whether forces generated by the growing retina participate in the folding of the eye cup, which finally separates from the ectoderm, is not known, but the current model proposes that it is simultaneous folding through polarized constriction of the two epithelia (neuroepithelium and ectoderm), coordinated by morphogen gradients, that causes the formation of the eye cup.[33] Thus, while the structure of the eye cup is seemingly similar to a spherical bud generated by membrane coats, the mechanism may be totally different.

Conclusion

New mechanisms of membrane remodeling or of epithelium morphogenesis are still being discovered, or proposed. Recent theoretical work has proposed membrane remodeling through nematic order of chiral molecules that can create gradients of stresses within the plane of the lipid membrane.[34] Recently, we showed that gradients of active stresses organized by topological defects in cell monolayers can concentrate stresses to form cell protrusions.[35] The future will say if among these two possibilities, common principles can be found.

Conflicts of interest

There are no conflicts to declare.

References

1 B. Sorre, PhD thesis, Université Paris-Diderot, 2010.
2 J. N. Israelachvili, S. Marcelja and R. G. Horn, *Q. Rev. Biophys.*, 1980, **13**, 121–200.
3 M. Luciano, S.-L. Xue, W. H. De Vos, L. R. Morata, M. Surin, F. Lafont, E. Hannezo and S. Gabriele, *Nat. Phys.*, 2021, DOI: 10.1038/s41567-021-01374-1.
4 T. Lecuit, P. F. Lenne and E. Munro, *Annu. Rev. Cell Dev. Biol.*, 2011, **27**, 157–184.
5 P. Gomez-Galvez, P. Vicente-Munuera, A. Tagua, C. Forja, A. M. Castro, M. Letran, A. Valencia-Exposito, C. Grima, M. Bermudez-Gallardo, O. Serrano-Perez-Higueras, F. Cavodeassi, S. Sotillos, M. D. Martin-Bermudo, A. Marquez, J. Buceta and L. M. Escudero, *Nat. Commun.*, 2018, **9**, 2960.
6 A. Roux, D. Cuvelier, P. Nassoy, J. Prost, P. Bassereau and B. Goud, *EMBO J.*, 2005, **24**, 1537–1545.
7 B. Sorre, A. Callan-Jones, J. B. Manneville, P. Nassoy, J. F. Joanny, J. Prost, B. Goud and P. Bassereau, *Proc. Natl. Acad. Sci. U. S. A.*, 2009, **106**, 5622–5626.
8 L. Rhumbler, *Arch. Entwicklungsmech. Org.*, 1902, **14**, 401–476.
9 N. Rashevsky, *Bull. Math. Biophys.*, 1940, **2**, 169–175.
10 M. Rauzi, A. Hocevar Brezavscek, P. Ziherl and M. Leptin, *Biophys. J.*, 2013, **105**, 3–10.
11 W. Romer, L. Berland, V. Chambon, K. Gaus, B. Windschiegl, D. Tenza, M. R. Aly, V. Fraisier, J. C. Florent, D. Perrais, C. Lamaze, G. Raposo, C. Steinem, P. Sens, P. Bassereau and L. Johannes, *Nature*, 2007, **450**, 670–675.
12 W. Romer, L. L. Pontani, B. Sorre, C. Rentero, L. Berland, V. Chambon, C. Lamaze, P. Bassereau, C. Sykes, K. Gaus and L. Johannes, *Cell*, 2010, **140**, 540–553.
13 R. Lakshminarayan, C. Wunder, U. Becken, M. T. Howes, C. Benzing, S. Arumugam, S. Sales, N. Ariotti, V. Chambon, C. Lamaze, D. Loew, A. Shevchenko, K. Gaus, R. G. Parton and L. Johannes, *Nat. Cell Biol.*, 2014, **16**, 595–606.
14 B. Antonny, *Curr. Opin. Cell Biol.*, 2006, **18**, 386–394.
15 M. P. Sheetz and S. J. Singer, *Proc. Natl. Acad. Sci. U. S. A.*, 1974, **71**, 4457–4461.
16 M. M. Kozlov, F. Campelo, N. Liska, L. V. Chernomordik, S. J. Marrink and H. T. McMahon, *Curr. Opin. Cell Biol.*, 2014, **29**, 53–60.
17 C. Bielmeier, S. Alt, V. Weichselberger, M. La Fortezza, H. Harz, F. Julicher, G. Salbreux and A. K. Classen, *Curr. Biol.*, 2016, **26**, 563–574.
18 M. Matejcic, G. Salbreux and C. Norden, *PLoS Biol.*, 2018, **16**, e2006018.

19 B. Antonny, S. Vanni, H. Shindou and T. Ferreira, *Trends Cell Biol.*, 2015, **25**, 427–436.

20 N. Rashevsky, *Bull. Math. Biophys.*, 1940, **2**, 109–121.

21 A. Trushko, I. Di Meglio, A. Merzouki, C. Blanch-Mercader, S. Abuhattum, J. Guck, K. Alessandri, P. Nassoy, K. Kruse, B. Chopard and A. Roux, *Dev. Cell*, 2020, **54**, 655–668.

22 J. Solon, J. Pécréaux, P. Girard, M. Fauré, J. Prost and P. Bassereau, *Phys. Rev. Lett.*, 2006, **97**, 1–4.

23 M. Lenz, D. J. Crow and J. F. Joanny, *Phys. Rev. Lett.*, 2009, **103**, 038101.

24 P. I. Hanson, R. Roth, Y. Lin and J. E. Heuser, *J. Cell Biol.*, 2008, **180**, 389–402.

25 A. G. Cashikar, S. Shim, R. Roth, M. R. Maldazys, J. E. Heuser and P. I. Hanson, *eLife*, 2014, **3**, e02184.

26 M. Vietri, M. Radulovic and H. Stenmark, *Nat. Rev. Mol. Cell Biol.*, 2020, **21**, 25–42.

27 N. Chiaruttini, L. Redondo-Morata, A. Colom, F. Humbert, M. Lenz, S. Scheuring and A. Roux, *Cell*, 2015, **163**, 866–879.

28 A. Bertin, N. de Franceschi, E. de la Mora, S. Maity, M. Alqabandi, N. Miguet, A. di Cicco, W. H. Roos, S. Mangenot, W. Weissenhorn and P. Bassereau, *Nat. Commun.*, 2020, **11**, 2663.

29 J. Moser von Filseck, L. Barberi, N. Talledge, I. E. Johnson, A. Frost, M. Lenz and A. Roux, *Nat. Commun.*, 2020, **11**, 1516.

30 A. K. Pfitzner, V. Mercier, X. Jiang, J. Moser von Filseck, B. Baum, A. Saric and A. Roux, *Cell*, 2020, **182**, 1140–1155.

31 M. Tozluoglu, M. Duda, N. J. Kirkland, R. Barrientos, J. J. Burden, J. J. Munoz and Y. Mao, *Dev. Cell*, 2019, **51**, 299–312.

32 T. Kirchhausen, *Nat. Rev. Mol. Cell Biol.*, 2000, **1**, 187–198.

33 S. Fuhrmann, *Curr. Top. Dev. Biol.*, 2010, **93**, 61–84.

34 R. C. Sarasij and M. Rao, *Phys. Rev. Lett.*, 2002, **88**, 088101.

35 P. Guillamat, C. Blanch-Mercader, K. Kruse and A. Roux, Integer topological defects organize stresses driving tissue morphogenesis, *Nat. Mater.*, 2021, in press.

Faraday Discussions

PAPER

Heterogeneity of cell membrane structure studied by single molecule tracking†

Gregory I. Mashanov, ⓘ *a Tatiana A. Nenasheva, ⓘ b
Alla Mashanova, ⓘ c Remigijus Lape,‡d Nigel J. M. Birdsall,a
Lucia Sivilottid and Justin E. Molloy ⓘ *a

Received 27th May 2021, Accepted 23rd July 2021

DOI: 10.1039/d1fd00035g

Heterogeneity in cell membrane structure, typified by microdomains with different biophysical and biochemical properties, is thought to impact on a variety of cell functions. Integral membrane proteins act as nanometre-sized probes of the lipid environment and their thermally-driven movements can be used to report local variations in membrane properties. In the current study, we have used total internal reflection fluorescence microscopy (TIRFM) combined with super-resolution tracking of multiple individual molecules, in order to create high-resolution maps of local membrane viscosity. We used a quadrat sampling method and show how statistical tests for membrane heterogeneity can be conducted by analysing the paths of many molecules that pass through the same unit area of membrane. We describe

aThe Francis Crick Institute, 1 Midland Road, London, NW1 1AT, UK. E-mail: Gregory.mashanov@crick.ac.uk; Justin.molloy@crick.ac.uk

bKoltzov Institute of Developmental Biology, 26 Vavilova Str., Moscow, 119334, Russia

cSchool of Life and Medical Sciences, University of Hertfordshire, College Lane, Hatfield, Hertfordshire, AL10 9AB, UK

dDepartment of Neuroscience, Physiology and Pharmacology, Division of Biosciences, University College London, Gower St., London, UK

† Electronic supplementary information (ESI) available: Movie 1: M_2 muscarinic receptors labelled with Cy3B-telenzepine moving on the plasma membrane of cultured HL1 cardiomyocyte and imaged at 33 fps (23 °C). Real time movie, Movie 2: M_2 muscarinic receptors labelled with Cy3B-telenzepine moving on the plasma membrane of primary, cultured, cardiomyocyte extracted from embryonic mouse heart and imaged at 33 fps (23 °C). Movie 3: M_2 muscarinic receptors labelled with Cy3B-telenzepine on the plasma membrane of cultured HL1 cardiomyocyte treated with 1% of paraformaldehyde and imaged at 20 fps (23 °C). Some patches of plasma membrane adjacent to the substrate were sealed off by PFA and retained low viscosity allowing individual molecules to move freely inside these patches. Movie 4: M_2 muscarinic receptors tagged with eGFP transfected into a HUVEC and imaged at 33 fps (23 °C). Real time movie. Movie 5: tracking individual muscarinic receptors on the surface of a heart slice from the zebrafish reveals the fine viscosity map of the plasma membranes when imaged at 50 fps (23 °C). The trajectory map is inserted at the end of Movie 1 to show all the trajectories detected in this record (×0.5 real time). Movie 6: model simulating the random movements ($D_{lat} = 0.2 \ \mu m^2 \ s^{-1}$) of fluorescent molecules on the surface of a rectangular cell (30 × 30 × 5 μm^3. TIRFM illumination mode). A trajectory map is inserted at the end of Movie 1 to show all the trajectories detected in this record (×0.5 real time). See DOI: 10.1039/d1fd00035g

‡ Current address: MRC LMB, Francis Crick Avenue, Cambridge, CB2 0QH, UK.

experiments performed on cultured primary cells, stable cell lines and *ex vivo* tissue slices using a variety of membrane proteins, under different imaging conditions. In some cell types, we find no evidence for heterogeneity in mobility across the plasma membrane, but in others we find statistically significant differences with some regions of membrane showing significantly higher viscosity than others.

Introduction

It is now thought that structural heterogeneity of the plasma membrane plays a critical role in a variety of cell functions.[1-2] This contrasts with our classical view of the plasma membrane as a fluid mosaic bilayer of phospholipids and other amphipathic molecules within which proteins can diffuse in an unhindered manner.[3] The path taken by a diffusing transmembrane protein should follow a random Brownian walk and a plot of its mean-squared displacement against time interval (MSD *vs.* dT) should be linear,[4] with a gradient determined by D_{lat}, the lateral diffusion coefficient. D_{lat} should be a linear function of membrane viscosity and thickness but should have only a weak dependence (*i.e.* log-function) on protein radius and viscosity of the bounding aqueous media (*i.e.* cytosol and extra cellular fluid).[5,6] Recently a revised picture of the plasma membrane has emerged in which it is thought to have a more heterogenous structure compared to the classical fluid mosaic model, and there is biochemical and biophysical evidence for submicron, phase-separated, domains[7] that are enriched with particular lipid species[8] ("lipid rafts") and regions of protein crowding or clustering. In addition, interaction of membrane proteins with the cortical cytoskeleton[9] or extracellular matrix can lead to obstructions to free-diffusion[10] ("picket-fences"). Together these features contribute to anomalous diffusive behaviour[11,12] whereby MSD *vs.* dT plots are non-linear and show distinct downward curvature. The features that contribute to membrane heterogeneity are thought to be transient in nature and to extend over a length scale <200 nm, making them notoriously difficult to study in live cells. However, the advent of super-resolution, single molecule imaging methods[2,13] allows "rafts", "picket fences", protein clustering and other heterogeneous properties to be studied in live cells and tissue slices in real-time.[14,15] An advantage of single molecule studies is that they eliminate problems of bulk averaging that can mask short-lived, sub-resolution structural heterogeneities. Individual transmembrane proteins have been used as probes of membrane structure because they are of uniform size, can be labelled with high-specificity and known stoichiometry, they target to known membrane compartments and genetic manipulation allows structural alterations. However, tracking of single fluorophores in live cells is technically challenging because the spatial resolution and temporal precision of tracking is limited by photon emission rate and photobleaching of the fluorescent moiety. A useful figure of merit for a given fluorophore is the average number of photons emitted before photobleaching, giving rise to the idea of a "photon budget".[16] For a given experiment, best use of the photon budget requires a compromise to be made between tracking precision and measurement bandwidth. To measure small-scale, rapid movements, high laser power is required so that single fluorophores are bright and can be tracked with high-precision at fast video frame rate, but the cost is they soon photobleach. At low laser power, fluorophores can be tracked for longer but tracking resolution is reduced because fewer photons are emitted per second. It is also important to consider fluorophore density across the sample field of view (see ESI of ref. 14†) because at high density, fluorophore tracking becomes difficult as paths overlap, whereas at low density fewer molecules pass through a given region of interest making it difficult to accumulate sufficient data for statistical analysis. In general, fluorescent fusion proteins (like eGFP) photobleach faster than synthetic organic fluorophores (like rhodamine or Cy-dyes) and offer a lower photon budget.

The aim of the current study has been to test whether live mammalian cell membranes exhibit heterogeneity in viscosity on the micron length-scale. We have explored whether heterogeneity in protein mobility is evident in immortal-ised cell lines, primary cell culture and live tissue samples using different protein probes. Total internal reflection fluorescence microscopy (TIRFM)[17] was used to image membrane proteins labelled with single fluorophores and studied their diffusive behaviour to infer local membrane properties. We used a quadrat sampling method that enables statistical tests to be performed on data obtained from different regions of the cell. Trajectories were segmented and diffusion coefficients computed from ≥ 10 data points (~ 0.5 second) sections of each trajectory. The image of the cell was then sub-divided into a checkerboard pattern of sample quadrats (~ 1–4 μm^2 area each), and local statistics were calculated for trajectory segments that resided within each quadrat so statistical tests could be performed across different regions of the cell membrane. For most of the speci-mens tested, protein mobility was homogenous across the cell surface but in some cells we found evidence for spatial heterogeneity. We conclude that lipid rafts (if they exist) are evenly distributed across the plasma membrane of most cells, but in some cell types, they might show heterogeneity in density forming a "raft of rafts".

Experimental

All chemicals were obtained from Sigma-Aldrich, UK, unless stated otherwise.

Cultured cell transfection and tissue labelling

All of the cells and tissue materials were imaged using a custom-made imaging chamber described previously.[18] Human umbilical vein endothelial cells (HUVECs) (PromoCell GmbH, Heidelberg, Germany) derived from a pooled, primary cell culture were transfected with M_2-eGFP muscarinic receptor fusion protein (kind gift of Prof. J. W. Wells, University of Toronto, Canada) using nucleofection according to the manufacturer's instructions (Amaxa GMBH). CHO-K1 cells were transfected with cDNAs encoding the mouse $\alpha 1$, $\beta 1$, and δ nicotinic receptor subunits subcloned within vector pRBG4 (kind gift of S. M. Sine, Mayo Clinic, MN, USA). The mouse γ-subunit was tagged with eGFP inserted into the cytoplasmic loop between transmembrane helices M3 and M4 and subcloned in pRK5 plasmid (kind gift of V. Witzemann, Max-Plank-Institute, Heidelberg, Germany). Primary cardiomyocytes were prepared as described previously[14] and allowed to settle on to the imaging coverslip before labelling with Cy3B-telenzepine (10 nM for 1 h at 23 °C). The HL1 cardiomyocyte cell-line was cultured for 24 h before labelling with Cy3B-telenzepine (1 nM for 3.5 h at 23 °C). Zebrafish heart was removed and washed in room temperature PBS (pH 7.2)

(Thermo Fisher Scientific, UK) solution supplemented with 10 U ml^{-1} heparin and 100 U ml^{-1} penicillin–streptomycin for 5 min. After washing, the heart was moved into an ice-cold "relaxing solution" consisting of PBS supplemented with; 1 mM EDTA, 2.5 mM KOH, and 3 mM MgCl$_2$ for 5 min. A cutting block, with a 2 mm diameter cavity, and a single cutting-slot was used to bisect the heart along the ventricle axis.[15] Mouse embryonic heart slices were prepared using procedures described previously.[14] The prepared tissue slices were placed in relaxing PBS solution containing 10 nM of Cy3B-telenzepine for 1 h at 4 °C. This procedure labelled >95% of M$_2$ muscarinic receptors with the tight-binding fluorescent ligand. The tissue slice was fixed against the coverslip with a fine nylon mesh grid ($\sim$0.5 × 0.5 mm^2) stretched in a stainless-steel tambour. The chamber was filled with Hank's Balanced-Salt solution supplemented with 20 mM HEPES (pH 7.4). All imaging was performed at 23 °C.

TIRF imaging

A custom-built TIRF microscope was used, as described previously.[19] Briefly, the beam from a 100 mW, 488 nm laser or 150 mW, 561 nm laser (Light HUB-6, Omicron, Germany) was expanded using a Galilean beam expander and focused at the back focal-plane of a high numerical aperture, oil-immersion, objective lens (PlanApo, 100×, NA 1.45, Olympus, Japan) using a small, aluminium-coated mirror (3 mm diameter, Comar Optics, UK) placed at the edge of the back-aperture of the objective lens. The average laser power at the specimen plane was adjusted to $\sim$0.5 µW µm^{-2} and the incident laser beam angle was adjusted to $\sim$63° to create the evanescent field at the glass–aqueous medium interface. A second small mirror was placed at the opposite edge of the objective lens back-aperture to remove the returning (internally-reflected) laser beam from the microscope and a narrow band-pass emission filter FF01-525/50 or FF01-593/40 (Semrock, Rochester, NY) was used to block the scattered 488/561 nm laser light and other unwanted light. An EMCCD camera (iXon897BV, Andor, UK) captured video sequences at a rate of 20–50 fps, and the data were stored on a computer hard drive for analysis.

Video data analysis

Video image sequences were analysed using GMimPro software (http://www.mashanov.uk), which employs an automatic single particle tracking algorithm (described previously[20]) to detect and track individual fluorophores. The position of each fluorescent spot was localised with sub-pixel resolution using a Gaussian fitting method and systematic drift was corrected *via* cross-correlation. Individual particle trajectories were output as a table of x,y coordinates measured at each video frame (*i.e.* each time point) for every fluorophore detected in the sample. From the tabulated data, the mean-squared displacement (MSD) of each fluorophore was computed over all possible time intervals (dT) (*e.g.* 1,2,3...n frames). MSD *vs.* dT plots were then generated to examine the particle motion; a linear-relationship is expected for simple Brownian motion, but curvature indicates anomalous diffusion (discussed later). To extract statistically meaningful mobility data from the trajectory analysis, we analysed trajectories where single fluorescent spots could be tracked for at least 10 consecutive frames with a permitted spot displacement of ≤0.7 µm between frames.[20]

Single molecule mobility maps

We created a map to depict the lateral diffusion coefficient of our labelled protein probes by taking sequential 10-frame time windows (~400 ms) along each single molecule trajectory and estimating the local D_{lat} value from the gradient of the MSD $vs.$ dT plot. For display purposes, tracks were dilated to be 5 pixels wide (0.5 μm) and coded to produce a pseudo-colour heat map of membrane viscosity. At positions where trajectories overlapped, the pixel intensity was averaged. This approach is similar, but not identical, to single particle velocimetry.

Measuring the local heterogeneity of cell membrane

We divided the image of each cell into a checkerboard pattern of sample quadrats and selected segments of single molecule trajectories that resided within each quadrat region. The local trajectory segments (at least 5 trajectories, ≥10 data points each per segment) were used to build an average MSD $vs.$ dT plot. The initial slope of the plot was used to estimate the local D_{lat} ($i.e.$ slope/4, for 2-dimensional diffusion). Local D_{lat} estimates were binned and histogrammed to build a global mobility map across the cell surface. The distribution of D_{lat} values of individual trajectories is described by a Gamma distribution due to the limited number of data points in each trajectory, but, according to the central limit theorem, the distribution of mean D_{lat} values measured at each quadrat sample should be normally distributed ($i.e.$ Gaussian). We tested for heterogeneity across the checkerboard pattern of quadrat samples against an expectation based on the global distribution. The normality of the distributions was checked using the Kolmogorov–Smirnov (KS) test. The pairwise comparisons were done using Student's t-test when both distributions were normal, and Mann–Whitney U (MW) test when at least one distribution was not normal.[21] Calculations were performed using open-source R software.[22]

Results

Comparison of membrane viscosity in HL1 cell-line $vs.$ primary cardiomyocytes

We first imaged HL1 cells, a cardiac muscle cell-line, that natively-express M_2 muscarinic acetylcholine receptors (Fig. 1A and Movie 1†). Receptors were labelled using the tight-binding fluorescent ligand Cy3B-telenzepine[14] under conditions that gave an optimal density of labelled receptors (~0.8 receptors per μm²) for single molecule imaging and tracking purposes. Under the conditions used here, the Cy3B photobleaching rate was 0.19 s^{-1}, (Fig. S1A†) which allowed us to track individual molecules for around 5 s. Many trajectories were prematurely truncated because fluorophores overlapped during a single frame. This caused one track to terminate and a new track to be generated when the fluorophores separated. A small number of, non-specifically, surface-bound Cy3B-telenzepine molecules were completely immobile throughout the video recording and these objects were removed from the dataset. The projected image of all single molecule trajectories shows the degree of coverage across the cell surface (Fig. 1A(ii) and S2†). The cell periphery has a higher track density because fresh (unbleached) fluorophores diffuse into this region from the apical cell surface which is beyond the evanescent field.

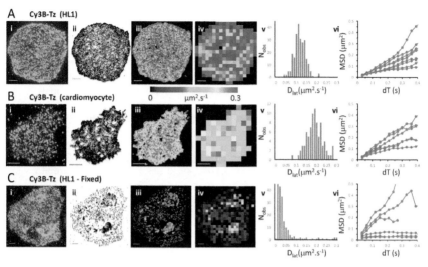

Fig. 1 M_2 muscarinic acetylcholine receptor diffusion on the plasma membrane of HL1 cells and primary cardiomyocytes. (A(i)) First video frame showing an HL1 cell-line with Cy3B-telenzepine labelled M_2 receptors. (ii) A projection of all single molecule trajectory paths. (iii) Heat-map showing locally averaged D_{lat} values; mean D_{lat} = 0.12 μm^2 s^{-1}, n = 7526 trajectories. (iv) Heat map of the quadrat-sampled (2 × 2 μm^2) D_{lat} values. (v) Distribution of D_{lat} measured in each quadrat region across the cell: 250 compartments, overall mean (of the local mean values) D_{lat} = 0.112 ± 0.028 μm^2 s^{-1} (±SD). (vi) MSD $vs.$ dT plots for typical fast (where D_{lat} > mean +1.5 × SD; red) and slow (where D_{lat} < mean −1.5 × SD; blue) quadrat regions. Video was recorded at 33 fps. (B) Panels as for A except: (i) primary cardiomyocyte. (iii) Mean D_{lat} = 0.203 μm^2 s^{-1}, n = 3133 trajectories. (v) Quadrats: 86 (2 × 2 μm^2); mean D_{lat} = 0.188 ± 0.043 μm^2 s^{-1}. (vi) See panel A. (C) Panels as for A except: (i) HL1 cell chemically fixed with 1% paraformaldehyde. (ii) Trajectory paths show that most molecules have very restricted motion following fixation. Some regions show receptor diffusion (dark areas where molecular trajectories greatly overlap). (iii) Most M_2 receptors are fixed, but some isolated regions show rapid receptor diffusion; average D_{lat} = 0.132 μm^2 s^{-1}, n = 6673 trajectories. (iv) Quadrats: 327 (2 × 2 μm^2). (v) Distribution of D_{lat} values with overall mean D_{lat} = 0.029 ± 0.05 μm^2 s^{-1}. Fixed regions indicate the tracking noise floor (see main text and Fig. S1 C and D†). (vi) See panel A.

Analysis of individual trajectories (see methods) leads to estimates of D_{lat} for each molecule tracked and these values can be plotted as a pseudo-colour map (Fig. 1A(iii)) indicating M_2 receptor mobility at different regions across the cell surface. To test for homogeneity, the mobility map is divided into a checker-board pattern of sample quadrats (Fig. 1A(iv)). Each quadrat is then colour-coded to represent the average D_{lat} value within that region. According to the central limit theorem the distribution of mean D_{lat} values should be Gaussian (Fig. 1A(v)), and in the following text "mean D_{lat}" refers to the "mean of the quadrat means". To investigate whether the diffusion of molecules in quadrats with extreme low or high D_{lat} values was anomalous we plotted MSD $vs.$ dT diagrams from those, selected, quadrats (Fig. 1A(vi), S3A, B† and Table 1).

We next imaged M_2 receptors in primary cultured mouse cardiomyocytes that had been labelled the same way as the HL1 cells (Fig. 1B and Movie 2†). We found receptor diffusion was significantly faster (D_{lat} = 0.2 μm^2 s^{-1}) compared to the HL1 cell-line (t-test with unequal variances (p < 0.0001), $t(111)$ = −15.4, p < 2.2 ×

Table 1 Kolmogorov–Smirnov analysis of the local viscosity distributions[a]

Figure	Sample	Mean ($\mu m^2\ s^{-1}$)	SD	n	K–S D-statistic	p-Value	Conclusion
Fig. 1	M$_2$-Cy3B HL1	0.112	0.028	250	0.045	0.704	Normal
	M$_2$-Cy3B primary myocyte	0.188	0.043	86	0.073	0.747	Normal
	M$_2$-Cy3B HL1 fixed	**0.029**	**0.050**	**327**	**0.283**	**<0.0001**	**Not-normal*****
	M$_2$-Cy3B inter-cell HL1	0.129	0.015	29	0.121	0.793	Normal
Fig. 2	**M$_2$-eGFP-HUVEC**	**0.204**	**0.086**	**150**	**0.178**	**0.0002**	**Not-normal*****
Fig. 3	M$_2$-Cy3B zebrafish cell 1 and cell 2	0.224	0.116	49	0.185	0.069	Normal
	Zebrafish cell 1	0.291	0.144	20	0.198	0.367	Normal
	Zebrafish cell 2	0.179	0.056	29	0.131	0.699	Normal
	M$_2$-Cy3B mouse slice (all cells)	**0.427**	**0.457**	**79**	**0.294**	**<0.0001**	**Not-normal*****
	Mouse slice cell 1	0.533	0.518	27	0.253	0.052	Normal
	Mouse slice cell 2	**0.534**	**0.590**	**18**	**0.370**	**0.010**	**Not-normal****
	Mouse slice cell 3	**0.285**	**0.228**	**34**	**0.260**	**0.020**	**Not-normal***
Fig. 4	Nicotinic-eGFP-CHO	0.114	0.078	181	0.191	<0.0001	Not-normal***
Fig. 5A	Model	0.188	0.034	227	0.046	0.721	Normal
Fig. 5B	**Model (both regions)**	**0.111**	**0.080**	**226**	**0.192**	**<0.0001**	**Not-normal*****
	Model (non-raft)	0.188	0.030	120	0.056	0.847	Normal
	Model (raft region)	**0.031**	**0.021**	**116**	**0.160**	**0.005**	**Not-normal****

[a] Significance testing against normal distribution: *p-value < 0.05 (*e.g.* >95% confidence); **p-value < 0.01 ***p-value < 0.001.

10^{-16}), implying membrane viscosity was lower in the primary cultured cells. The molecular tracks were analysed as before and binned to give a checkerboard pattern of quadrat samples across the cell (Fig. 1B(i–vi) and Table 1).

In order to measure localisation and tracking errors (*i.e.* the "noise floor" of our measurements), we imaged HL1 cells that had been chemically fixed using 1% paraformaldehyde. Interestingly, after several minutes treatment with fixative, many cells still had isolated regions, or "islands", of membrane that appeared sealed off from the fixative and where the Cy3B-telenzepine-labelled receptors continued to diffuse (Fig. 1C and Movie 3†). This fixation artefact gave the opportunity to generate diffusion maps in cells that showed extreme heterogeneity in receptor mobility due to incomplete fixation. Cell regions where chemical fixation was effective allowed individual molecules to be selected and tracked in order to give an estimate of the localisation and tracking error measured under our imaging conditions. The sum of all noise sources gave 33 nm root mean squared deviation (rms) (*i.e.* MSD = $1 \times 10^{-3}\ \mu m^2$) (Fig. S1C and D).† However, the quadrat sampling method gave a three-fold higher estimate of the noise floor: rms ∼100 nm (MSD = $1 \times 10^{-2}\ \mu m^2$) (Fig. 1C(v)). This difference arose because mobile molecules were often present within a given quadrat region. The quadrat mapping of D_{lat} values showed a heterogeneous distribution of values which is readily explained by the incomplete fixation, whereby some islands of plasma membrane seemed to be isolated and protected from the paraformaldehyde treatment. The mean D_{lat} values determined for each of the quadrat sample regions were histogrammed as before. The D_{lat} values showed an exponential distribution (Fig. 1C(iv)). Typical slow-moving quadrat regions (blue lines in Fig. 1C(vi)) exhibited MSD vs. dT plots that were highly anomalous, consistent with the molecules being either totally immobile or trapped in a confined area. The faster moving, outlier, D_{lat} values were from unfixed regions of membrane and MSD vs. dT plots (red lines in Fig. 1C(vi)) showed a variety of behaviours, with some quadrat regions showing a relatively linear slope of MSD vs. dT plots and others showing a distinct downward curvature consistent with anomalous diffusive behaviour (Table 1).

Mapping membrane viscosity using eGFP-tagged M$_2$ receptors

We next examined the mobility of a C-terminally-tagged M$_2$-receptor eGFP-fusion protein where the fluorophore is positioned on the intracellular-side of the molecule in contrast to the extrinsic, synthetic labelling with Cy3B-telenzepine used in the previous experiments. HUVECs were transiently transfected with a recombinant M$_2$-eGFP fusion construct in a mammalian cell expression vector and cells were then selected for imaging based on their receptor expression level (target level ~0.8 receptors per µm^2). The eGFP-tagged receptors moved rapidly and in a seemingly unhindered fashion across the plasma membrane (Fig. 2 and Movie 4†). The photobleaching rate of the eGFP fluorophore was 0.4 s^{-1} (Fig. S1B†) which is twice as fast as the Cy3B fluorophore. In addition to that, fluorophores could only be tracked for ~0.5 s, because the eGFP signal-to-noise intensity ratio is lower than for Cy3B and this leads to premature track termination (i.e. tracking can fail before the fluorophore bleaches). Individual static molecules were selected and tracked in order to give an estimate of the localisation and tracking error measured under our imaging conditions. The sum of all noise sources gave 26 nm root mean squared deviation (rms) (i.e. MSD = 1 × 10^{-3} µm^2) (Fig. S1C and D†), similar precision to Cy3B. The average rate of receptor diffusion (D_{lat} = 0.2 µm^2 s^{-1}) was similar to the primary cardiomyocytes (MW, U = 6265, p = 0.715) but the trajectories appear more compact (Fig. 2(ii) vs. Fig. 1A and B panel (ii)) simply because they are of shorter duration. The short

HUVEC – eGFP-tagged M$_2$ receptors

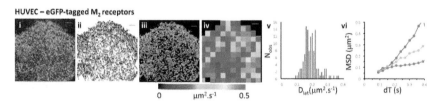

Fig. 2 eGFP-tagged M$_2$ muscarinic acetylcholine receptor diffusing at the plasma membrane of HUVECs. (i) Single image from the beginning of the record shows individual eGFP-tagged M$_2$ receptors at the plasma membrane. (ii) Trajectory map; 5031 trajectories. (iii) Heat-map of D_{lat}; mean D_{lat} = 0.197 µm^2 s^{-1}. (iv) D_{lat} quadrat map; 129; 4 × 4 µm^2 quadrats (black indicates <5 trajectories, removed from later analysis). (v) Distribution of D_{lat} values; D_{lat} = 0.204 ± 0.086 µm^2 s^{-1} (mean ± SD). (vi) MSD vs. dT plots for typical fast (where D_{lat} > mean +1.5 × SD; red), intermediate (where D_{lat} ≈ mean; green) and slow (where D_{lat} < mean −1.5 × SD; blue) quadrat regions. Video was recorded at 33 fps.

tracks mean that there is less coverage (*i.e.* fewer unique tracks per unit area) when membrane viscosity maps are generated using a 4×4 µm^2 quadrat size (Table 1).

Comparison of Cy3B-telenzepine labelled M$_2$ receptors in zebrafish tissue slice *vs.* mouse tissue slice

We investigated the M$_2$ muscarinic receptor mobility in live tissue slices using two model cardiac systems (zebrafish and mouse), using Cy3B-telenzepine to fluorescently label the proteins. An initial observation was that the spread-area of the cells was noticeably smaller than for isolated primary cardiomyocytes and the HL1 cell-line; presumably because cells are held within the tissue and are less able to spread across the coverslip surface. Receptor density was similar in the tissue slices compared to the isolated cultured cells ($\sim$1 receptor per µm^2) (Fig. 3A, B and Movie 5† for comparison), but, the rate of diffusion was significantly faster in mouse tissue slices (mean $D_{lat} = 0.43$ µm^2 s^{-1}) compared to primary myocytes (MW, $U = 1683, p = 2.3 \times 10^{-8}$) and HL1 cell-line (MW, $U = 2135, p < 2.2 \times 10^{-16}$) and was similarly fast in the zebrafish tissue (mean $D_{lat} = 0.33$ µm^2 s^{-1}), but the median values were significantly different (MW, $U = 1261, p = 0.001$). Rapid receptor movement in both specimens produced a dense network of single molecule tracks so a small quadrat size (1×1 µm^2) could be used and this gave a higher-resolution map of membrane viscosity (Fig. 3(iv)). A histogram of mean

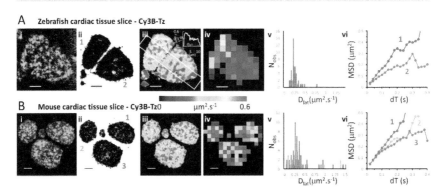

Fig. 3 Cy3B-telenzepine labelled M$_2$ muscarinic receptors in zebrafish and mouse cardiac tissue slices. (A(i)) First video frame showing individual Cy3B-telenzepine labelled M$_2$ receptors on the plasma membrane of zebrafish cardiomyocytes in a tissue slice. (ii) Map showing the trajectory paths of individual molecules (see Movie 3).† (iii) Pseudo-colour heat-map showing locally averaged M$_2$ receptor D_{lat} values. The overall mean $D_{lat} = 0.33$ µm^2 s^{-1}, 2518 trajectories. White rectangle shows profile plotted on the insert in the right-top corner. (iv) Heat map of the quadrat-sampled D_{lat} values; 49 quadrats (1×1 µm^2) (black indicates <5 trajectories, removed from later analysis). (v) Distribution of D_{lat} measured in each quadrat region (red = "cell 1", blue = "cell 2"): overall mean $D_{lat} = 0.224 \pm 0.116$ µm^2 s^{-1} ($\pm$SD): "cell 1" $D_{lat} = 0.291 \pm 0.144$ µm^2 s^{-1}, "cell 2" $D_{lat} = 0.179 \pm 0.056$ µm^2 s^{-1}. (vi) MSD *vs.* dT plots for typical "cell 1" (red), "cell 2" (blue) quadrat regions. (B) As for Panel A except: (i) mouse cardiomyocytes in a tissue slice. (iii) Mean $D_{lat} = 0.428$ µm^2 s^{-1}, 2336 trajectories. (iv) D_{lat} map: 83 quadrats (1×1 µm^2). (v) Distribution of D_{lat} ("cell 1"(red); "cell 2" (green); "cell 3" (blue): mean $D_{lat} = 0.427 \pm 0.457$ µm^2 s^{-1}). (vi) MSD *vs.* dT plots for typical quadrat regions (where $D_{lat} \approx$ global mean D_{lat}); within "cell 1" (red); "cell 2" (green); "cell 3" (blue).

D_{lat} values derived from each quadrat sample was normally distributed (Fig. 3(v)). MSD $vs.$ dT plots for three of the "slower" quadrat samples compared to 3 of the "faster" regions showed that MSD increased linearly with dT for both fast and slow samples and did not show substantial curvature (that would indicate anomalous diffusion) (Fig. 3(vi)).

Although the global average D_{lat} was significantly faster in tissue slices (D_{lat} = 0.33 μm^2 s^{-1}) than isolated cultured cells, the difference was smaller in the D_{lat} maps because faster trajectories were underrepresented and the mean D_{lat} value fell to 0.22 μm^2 s^{-1} for the zebrafish tissue slices. Consistent with this, when quadrat size was increased to 2 $\times$ 2 μm^2, the mean D_{lat} value increased to 0.27 μm^2 s^{-1}. Interestingly, we found significant differences in mobility between neighbouring cells (Fig. 3(iii)), implying that there are significant differences in membrane composition between cardiomyocytes within a given tissue sample. In the example shown here, one cell gave D_{lat} = 0.29 μm^2 s^{-1} the other, D_{lat} = 0.18 μm^2 s^{-1} (t-test with unequal variances, $t(23)$ = 3.224, p = 0.004).

Mapping membrane viscosity using nicotinic acetylcholine ion channels

Our expectation was that the nicotinic acetylcholine receptor, which is a hetero-pentameric ion channel, should exhibit unrestricted diffusive motion at the plasma membrane. However, we found it did not localise strongly to the plasma membrane and most molecules were retained in the endoplasmic reticulum (ER) following transfection. This might arise because of its hetero-pentameric nature, comprising four different polypeptides; β1, δ, γ and two α1 subunits, that must assemble correctly following co-transfection of the α1, β1, and δ subunits, together with the eGFP-γ-subunit. It is likely that a fraction of eGFP-γ subunits failed to assemble correctly into the heteromeric complex and are retained in the ER. So, we see a mixture of fluorophores at the ER network and also at the plasma membrane. Receptors that had localised correctly to the plasma membrane were most evident beneath the cell nucleus where the ER was excluded. We tracked all fluorophores as before and generated mobility maps to investigate mobility across the cell (Fig. 4). Our data show, perhaps unsurprisingly, that the histogram of D_{lat} values derived from the quadrat maps is not normally distributed, and while MSD $vs.$ dT plots produced from "fast-moving" quadrat regions (plasma membrane) are consistent with an unconstrained Brownian walk, "slow-moving" regions show anomalous diffusive behaviour (MSD $vs.$ dT plots show a distinct downward curvature) because γ subunits are confined to the ER membrane network.

Monte Carlo simulation of membrane protein random walks

We used an object-based, Monte Carlo stochastic model[23] to generate sequences of simulated TIRFM videos (Fig. 5 and Movie 6†). The random movement of single molecules were realistically reproduced with known intensity levels, stochastic shot-noise, background noise and photobleaching behaviour consistent with our experimental TIRFM imaging modality. The simulated video data sets were analysed with the same image processing software used for our real data sets. We first checked that our localisation and tracking software gave consistent results over a wide range of simulated D_{lat} values (see ESI Fig. S5†). The example simulation in Fig. 5A shows trajectories detected in a sequence of images simulating molecules

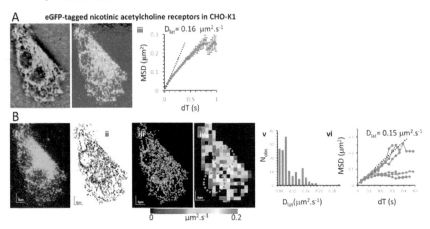

Fig. 4 Nicotinic acetylcholine receptors were present at the plasma membrane and the ER of CHO-K1 cells: diffusion of nicotinic receptors was highly anomalous because the eGFP tagged γ-subunit only partially targeted the plasma membrane and the majority was found in the ER. (A(i)) Average projection of the video stack. (ii) Standard-deviation projection shows the reticulated network and also darker regions (presumably below the nucleus) where the ER was excluded and receptors could be seen moving at the plasma membrane. (iii) The overall MSD *vs.* dT plot showed evidence of anomalous diffusion with initial gradient (determined by least-squares' linear regression) indicating $D_{lat} = 0.16 \ \mu m^2$ s^{-1}. (B(i)) First video frame showing single fluorophores. (ii) Projection of all the single fluorophore trajectories. (iii) D_{lat} map from trajectory segments. (iv) D_{lat} quadrat map (black indicates <5 trajectories, removed from later analysis). (v) Distribution of quadrat D_{lat} means (note that the histogram is not normally distributed). (vi) MSD *vs.* dT for fast (where D_{lat} > mean +1.5 × SD; red) and slow (where D_{lat} < mean −1.5 × SD; blue) regions. The fast-moving regions were located under the cell nucleus where receptors diffused freely at the plasma membrane (MSD *vs.* dT plots were linear with initial gradient (determined as in A) giving $D_{lat} = 0.15 \ \mu m^2 \ s^{-1}$). The slow-moving regions were from the network (ER) regions where receptors showed anomalous diffusion (note distinct downward curvature).

moving at $D_{lat} = 0.2 \ \mu m^2 \ s^{-1}$. The D_{lat} map (Fig. 5A(iii)) appears similar to our cell imaging data (Fig. 1A, B, 2 and 3). The local D_{lat} map calculated for a 2 × 2 μm^2 quadrat size showed a normal distribution of diffusion coefficients (Fig. 5A(v)). Note here, that any superficial appearance of heterogeneous behaviour is simply due to statistical variation across sample quadrats, and demonstrates the necessity for statistical tests of heterogeneity. In fact, there is no deviation from normality (Table 1).The MSD *vs.* dT plots (Fig. 5A(vi)) were linear for the "fast", "intermediate" and "slow" quadrat samples (note: we define "fast" as D_{lat} > mean +1.5 × SD, "slow" D_{lat} < mean −1.5 × SD and "intermediate" as $D_{lat} \approx$ mean). We also simulated the presence of lipid raft regions 300 × 300 nm^2, dispersed randomly across the membrane (Fig. 5B, on the left-hand side of the modelled membrane) in which molecules were confined ($D_{lat} = 0.02 \ \mu m^2 \ s^{-1}$) and these regions adjoined others where molecules diffused freely with $D_{lat} = 0.2 \ \mu m^2$ s^{-1} (on the right-hand side of the membrane). Now, the distribution of quadrat mean D_{lat} values were no longer normally distributed (Fig. 5B(v)). "Slow" regions showed anomalous diffusion with downward curvature, because molecules were confined within the modelled "lipid rafts", whereas "fast" regions had linear MSD *vs.* dT plots.

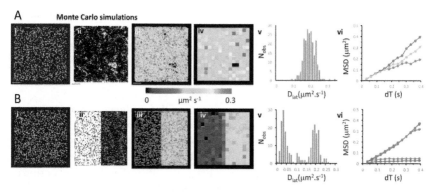

Fig. 5 Monte Carlo simulation of single, fluorescently-tagged molecules moving at the plasma. (A(i)) Single image from the beginning of the simulated video recording shows individual fluorescent molecules (initial density 0.8 molecules per μm^2). The simulated molecules are free to move in an unrestricted manner. (ii) Trajectory map consisting of 4434 trajectories. Fluorophore intensity, noise levels and photobleaching rate (0.2 s^{-1}) were chosen to closely mimic our real data sets. (iii) D_{lat} map. (iv) Quadrat sampled D_{lat} map, 227 quadrats; 2 × 2 μm^2, black squares contain <5 trajectories rejected from further analysis. (v) Histogram of quadrat D_{lat} values, is normally distributed and consistent with a single population. (vi) MSD vs. dT plots for typical fast (where D_{lat} > mean +1.5 × SD; red), intermediate (where D_{lat} ≈ mean; green) and slow (where D_{lat} < mean −1.5 × SD; blue) quadrat regions. (B) As for Panel A except: (i) the left-hand side of the simulation molecules are confined to lipid rafts; right-hand side molecules are free to move in an unrestricted manner, (ii) trajectory map, (iii) D_{lat} map, comprises two distinct regions. (iv) Quadrat sampled D_{lat} map. (v) Histogram of quadrat D_{lat} values, is now clearly bi-modal and not consistent with a single population. (vi) MSD vs. dT plots for typical fast (where D_{lat} > mean +1.5 × SD; red) and slow (where D_{lat} < mean −1.5 × SD; blue) quadrat regions. The "slow" (blue curves) regions show anomalous diffusive behaviour because molecules are trapped within modelled rafts of 300 × 300 nm^2 area.

Discussion

In the spirit of Faraday meetings our intention is to provoke discussion on the structure of biological membranes: the classical view of the plasma membrane is that it comprises a fluid-mosaic of lipids, proteins and other amphipathic molecules.[3] This view has been challenged and it is now proposed that there is structural heterogeneity arising from lipid de-mixing and phase-separation which results in the formation of sub-micron sized microdomains or "lipid rafts", enriched with saturated lipids, sphingolipids and cholesterol.[24] Our view of the plasma membrane has therefore changed from a fully-mixed fluid-mosaic sheet to a cholesterol- and protein-stabilised, oil-in-oil, emulsion that resembles a two-dimensional "mayonnaise". The small-size and short-lived, nature of lipid microdomains has led to controversy[25–27] and as technologies have improved, microdomain size has fallen from 1 micrometre to as small as 10 nm diameter. Although phase separation has been demonstrated in model lipid systems[7,28] there are few, direct observations of heterogeneity in live mammalian cells at physiological temperature.[29,30] In a previous study, we found the mobility of M_2 receptors in HL1, CHO-K1 and primary cardiomyocytes showed no evidence of a phase transition over a −5 °C to +45 °C temperature range.[14] Stability of lipid microdomains may be impacted by in-plane, protein–protein and protein–lipid interactions and there may be additional interactions that extend out of the membrane plane onto the intracellular cytoskeleton and/or extracellular matrix. These interactions may act to stabilise raft structures and corral or otherwise interfere with the free diffusion of membrane proteins[10] which may either partition into the raft or the surrounding isotropic lipid, or may "hop" between the two regions.[31]

Notwithstanding the controversy surrounding the existence of lipid rafts and membrane microdomains, their proposed presence is thought to affect both the distribution and mobility of transmembrane proteins, with important consequences in neurobiology, virology, immunology and membrane–peptide interactions. One manifestation of membrane heterogeneity is that diffusive motion of transmembrane proteins becomes anomalous and MSD $vs.$ dT plots are non-linear and show distinct downward inflection, indicating that molecules diffuse rapidly over short time and length scales, but more slowly over longer time intervals and distances.[12,28,32] A significant problem with use of MSD $vs.$ dT plots is that raw data derived from single molecule imaging experiments rarely enables meaningful analysis to be made on a single molecular trajectory.[11,33] In earlier work, we found that muscarinic acetylcholine receptors (a class of membrane-spanning, 7-helix, G-protein coupled receptor) undergoes unrestricted diffusion at the plasma membrane.[34] So, we chose this protein as a prototype probe of membrane viscosity and structural heterogeneity. Here, we have investigated whether molecules diffuse at different speeds in different regions of the plasma membrane and whether in regions where molecules appear to move more slowly they also exhibit anomalous diffusive behaviour. We divided the cell membrane into 1×1 µm^2 or 2×2 µm^2 quadrats and analysed (≥ 5) independent molecular trajectories in each quadrat region so we could test for significant variation between cell membrane regions. We also tested if the mean D_{lat} values estimated at each quadrat were homogeneous and normally distributed, as expected by the central-limit theorem. We then specifically compared MSD $vs.$ dT plots from "slow" and "fast" moving regions to see if the plots were linear or showed evidence of anomalous diffusive behaviour.

Measurement precision to some extent depends on imaging conditions, including the type of cell, choice of membrane protein and the fluorescent tag that has been employed. The density of single molecule trajectories (or "tracks") and the spatial resolution of local D_{lat} maps, depends on the lateral diffusion (D_{lat}) and track duration; which are limited mainly by the fluorophore photobleaching rate. To generate a local D_{lat} estimate we constructed MSD $vs.$ dT plots from single particle tracks that extended for ≥ 10 consecutive video frames.[35] Excitation laser power and imaging rate were optimised to allow molecules to be unambiguously tracked with high spatial and temporal resolution over a sufficient number of video frames. The density of unique single molecule trajectories that were accumulated over the entire imaging period (~ 1 min) varied between samples even though the starting fluorophore density was similar (~ 1 molecule per µm^2). For most specimens a quadrat size of 2×2 µm^2 was used, but some (with high track density) allowed the use of a smaller quadrat size (1×1 µm^2). For randomly moving molecules (free diffusion), a histogram of D_{lat} values obtained by fitting all of the individual single molecule MSD $vs.$ dT plots over the entire cell surface is best described by a gamma distribution.[11,36] However, we show here that a histogram of mean D_{lat} values, obtained by averaging individual tracks within each quadrat sample area, obeys the central limit theorem and shows a normal distribution. Departure from normality implies heterogeneity in the data set at the level of the quadrat sample size; this can be assessed by Kolmogorov–Smirnov analysis (summarised in Table 1).

We found that the diffusive motion of M_2 receptors at the plasma membrane of HL1 cells, primary cardiomyocytes and zebrafish cardiac tissue slices all gave pooled MSD *vs.* dT plots (*i.e.* taking all molecular trajectories) that were linear (Fig. S4†), with no obvious evidence of anomalous diffusion. When trajectories were segmented and subdivided into checkerboards of quadrat samples, the histograms of mean D_{lat} values were found to be normally distributed. Thus, there was also no evidence for heterogeneity in receptor mobility across the plasma membrane of these specimens. When we further analysed the data by examining MSD *vs.* dT plots on quadrat sample trajectories drawn from seemingly "fast" and "slow" moving regions we found the plots have no obvious downward curvature. Together these findings are consistent with variation in individual D_{lat} values, being the result of random sampling of a homogeneous population. So, in these cases, lipid rafts must not cause anomalous diffusive motion and/or must be monodisperse and evenly distributed across the plasma membrane (see Table 1 and Fig. 1–3).

We found three specimens that convincingly exhibited D_{lat} histograms that were not normally distributed; implying that the quadrat samples were not drawn from a simple homogeneous population. We were interested to compare our findings for nicotinic acetylcholine receptors with another recent study[37] in which ion channel motion was tracked in frog embryonic muscle fibres using quantum dot labels and where MSD *vs.* dT plots were essentially linear once immobile objects were removed from the data sets and MSD *vs.* dT plots were essentially linear. However, an increasing deviation from unity gradient for log(MSD) *vs.* log(dT) plots at shorter times and smaller distances, indicated the receptors were diffusing in an anomalous manner that was best fit by an exponential distribution of diffusion coefficients. The authors point out that a distribution of diffusion coefficients is inconsistent both with simple lipid raft or cytoskeletal picket fence model. The same type of effect was seen in another high-resolution, quantum-dot tracking study.[38]

In the current study, we found downward curvature of the MSD *vs.* dT plots when we analysed data collected across the whole CHO-K1 cells transfected with the eGFP-tagged γ-subunit and all other component polypeptides of the nicotinic receptor. This indicates a highly anomalous diffusive behaviour of the ion channel. However, the explanation for our finding is rather simple because a standard-deviation projection of the video data revealed that many of the eGFP-tagged subunits were associated with the ER, and when regions that were rich in ER were examined separately (using our quadrat sampling method), diffusion was slow and highly anomalous. However, regions of the cell where the ER was excluded (*e.g.* beneath the nucleus) the ion channels moved rapidly and exhibited linear MSD *vs.* dT plots. The rate of diffusion in the fast-moving regions was 10× greater than values reported in the earlier study using frog muscle cells.[37]

We show that data averaging between the cells should be applied with caution, or avoided altogether, because variation in viscosity between the cells can be higher than variations within a single cell (see the viscosity map profile values on Fig. 3A(iii)). The D_{lat} quadrat distributions found between two neighbouring myocytes have very different K–S test scores when considered either separately or together (Fig. 3A(iv) and Table 1). Also, membrane heterogeneity within an individual endothelial cell expressing eGFP-tagged M_2 receptors was evidenced by the "non-normal" distribution of local viscosity values (Fig. 2(v) and Table 1). Visual inspection of the quadrat map (Fig. 2(iv)) revealed that mobility in the lower-left region of the cell was significantly lower than for the rest of the cell.

Conclusion

We have presented a method to analyse single fluorophore tracking data, which employs a quadrat sampling technique to partition data into localized maps of transmembrane protein mobility. We have shown that, in many cases, the plasma membrane has a uniform viscosity and homogeneous structure, and quadrat mean values are normally distributed as expected by the central limit theorem. However, and perhaps as expected, proteins that localised to two different membrane systems (nicotinic receptors found at the ER and plasma membrane or CHO-K1 cells) or following incomplete chemical fixation (M_2 receptors in HL1 cell) showed distinct heterogeneity. In one specimen, M_2 receptors in an endothelial cell, we found spatial variation in membrane viscosity across an individual cell, and while the bulk MSD *vs.* dT plot was perfectly linear (Fig. S3B†) the quadrat mean D_{lat} values were not normally distributed (Table 1). Closer inspection of the MSD *vs.* dT plots for "fast" and "slow" diffusing regions showed the difference in mobility was not due to anomalous diffusion (*i.e.* microscopic variation in structure) but more likely to a bulk variation in membrane viscosity (a "raft of rafts") that impacts the macroscopic D_{lat} value but has minimal effect on microscopic, anomalous diffusive motion of individual M_2 receptors.

Conflicts of interest

There are no conflicts to declare.

Acknowledgements

We thank Mr Alan Ling (Mechanical Engineering Workshop, Francis Crick Institute) for manufacturing some of the equipment used in this study, and Dr Qiling Xu (Neural Development Laboratory, Francis Crick Institute) for providing advice on zebrafish preparation. This work was supported by the Francis Crick Institute which receives its core funding from Cancer Research UK (FC001119), the UK Medical Research Council (FC001119), and the Wellcome Trust (FC001119). For the purpose of Open Access, the author has applied a CC BY public copyright licence to any Author Accepted Manuscript version arising from this submission. RL and LC were supported by Leverhulme Trust, project grant RPG-2016-407.

References

1 K. Jacobson, O. G. Mouritsen and R. G. W. Anderson, *Nat. Cell Biol.*, 2007, **9**, 7–14.

2 E. Sezgin, I. Levental, S. Mayor and C. Eggeling, *Nat. Rev. Mol. Cell Biol.*, 2017, **18**, 361–374.

3 S. J. Singer and G. L. Nicolson, *Science*, 1972, **175**, 720–731.

4 H. C. Berg, *Random walks in biology*, Princeton University Press, Chichester, 1993.

5 P. G. Saffman and M. Delbruck, *Proc. Natl. Acad. Sci. U. S. A.*, 1975, **72**, 3111–3113.

6 Y. Gambin, R. Lopez-Esparza, M. Reffay, E. Sierecki, N. S. Gov, M. Genest, R. S. Hodges and W. Urbach, *Proc. Natl. Acad. Sci. U. S. A.*, 2006, **103**, 2098–2102.

7 T. Baumgart, A. T. Hammond, P. Sengupta, S. T. Hess, D. A. Holowka, B. A. Baird and W. W. Webb, *Proc. Natl. Acad. Sci. U. S. A.*, 2007, **104**, 3165–3170.

8 K. Simons and E. Ikonen, *Nature*, 1997, **387**, 569–572.

9 M. Edinin, S. C. Kuo and M. P. Sheetz, *Science*, 1991, **254**, 1379–1382.

10 A. Kusumi, C. Nakada, K. Ritchie, K. Murase, K. Suzuki, H. Murakoshi, R. S. Kasai, J. Kondo and T. Fujiwara, *Annu. Rev. Biophys. Biomol. Struct.*, 2005, **34**, 351–378.

11 H. Qian, M. P. Sheetz and E. L. Elson, *Biophys. J.*, 1991, **60**, 910–921.

12 M. J. Saxton and K. Jacobson, *Annu. Rev. Biophys. Biomol. Struct.*, 1997, **26**, 373–399.

13 D. M. Owen, A. Magenau, D. Williamson and K. Gaus, *BioEssays*, 2012, **34**, 739–747.

14 T. A. Nenasheva, M. Neary, G. I. Mashanov, N. J. M. Birdsall, R. A. Breckenridge and J. E. Molloy, *J. Mol. Cell. Cardiol.*, 2013, **57**, 129–136.

15 G. I. Mashanov, T. A. Nenasheva, T. Mashanova, C. Maclachlan, N. J. M. Birdsall and J. E. Molloy, *J. Gen. Physiol.*, 2020, **153**, e202012657.

16 S. Shashkova and M. C. Leake, *Biosci. Rep.*, 2017, **37**(4), BSR20170031.

17 D. Axelrod, *Methods Cell Biol.*, 1989, **30**, 245–270.

18 T. A. Nenasheva, T. Carter and G. I. Mashanov, *J. Microsc.*, 2012, **246**, 83–88.

19 G. I. Mashanov, D. Tacon, A. E. Knight, M. Peckham and J. E. Molloy, *Methods*, 2003, **29**, 142–152.

20 G. I. Mashanov and J. E. Molloy, *Biophys. J.*, 2007, **92**, 2199–2211.

21 M. J. Crawley, *Statistics: An Introduction Using R*, John Wiley & Sons, Chichester, 2nd edn, 2015.

22 R_Core_Team, *R: A language and environment for statistical computing*, http://www.R-project.org/.

23 G. I. Mashanov, *J. R. Soc. Interface*, 2014, **11**, 20140442.

24 D. Lingwood and K. Simons, *Science*, 2010, **327**, 46–50.

25 S. Munro, *Cell*, 2003, **115**, 377–388.

26 J. F. Hancock, *Nat. Rev. Mol. Cell Biol.*, 2006, **7**, 456–462.

27 B. Nichols, *Nature*, 2005, **436**, 638–639.

28 R. Metzler, J. H. Jeon and A. G. Cherstvy, *Biochim. Biophys. Acta*, 2016, **1858**, 2451–2467.

29 C. Eggeling, C. Ringemann, R. Medda, G. Schwarzmann, K. Sandhoff, S. Polyakova, V. N. Belov, B. Hein, C. von Middendorff, A. Schonle and S. W. Hell, *Nature*, 2009, **457**, 1159–1162.

30 D. M. Owen, D. J. Williamson, A. Magenau and K. Gaus, *Nat. Commun.*, 2012, **3**, 1256.

31 K. Suzuki, K. Ritchie, E. Kajikawa, T. Fujiwara and A. Kusumi, *Biophys. J.*, 2005, **88**, 3659–3680.

32 K. Ritchie, X. Y. Shan, J. Kondo, K. Iwasawa, T. Fujiwara and A. Kusumi, *Biophys. J.*, 2005, **88**, 2266–2277.

33 M. J. Saxton, *Biophys. J.*, 1993, **64**, 1766–1780.

34 J. A. Hern, A. H. Baig, G. I. Mashanov, B. Birdsall, J. E. T. Corrie, S. Lazareno, J. E. Molloy and N. J. M. Birdsall, *Proc. Natl. Acad. Sci. U. S. A.*, 2010, **107**, 2693–2698.

35 X. Michalet, *Phys. Rev. E: Stat., Nonlinear, Soft Matter Phys.*, 2010, **82**, 041914.

36 M. J. Saxton, *Biophys. J.*, 1997, **72**, 1744–1753.

37 W. He, H. Song, Y. Su, L. Geng, B. J. Ackerson, H. B. Peng and P. Tong, *Nat. Commun.*, 2016, **7**, 11701.

38 A. V. Weigel, B. Simon, M. M. Tamkun and D. Krapf, *Proc. Natl. Acad. Sci. U. S. A.*, 2011, **108**, 6438–6443.

Faraday Discussions

DISCUSSIONS

Behaviour and interactions of proteins and peptides with and within membranes; from simple models to cellular membranes: general discussion

Mibel Aguilar, Kareem Al Nahas, Francisco Barrera, Patricia Bassereau, Margarida Bastos, Paul Beales, Burkhard Bechinger, Boyan Bonev, Izabella Brand, Amitabha Chattopadhyay, William DeGrado, Patrick Fuchs, Ana J. Garcia Saez, Bart Hoogenboom, Shobhna Kapoor, Paula Milán Rodríguez, Justin Molloy, Paul O'Shea, Georg Pabst, Sreetama Pal, Amy Rice, Aurelien Roux, John Sanderson, John Seddon, Lukas K. Tamm and Aishwarya Vijayakumar

DOI: 10.1039/d1fd90067f

Aishwarya Vijayakumar opened the discussion of the paper by Shobna Kapoor: Does the charge of the lipids on the membrane influence its interaction with the cytoskeleton?

Shobhna Kapoor responded: Yes, the charge of the lipids on the membrane does influence its interaction with the cytoskeleton. In our work (DOI: 10.1039/d0fd00051e), some mycobacterial lipids had no net charge at physiological conditions and these lipids did not affect the membrane adhesion energy (implying no changes to the membrane–cytoskeletal interactions). However, these did influence the actin morphology re-distribution. Most likely, the electrostatic interactions between bacterial lipids and the host cell membrane/actin network underneath it play a role in the selective recruitment of host proteins and lipids, impacting the membrane–cytoskeletal interactions.

Paula Milán Rodríguez asked: What is the lipid transfer mechanism from the bacterial membrane to the host membrane?

Shobhna Kapoor replied: There are two possible mechanisms of lipid transfer from the bacterial surface to the host cell membrane. First is *via* the transfer of exogenous lipids within extracellular vesicles (EVs) released by the bacteria. The second is *via* direct lipid transfer upon direct contact of the bacteria with the host cell. At the moment, which of these is at play for the transfer of mycobacterial lipids to the host cell is not known. One way to approach this issue is by the use of double labeled bacteria, wherein the bacteria are labeled with a fluorophore and the lipids are also labelled with an orthogonal fluorophore. The absence of the lipid signal in the bystander cells (which have just the bacteria signal) would argue in favor of the EV mediated transfer of lipids from the bacterial surface to the host cell.

Patricia Bassereau commented: Which type of actin-related process is triggered eventually?

Shobhna Kapoor answered: Migration and phagocytosis/endocytosis are among the most applicable actin-related host processes impacted eventually by exogenous bacterial lipids. This has indeed been shown with some mycobacterial lipids. Other processes that could be implicated are autophagosome maturation and formation,[1] wherein actin has been recently shown to play a critical role. Thus, modulation of these processes would work in the favor of the pathogen to either foster its host uptake or enhance its survival by blocking the autophagosome maturation.

1 K. Zientara-Rytter and S. Subramani, *Autophagy*, 2016, **12**, 2512–2515.

Patricia Bassereau remarked: Could it trigger some activation of CDC42?

Shobhna Kapoor replied: There is a possibility for the activation/modulation of the activity of CDC42 GTPase in mycobacterial host cross talk, as CDC42 is implicated in actin nucleation *via* the ARP3–WASP pathway. However, direct proof for the same by mycobacterial lipids has not been shown. We think that global host cell proteomics in the presence of mycobacterial lipids might shed some light on these issues.

Shobhna Kapoor commented: The changes in the cellular actin cytoskeleton are associated with tether force in general and we see the same with our exogenously added mycobacterial lipids. Specifically, we have observed the host actin to be re-distributed between patches, filaments and punta in presence of mycobacterial lipids, impacting tether force distribution. However, the contribution of each actin morphology on the tether force is not known. For this, we would have to move to model systems.

William DeGrado opened a general discussion of the paper by John Sanderson: Have you looked at many histidine containing peptides? Can histidine residues in peptides serve as catalysts of the reaction by forming acyl imidazoles that then transfer to Lys?

John Sanderson answered: We haven't looked at many, but we have some circumstantial evidence that His is acylated so it is likely that there is some transfer onwards. It will be very structure dependent though. When we look at the lipidation of propranolol, which has both a secondary alcohol and a secondary amine which are well placed for intramolecular transfer (the structure is presented below) we find that lipidation on the oxygen forms an ester, but the process stops at that point as long as the O-lipidated propranolol is embedded in the membrane. If we chemically synthesise an authentic sample of *O*-palmitoyl or

O-oleoyl propranolol, if we are not careful the acyl group migrates to the nitrogen before we get the material into the NMR spectrometer – it is very facile, as you would imagine. As the *N*-acyl product is an amide, it is effectively irreversible in these conditions. It therefore appears that propranolol is O-acylated in the membrane by virtue of its penetration depth and orientation favouring this nucleophilic site. The O-acyl product is isolated by virtue of being embedded in the membrane, which prevents the migration from occurring, presumably for steric reasons. It is reasonable to expect that similar arguments would apply for a lipidated peptide, with the caveat that there is more conformational space to explore for a potential acyl group acceptor, so there may be some scenarios where acyl group transfer is possible.

(±)-Propranolol

William DeGrado asked: Do you think it is possible that the fatty acid acyl group might transfer from the acylated imidazole to the hydroxyl of cholesterol?

John Sanderson replied: Yes, that is certainly possible in theory. We have seen lipidated products that we think are lipidated on His, but these have been difficult to characterise. They have the retention time properties expected for a lipidated peptide, but fragment in-source very easily in the mass spectrometer. However, when we look at all the potential lipidation sites, we can rule out many of them apart from His because we have seen them elsewhere (identified by tandem MS approaches). So we have circumstantial evidence that His is acylated, but not concrete proof. As for the transfer to cholesterol, I have looked through old data sets to try to see whether there is any evidence of cholesteryl esters. I have not been able to find any, but then neither the chromatography nor the MS conditions were optimised for cholesteryl esters (lipidated peptides elute earlier than lipids, and cholesteryl esters are liable to in-source fragmentation). It is something we will look for at some point, because we also want to rule out (or in) transfer from a cholesteryl ester to a peptide, which may be something that happens in lipid droplets or lipoproteins. There is some evidence that amyloid peptides can associate with lipoproteins, and many real (*ex vivo*) amyloid deposits contain cholesterol and cholesteryl esters, so this might be a relevant route for them to be lipidated.

Amy Rice asked: Is there evidence of the back transfer of acyl chains from the lipidated peptides to lysolipids or is this transfer mostly irreversible?

John Sanderson responded: It is certainly irreversible for the transfer to amino groups (Lys, N-terminus) as the products are amides. Transfer to Ser is reversible in principle, but we have not examined it to date. I have looked over some of our old data to see whether there is any evidence of *e.g.* acyl group scrambling or formation of cholesterol esters, but I have not found any evidence – although those experiments were not optimised for detecting those products.

Mibel Aguilar remarked: Melittin acts very quickly – and you observed the lipidation over 24 hours. Could this be a reversible reaction which occurs in cells transiently?

John Sanderson responded: The rate is indeed slow relative to the membrane lytic activity. Lipidation could be transient, especially if there is an acylase in the cell that can reverse lipidation. This kind of acylase activity is something that we are actively looking for at the moment. Transient lipidation might also feature in enabling some peptides to cross membrane boundaries more easily.

Francisco Barrera queried: Is it known if the acylation process that you describe occurs in cellular membranes? If that was the case, it would maybe imply that membrane proteins get slowly acylated, in a process that could constitute a sort of "membrane protein aging". However, maybe cellular enzymes (deacylases) could selectively revert this process?

John Sanderson answered: We have not looked at peptides – there are some challenges to doing that – but we have looked at small organic molecules, including the drug propanolol, a beta blocker.[1] Propranolol has two potentially reactive sites – a secondary alcohol and a secondary amine. We conducted initial experiments in simple single-component liposomes and established that there was indeed lipidation of this drug. We synthesised four authentic lipidated propranolol analogues (palmitoyl and oleoyl derivatives at each of the alcohol and the amine groups) and used these to find the best conditions for extracting the lipidated propranolol from the membrane mixture. We then incubated propranolol with liposomes made from commercial liver extracts and verified that we saw a series of lipidated products, and verified that we could extract them. Finally we administered propranolol to liver cells and extracted the cells after 24 h and saw a similar series of lipidated products. The challenges with doing this with peptides like melittin include the potential for degradation by peptidases, but also the problem with handling a large number of controls. We know that melittin lipidation occurs on at least 5 locations, so to repeat the process that we used for propranolol, we would need 10 different synthetic analogues as controls. There are also challenges to proving where the acyl group comes from *in vivo* because you have other good sources of acyl groups, most notably coenzymes (thioesters). We were on the point of doing experiments to try to resolve this when the first lockdown happened. We had loaded cells with isotopically labelled oleate and planned to used CoA synthetase inhibitors to do isotope chasing experiments.

We have also looked at proteins. There are some proteins in locations where you might expect this kind of acyl modification to accumulate because the proteins are not turned over. An example of this is aquaporin 0 (AQP0) in lens fiber cells. Because of the role they perform, these post-mitotic cells have many of their organelles removed (because they scatter light) and proteins such as AQP0 are not turned over – the AQP0 in the nucleus of your lens has been there since birth. When we look at the lipidation of AQP0,[2] we find all the hallmarks of this lipidation process: lipidation occurs at two sites that are proximal to the membrane interface and not predicted as lipidation sites by known consensus sequences. Lipidation is also complete, and there is a series of lipidated products found, with a relative abundance in line with the cell membrane fatty acid composition. So, we think that AQP0 is lipidated by this process, but the issue raised above – potential lipidation by CoA derivatives – has not been ruled out yet. There are other proteins that have caught our attention, including surfactant protein C in lung surfactant – again a location where there may not be means to recycle or correct lipidated proteins. SP-C is lipidated on Cys residues, again with a pattern that is unusual. In this case, the product is mostly palmitoyl – which is expected as DPPC is a major component of lung surfactant.

The typical half-life for a protein is about 100 h, so there is time for acyl modifications to accumulate. So, in principle, the cell may correct for these by recycling the protein, or by reversing the acylation. There are some enzymes, such as some sirtuins for example, that have very broad substrate specificity and may fulfil this role. This hypothesis is developed more fully in the *BioEssays* article.[3] In that article, I have also hypothesised that the lipidation of amyloid peptides is a mechanism to drive these peptides into conformations that are on-pathway for fibril formation. There are some arguments in favour of this: in parallel to our work with melittin, many amyloid peptides adopt amphipathic helices in the membrane; experiments with amyloid peptides are typically done with similar peptide to lipid ratios to ours; the lag phase has kinetics similar to our lipidation kinetics; and lipidation would be expected to be sensitive to changes in the lipid profile and/or peptide sequence, which can account for the spatial and temporal difference in amyloid nucleation.

1 H. M. Britt, C. A. García-Herrero, P. W. Denny, J. A. Mosely and J. M. Sanderson, *Chem. Sci.*, 2019, **10**, 674–680.
2 V. S. Ismail, J. A. Mosely, A. Tapodi, R. A. Quinlan and J. M. Sanderson, *Biochim. Biophys. Acta Biomembr.*, 2016, **1858**, 2763–2768.
3 J. M. Sanderson, *BioEssays*, 2020, **42**, 1900147.

Sreetama Pal asked: Do you expect the melittin lipidation events to mature or coalesce into the peptide acting as a lipid-solubilizing detergent (as reported earlier[1,2]), especially if you keep decreasing the lipid-to-peptide molar ratio?

1 A. Therrien, A. Fournier and M. Lafleur, *J. Phys. Chem. B*, 2016, **120**, 3993–4002.
2 A. Therrien and M. Lfleur, *Biophys. J.*, 2016, **110**, 400–410.

John Sanderson responded: There are two points here. First, we can measure an apparent CMC for the peptide, so there is every reason to believe that it will have detergent-like properties. But at some point you will fall below the CMC if you decrease the concentration, and at that point I would expect the peptide to remain as a peptide anchored in the membrane by the acyl group. It is also worth remembering that the lipidation by-product is a lysolipid, so the situation with regard to detergent activity is a little more complex. Second: our peptide to lipid ratios are quite high for pragmatic reasons. If we use a lower P : L, it becomes harder to characterise the peptide in the excess of lipid. We have, however, run experiments at 1 : 100, below the critical concentration for toroidal formation, and seen a similar lipidation process. I should add that we generally load our sample onto the LC column without any treatments so that we know we are not losing any lipidated material during chemical extraction.

Paul Beales remarked: Could you take a peptide and predict where lipidation might occur or not? Or is this still an empirical observation on a case by case basis?

John Sanderson responded: We found this process because we doing some work with Alison Rodger on the kinetics of melittin binding to liposomes and found that equilibrium was hard to attain.[1] This lipidation emerged from trying to understand that. Prediction is currently challenging, but I have just started a project to examine this. At the moment, the location at the hydrophobic/hydrophilic boundary seems to be critical. This has not generally been picked up for a couple of reasons: first, no-one has been looking for it, and second, you cannot predict the process on the basis of a consensus sequence – it depends on how something sits in the membrane. Another thing we know is that when you lipidate Lys21 or Lys23, the peptide no longer digests with trypsin – so if you were doing a trypsin digest of a membrane protein, you have a transmembrane sequence that is lipidated, making it more hydrophobic, and that will either form a large insoluble fragment, or elute off the end of your LC gradient.

1 A. Damianoglou, A. Rodger, C. Pridmore, T. R. Dafforn, J. A. Mosely, J. M. Sanderson and M. R. Hicks, *Protein Pept. Lett.*, 2010, **17**, 1351–1362.

Sreetama Pal asked: Do you observe any lipid dependence in the lipidation-mediated peptide folding? Since peptide folding during or after membrane adsorption would include favorable hydrogen-bonding interactions of the peptide, membrane, and surrounding water molecules, it might be interesting to specifically explore the role of lipids that are prone to participating in hydrogen bonds (such as PE) or influencing hydration dynamics (such as PG or other negatively charged lipids).

John Sanderson replied: In terms of the folding, that is not something that we have looked at. Our work (DOI: 10.1039/d1fd00030f) used a synthetic peptide obtained commercially with fatty acyl chains at either the N-terminus or the side chain of Lys23. We looked at the folding of this in the absence of membranes, but it would be interesting to know whether the folding changes if the peptide is anchored in the membrane by the fatty acyl group. It is notable, though, that we do see transfer of a second and sometimes a third acyl group to the peptide in some circumstances, and these transfers occur preferentially at the same locations (N-terminus, Lys23), so my guess is that the membrane-associated singly lipidated peptide is still helical. We have looked at the effects of components like PE on lipidation, and generally we see transfer from both PE and PC in the ratio you would expect based on the composition, but with an enhanced rate. In some cases, such as DOPC/DPPS, we have only observed transfer from one component. In PC/PG mixtures, we see transfer from both components.[1]

1 R. H. Dods, J. A. Mosely and J. M. Sanderson, *Org. Biomol. Chem.*, 2012, **10**, 5371–5378.

Paul Beales said: Could there be specific nearest neighbours that would increase the likelihood of an amine being lipidated? Normally, the amine would be in a hydrophilic residue that often sits away from the membrane. For it to sit at the hydrophilic–hydrophobic interface, are you looking for a lysine to be close to a few hydrophobic amino acids? While it is not possible to fully predict, might local residues be used to score a likelihood that a lysine (or other amine residue) might get lipidated? This could then be used in bioinformatics approaches to seek other peptides of potential interest.

John Sanderson replied: It is possible that some neighbouring residues could transfer on the acyl group. So, for example, initial transfer from the lipid to a His or Ser may then lead to subsequent intramolecular transfer to a second residue. As this would be through space you would need to know both the sequence relationship between the two groups, and the structure type (helix *vs.* sheet). It should be possible to predict those, however. Propranolol is an interesting case here, as if we make *O*-acyl propranolol *in vitro*, intramolecular O to N acyl migration occurs before we can analyse it – it is very fast. In membranes, transfer from the lipid to the alcohol occurs, but there is not subsequent transfer to the nitrogen – the intramolecular transfer is inhibited, presumably for steric reasons within the membrane. This highlights that local structural effects, including how the lipidated molecule partitions, are likely to be crucial.

Izabella Brand enquired: May lipopolysaccharides undergo diacylation and trigger acylation of a peptide? Does the acylation of the peptide depend on the pH? Are any amino acids/amino acid sequences particularly sensitive to acylation?

John Sanderson answered: The acyl group transfer process occurs from ester groups on the lipid. So, to use lipid A as an example, in principle there are four ester groups that could serve as the source of an acyl group. It is also possible – again in principle – that an alcoholic group of lipid A is transiently lipidated before passing the acyl group on to a suitable acceptor. Transfer from the amides of lipid A would not be expected to occur. We haven't addressed the pH dependence of the process – we've generally used physiological pH for the bulk medium. Changes in pH could have complex effects on the process. First, the charged ammonium forms of the Lys side chain and the N-terminal amino group are not reactive, so you might expect the rate of transfer to these sites to slow down at a low pH, particularly for the N-terminal amino group, which has a pK_a in bulk solution closer to the physiological pH. However, the effective pH in the membrane interface is not the same as that of the bulk solution, so it is not trivial to understand the effects of changes in the bulk pH on the ionisation state of the peptide and its net orientation in the membrane interface. You might also expect there to be changes in the rate determining step of the process at low and high pH (based on studies of aminolysis and transesterification in bulk solution, mostly by Jencks and Bruice[1,2] in the 1960s to 1980s), so it may well be the case that the overall rate of transfer to all sites (including the serine side chain) changes fundamentally at extremes of pH. The effects of pH are on our list of things to look at, specifically with regard to the reactivity in endosomes, as it looks like some drugs get trapped in these organelles, and lipidation may be a cause, or outcome, of that.

So far, most of the lipidation that we have seen has been to amino groups (N-terminus, Lys side chain) and hydroxyl groups (Ser). We also think that lipidation of the His imidazole occurs, as well as possibly also the guanidinium of Arg. In the latter two cases, the evidence comes from the retention times of the products, combined with a high tendency to fragment in-source in MS analyses, alongside other sequence considerations (*i.e.* we have identified all the other possible products already). We also anticipate that the thiol of Cys will be lipidated, but we have not been able to see that yet.

1 A. C. Satterthwait and W. P. Jencks, *J. Am. Chem. Soc.*, 1974, **96**, 7018–7031.
2 T. C. Bruice and S. M. Felton, *J. Am. Chem. Soc.*, 1969, **91**, 2799–2800.

Paul Beales queried: To follow on from the sensitivity to the lipid composition, does that mean melittin will get lipidated in some cell membranes but not others? Could this have a role in selectivity for different cell types?

John Sanderson replied: For interfacial helices such as melittin, it may be possible to develop methods to predict likelihood based on neighbouring residues and predictions of amphipilicity. These will need to be nuanced though if they are to predict the difference in reactivity according to membrane composition. We have seen variations in melittin reactivity. The reactivity is higher in PC membranes containing cholesterol. In mixtures of PC with PS, PG or PE, reactivity is generally higher than that in PC alone, but there are some quirks. So, for example, in a membrane composed of DOPC and DMPG (4 : 1), we see transfer from both lipids. In DOPC/DPPS (4 : 1), we only see transfer from the oleoyl component.[1] This could lead to selectivity in different cell types according to differences in the lipid composition, or even in one cell type under different stress conditions. We have proposed this as one method by which amyloid peptides may be lipidated under specific circumstances if the membrane composition permits it.

1 R. H. Dods, J. A. Mosely and J. M. Sanderson, *Org. Biomol. Chem.*, 2012, **10**, 5371–5378.

Georg Pabst asked: Can you comment on the role of Mel partitioning in different membrane systems? Your experiments (DOI: 10.1039/d1fd00030f) used lipid concentrations (50 μM) where this should be important. In other word, have you repeated the experiments at different lipid concentrations? For example, you could go to high lipid concentrations (∼ > a few mM) and repeat the experiments at the same peptide to lipid ratio. Under such conditions, effects from differential partitioning between your lipid systems should be insignificant (all Mel should be bound). Possibly, this could even speed up peptide lipidation in the time range, and maybe then this effect becomes relevant for the biological activity.

John Sanderson answered: We have looked at the effect of different peptide to lipid ratios, but not the effect of absolute concentrations. However, the data we have for cholesterol are very interesting.[1] Melittin binds to POPC/cholesterol with a lower affinity than POPC, yet the rate of lipidation increases, alongside a change in selectivity away from N-terminal lipidation towards lipidation on the side chain of Lys23. This is ultimately because the "on" rate is not rate-limiting in the process, even when the binding is weak. I do agree though that increasing the absolute concentration will lead to a high percentage of the peptide being bound, which will simplify analysis of the kinetics.

1 H. M. Britt, J. A. Mosely and J. M. Sanderson, *Phys. Chem. Chem. Phys.*, 2019, **21**, 631–640.

Boyan Bonev enquired: Does the lipidation of melittin benefit from PLA release from lipids?

John Sanderson responded: Probably. It is certainly the case that any formation of a lysolipid by PLA in an otherwise perfect membrane will promote initial lipidation. We looked at bee venom melittin, which is contaminated with PLA2, and found that there was significant reactivity with all components of the membrane. The lysolipid formed by PLA2 dominated though, as the rate of the phospholipase reaction is much faster than that of the lipidation reaction.

Lukas K. Tamm said: It is well known that peptides such as melittin change orientation in the membrane depending on their concentration or the degree of hydration of the lipid bilayer. Could these effects also influence the degree of acylation as a result of the position and orientation of the peptide in the membrane?

John Sanderson responded: Yes, indeed that may be the case. To date, we have not systematically studied the effects of hydration and concentration. Our concentrations tend to be defined by what we can study analytically, so if our peptide to lipid ratio drops much below 1 : 100 the analysis becomes more challenging because the peptide signals are swamped by the lipid. We do not perform any extraction on the peptide–lipid mixtures to avoid components (*i.e.* lipidated peptides) being lost during the process. The untreated mixture is loaded onto the LC column. However, you make a good point and we will look at the effects of absolute concentration in the future, which will affect the bound melittin ratios. There is some empirical evidence that hydration levels alter lipidation, as samples prepared at low hydration between glass slides for solid state NMR appear to be stable when refrigerated for long periods, whereas preparations with liposomes in excess water are less stable.

Sreetama Pal commented: Regarding the influence of cholesterol in the membrane interaction of melittin behavior, there have been some studies showing that although melittin binding to membranes decreases in the presence of cholesterol, cholesterol preferentially interacts with the tryptophan residue of melittin.[1,2] This could be one reason behind your observations that melittin lipidation in the presence of PC/Chol membranes is higher than that in PC membranes.

Having said that, I would expect any effect of cholesterol on melittin lipidation to not be dependent solely on the binding/affinity parameters, since cholesterol is also known to influence a range of local and global membrane physical properties and there could be a combinatorial effect of these factors.

1 H. Raghuraman and A. Chattopadhyay, *Biophys. J.*, 2004, **87**, 2419–2432.
2 P. Wessman, A. A. Strömstedt, M. Malmsten and K. Edwards, *Biophys. J.*, 2008, **95**, 4324–4336.

John Sanderson responded: Indeed, it is known that cholesterol decreases the affinity of the peptide for the membrane and changes the penetration depth of the peptide – there are red edge excitation shift data that demonstrate this, which we cited in our paper.[1] The penetration depth, and probably also orientation, are responsible for the shift in selectivity away from the N-terminus towards Lys23. The enhanced rate is most likely a consequence of increased water penetration to the carbonyl region where the reaction occurs. Ultimately, several things are contributing to the process – some are mechanistic (physical organic chemistry) relating to water involvement in the rate determining step; others relate to the biophysics of peptide absorption into the membrane interface – the preferred depth and orientation *etc.* Binding affinity is not really a factor, exemplified by the finding that even low molecular weight organic molecules with little membrane affinity can be lipidated.[2]

1 H. M. Britt, J. A. Mosely and J. M. Sanderson, *Phys. Chem. Chem. Phys.*, 2019, **21**, 631–640.
2 H. M. Britt, A. S. Prakash, S. Appleby, J. A. Mosely and J. M. Sanderson, *Sci. Adv.*, 2020, **6**, eaaz8598.

Boyan Bonev asked: Perhaps the proton chemical shift may turn in something from carbon 2.

John Sanderson replied: Yes, that is not something that we have considered, but there should be a significant difference between ester *vs.* amide *vs.* acid. It is known that the ^{13}C shift of the free fatty acid varies with pH, which is a tool we could use to examine the pH close to the interface. I am unsure whether the shift at the 2-position also has pH dependent shifts.

Paul Beales commented: Some venoms have phospholipases and membrane active peptides. Perhaps they work in synergy in some cases?

John Sanderson replied: This may be the case. Most of our melittin work has been done with synthetic melittin for this reason, but in this paper (DOI: 10.1039/d1fd00030f) we did look at bee venom melittin. We saw the lipidation we expected, but also significantly raised quantities of lysolipid, mostly because the hydrolysis by phospholipase A is much faster than the lipidation process.

Boyan Bonev commented: Yes, this is an interesting conclusion and tangent from your study.

John Sanderson said: Magainin II and PGLa also have this lipidation activity,[1] but there is no evidence that it is related to their antimicrobial activity.

1 R. H. Dods, B. Bechinger, J. A. Mosely and J. M. Sanderson, *J. Mol. Biol.*, 2013, **425**, 4379–4387.

Margarida Bastos commented: Thanks a lot, for the answer and reference, which I already downloaded and read. The lipidation is a interesting aspect.

Bart Hoogenboom opened a general discussion of the papers by Boyan Bonev: I am not sure if I have fully understood this – do you have vesicles with PC on the inner leaflet and LPS on the outer leaflet? Could you elaborate a bit more on how you prepare these and on how you ensure the asymmetry? Apologies if I have misunderstood this.

Boyan Bonev responded: No, the LUVs are symmetric but LPS is presented as a polymyxin target on the outer leaflet only.

Patricia Bassereau asked: Do resistant strains have different LPS? Would it be possible to target them with other drugs that use other LPS?

Boyan Bonev answered: Resistance is a very complex and multifaceted process. LPS adaptations can contribute to resistance, specifically *via* the alkylation or Etn-ation of LPS phosphates or pyrophosphates. An antibiotic combination approach or the use of adjuvants is certainly a very appropriate way forward.

Burkhard Bechinger commented: The spectra shown in Fig. 3 of the paper (DOI: 10.1039/d1fd00036e) show the ^{31}P powder patterns of DMPC lipids. In the presence of polymyxin, the discontinuities/'edges' at about 30 ppm and −15 ppm are less steep and thereby less well defined; is there heterogeneity in the lipid population, problems with ^{1}H decoupling or what else could explain the difference in the ^{31}P NMR spectra in the absence of the peptide?

Boyan Bonev replied: Wideline ^{31}P NMR spectra from fluid lipid membranes are inhomogeneous statistical sums of contributions from the effective CSA (CSAeff) of lipid phosphates, offset by the statistically weighted contribution from molecular populations with different orientations with respect to the external magnetic field. In the fluid phase, the magnitude of this CSAeff is determined by the residual anisotropy after the full ^{31}P CSA (*ca.* 200 ppm) is averaged by the fast (GHz) axial rotation of the lipid molecules to approximately 40–45 ppm, as seen for DMPC. Clean spectral outliers are observed, as the lateral molecular librations within the bilayer are slow compared to the NMR timescale (100s of MHz).

In the presence of a large molar fraction of membrane-perturbing compounds, such as LPS or polymyxin, and particularly in the presence of both, lipid packing can be disrupted, which increases the librational freedom, frequency and magnitude. As a result of this, the overall CSAeff is reduced significantly. In addition, the individual contributions to the inhomogeneous distribution are broadened homogeneously due to angular excursions caused by molecular librations (rocking) that introduce an additional partial averaging mechanism onto the axial lipid rotation. The superposition of such significantly broader homogeneous lines leads to rounding of the 0 and 90° "edges" of the powder distribution. This enhanced mobility on the NMR timescale can also partially interfere with proton decoupling. This latter effect, however, is not significant for phosphates, in which proton coupling is weak as there are no protons directly bonded to the phosphorus atom, but all proton bonding is at least two bonds away and mediated by oxygen atoms.

Kareem Al Nahas enquired: Would the use of LPS free anionic membranes (PC : PG, 3 : 1) instead of zwitterionic (only PC) yield membranolytic activity similar to those of polymyxin B and LPS membranes? In other words, is the observed membrane activity of polymyxin B specific to the presence of LPS or is the enhanced activity a result of general electrostatic interactions?

Boyan Bonev responded: There is no difference between polymyxin B-induced leakage in PG vesicles with or without LPS. The presence of the PG charge overwhelms the system and sequesters most of the polymyxin in non-specific interactions, which masks the LPS-mediated ones. This, however, is not the composition of the outer leaflet of bacterial outer membranes, which consists almost entirely of LPS, and for this reason the PG/LPS model is inappropriate. The use of a zwitterionic lipid allows the teasing out of the LPS-specific effects.

Bart Hoogenboom asked: Can you relate your findings to recent evidence[1] that *E. coli* may develop resistance to polymyxin B by modification of the inner membrane, pretty much ignoring what happens at the outer membrane (and hence to LPS)?

1 G. Benn, I. V. Mikheyeva, P. G. Inns, J. C. Forster, N. Ojkic, C. Bortolini, M. G. Ryadnov, C. Kleanthous, T. J. Silhavy and B. W. Hoogenboom, *Proc. Natl. Acad. Sci. U. S. A.*, 2021, **118**(44), e2112237118, DOI:10.1073/pnas.2112237118.

Boyan Bonev replied: There are two obstacles to breaching the Gram-negative envelope, the OM and the IM. Each contributes independently to the MIC for a particular antibiotic. While we focus on the difference between the OM and IM, stress-related adaptations in the IM contribute to an increase in the MIC. That also depends on the growth phase point.

Bart Hoogenboom remarked: To clarify my previous question about resistance: clearly polymyxin disrupts the outer membrane, but I wonder if you can say something about the differences by which it disrupts the outer and inner membranes, with possible ramifications for resistance.

Boyan Bonev responded: In our study (DOI: 10.1039/d1fd00036e), we hypothesise that the presence of rLPS in lipid membranes facilitates polymyxin-mediated membrane disruption. In other words, we compare non-specific membrane breaches (no rLPS) to LPS-mediated membrane disruption. So, yes, this is our model of the rLPS-aided differential disruption of outer *vs.* inner membranes. LPS P-Etn-ation interferes with CAP-mediated OM disruption and leads to resistance.

Aishwarya Vijayakumar opened the discussion of the paper by Sreetama Pal: If lysine has a role in hydrophobic mismatching, have the roles of arginine and histidine been studied? Which type of NMR experiment is used to study the hydrophobic mismatching?

Sreetama Pal responded: The incorporation of arginine and histidine into the WALP scaffold has led to the identification of some interesting peptide behavior as a function of both hydrophobic mismatching and pH. For example, arginine–tryptophan interactions have been shown to control helical fraying at the edges of the WALP helix, with important consequences for the stability of transmembrane peptides and proteins.[1] In peptides with a single arginine acting as a membrane interfacial anchor, the introduction of a glutamate near the arginine has been shown to induce multiplicity in peptide conformational states.[2] Similarly, the introduction of histidine residues has been shown to modulate the degeneracy in peptide conformational and orientational states, depending on hydrophobic mismatching and pH.[3] More importantly, the pH responsiveness of WALP analogs has been found to depend on the precise position of histidine residues.[4] It is worth mentioning here that, in spite of decades of research into peptide–membrane interactions, these aspects of peptide behavior remain enigmatic and are therefore worth investigating from the perspective of tryptophan fluorescence (on account of the central role of tryptophans in membrane protein organization, evolution and biology[5,6] and the substantial overlap of fluorescence timescales with a range of membrane-associated phenomena[7]). However, a robust analysis and meaningful interpretation of tryptophan dynamics in the context of such nuanced behavior would require 'basis sets' of tryptophan (fluorescence) signatures from simpler WALP variants, which remain limited.[8] These considerations informed our choice of KWALP and GWALP as suitably minimalistic systems for exploring the consequences of near-neighbor interactions between tryptophan and nonaromatic interfacial amino acids, such as lysine and glycine. The type of NMR experiments used to study hydrophobic mismatch would depend, among other things, on whether protein/peptide or lipid responses to mismatch conditions are being investigated. For example, since hydrophobic mismatching could induce the formation of non-bilayer (such as cubic or reversed hexagonal) lipid phases, lipid responses to mismatching can be tracked using ^{31}P-NMR (based on the fact that the environment of the phosphate in the phospholipid headgroup would be distinct in different membrane phases[9]). More subtle lipid responses, such as changes in the lipid acyl chain order in positive or negative mismatching conditions, could also be identified based on order parameters calculated using ^{2}H-NMR. On the other hand, peptide responses to mismatching range from localized changes in the conformational dynamics of the peptide backbone and side chains to more global changes in the peptide orientation (transmembrane or surface-adsorbed) and even aggregation. These aspects have been explored using a combination of ^{2}H-NMR and other specialized NMR-based methodologies, such as GALA (geometric analysis of labeled alanines[10]) and PISEMA (polarization inversion with spin exchange at magic angle[11]).

1 S. J. Sustich, F. Afrose, D. V. Greathouse and R. E. Koeppe II, *Biochim. Biophys. Acta, Biomembr.*, 2020, **1862**, 183134.
2 J. R. Price, F. Afrose, D. V. Greathouse and R. E. Koeppe II, *ACS Omega*, 2021, **6**, 20611–20618.
3 F. Afrose, A. N. Martfeld, D. V. Greathouse and R. E. Koeppe II, *Biochim. Biophys. Acta, Biomembr.* 2021, **1863**, 183501.
4 F. Afrose and R. E. Koeppe II, *Biomolecules*, 2020, **10**, 273.
5 D. A. Kelkar and A. Chattopadhyay, *J. Biosci.*, 2006, **31**, 297–302.
6 R. E. Koeppe II, *J. Gen. Physiol.*, 2007, **130**, 223–224.
7 S. Pal and A. Chattopadhyay, in *Membrane Organization and Dynamics*, ed. A. Chattopadhyay, Springer, Heidelberg, 2017, pp. 1–9.
8 S. Pal, R. E. Koeppe II and A. Chattopadhyay, *J. Fluoresc.*, 2018, **28**, 1317–1323.
9 G. Gröbner and P. Williamson, in *Solid-State NMR: Applications in Biomembrane Structure*, ed. F. Separovic and M. A. Sani, IOP Publishing, 2020, pp. 1–30.
10 P. C. A. van der Wel, E. Strandberg, J. A. Killian and R. E. Koeppe II, *Biophys. J.*, 2002, **83**, 1479–1488.

11 V. V. Vostrikov, C. V. Grant, A. E. Daily, S. J. Opella and R. E. Koeppe II, *J. Am. Chem. Soc.*, 2008, **130**, 12584–12585.

Paul O'Shea enquired: Regarding your use of TCSPC to probe indole lifetimes, have you also looked at the polarisation of Trp, as the anisotropies can also be very revealing about the microenvironments? In an earlier paper, for example ref. 1, we were able to show that the Trp environment changed significantly within the interior of a protein using time-resolved anisotropy measurements, I think in your system it would work well.

1 N. Chadborn, J. Briant, A. J. Bain and P. O'Shea, *Biophys. J.*, 1999, **76**, 2198–2207.

Sreetama Pal answered: That's a great point! We did, in fact, measure trends in the fluorescence anisotropy of tryptophan residues in WALP analogs with excitation (Fig. 2c of the paper, DOI: 10.1039/d0fd00065e) and emission (Fig. 3c of the paper) wavelengths. The magnitude of change in the tryptophan fluorescence anisotropy with excitation (Fig. 2d of the paper) or emission (Fig. 3d of the paper) wavelengths did not show any significant differences across the three peptide analogs or in the presence/absence of negatively charged lipids. However, the absolute values of the fluorescence anisotropy for WALP tryptophans (the blue traces in Fig. 2c and 3c of the paper) were markedly lower than those for KWALP/GWALP (the green/maroon traces in Fig. 2c and 3c of the paper). This is possibly due to homo-fluorescence resonance energy transfer (homo-FRET) among the tryptophan pairs present at each end of WALP, but not KWALP/GWALP. Of course, this does not rule out contributions from specific near-neighbor interactions (for example, cation–π interactions among lysine and tryptophan residues in KWALP) in dictating the fluorescence anisotropy values. Time-resolved anisotropy measurements (as suggested) could be expected to provide more spatiotemporally resolved information about these aspects of tryptophan dynamics in WALP analogs.

Burkhard Bechinger asked: In the paper you proposed to use Trp fluorescence for tilt angle determination. However, in other membrane proteins there may be many other factors that influence the local environment of the Trp, including *e.g.* oligomerisation or membrane deformations. Therefore, the comparison with WALP may not be straightforward. In solid-state NMR, we have taken many years to elaborate the technique to determine tilt angles quite accurately, also including wobbling and rocking motions of the peptides (*e.g.* reviewed in ref. 1). In your opinion, where are the limitations of the fluorescence approach?

1 B. Bechinger, J. M. Resende and C. Aisenbrey, *Biophys. Chem.*, 2011, **153**, 115–125.

Sreetama Pal replied: I agree that the analysis of fluorescence readouts, such as the emission maximum, fluorescence anisotropy or fluorescence lifetime, would suffer from ambiguity of interpretation due to the multiplicity of factors that could influence these values. This is especially true for the spectral analysis of tryptophan fluorescence. However, since the red edge excitation shift approach employed here reports specifically on the solvation environment of the fluorophore (and not the fluorophore itself), the magnitude of REES (defined as the red shift in the fluorescence emission maximum on increasing the excitation wavelength) is known to be a very faithful marker of the membrane penetration depth of a fluorophore. Previous work from our group has established this working relationship for a range of fluorophores known to preferentially partition into specific membrane depths.[1,2] These considerations, along with reports on the scaling of tilt angles with positions of tryptophan residues along the helix,[3] led us to propose the use of REES signatures for interfacial tryptophans as an indirect measurement of the WALP tilt angles. This is validated by our observations of a higher magnitude of REES in KWALP/GWALP tryptophans, relative to that in WALP (Fig. 2b of the paper, DOI: 10.1039/d0fd00065e), since the tilt angles of KWALP/GWALP are known to be higher than those in WALP.[4,5] However, your point is well-taken in that the proportionality between the REES signatures of tryptophans and peptide tilt angles must be tested and validated for a wide range of peptides and proteins (including more complex WALP variants) before this could be treated as a universal working rule applicable for all membrane proteins. Regarding the comparative usefulness of NMR and fluorescence methodologies for tilt angle determination, NMR approaches have definitely emerged as one of the most effective. However, to the best of my understanding, resolving the differences in tilt angles estimated by NMR and fluorescence remains a challenge. As suggested elsewhere,[6,7] this could be due to the differential overlap of NMR and fluorescence timescales with that of peptide motion, which could lead to underestimation of the tilt angles by NMR on account of motional averaging. Yet another source contributing to this discrepancy could be the differences in experimental conditions required for NMR and fluorescence, translating to different lipid-to-peptide ratios, which could influence peptide dynamics beyond a threshold value. On the other hand, fluorescence-based approaches could introduce complexities arising from changes in peptide organization and dynamics due to incorporation of extrinsic fluorescent labels (in the absence of tryptophans). In my opinion, one way forward could be to use a judicious combination of NMR and fluorescence approaches to define the lower and upper bounds of the peptide tilt angles, followed by the use of directed analytical and simulation approaches for further refinement.

1 A. Chattopadhyay and S. Mukherjee, *Langmuir*, 1999, **15**, 2142–2148.
2 H. Raghuraman, S. Shrivastava and A. Chattopadhyay, *Biochim. Biophys. Acta, Biomembr.*, 2007, **1768**, 1258–1267.
3 V. V. Vostrikov and R. E. Koeppe, *Biochemistry*, 2011, **50**, 7522–7535.
4 A. E. Daily, D. V. Greathouse, P. C. A. van der Wel and R. E. Koeppe, *Biophys. J.*, 2008, **94**, 480–491.
5 V. V. Vostrikov, C. V. Grant, A. E. Daily, S. J. Opella and R. E. Koeppe, *J. Am. Chem. Soc.*, 2008, **130**, 12584–12585.
6 S. Özdirekcan, C. Etchebest, J. A. Killian and P. F. Fuchs, *J. Am. Chem. Soc.*, 2007, **129**, 15174–15181.
7 A. Holt, R. B. M. Koehorst, T. Rutters-Meijneke, M. H. Gelb, D. T. S. Rijkers, M. A. Hemminga and J. A. Killian, *Biophys. J.*, 2009, **97**, 2258–2266.

John Sanderson remarked: I noted that the intensities for KWALP are lower than WALP. Is there any effect of the neighbouring lysine on the tryptophan excitation or emission?

Sreetama Pal replied: That would have been neat, but we did not find any obvious spectral shape changes in fluorescence excitation spectra for KWALP relative to WALP, although there was a small decrease in fluorescence intensity at the excitation maximum (the left panel in the Fig. 1) and a blue shift in the emission maxima (the right panel in Fig. 1), as also seen in Fig. 2a of the paper

(DOI: 10.1039/d0fd00065e). Interestingly, the fluorescence intensity for both WALP and KWALP tryptophans showed a modest increase in negatively charged POPC/POPG (dashed lines in Fig. 1) membranes relative to that in zwitterionic POPC membranes (solid lines). For GWALP, a reverse trend of decreased fluorescence intensity in negatively charged membranes was observed. The reason for the opposing trend for KWALP and GWALP is not apparent at this point in time, but could indicate that the modulation of the tryptophan microenvironment by interfacial lysines (in KWALP) occurs in a fundamentally different manner than that by interfacial glycines (in GWALP). In addition, it is worth mentioning here that the absorption spectroscopy based protein charge transfer spectra (Pro-CharTS) method could be a good alternative for specifically probing the interaction of charged and aromatic amino acids in peptides, since proximity to tryptophan residues has been reported to reduce ProCharTS absorbance in soluble peptides.[1] Of course, successful adaptation of ProCharTS for membrane-interacting peptides and proteins would first require the (rather non-trivial) development of analytical tools to identify and correct for membrane-induced scattering artifacts in absorption spectra.

1 M. Z. Ansari, A. Kumar, D. Ahari, A. Priyadarshi, P. Lolla, R. Bhandari and R. Swaminathan, *Faraday Discuss.*, 2018, **207**, 91–113.

Patrick Fuchs addressed Sreetama Pal and Burkhard Bechinger: Adding to the conversation of Sreetama and Burkhard: one difficulty is also to have a good model of the motion of the peptide over the time scale of the measurement.

Sreetama Pal replied: Yes, absolutely, and this is where computational approaches can provide some robust insights!

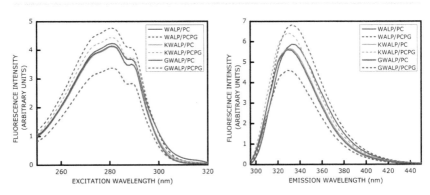

Fig. 1   Fluorescence excitation (left) and emission (right) spectra of WALP (blue), KWALP (green), and GWALP (maroon) tryptophan residues in zwitterionic POPC (solid lines, PC) and negatively charged POPC/POPG (dashed lines, PCPG) membranes. Excitation spectra were acquired with the emission wavelength set to the emission maximum of each peptide observed upon excitation at 280 nm (i.e., 334 nm for WALP and 331 nm for KWALP/GWALP), while emission spectra were collected at an excitation wavelength of 280 nm. The spectra represent averaged traces from at least three independent measurements, with each spectrum recorded in the corrected spectrum mode. All experimental conditions are the same as in Fig. 2a of the paper (DOI: 10.1039/d0fd00065e).

Burkhard Bechinger answered: Indeed, extended sets of complementary solid-state NMR measurements were needed to develop models which describe the tilt and pitch angles as well as the wobbling and rocking motions of helical peptides (*e.g.* for surface oriented helices see ref. 1; for transmembrane helices see ref. 2).

1 M. Michalek, E. S. Salnikov and B. Bechinger, *Biophys. J.*, 2013, **105**, 699–710.
2 E. S. Salnikov, C. Aisenbrey, B. Pokrandt, B. Brügger and B. Bechinger, *Front. Mol. Biosci.*, 2019, **6**, 83.

Burkhard Bechinger said: Apart from motions of the whole peptide and/or changes in the bilayer properties and morphology, the Trp side chain can wobble/rotate around the Cβ–Cγ bond. How does this affect your analysis?

Sreetama Pal replied: That is a good point. These localized tryptophan motions could reflect subtle changes in the fluorophore microenviroment. However, the red edge excitation shift analysis employs steady state fluorescence anisotropy (Fig. 2c of the paper, DOI: 10.1039/d0fd00065e) and therefore, we have no direct handle on the tryptophan side chain dynamics. However, the apparent rotational correlation times calculated for tryptophan (Fig. 4 of the paper, DOI: 10.1039/d0fd00065e) could be more sensitive to tryptophan side chain motions, although deconvoluting those trends to identify the contribution of tryptophan side chain dynamics would require data acquisition in a time-resolved setup.

Patrick Fuchs remarked: Do you use also tilt/azimuthal rotation fluctuations to determine peptide orientation by fluorescence?

Sreetama Pal answered: Great question! Different types of fluorescence spectroscopy and microscopy have been used to estimate the tilt and azimuthal angles of membrane-interacting peptides and proteins. In the case of WALP, incorporation of the BADAN fluorophore at different positions along the helix was employed to construct a calibration plot for emission maximum as a function of the BADAN position, which was then used to calculate the tilt and rotation angles.[1] The fluorescence quenching based parallax approach[2] has been employed to calculate the tilt angle of the bee venom peptide melittin in its membrane-bound form, based on simple geometric considerations of the membrane penetration depth of two distinct fluorescent moieties in the peptide.[3] In addition, FRET efficiencies of membrane proteins have been correlated to changes in azimuthal angles.[4]

1 A. Holt, R. B. M. Koehorst, T. Rutters-Meijenke, M. H. Gelbe, D. T. S. Rijkers, M. A. Hemminga and J. A. Killian, *Biophys. J.*, 2009, **97**, 2258–2266.
2 A. Chattopadhyay and E. London, *Biochemistry*, 1987, **26**, 39–45.
3 S. Haldar, H. Raghuraman and A. Chattopadhyay, *J. Phys. Chem. B*, 2008, **112**, 14075–14082.
4 S. E. D. Webb, D. J. Rolfe, S. R. Needham, S. K. Roberts, D. T. Clarke, C. I. McLachlan, M. P. Hobson and M. L. Martin-Fernandez, *Opt. Express*, 2008, **16**, 20258–20265.

Patricia Bassereau opened a general discussion of the paper by Aurelien Roux: In the case of tissue bending, there are many internal forces involved. It is an active process. The bending of membranes due to lipid shapes that you mentioned occurs at equilibrium. Can you comment and is the analogy justified?

Aurelien Roux answered: I would say that at a short timescale, epithelial cells are responding elastically, which may indicate that they are in a sort of mechanical equilibrium. So in that short timescale, the analogy would be arguable, I think. However, on long time scales, above ten minutes, this may be less relevant, as cells would actively change their shape on that time scale.

John Seddon said: I like your idea of comparing lipid membranes with the monolayers of cells, in terms of their surface properties. It is clear that a layer of epithelial cells can develop a local negative Gaussian curvature. Could this curvature be extended through space to form a lattice of saddle-like structures, similar to an ordered sponge, that could play a role in tissue formation or tissue engineering?

Aurelien Roux answered: This is certainly a good idea. I can surely see that the local lattice of cells can create negative Gaussian curvature, and certainly, the negative Gaussian curvature could then extend, as in the lipid sponge phase, to form a foam like material.

Ana J. Garcia Saez enquired: How do cell/cell adhesion forces between neighbouring cells in the monolayer affect the bulging process? Can this be compared to lipid packing?

Aurelien Roux responded: Dear Ana, thank you for the excellent question. First, adhesion energy is associated with contractility, so cells with higher adhesion energy are also more contractile along the lateral parts of the cells. Thus, they usually are taller and thinner. This can be compared to the lipids with long, saturated acyl chains, which will have a higher density and will make thicker bilayers. The big difference is the dynamics, as cells can change their adhesion/contractility, allowing them to change the overall curvature of the surface (as proposed in Fig. 2 of my article, DOI: 10.1039/d1fd00040c), whereas lipids cannot change shape directly.

Justin Molloy asked: I am interested in whether the direction of the membrane deformation caused by the spiral protein assemblies can occur either way, *e.g.* positive or negative curvature relative to the leaflet to which the protein binds?
More specifically, do you think that the lipid composition of the inner and outer leaflets might affect the probability of the buckling direction or is it totally dominated by the face to which the protein binds?

Aurelien Roux replied: This is an excellent question. The buckling mechanism, at least theoretically, does not predict direction. The probability of breaking symmetry in one direction or another is 50% for both directions. But anything that would make the system asymmetric, including the fact that the protein is binding only one side, or that the lipid composition is different in the two leaflets of the membrane, could bias this 50/50 probability. More work needs to be done on this though.

Paula Milán Rodríguez queried: So your initial information is the shape and then from the shape you try to see which mechanism could provoke it?

Aurelien Roux responded: Exactly, but we need another information, which is force. And the exact mechanism of shape generation can be established from the quantitative relation between the force and the shape.

Paula Milán Rodríguez enquired: About the buckling of surface through growth under confinement: is this mechanism present in living organisms?

Aurelien Roux responded: This is an essential, important question that developmental biologist have tackled for about a century. Rhumbler, in the early 1900s, had developed a purely mechanical model (made of springs, ropes, *etc.*) that reproduced the shape taken by simple embryos, such as the sea anemone, during gastrulation. The theory about buckling by growth pressure arose at that time, and many experimental works associated with computational work showed that shapes of embryos can be reproduced by buckling-based algorithms. The main problem of measuring the force associated with the deformation remained, which is practically impossible *in vivo*. We used an *in vitro* approach to quanti-tatively link the shape of the epithelium and compressive stress to show that growth under confinement can generate sufficient compressive stresses to buckle the epithelium (see ref. 1). I should mention the experimental tour-de-force of Guillaume Charras, who buckled the epithelium using micro-levers, showing that the buckling can be compensated by cell contractility to flatten the epithelium after buckling for compressive rates below 35%.[2]

1 A. Trushko, I. Di Meglio, A. Merzouki, C. Blanch-Mercader, S. Abuhattum, J. Guck, K. Alessandri, P. Nassoy, K. Kruse, B. Chopard and A. Roux, *Dev. Cell*, 2020, **54**, 655–668.
2 T. P. J. Wyatt, J. Fouchard, A. Lisica, N. Khalilgharibi, B. Baum, P. Recho, A. J. Kabla and G. T. Charras, *et al.*, *Nat. Mater.*, 2020, **19**, 109–117.

Paul Beales commented: One difference with lipid bilayers is that they are fluid, whereas cell monolayers usually aren't (excepting metastatic effects!). Could you comment on how this lack of fluidity may give rise to differences between lipid and cell layers in your packing parameter model analogy? Instead, the cells might be better approximated by colloidal packing theories. Does the physics of topological defects in liquid crystals or colloidal monolayers have a role to play in understanding curvature effects in cell monolayers? I am particularly thinking of the work of the likes of David Nelson[2] on packing and defects on curved surfaces and the interactions that occur between the topological defects that must arise in these geometries. Have you considered these models in the context of cell packing and the curvature of cell layers?

Aurelien Roux replied: This is an excellent question. Indeed, it is essential for the material not to be completely fluid in order to have propagation of long range forces in cellular tissues. And yes, many people have used the Nelson description for studying how order and constraints propagates in active matter. We actually show that topologies with charge +1 (spirals, asters, vortices) can concentrate cellular forces and deform cell monolayers into protrusions, resulting in cellular tornadoes (see ref. 1).

1 P. Guillamat, C. Blanch-Mercader, K. Kruse and A. Roux, *bioRxiv*, 2020, 2020.06.02.129262, https://www.biorxiv.org/content/10.1101/2020.06.02.129262v1.
2 M. J. Bowick, D. R. Nelson and A. Travesset, *Phys. Rev. B*, 2000, **62**, 8738.

Paul Beales remarked: Regarding your model of ESCRT filaments as a spring, you predict that on an unsupported membrane they will remodel the membrane by pushing out towards the membrane, but if ESCRT spirals grow on a solid-supported membrane (as in your AFM studies), why do they not buckle away from the membrane due to elastic stress in this case since the solid support prevents them from buckling in the "usual" direction?

Aurelien Roux answered: Again, an excellent question Paul. First, the spring model does not define any particular direction for buckling. But any asymmetry, such as different lipid compositions in the two leaflets, or just the fact that ESCRT-III is binding only on one side of the membrane, may force the system to choose one direction over the other. But in both cases, when the membrane is attached to a solid substrate, the adhesion energy is so high (in particular on mica for AFM, and in clean glass experiments) that it prevents the membrane from any deformation, whatever direction it is. Thus, the spirals stay flat. Recent work from our colleagues (Simon Scheuring's group) imaging the spirals on soft substrates show spontaneous and reversible buckling in the center of the spirals (soon to be published). Also, buckling is aided by the addition of Vps2/Vps24 to the spirals, which increases the rigidity and twisting of filaments (see ref. 1 and 2).

1 J. Moser von Filseck, L. Barberi, N. Talledge, I. E. Johnson, A. Frost, M. Lenz and A. Roux, *Nat. Commun.*, 2020, **11**, 1516.
2 A.-K. Pfitzner, V. Mercier, X. Jiang, J. Moser von Filsek, B. Baum, A. Šarić and A. Roux, *Cell*, 2021, **182**, 1140–1155.

Patricia Bassereau opened a general discussion of the paper by Justin Molloy: You use your tracking system to probe the local viscosity. But, as you know, the probe mobility can be influenced by interactions with the cytoskeleton. There are many membrane components that the receptor could interact with, not just the surrounding lipids. Thus, I was surprised that you claimed to measure primarily the viscosity of the membrane with such experiments.

Justin Molloy responded: We make that claim because the M_1 and M_2 muscarinic acetylcholine receptors that we selected for this study exhibit MSD *vs.* dT plots which are linear and show no evidence of sub-diffusion or super-diffusion. However, other membrane proteins that we[1] and other groups have studied show distinctly non-linear MSD *vs.* dT plots. The experimental work of Kusumi, Jacobson and many others and the theory of Saxton (see the references in our article, DOI: 10.1039/d1fd00035g) indicate that these more complex diffusive behaviors might arise from interactions with cytoskeletal networks or other obstructions to free-diffusion.

1 G. I. Mashanov, M. Nobles, S. C. Harmer, J. E. Molloy and A. Tinker, *J. Biol. Chem.*, 2010, **285**, 3664–3675.

Patricia Bassereau said: People have been using single particle tracking for some time to reveal the organization of cell membranes, but it remains a very challenging task considering their complexity and the interaction with the cytoskeleton. What I find quite exciting in your results is the difference in probe mobility in the plasma membrane of the same cell types, depending if they are primary cells, cells in culture or in a tissue.

Justin Molloy replied: Yes, we find that cell-lines and primary cell-cultures show little cell-to-cell variation within a cell type; however, the differences between a stable cardiac cell-line (HL1) and to primary cardiomyocytes are significant and the differences are greater still between cells embedded in tissue slices compared to those studied in a primary cell-culture. It is curious that even when we see a significant difference between adjacent cells within a tissue slice, our quadrat sampling method indicates that the plasma membrane of a given cell is homogeneous and the variation we see can not be explained by sampling statistics.

John Seddon asked: Is it clear that Monte Carlo is the best approach to simulating membrane protein self-diffusion?

Justin Molloy replied: One great advantage of Monte Carlo (*i.e.* single molecule stochastic models) simulations is that we can generate mock single molecule video datasets and test our analytical tools (*i.e.* test our image analysis methods against a known truth). Another advantage is that we can make the model system as complex or simple as we like. Sophisticated, closed analytical solutions require a lot of brain-power, whereas Monte Carlo simulations just require computing time (usually enough for a quick cup of coffee). Numerical simulations (*e.g.* networks or sets of ODEs) are not so easily adapted for 2D or 3D geometrical systems. So, we prefer Monte Carlo stochastic models, because each simulated molecule can take on its own unique properties, and we can build a physical 2D or 3D framework (or system) of the "cell" and "plasma membrane" and then give the system different viscosities, include cytoskeletal networks or protein binding sites *etc.* Then, we let the simulated molecules meander under thermal force with motional probabilities determined by the local physical environment.

Paul O'Shea enquired: The approach is very interesting, but could you clarify that although you indicate that although you rely on random sampling for this to behave, do you bias the initial starting conditions/point? Monte Carlo assumes a random sample/seed but if it is not random you can get spurious results. Are you able to confirm a random start? The problem is loosely analogous to that of the atomistic modeling of membrane lipids in which a starting condition of 'equilibrium' is desirable.

Justin Molloy responded: Yes, we get your point. The single fluorophore simulations are initialized by randomly seeding "virtual" molecules over the "virtual" membrane. So, they have random starting positions and any state variables are also randomly assigned. We then let the model run for a hundred (or a thousand) cycles so that the "molecules" and any state variables relax towards equilibrium. We then start to output simulation data and either simply watch the system fluctuate about the steady-state or, if we wish, perturb the system and observe it relax to a new steady-state.

Amitabha Chattopadhyay commented: I want to share some general remarks on the measurement of membrane diffusion. One needs to keep in mind that, in membranes, diffusion follows the Saffman–Delbrück model, which means that diffusion is proportional to the mass of the diffusing body in a logarithmic fashion (weak function of mass). Therefore, questions such as receptor dimerization in membranes are difficult to address using membrane diffusion measurements. Having said that, we were able to monitor the activation of G protein-coupled receptors by measuring the diffusion of the receptor using FRAP (since G-proteins are dissociated upon signaling, there is a mass difference).[1] We have also carried out the measurement of GPCR diffusion in membranes using FRAP, z-FCS, and SPT. Although we used the same construct and cell type in these experiments, the diffusion coefficients exhibited some variation due to differences in the sampling time in these measurements. Interestingly, in z-FCS measurements, it is possible to dissect out diffusion modes (such as random diffusion, anomalous diffusion, coralled diffusion) by the application of diffusion laws.[2] In recent times, these measurements have gained spatial resolution by using diffraction-limited microscopy. By using SPT measurements, we could gain further insights on relative distributions of receptors with various modes of diffusion.[3]

For a recent Perspective, see ref. 4.

1 T. J. Pucadyil, S. Kalipatnapu, K. G. Harikumar, N. Rangaraj, S. S. Karnik and A. Chattopadyay, *et al.*, *Biochemistry*, 2004, **43**, 15852–15862.
2 S. Ganguly and A. Chattopadhyay, *Biophys. J.*, 2010, **99**, 1397–1407.
3 S. Shrivastava, P. Sarker, P. Preira, L. Salomé and A. Chattopadhyay, *Biophys. J.*, 2020, **118**, 944–956.
4 P. Sarkar and A. Chattopadhyay, *Phys. Chem. Chem. Phys.*, 2019, **21**, 11554–11563.

Justin Molloy answered: Thank you for raising this Amit and I apologize that we did not do justice to your previous work nor indeed the work of many others in the field in our short manuscript. This is a good opportunity to refer the general reader to your papers and the references therein.

Patricia Bassereau said: We did *in vitro* experiments to show there are limitations to Saffman–Delbrück (SD). We were using single particle tracking to measure the diffusion of voltage-gate potassium channels (KvAP) and of aquaporin 0 in giant liposomes. According to SD, these proteins should have similar diffusion coefficients since they have the same size. We showed that if you change the liposome tension, the diffusion of aquaporin was almost not changed but we observed a reduction by 50% for the lowest tension with KvAP, which is not predicted by SD. We propose that if a protein deforms a membrane (like KvAP), it affects its mobility in the membrane. This might also occur when membrane proteins change conformation, and shape due to some activity. But the question is whether or not the timescale of this change allows us to capture this effect in the diffusion. Experiments have been done *in vitro* to investigate this question.

Justin Molloy commented: If a protein causes local membrane deformation then, in terms of its mobility within the lipid bilayer, can you consider the effect as a (perhaps, tension-dependent) change in the effective radius of the protein *i.e.* otherwise consistent with the Saffman–Delbruck analysis?

Patricia Bassereau responded: This was our first guess that what is effectively diffusing is a "bump" with a size that depends on the membrane tension. But if you evaluate the effect, in particular the intrinsic curvature of the protein, you find that you would need a much higher curvature to explain our results, than the value that we had measured with separate experiments.[1]

1 F. Quemeneur, J. K. Sigurdsson, M. Renner, P. J. Atzberger, P. Bassereau and D. Lacoste, *Proc. Natl. Acad. Sci. U. S. A.*, 2004, **111**, 5083–5087.

Paul O'Shea said: Justin, in order to disentangle the various structures that may underlie the different diffusion processes of large membrane components (receptors), if you look at purely membrane surface indicators that can float *versus* a transmembrane receptor protein, the 2D diffusion coefficients of something that is just on the surface would be different in form *vs.* something embedded in the membrane and may offer a way of discriminating between the various diffusion/structural models. Personally, I think all of these structures probably exist in membranes at the same time but with different dynamics.

Justin Molloy responded: Yes, I agree Paul, the lateral diffusion of transmembrane proteins is likely to be very different from that of monotopic membrane proteins. We chose acetylcholine receptors as a model system because we are able to label them in different ways, express them in different cell types and because their molecular structures have been solved. In our earlier work with muscarinic receptors (M_1 and M_2 classes), we found they showed near-perfect Brownian motion in the lipid bilayer (linear MSD *vs.* dT plots). In the current study (DOI: 10.1039/d1fd00035g), we used these proteins as surrogate probes of the membrane bilayer structure; specifically testing if they show the same lateral diffusion constant across different cells types (primaries *vs.* cell lines, *vs.* tissue culture); between neighboring cells and different regions of membrane across an individual cell. Membrane heterogeneity might well exist on length and time scales that are outside our measurement window (30 ms to 2 seconds on membrane regions with a length scale of $\sim$1 μm^2) and indeed our findings would be strengthened if we had an additional, independent measure of membrane structure that we could cross-correlate with our single molecule tracking data. I think Patricia makes a similar point and we agree with both of you, but it is sadly beyond the scope of our current study.

Paul O'Shea said: Justin – you are suggesting that all the anomalous diffusion is driven by molecular motors located in the cytoskeleton? Or are there other mechanisms, perhaps a bulk lipid flow?

Justin Molloy replied: Anomalous diffusion can manifest either as an upward or downward curvature of mean squared displacement *vs.* time interval plots (MSD *vs.* dT) for super-diffusion and sub-diffusion, respectively. Anomalous sub-diffusion usually arises from passive effects, such as obstacles (molecular crowding), cages (cytoskeletal or extracellular matrix networks) or transient confinement (stochastic binding/unbinding) which all tend to impede the path of a free Brownian walk (which would show a perfectly linear MSD *vs.* dT plot).

Conversely, anomalous super-diffusion requires energy input, arising from *e.g.* convective flow, systematic drift or the action of molecular motors. To some extent, one can distinguish super-diffusion mechanisms by looking at the autocorrelation and cross-correlation of molecular trajectories: convective flow and systematic drift both tend to show strong cross-correlation between different molecules (*i.e.* all (or many) molecules will "drift" in the same direction), whereas the direction of super-diffusion that is driven by individual molecular motors is usually uncorrelated between molecular paths; unless cytoskeletal tracks are aligned (*e.g.* cytoplasmic streaming in plant and amoeboid cells). Matters become more complicated when short-lived single molecule trajectories are studied, because then one finds that only a minority of molecules show perfectly straight-line MSD *vs.* dT plots because of simple statistical variation and this is easy to show by Monte Carlo simulations.

We have tried to address these sampling (perhaps better termed "under-sampling") problems in our paper (DOI: 10.1039/d1fd00035g).

Conflicts of interest

William DeGrado is an advisor to Innovation Pharmaceuticals, and there are no other conflicts to declare.

Faraday Discussions

The impact of antibacterial peptides on bacterial lipid membranes depends on stage of growth†

Tzong-Hsien Lee,[a] Vinzenz Hofferek,[b] Marc-Antoine Sani, [b]
Frances Separovic, [b] Gavin E. Reid[bc] and Marie-Isabel Aguilar [*a]

Received 4th May 2020, Accepted 24th June 2020
DOI: 10.1039/d0fd00052c

The impact of maculatin 1.1 (Mac1) on the mechanical properties of supported lipid membranes derived from exponential growth phase (EGP) and stationary growth phase (SGP) *E. coli* lipid extracts was analysed by surface plasmon resonance and atomic force microscopy. Each membrane was analysed by quantitative nanomechanical mapping to derive measurements of the modulus, adhesion and deformation in addition to bilayer height. Image analysis revealed the presence of two domains in the EGP membrane differing in height by 0.4 nm. Three distinct domains were observed in the SGP membrane corresponding to 4.2, 4.7 and 5.4 nm in height. Using surface plasmon resonance, Mac1 was observed to bind strongly to both membranes and then disrupt the membranes as evidenced by a sharp drop in baseline. Atomic force microscopy (AFM) topographic analysis revealed the formation of domains of different height and confirmed that membrane destruction was much faster for the SGP derived bilayer. Moreover, Mac1 selectively disrupted the domain with the lowest thickness, which may correspond to a liquid ordered domain. Overall, the results provide insight into the role of lipid domains in the response of bacteria to antimicrobial peptides.

Introduction

Antibiotic resistance continues to evolve and intensify and in the absence of new effective drugs, the emerging global healthcare crisis is undeniable.[1] Antimicrobial peptides (AMPs) are a promising class of antibiotics that generally act by destroying the lipid membrane. However, as seen for all other classes of antibiotics, bacteria have evolved a range of resistance mechanisms to AMPs by changing the lipid structure of their membranes.[2–4] It has been proposed that

[a]Department of Biochemistry and Molecular Biology, Monash University, Clayton, VIC 3800, Australia. E-mail: mibel.aguilar@monash.edu

[b]School of Chemistry, Bio21 Molecular Science and Biotechnology Institute, University of Melbourne, VIC 3010, Australia

[c]Department of Biochemistry and Molecular Biology, University of Melbourne, Parkville, VIC 3010, Australia

† Electronic supplementary information (ESI) available. See DOI: 10.1039/d0fd00052c

AMPs and the emerging AMP-resistance mechanisms have co-evolved to support the delicate host–pathogen balance.[5] Understanding the molecular basis of AMP-resistance mechanisms could, therefore, lead to the development of more sustainable antibiotics. However, while much is known about the genetic changes involved in resistance mechanisms,[6,7] very little is known about how membrane structure influences the ability of microbes to resist AMP action.

There are a number of approaches that have been developed to measure the changes in membrane bilayer structure during the binding of antimicrobial peptides. These include optical biosensors,[8] solid-state nuclear magnetic resonance (ss-NMR)[9] and neutron reflectivity.[10] These studies have revealed multiple stages in bilayer structure and have demonstrated that the mechanism of binding, insertion and disruption is more complex than can be described by simple pore-forming or carpet mechanisms.

We have previously developed optical biosensor techniques to measure the binding to membrane bilayers and also the impact of the peptide binding on bilayer structure.[11,12] These studies demonstrated that the bilayer structure changes substantially throughout the binding event and different peptides cause effects that are reversible (in which the bilayers are able to recover) or result in irreversible damage to the membrane where recovery of the bilayer structure was limited. Overall, these earlier studies clearly demonstrated that lipid membrane structure undergoes significant change and the quantitative analysis of these changes in terms of a range of physical properties can provide significant insight into these complex binding events. We hypothesise that the lipid membrane properties play a central role in bacterial response to AMPs and that changes in bacterial membrane components modulate the potency of AMP binding and membrane disruption.

In this study we have investigated the impact of the antimicrobial peptide maculatin 1.1 (Mac1) on supported membrane bilayers formed from total lipid extracts of *E. coli* BL21 (K-12 MG1655) obtained at different growth phases. We used first mass spectrometry based lipidomics[13] to identify and quantify the lipid species present in lipid extracts from two different growth phases. We then used surface plasmon resonance (SPR) to measure peptide binding and analysed the membrane structural properties using atomic force microscopy (AFM) with quantitative nanomechanical mapping to monitor *in situ* changes in membrane bilayer properties during the interaction of the AMPs with each membrane. The results showed the formation of specific domains of different height and 'packing', and that Mac1 selectively disrupts the lowest (or thinnest) domain, clearly demonstrating the important role that membrane lipid composition and structure plays in the defence of bacteria against AMPs.

Results

Exponential growth phase (EGP) and stationary growth phase (SGP) *E. coli* lipid extracts

Bacteria modify the lipid composition of their membranes during growth.[14] The aim of this study was to determine the effects of these different compositions on the physical properties of the membranes and the subsequent binding of AMPs.

Semi-quantitative mass spectrometry analysis was used to compare the levels of phosphatidylethanolamine (PE), phosphatidylglycerol (PG) and phosphatidic

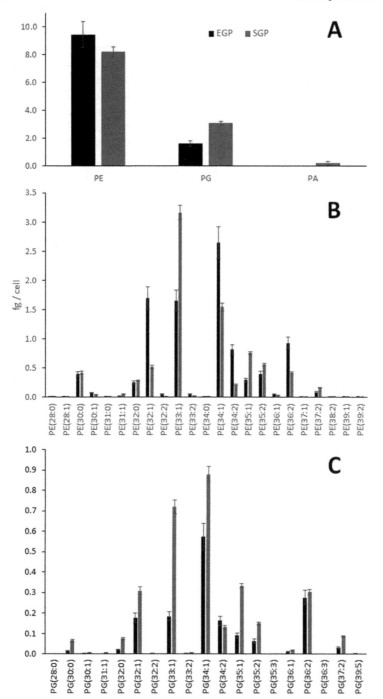

Fig. 1 (A) Identified lipid from EGP and SGP *E. coli* total lipid extracts. (B) Comparison of phosphatidylethanolamine (PE) lipid species within EGP and SGP *E. coli* lipid extracts. (C) Comparison of phosphatidylglycerol (PG) lipid species within EGP and SGP *E. coli* lipid extracts. The nomenclature used is exemplified by dimyristoylphosphatidylglycerol (DMPG) which corresponds to PG(28:0) and palmitoyloleoylphosphatidylglycerol (POPG) corresponds to PG(34:1).

acid (PA) lipids between the SGP and EGP *E. coli* total lipid extracts and the results are shown in Fig. 1. The nomenclature used in Fig. 1 is exemplified by dimyristoylPG (DMPG, (*i.e.*, PG(14:0/14:0))) which corresponds to PG(28:0) and palmitoyloleoylPG (POPG, (*i.e.*, PG(16:0/18:1))) which corresponds to PG(34:1). The PE lipid class was found to be significantly decreased, while the PG and PA lipid classes were significantly increased ($p < 0.02$) (see Fig. 1A). PE lipids are zwitterionic and PG and PA lipids are anionic.[15] Therefore, the change from PE to PG or PA in SGP *E. coli* is suggestive of a change in membrane charge. The lipid composition was also found to change at the individual lipid species level. Notably, odd-carbon numbered (predominantly cyclopropyl-containing)[16] species were found to increase in both the PE and PG lipid class in SGP *E. coli* compared to EGP, with a corresponding decrease in even-carbon numbered lipids (Fig. 1B and C). This resulted in a change in the ratio of odd-chain to even-chain containing lipids within the EGP and SGP *E. coli* samples for the PE lipid class from 0.37 to 1.37 (*i.e.*, a 3.7 fold increase), and for the PG lipid class from to 0.3 to 0.73 (*i.e.*, a 2.4 fold increase).

AFM analysis of supported lipid bilayers (SLBs) from *E. coli* lipid extracts

AFM has been widely used to image the surface of membranes and has provided insight into the nanoscale changes in the topography of bilayer formation. More recently, nanomechanical mapping using AFM allows the measurement of different material properties such as adhesion, modulus, dissipation and deformation.[17] In this study, we have applied this technique to the nanomechanical mapping of *E. coli* lipid extracts to determine the bilayer properties as a function of growth phase.

Liposomes of 100 nm diameter were prepared and deposited onto fresh mica and then imaged by AFM. Fig. 2 shows characteristic images of the deposited membranes from the EGP (A) and SGP (B) lipid extracts. For each lipid extract, clear domains of different height were observed. Fig. 2A shows the presence of two different domains in the EGP membrane which differed in height by approximately 0.4 nm (Table 1) and designated Domain 1 (corresponding to 4.1 nm in height) and Domain 2 (corresponding to height of 4.5 nm). In comparison, Fig. 2B shows the presence of three domains in the SGP extract designated Domain 1 (4.2 nm in height), Domain 2 (4.7 nm height) and Domain 3 (5.4 nm height) (Table 1). Fig. 2C and D show the height distribution for each SLB, and show a similar distribution of Domains 1 and 2 for both the EGP and SGP membranes, while Domain 3 represents a small proportion of the total membrane surface.

The co-existence of different domains in lipid bilayers deposited onto mica has been observed in a number of studies and most likely represents ordered regions of different lipid composition and fluidity.[18–21] We, therefore, performed quantitative nanomechanical mapping (QNM) analysis of each membrane bilayer to derive measurements of the modulus, adhesion and deformation, and the values for the EGP and SGP lipid extract bilayers are listed in Table 1. The modulus relates to the stiffness of the bilayer, the adhesion refers to the adhesion force between the tip and the surface and the deformation is defined as the penetration of the tip into the surface at the peak force. For the EGP membrane, while there was a significant difference between the bilayer heights for Domains 1 and 2, the

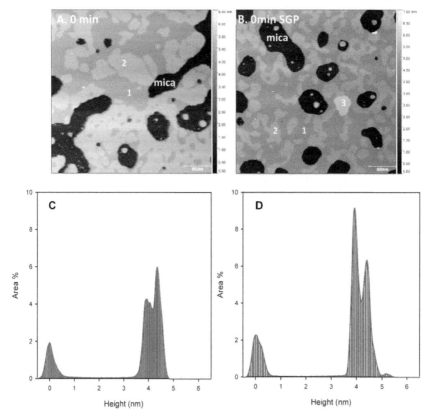

Fig. 2 Topographical analysis of SLBs formed from EGP and SGP lipid extracts. AFM image of (A) EGP and (B) SGP *E. coli* extracts. Images show the presence of Domains 1 and 2 in (A) and Domains 1, 2 and 3 in (B). The height distribution of SLBs for membrane lipid extract from *E. coli* at (A) the exponential growth phase (EGP) and (B) the stationary growth phase (SGP). (C) The peaks centred at 4.1 and 4.5 nm are the height (thickness) for Domain 1 and Domain 2, respectively. (D) The peaks centred at 4.2, 4.7 and 5.4 nm are the height (thickness) for Domain 1, Domain 2 and Domain 3, respectively.

values for the modulus, adhesion and deformation were similar for both domains. The corresponding data for the SGP membrane also revealed similar values for the modulus, adhesion and deformation for Domains 1 and 2, which exhibited different bilayer heights. However, Domain 3 exhibited significantly

Table 1 Physical properties of membrane bilayers prepared from each *E. coli* lipid extract

Property	Exponential Growth Phase (EGP)		Stationary Growth Phase (SGP)		
	Domain 1	Domain 2	Domain 1	Domain 2	Domain 3
Height (nm)	4.1 ± 0.0	4.5 ± 0.0	4.2 ± 0.1	4.7 ± 0.1	5.4 ± 0.0
Modulus (MPa)	11.4 ± 5.5	15.6 ± 6.2	15.3 ± 9.8	27.7 ± 4.3	68.4 ± 27.6
Adhesion (pN)	13.2 ± 8.3	22.5 ± 9.1	12.2 ± 4.4	6.7 ± 2.6	0.9 ± 1.6
Deformation (nm)	3.2 ± 2.4	2.9 ± 1.2	2.2 ± 0.4	1.9 ± 0.2	1.1 ± 0.2

different properties. In addition to an increased height of 5.4 nm, the modulus higher and the adhesion and deformation were lower than the corresponding values for Domains 1 and 2.

The interaction of Mac1 on membrane bilayers of *E. coli* exponential growth phase (EGP) lipid extracts

SPR analysis of Mac1 binding. The interaction of Mac1 with the bilayers formed from each *E. coli* extract was investigated using SPR. Fig. 3 shows the sensorgrams for the binding of Mac1 to (A) EGP and (B) stationary growth phase (SGP) *E. coli* lipid extracts between 0–20 µM peptide. On the EGP membrane, there was a concentration-dependent increase in binding between 0–5 µM. However, at 10 µM and 20 µM, there was an initial large increase in response units followed by a drop in response units (RU) during the association phase, which most likely corresponds to removal of some of the membrane from the chip surface. This result has been previously observed for AMPs with a range of lipids.[19,22,23] A similar phenomenon was observed on the SGP membrane (Fig. 3B), except that the decrease in RU was evident at much lower concentrations, beginning at 2.5 µM Mac1. These results indicate that the SGP membrane is more susceptible to disruption by the binding of Mac1, and suggests that the increase in anionic PG and cyclopropyl-containing lipids influences bilayer structure and the ability of a membrane to resist AMP-mediated disruption. Based on the lipidomic analysis results comparing EGP and SGP membranes, where an increase in anionic PG lipids and an increase in odd-chain, cyclopropane-containing lipids were observed in the SGP, this suggests that a cause for the increased susceptibility of the membrane to Mac1 directly results from alterations in both the charge and 'packing' or fluidity of the membrane bilayer.

AFM analysis of the binding of Mac1 to SLBs formed from *E. coli* lipid extracts. The interaction of Mac1 with the *E. coli* EGP membrane surface was then studied by injecting a 10 µm solution. Fig. 4 contains images corresponding to specific times after injection of Mac1 and a video depicting the changes over time is included in the ESI (Video S1†). In Fig. 4B it can be seen that regions corresponding to Domain 1 are impacted after 21 min, as evident from the appearance of small dark brown patches corresponding to bare mica. In addition, a number of small holes are evident in the regions corresponding to Domain 1. At 68 min

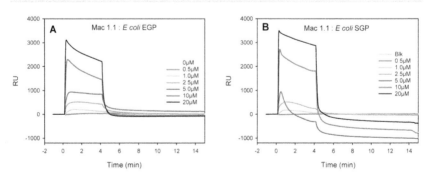

Fig. 3 SPR analysis of Mac1 binding to SLBs formed from *E. coli* lipid extracts at (A) exponential growth phase and (B) stationary growth phase.

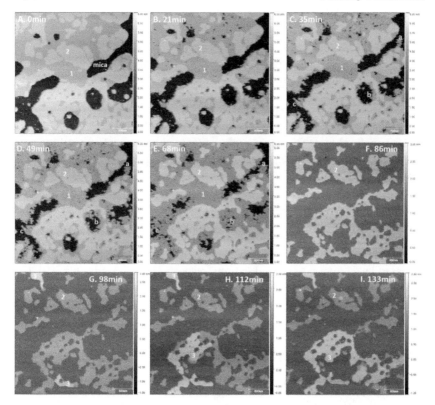

Fig. 4 AFM images of SLBs formed from *E. coli* EGP lipid extracts in the presence of 20 μM Mac1. Heights of each domain and edges labelled (a–c) are listed in Table 2.

Table 2 Height values (nm) derived from AFM images for the binding of Mac1 to the SLBs formed from EGP lipid extracts[a]

Time (min)	Height (nm)			Height difference (nm)	
	Domain 1	Domain 2	Domain 3	$\Delta(2-1)$	$\Delta(3-2)$
0	4.1 ± 0.0	4.5 ± 0.0	—	0.5 ± 0.1	—
21	4.1 ± 0.1	4.6 ± 0.1	—	0.5 ± 0.1	—
35	4.0 ± 0.1	4.6 ± 0.1	—	0.6 ± 0.1	—
49	3.9 ± 0.1	4.5 ± 0.1	—	0.6 ± 0.2	—
68	3.5 ± 0.1	4.4 ± 0.1	—	0.9 ± 0.1	—
86	—	1.1 ± 0.1	—	—	—
98	—	1.1 ± 0.1	1.6 ± 0.1	—	0.5 ± 0.1
112	—	1.0 ± 0.1	1.6 ± 0.0	—	0.6 ± 0.1
133	—	1.0 ± 0.0	1.5 ± 0.0	—	0.5 ± 0.0

[a] The measured height for Domain 1 and Domain 2 from 0 to 86 min using bare mica as reference. The lipid molecules spread and fully cover all mica surface as shown at 86 min. The height measured from 86 min onward becomes the height difference to Domain 1 which has increased from 0.5 nm to 1.0 nm. The heights of newly formed edges (a, b, & c) observed at 35 min (C) are 3.8, 3.8, & 4.0 nm, respectively. There is a new area 3 formed from 98 min (G) onward and this area 3 expands in size as shown in (H) & (I).

(Fig. 4E), the height of Domain 1 decreased by approximately 0.5 nm to 3.5 nm (Table 2), and by 86 min after the addition of Mac1, Domain 1 had disappeared. At the same time, Domain 2 decreased to a height of 1.0–1.1 nm, while Domain 3 appears to persist with a height of 1.5 nm. In summary, the addition of Mac1 caused the dissolution of Domain 1, the contents of which appear to spread onto the bare mica, resulting in a layer of unstructured lipid, while Domains 2 and 3 are more resistant to disruption.

The impact of Mac1 on membranes prepared from the *E. coli* SGP lipid extract was also analysed by AFM. Height profiles obtained before and after addition of 20 μM Mac1 are shown in Fig. 5 and a video depicting the changes over time is included in the ESI (Video S2†). At time = 0 min (Fig. 5A – no Mac1), three domains of different height were observed corresponding to 4.2 nm (Domain 1), 4.7 nm (Domain 2) and 5.4 nm (Domain 3) (Table 3). 25 min after the addition of Mac1 (Fig. 5B), the edges surrounding the bare mica started to become diffuse. After 30 min (Fig. 5C), the mica areas contained significant amounts of material, presumably largely dispersed lipid and the heights of Domains 1, 2 and 3 decreased by 0.5 nm, 0.4 nm and 0.3 nm, respectively. By 35 min (Fig. 5D), the bare mica regions had almost completely disappeared and by 40 min (Fig. 5E), Domains 2 and 3 were 1.0 nm and 1.5 nm higher than the lipid filled regions. In

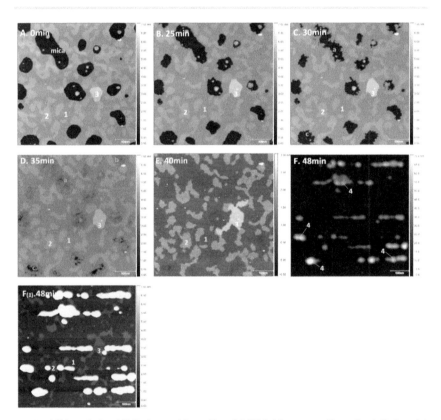

Fig. 5 AFM analysis of SLBs formed from *E. coli* SGP lipid extracts. Domains 1, 2, 3 and 4 and bare mica are indicated. Heights of each domain and edges labelled a–c are listed in Table 3.

Table 3 Height values (nm) derived from AFM images for the binding of Mac1 to the SLBs formed from SGP lipid extract

Time (min)	Height (nm)				Height difference (nm)			Height (nm)		
	Domain 1	Domain 2	Domain 3	Domain 4	$\Delta(2-1)$	$\Delta(3-1)$	$\Delta(3-2)$	Edge a	Edge b	Edge c
0	4.2 ± 0.1	4.7 ± 0.1	5.4 ± 0.0	—	0.5 ± 0.1	1.2 ± 0.1	0.7 ± 0.1	—	—	—
25	4.2 ± 0.0	4.7 ± 0.1	5.4 ± 0.1	—	0.5 ± 0.1	1.2 ± 0.0	0.7 ± 0.1	—	—	—
30	3.7 ± 0.0	4.3 ± 0.0	5.1 ± 0.0	—	0.6 ± 0.1	1.4 ± 0.0	0.8 ± 0.1	3.4 ± 0.2	3.3 ± 0.3	2.9 ± 0.2
35	3.8 ± 0.1	4.4 ± 0.1	5.0 ± 0.1	—	0.6 ± 0.2	1.3 ± 0.1	0.7 ± 0.1	2.9 ± 0.1	2.8 ± 0.1	2.7 ± 0.1
40	—	1.0 ± 0.0	1.5 ± 0.0	—	a	b	0.5 ± 0.1	—	—	—
48	—	1.2 ± 0.1	1.4 ± 0.1	9.8~79.6	a	b	0.2 ± 0.1	—	—	—

[a] The same value as the height of Domain 2. [b] The same value as the height of Domain 3.

addition to the changes in Domains 1, 2 and 3, Fig. 5F and $F_{(1)}$ also reveal the formation of larger spherical clusters 10–70 nm in height, which was not observed for the EGP lipid extract. In addition, the small holes observed in the Domain 1 in the EGP lipid extract were not evident in Domain 1 in the SGP lipid extract.

Overall, Mac1 exerted similar changes in the Domains 1 and 2 structure of both lipid extracts but acted much faster on the SGP lipid extract where significant changes were observed by 35 min compared to 85 min for the EGP lipid extract. Moreover, Mac1 led to the formation of larger spherical particles in the membrane derived from SGP lipid extract.

PF-QNM analysis of *E. coli* EGP lipid SLBs with 20 μM Mac1

PF-QNM analysis was performed on each lipid membrane over the time course of Mac1 binding. Fig. 6 shows the adhesion, deformation and modulus imaged at 0, 49, 86 and 112 min after the addition of Mac1 to the EGP membrane and the corresponding values for each parameter are listed in Table 4. Each image is the

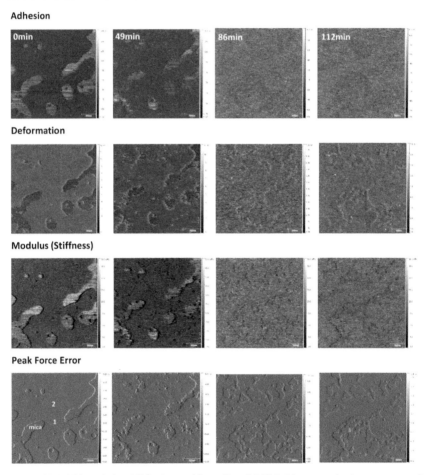

Fig. 6 PF-QNM analysis of SLBs formed from *E. coli* EGP lipid extracts in the presence of 20 μM Mac1. Heights for Domains 1, 2 and 3 are listed in Table 2 and the corresponding QNM parameters are listed in Table 4.

Table 4 Nanomechanical properties of SLBs formed from *E. coli* EGP lipid extracts at various time points after the addition of 20 μM Mac1

	Time (min)			
	0	49	86	112
Adhesion (pN)				
Domain 1	13.2 ± 8.3	2.5 ± 3.0	4.2 ± 5.7	10.7 ± 8.3
Domain 2	22.5 ± 9.1	0.1 ± 3.9	0.2 ± 3.2	5.2 ± 9.7
Domain 3	—	—	—	0.9 ± 5.2
Deformation (nm)				
Domain 1	3.2 ± 2.4	3.2 ± 1.3	3.1 ± 1.6	3.9 ± 1.8
Domain 2	2.9 ± 1.2	2.8 ± 1.2	2.7 ± 1.7	3.8 ± 1.7
Domain 3	—	—	—	4.1 ± 1.7
Modulus (MPa)				
Domain 1	11.4 ± 5.5	11.5 ± 5.6	15.2 ± 7.6	18.2 ± 6.5
Domain 2	15.6 ± 6.2	14.6 ± 7.4	16.9 ± 8.8	15.7 ± 8.5
Domain 3	—	—	—	15.7 ± 9.2

same area scanned in Fig. 6. The adhesion for Domains 1 and 2 decreased significantly at 49 and 86 min after Mac1 addition, and increased somewhat at 112 min while the deformation and modulus for both domains did not change over the 112 min time course. However, given the significant restructuring of the membrane after 49 min shown in Fig. 4, the values for adhesion, deformation and modulus after 49 min represent the properties of a disrupted membrane structure.

The corresponding PF-QNM analysis of the SGP lipid extract derived membrane is shown in Fig. 7 and the data listed in Table 5 at 0, 30, 40 and 48 min after the addition of Mac1. The adhesion for Domains 1 and 2 showed a similar trend to the EGP lipid extract derived membrane, fluctuated slightly and then increased significantly at 48 min. The adhesion for Domain 3 was much lower than for Domains 1 and 2, but it also increased over the time course for Mac1 binding. The deformation for Domains 1, 2 and 3 did not change significantly over 48 min, similar also to the trend observed for the EGP lipid extract derived membrane. The most significant change was observed for the modulus or stiffness values. While the values for Domains 1 and 2 decrease over time, they exhibit a large increase at 48 min. Moreover, the modulus for Domain 3 is much higher than for Domains 1 and 2 and then decreased substantially after 30 and 40 min but then increased to a final value that is similar to that for Domains 1 and 2.

In summary, at $t = 0$ min, membranes derived from EGP lipid extracts are more adhesive and deformable while those from SGP extracts were stiffer. The addition of Mac1 to the EGP and the SGP lipid extract derived membranes caused the destruction of Domain 1 in both membranes, but this occurred much faster on the EGP than the SGP membrane.

Discussion

Topographic analysis of SLBs

The aim of this study was to investigate the influence of changes in the *E. coli* membrane lipid composition at different growth phases on the binding of the antimicrobial peptide Mac1. Lipidomic analysis revealed a significant increase in

Adhesion

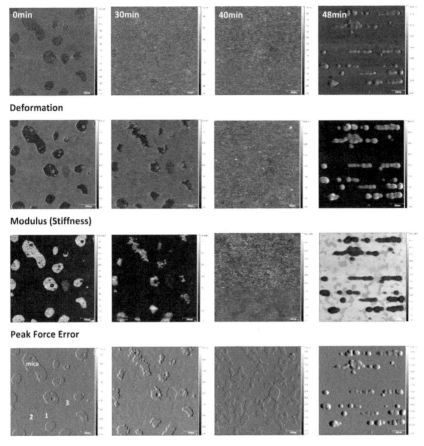

Deformation

Modulus (Stiffness)

Peak Force Error

Fig. 7 PF-QNM analysis of SLBs formed from *E. coli* SGP lipid extracts in the presence of 20 μM Mac1. Heights for Domains 1, 2 and 3 are listed in Table 3 and the corresponding QNM parameters are listed in Table 5.

PG lipids under SGP conditions compared to the EGP, while PE lipids were decreased along with an increase in the ratio of odd-numbered carbon chain length to even-numbered carbon chain length containing lipids in both the PE and PG lipid classes. This is consistent with previous reports in the literature.[14]

Liposomes were then prepared from lipid extracts of each growth phase and deposited onto mica to generate SLBs. Topographic analysis was then performed by AFM. The results revealed the formation of domains of different height. Domains of two different thickness values were observed in the EGP membrane while three different thickness values were found for the SGP membrane. These topographic differences are indicative of separated phases related to molecular packing order and have previously been observed for a wide range of model membranes comprised of different, albeit simple, lipid compositions.[18-21]

This study is amongst the first to report the SPR and AFM analysis of lipids derived from endogenous *E. coli* lipid extracts. Furthermore, to our knowledge, this is the first report of the comparative analysis of lipid extracts from two

Table 5 Nanomechanical properties of SLBs formed from *E. coli* SGP lipid extracts at various time points after the addition of 20 μM Mac1

	Time (min)			
	0	30	40	48
Adhesion (pN)				
Domain 1	12.2 ± 4.4	10.1 ± 4.8	18.3 ± 4.3	16.6 ± 5.7
Domain 2	6.7 ± 2.6	10.1 ± 4.8	18.7 ± 4.6	17.8 ± 6.1
Domain 3	0.9 ± 1.6	1.1 ± 1.8	0.8 ± 2.7	0.7 ± 2.5
Domain 4	—	—	—	459.1 ± 196.5
Deformation (nm)				
Domain 1	2.2 ± 0.4	2.0 ± 0.3	1.6 ± 0.6	1.2 ± 0.8
Domain 2	1.9 ± 0.2	2.0 ± 0.2	1.8 ± 0.7	1.3 ± 0.7
Domain 3	1.1 ± 0.2	0.6 ± 0.2	0.8 ± 0.6	1.2 ± 0.7
Domain 4	—	—	—	19.6 ± 10.2
Modulus (MPa)				
Domain 1	15.3 ± 9.8	8.9 ± 2.6	11.2 ± 6.2	51.3 ± 12.6
Domain 2	27.7 ± 4.3	10.5 ± 3.6	11.2 ± 6.2	51.3 ± 12.6
Domain 3	68.4 ± 27.6	41.6 ± 9.1	39.8 ± 9.5	42.8 ± 8.9
Domain 4	—	—	—	5.6 ± 7.5

different growth phases of *E. coli*. Moreover, the coexistence of three phases for *E. coli* stationary growth phase membranes has not been previously reported.

Comparison of the bilayer thickness values can also provide insight into the packing of each domain. For example, PE lipids have been reported to have higher packing order than PG molecules and mixed PE/PG SLBs showed separate domains corresponding to liquid disordered and gel phase for lower and higher domains, respectively.[20,24–27]

In the present study, it is possible that Domain 1 corresponds to a liquid disordered phase, Domain 2 is a liquid ordered phase while Domain 3 is a gel or liquid ordered (or 'raft') phase. The QNM data revealed no significant difference between the modulus, adhesion and deformation values for Domains 1 and 2 in the EGP and also for Domains 1 and 2 for the SGP membranes. These results, therefore, suggest that the fluidity or packing of the domains may not be related to the modulus or stiffness as measured by QNM.

Maculatin binding to membranes of EGP and SGP lipid extracts

Clear differences were observed in the binding of Mac1 to SLBs of *E. coli* lipid extracts obtained from different growth phases. SPR showed that Mac1 bound more to the SGP derived SLB than the EGP derived SLB at a given concentration and caused loss of the SGP derived lipid SLB at a lower concentration than on the EGP. The corresponding analysis by AFM revealed that Mac1 selectively disrupts Domain 1 in both the EGP and SGP derived SLBs, but disrupts the SGP derived bilayer much faster, consistent with the SPR results. The contents of Domain 1 in both systems appear to spread onto the bare mica while Domain 2 apparently remains intact. At the same time, Domain 3 in the SGP persisted while a new Domain 3 emerged around 100 min in the EGP membrane.

Another difference between the effect on the two membranes is that Mac1 causes the formation of small holes in Domain 1 on the EGP derived bilayer, which grow larger over time, until complete disruption is observed. These small holes were not observed in the SGP derived bilayer, suggesting a different mechanism of binding and disruption to the SGP. A third difference was the appearance of 'blebbing' on the SGP (and not the EGP) at 48 min, which further demonstrates distinctive behaviour of the two lipid extracts.

Taking together the results of this study, we propose the following model as a basis for discussion of the mechanism of Mac1 binding and disruption to both bilayer membranes. The schematic shown in Fig. 8 shows the differences between disordered or fluid acyl chain and gel phase or ordered acyl chain domains with different thickness values. Cyclopropyl-containing lipids have been shown to disrupt lipid packing and increase bilayer fluidity[28] and in the case of the EGP-derived bilayer, Domain 1 is depicted as a liquid disordered state and likely to contain a relatively higher proportion of the cyclopropyl-containing lipids than Domain 2. Following the addition of Mac1, holes appear in Domain 1 but not Domain 2, and the edges of Domain 1 diffuse into the empty mica space. This suggests that Mac1 binds and penetrates the bilayer to form large holes and also causes the spreading of the lipid chains leading to a lateral expansion of the bilayer material.[29] In the final stages of the Mac1 incubation, the pressure of this lateral expansion causes the conversion of Domain 2 into Domain 3 with an increased bilayer thickness and potentially tighter packing.

By comparison, Domain 1 in the SGP bilayer is more rapidly impacted by the addition of Mac1. The higher levels of anionic PG lipids and the cyclopropyl-containing lipids in the SGP lipid extract, therefore, have exerted a dual effect. Increased PG levels mediate a stronger affinity of Mac1 for the membrane[30–32] and increases in both PG lipids and the cyclopropyl-containing lipids further increase the membrane fluidity, again facilitating the binding and insertion of Mac1. While it may be anticipated that the mechanical properties of each domain would exhibit significant differences, this was only apparent prior to Mac1 addition for Domain 3 in SGP. However, the differences in properties for each domain became readily apparent following the addition of Mac1 to each membrane. Overall, the changes in the lipid composition

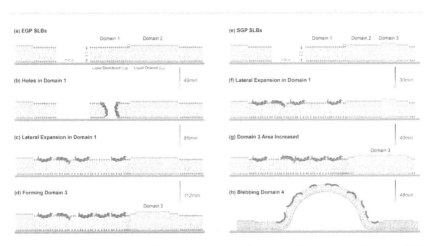

Fig. 8   Schematic proposing differences in the mechanisms of disruption of Mac1 on EGP and SGP membranes in terms of changes in bilayer thickness and lipid packing. Note that the exposed lipid chains would not exist, but more likely to resemble the edges of a bicelle.

dramatically changed the interactive properties of each membrane and what differentiated SGP from EGP was (1) the rate of membrane disruption, (2) the absence of holes, and (3) the appearance of blebbing.

Conclusions

In this study, we have presented a lipidomic analysis of *E. coli* lipids and revealed significant differences between the composition of the exponential phase and stationary phase lipids. We have also demonstrated that these differences in lipid composition influence the properties of the membrane and its ability to respond to the binding of the antimicrobial peptide Mac1. In particular, the changes in the amounts of anionic lipid and cyclopropyl-containing lipids altered the susceptibility and rate of disruption by Mac1. While further detailed experiments will be necessary to dissect the specific composition of each of the domains observed, our results provide significant new insight into the relationship between membrane composition, structure and function. Since bacteria have been shown to modify their membrane composition in response to exposure to different environmental stressors,[2,15] these results have important implications for antimicrobial drug development and the evolution of microbial resistance to membrane-disrupting compounds.

Experimental

Chemicals

4-(2-Hydroxyethyl)-1-piperazineethanesulfonic acid (HEPES), 3-[(3-cholamidopropyl)-dimethylammonio]-1-propanesulfonate (CHAPS), sodium hydroxide, sodium chloride, calcium chloride dihydrate, disodium ethylenediaminetetraacetate dihydrate (EDTA-Na2), sodium dodecylsulfonate (SDS) were purchased from Sigma-Aldrich (St. Louis, USA). HPLC grade solvent, chloroform, methanol and isopropanol, were obtained from Merck (Kilsyth, Australia). Milli-Q water was used to prepare all buffers and regeneration solutions. Maculatin 1.1 (Mac1) (H_2N-GLFGVLAKVAAHVVPAIAEHF-$CONH_2$) was synthesised with Fmoc chemistry. The peptide was purified with semi-prep C18 column (22 mm ID × 30 cm length) using water–acetonitrile gradient. The purity was 99% as determined by analytical C18 HPLC and the mass was confirmed by ESI-MS.

Bacterial growth and harvest

Overnight cultures of *Escherichia coli* (*E. coli*) K-12 MG1655 strain were grown at 37 °C, 250 rpm in lysogeny broth (LB: 1.25 g of NaCl, 1.25 g of yeast extract, 2.5 g of tryptone, 5 g of yeast extract in 250 mL distilled water). Experimental cultures were inoculated with 1% [v/v] of overnight cultures and cultured under the same growth conditions (37 °C, 250 rpm, LB). For exponential growth phase (EGP) and stationary growth phase (SGP) *E. coli*, cells were harvested when an optical density (OD) of 1 at 600 nm ($OD_{600} = 1$) and $OD_{600} = 8$ was reached, respectively.

Harvested cells were washed 3 times with 0.01 M phosphate buffered saline (1 × PBS: NaCl 0.138 M, KCl 0.0027 M, pH 7.4, at 25 °C). The final cell pellets were snap-frozen in liquid nitrogen, dried using a freeze-dryer, then weighed.

Lipid extraction

Lipids were extracted from freeze-dried cell pellets using a monophasic extraction method previously reported by Lydic et al.[33] In addition, a customised lipid standard mix was added prior to lipid extraction for lipid quantitation. For EGP and SGP cell samples, 50 μL and 100 μL customised lipid standard mix were added to an equivalent of 2 mL cell culture ($n = 5$ replicates). The customised lipid standard mix contained 843.9 μM PE(15:0/18:1(d7)), 209.4 μM PG(15:0/18:1(d7)), 87 μM PA(15:0/18:1(d7)), 10.3 μM LPE(18:1(d7)), and 21.5 μM LPG(13:0). Final lipid extracts were dissolved in 500 μL isopropanol : methanol : chloroform 4 : 2 : 1 [v/v/v] containing 0.01% [w/v] butylated hydroxytoluene.

Mass-spectrometry based lipidome analysis

Lipid extracts from the EGP and SGP cell samples were dried and diluted 1 : 10 [v/v] or 1 : 20 [v/v], respectively, in isopropanol : methanol : chloroform 4 : 2 : 1 [v/v/v] containing 20 mM ammonium formate (AF), then introduced to a Orbitrap Fusion Lumos mass spectrometer (Thermo Scientific, San Jose, CA) using a Triversa Nanomate nano electrospray ionization (nESI) source (Advion, Ithaca, NY). The ion source gas pressure was set at 0.3 psi and the spray voltage at 1.1 kV and 1.3 kV in negative and positive modes, respectively. The ion transfer capillary temperature was set at 150 °C, the RF-value at 10%, and an AGC target of 2×10^5. Spectra were recorded in positive and negative ionization modes for 2 minutes each at a mass resolving power of 500 000 (at 200 m/z) over a mass range of 350–1600 m/z.

Lipid identification and semi-quantitative analysis

Lipid identification was achieved using the Lipid Search 5.0a Software (Mitsui Knowledge Industry (MKI), Japan, and Thermo Fisher Scientific, USA), using a parent mass tolerance of 2 ppm, a correlation threshold (%) of 0.3, isotope threshold (%) of 0.1, max isotope number of 2, parent threshold of 50, and recalc intensity threshold of 10. Peak detection was set to profile and merge mode to average. Lipid 'sum composition'[34] assignments were achieved by searching against an accurate mass-based, user-defined database. Lipid nomenclature used is in accordance with the LIPID MAPS consortium proposed naming scheme.[35,36] Semi-quantification of lipid amounts were achieved by comparing the identified lipid ion peak areas to the internal standard of the same lipid subclass,[37] then normalized relative to the number of cells based on their dry weight, as previously described.[38] Additional processing for data visualization and quantification was performed using an in-house developed R-script.

Liposome preparation

The dried membrane lipids of E. coli exponential growth phase (EGP) $OD_{600} = 1$ and stationary growth phase (SGP) $OD_{600} = 8$ were dissolved in isopropanol : methanol : chloroform (volume ratio of 4 : 2 : 1) to 2 mg mL^{-1} as lipid stock solutions. Dried lipid films were prepared by adding the EGP and SGP lipid stock solutions to the bottom of clean, dried glass test tubes. The organic solvent was evaporated with a gentle stream of dried N_2 at 40 °C. The lipid films were further dried under high vacuum for 16 h to completely remove the residual organic solvent. The dried lipid films were flushed with argon gas, sealed and kept at $-78\ °C$ till use. For the liposome preparation, 10 mM HEPES, 150 mM NaCl, pH 7.2 (HBS) was used to hydrate the lipid films at a final concentration of 1 mg mL^{-1} with constant vortex at high speed for 5 min. The resulting liposome suspensions were incubated at 37 °C for 1 hour while shaking at 120 rpm min^{-1}, followed by bath-sonication for 30 min till full transparency. To make the liposome homogeneous in size, the clear liposome–HBS solutions were pass through a polycarbonate membrane (ATA scientific, Lucas Heights, Australia) with 100 nm pore diameter 31 times with AVESTIN Liposofast extruder (Avestin, Canada).

Dynamic light scattering

The hydrodynamic diameter of the liposomes was measured at 25 °C in HEPES buffer using a Zetasizer NanoZS (Malvern Instruments, Worcestershire, UK) collecting the scattering light at an angle of 173° at a wavelength of 633 nm using a 5 mW He–Ne laser. The mean diameter of liposomes to the light intensity was calculated using the Stokes–Einstein relation for spherical particles. The refractive index (RI) of HEPES buffer was 1.3345 and the viscosity of the buffer was 1.003 mPa s at 25 °C. The mean diameters of liposomes for *E. coli* EGP and SGP were 106 ± 12 nm and 103 ± 17 nm, respectively.

SPR analysis

SPR experiments were carried out with a Biacore T100 analytical system on L1 sensor chip (S-series, Biacore, Uppsala, Sweden). Prior to the experiment, the system was cleaned using the 'Desorb and Sanitize' protocol with a maintenance chip using water as bulk. 10 mM HEPES, 150 mM NaCl, pH 7.2 (HBS) was replaced for water as bulk and the chip temperature was controlled at 25 °C. The L1 chip was docked and system was primed with HBS. The chip surface was pre-cleaned twice with 10 μL of 40 mM CHAPS at 5 μL min^{-1} followed by 10 μL of isopropanol containing 50 mM NaOH (v/v = 3 : 2) at 5 μL min^{-1} through all 4 flow cells. The chip surface was equilibrated in HBS running buffer at 10 μL min^{-1} for 20 min till no baseline drift. The liposome solution for SPR analysis was diluted to 0.2 mg mL^{-1} with HBS and a solution of 100 mM CaCl$_2$–HBS was added to reach a final concentration of 2 mM Ca^{2+}. The diluted liposome solutions of each lipid composition in 2 mM CaCl$_2$–HBS were then injected to the chip surface at 2 μL min^{-1} for 60 min. At the end of the liposome injection, the deposited lipid surfaces were pulse-rinsed twice with 30 mM EDTA–HBS at 30 μL min^{-1} for 1 min to remove Ca^{2+} and multi-lamellar structures from the lipid surface and to stabilize the baseline. The responses for each EGP and SGP lipids were about 6500–7000 RU. The interaction of Mac1 with *E. coli* EGP and SGP lipid bilayers was examined at peptide concentrations 0.5, 1, 2, 5, 10, and 20 μM. All binding experiments were carried out at 25 °C. The peptide–HBS solutions of each concentration were injected at 30 μL min^{-1} with a total injection time of 200 s followed by a dissociation time of 400 s. The surface of L1 chip was regenerated by injecting twice with 10 μL of 40 mM CHAPS at 5 μL min^{-1} followed by 10 μL of isopropanol containing 50 mM NaOH (v/v = 3 : 2) at 5 μL min^{-1}. The sensorgrams for lipid deposition and each peptide–membrane interaction were analysed with BIAevaluation 4.0 software (Biacore, GE Health).

Complete kinetic analysis of the peptide–membrane interaction can be performed by numerical integration curve-fitting algorithm for different reaction models. On the basis of the possible membrane binding mechanisms of peptides, a two-state reaction model was chosen for the estimates for kinetic constants including the association (k_a) and dissociation (k_d) rate constants and affinity constants. The resultant sensorgrams for Mac1 to *E. coli* EGP and SGP extracts were fit globally at 0.5–20 μM. The affinity of peptides for each membrane was determined from a series of response curves at different peptide concentrations over each different lipid surface.

Supported lipid bilayer formation

The supported lipid bilayers (SLBs) were prepared *ex situ* with vesicle-fusion methods modified from previously described.[19] Liposome solution (1 mg mL^{-1}) was diluted with HBS to 0.4 mg mL^{-1} and 100 mM CaCl$_2$–HBS was added to reach a final concentration of 4 mM Ca^{2+}. 200 μL of the liposome solution with 4 mM Ca^{2+} was added onto the surface of freshly cleaved muscovite mica (grade V-1, 12 mm diam) (Ted Pella Inc, CA, USA) glued to a parafilm-coated metal disk. The samples were placed in saturated humidity chambers and incubated at 30 °C for 2–3 h in a programmable incubator. The SLBs were carefully and thoroughly rinsed with Ca^{2+}-free HBS. The final SLBs on mica were kept under aqueous environment by adding 200 μL HBS to the surface and allowed to equilibrate to room temperature before imaging.

Atomic force microscopy (AFM)

The topography and mechanical properties of the SLBs were characterised on FastScan Bio AFM (Bruker AXS, CA, USA) using PeakForce Mapping in Fluid in Nanomechanical Mapping. The instrument was controlled by NanoScope 9.1 software. The PeakForce QNM mode measures the adhesion, elastic modulus, deformation and dissipation simultaneously. A triangular ScanAsyst-Fluid+ probe (Bruker, CA, USA) with a nominal tip radius of 2 nm and a nominal spring constant of 0.7 N m^{-1} was used for imaging in fluid condition. To obtain more accurate Young's modulus (E) values, the deflection sensitivity of the probe was calibrated on a sapphire reference sample in PeakForce QNM sample kit (Bruker, CA, USA). A mean deflection sensitivity of ~23.5 nm V^{-1} obtained from 3 measurements was entered manually. The spring constant of ~0.7 N m^{-1} was determined using thermal tuning on simple harmonic oscillator model in fluid. The tip radius was calibrated on a RS Ti roughness sample using Tip Qualification function in NanoScope Analysis software. The SLBs samples were loaded to the scanner and scanned with a droplet method where the probe loaded onto the scanner was pre-wet with 30 μL HBS followed by engaging the sample. The force setpoint was manually maintained at ~500 pN with the feedback gain automatically adjusted by software. The amplitude and frequency of peak force were set at 50 nm and 2 kHz, respectively. Each of the *E. coli* EGP and SGP SLBs without peptide were scanned at 1 kHz with 512-line resolution at discrete area of 3 × 3, 5 × 5 and 10 × 10 μm in size. Prior to the addition of Mac1, the SLBs remain stable for at least 40 min during constant scanning. The real-time changes in the topography and mechanical properties of *E. coli* SLBs EGP and SGP induced by Mac1 were tracked at a scan rate of 2 kHz. The imaging size was about 3 × 3 μm and with a line resolution of 256. The topographic and nanomechanical images were analysed with NanoScope Analysis software and processed in Gwyddion 2.51 software.

Conflicts of interest

There are no conflicts to declare.

Acknowledgements

The authors acknowledge the support of the National Health and Medical Research Council Project grant APP1142750.

References

1 E. Tacconelli, *WHO publishes list of bacteria for which new antibiotics are urgently needed*, World Health Organisation, 2017, 27 February, https://pubmed.ncbi.nlm.nih.gov/29276051/.

2 H.-G. Sahl and Y. Shai, *Biochim. Biophys. Acta, Biomembr.*, 2015, **1848**, 3019–3020.

3 N. N. Mishra and A. S. Bayer, *Antimicrob. Agents Chemother.*, 2013, **57**, 1082–1085.

4 S. Omardien, S. Brul and S. A. Zaat, *Front. Cell Dev. Biol.*, 2016, **4**, 111.

5 A. Peschel and H. G. Sahl, *Nat. Rev. Microbiol.*, 2006, **4**, 529–536.

6 M. Arzanlou, W. C. Chai and H. Venter, *Essays Biochem.*, 2017, **61**, 49–59.

7 R. C. MacLean and A. S. Millan, *Science*, 2019, **365**, 1082–1083.

8 T. H. Lee, D. J. Hirst, K. Kulkarni, M. P. Del Borgo and M. I. Aguilar, *Chem. Rev.*, 2018, **118**, 5392–5487.

9 M. A. Sani and F. Separovic, *J. Magn. Reson.*, 2015, **253**, 138–142.

10 R. D. Barker, L. E. McKinley and S. Titmuss, *Adv. Exp. Med. Biol.*, 2016, **915**, 261–282.

11 T. H. Lee, K. N. Hall and M. I. Aguilar, *Curr. Top. Med. Chem.*, 2016, **16**, 25–39.

12 T. H. Lee, D. J. Hirst and M. I. Aguilar, *Biochim. Biophys. Acta*, 2015, **1848**, 1868–1885.

13 E. Ryan and G. E. Reid, *Acc. Chem. Res.*, 2016, **49**, 1596–1604.

14 J. Gidden, J. Denson, R. Liyanage, D. M. Ivey and J. O. Lay, *Int. J. Mass Spectrom.*, 2009, **283**, 178–184.

15 T. H. Lee, V. Hofferek, F. Separovic, G. E. Reid and M. I. Aguilar, *Curr. Opin. Chem. Biol.*, 2019, **52**, 85–92.

16 M. S. Blevins, D. R. Klein and J. S. Brodbelt, *Anal. Chem.*, 2019, **91**, 6820–6828.

17 A. Mularski and F. Separovic, *Aust. J. Chem.*, 2017, **70**, 130–137.

18 A. Aufderhorst-Roberts, U. Chandra and S. D. Connell, *Biophys. J.*, 2017, **112**, 313–324.

19 K. Hall, T. H. Lee, A. I. Mechler, M. J. Swann and M. I. Aguilar, *Sci. Rep.*, 2015, **4**, 5479.

20 D. Konarzewska, J. Juhaniewicz, A. Guzeloglu and S. Sek, *Biochim. Biophys. Acta, Biomembr.*, 2017, **1859**, 475–483.

21 T. H. Lee, K. N. Hall and M. I. Aguilar, *Aust. J. Chem.*, 2020, **73**, 195–201.

22 J. D. Gehman, F. Luc, K. Hall, T. H. Lee, M. P. Boland, T. L. Pukala, J. H. Bowie, M. I. Aguilar and F. Separovic, *Biochemistry*, 2008, **47**, 8557–8565.

23 T. H. Lee, C. Heng, M. J. Swann, J. D. Gehman, F. Separovic and M. I. Aguilar, *Biochim. Biophys. Acta*, 2010, **1798**, 1977–1986.

24 L. Picas, C. Suarez-Germa, M. Teresa Montero and J. Hernandez-Borrell, *J. Phys. Chem. B*, 2010, **114**, 3543–3549.

25 C. Suarez-Germa, M. T. Montero, J. Ignes-Mullol, J. Hernandez-Borrell and O. Domenech, *J. Phys. Chem. B*, 2011, **115**, 12778–12784.

26 C. Suarez-Germa, O. Domenech, M. T. Montero and J. Hernandez-Borrell, *Biochim. Biophys. Acta*, 2014, **1838**, 842–852.

27 L. Picas, C. Suarez-Germa, M. T. Montero, O. Domenech and J. Hernandez-Borrell, *Langmuir*, 2012, **28**, 701–706.

28 D. Poger and A. E. Mark, *J. Phys. Chem. B*, 2015, **119**, 5487–5495.

29 J. M. Henderson, A. J. Waring, F. Separovic and K. Y. C. Lee, *Biophys. J.*, 2016, **111**, 2176–2189.

30 D. I. Fernandez, A. P. Le Brun, T. H. Lee, P. Bansal, M. I. Aguilar, M. James and F. Separovic, *Eur. Biophys. J.*, 2013, **42**, 47–59.

31 D. I. Fernandez, T. H. Lee, M. A. Sani, M. I. Aguilar and F. Separovic, *Biophys. J.*, 2013, **104**, 1495–1507.

32 M. A. Sani, E. Gagne, J. D. Gehman, T. C. Whitwell and F. Separovic, *Eur. Biophys. J.*, 2014, **43**, 445–450.

33 T. A. Lydic, J. V. Busik and G. E. Reid, *J. Lipid Res.*, 2014, **55**, 1797–1809.

34 G. Liebisch, J. A. Vizcaino, H. Kofeler, M. Trotzmuller, W. J. Griffiths, G. Schmitz, F. Spener and M. J. Wakelam, *J. Lipid Res.*, 2013, **54**, 1523–1530.

35 E. Fahy, S. Subramaniam, H. A. Brown, C. K. Glass, A. H. Merrill Jr, R. C. Murphy, C. R. Raetz, D. W. Russell, Y. Seyama, W. Shaw, T. Shimizu, F. Spener, G. van Meer, M. S. VanNieuwenhze, S. H. White, J. L. Witztum and E. A. Dennis, *J. Lipid Res.*, 2005, **46**, 839–861.

36 E. Fahy, S. Subramaniam, R. C. Murphy, M. Nishijima, C. R. Raetz, T. Shimizu, F. Spener, G. van Meer, M. J. Wakelam and E. A. Dennis, *J. Lipid Res.*, 2009, **50**(suppl), S9–S14.

37 Y. H. Rustam and G. E. Reid, *Anal. Chem.*, 2018, **90**, 374–397.

38 F. C. Neidhardt and H. E. Umbarger, in *Escherichia coli and Salmonella: Cellular and Molecular Biology*, American Society of Microbiology (ASM) Press, 2nd edn, 1996, ch. 3, vol. 1.

Faraday Discussions

PAPER

Antimicrobial peptides: mechanism of action and lipid-mediated synergistic interactions within membranes

Dennis W. Juhl, [a] Elise Glattard, [a] Christopher Aisenbrey [a] and Burkhard Bechinger [*ab]

Received 20th April 2020, Accepted 8th September 2020

DOI: 10.1039/d0fd00041h

Biophysical and structural studies of peptide–lipid interactions, peptide topology and dynamics have changed our view of how antimicrobial peptides insert and interact with membranes. Clearly, both peptides and lipids are highly dynamic, and change and mutually adapt their conformation, membrane penetration and detailed morphology on a local and a global level. As a consequence, peptides and lipids can form a wide variety of supramolecular assemblies in which the more hydrophobic sequences preferentially, but not exclusively, adopt transmembrane alignments and have the potential to form oligomeric structures similar to those suggested by the transmembrane helical bundle model. In contrast, charged amphipathic sequences tend to stay intercalated at the membrane interface. Although the membranes are soft and can adapt, at increasing peptide density they cause pronounced disruptions of the phospholipid fatty acyl packing. At even higher local or global concentrations the peptides cause transient membrane openings, rupture and ultimately lysis. Interestingly, mixtures of peptides such as magainin 2 and PGLa, which are stored and secreted naturally as a cocktail, exhibit considerably enhanced antimicrobial activities when investigated together in antimicrobial assays and also in pore forming experiments applied to biophysical model systems. Our most recent investigations reveal that these peptides do not form stable complexes but act by specific lipid-mediated interactions and the nanoscale properties of phospholipid bilayers.

Introduction

Antimicrobial peptides (AMPs) provide a first line of defense against a multitude of pathogenic microorganisms and can be released rapidly when infections occur.[1,2] They are part of the innate immunity of a wide variety of species from the plant and animal kingdoms, including humans.[3] The corresponding databases

[a]Université de Strasbourg/CNRS, UMR7177, Institut de Chimie, 4, rue Blaise Pascal, 67070 Strasbourg, France. E-mail: bechinge@unistra.fr

[b]Institut Universitaire de France, France

list thousands of sequences and many are continuously added.[4,5] Understanding their mechanism of action allows one to design molecules with favorable properties that mirror the essential characteristics of the template compounds. They have also been shown to modulate the immune response of the host organisms.[6] Although peptides have a short half-life in natural environments they can be developed into therapeutics by modification of their composition, by nanostructure formulations that protect them from proteolysis until they reach their target or by fixation to surfaces.[7–9]

In this paper we focus on linear cationic antimicrobial peptides such as magainins which were first discovered in frogs.[2] These antimicrobial peptides are membrane active and they act by interfering with the barrier function of bacterial lipid bilayers.[10] Because it is more difficult for pathogens to adjust the physicochemical properties of their cell membranes compared to adjusting the sequence of proteinaceous receptors, the development of multi-resistance is less likely.[2,11] This, in an era where multi-resistance is a major problem for human health, bears great promise for the development of new lines of antibiotics.

Structural and mechanistic studies of membrane-associated antimicrobial peptides

Membrane-associated peptides exhibit large conformational and topological freedom. Thus, biophysical investigations reveal that hydrophobic sequences such as alamethicin preferentially, but not exclusively, adopt transmembrane alignments and form pores made of transmembrane helical bundles.[12] In contrast, linear cationic peptides such as magainins, cecropins or LL37 preferentially adopt alignments parallel to the membrane interface and work by different mechanisms that could only be established after decades of research and analysis.[13] Indeed, guidelines for the design of new compounds can be established from the models that arose from the biophysical and structural studies of cationic amphipathic antimicrobial peptides.[14] As a consequence, a number of antimicrobial small molecule mimetics,[15,16] pseudopeptides[17,18] and polymers[19] have been introduced.

Once the potential of linear cationic amphipathic sequences as antimicrobials was described, considerable research efforts were dedicated to understanding in detail their mechanisms of action and the underlying structural prerequisites.[2,20] Upon contact with membranes, linear cationic peptides adopt amphipathic folds, specifically disrupt the integrity of bacterial and fungal membranes,[2] and/or enter into the cell interior[10] where they can interact with and flocculate anionic macromolecules.[18]

Whereas magainins adopt random coil structures in aqueous buffer, they exhibit helical conformations in membrane environments. Membrane association is driven by electrostatic interactions and is reversible. The helices of magainin and other linear cationic antimicrobial peptides are predominantly oriented parallel to the membrane surface.[14] Whereas this surface orientation of magainin 2 has been observed under all conditions investigated,[21,22] PGLa has a more dynamic character. Indeed, in fully saturated membranes PGLa can adopt a broad range of tilt angles, a feature which has also been discussed for mixtures with magainin 2 (cf. below). However, when interacting with phospholipid bilayers carrying unsaturated groups, like magainin 2, the PGLa orientation is stable and close to perfectly parallel to the bilayer surface.[22–24] In contrast, the much more hydrophobic alamethicin preferentially adopts transmembrane helical arrangements, although depending on the conditions orientations along the membrane surface have also been observed.[25,26]

An amphipathic peptide helix that resides in the bilayer interface at an alignment parallel to the membrane surface uses more space at the level of the head group and glycerol regions when compared to the hydrophobic interior.[14] This topology thereby causes substantial disordering at the level of the hydrophobic region concomitant with membrane thinning.[27–29] The bilayer disruption has been estimated to extend over a radius of 50 Å.[30,31] Molecular dynamics (MD) simulations provide a view of possible magainin membrane arrangements, where pores form through stochastic rearrangements of peptides and lipids rather than well-defined channel structures. In some simulations side chains reach the opposite bilayer leaflet of a thinned membrane, thus water filled channels appear.[32,33]

From the ensemble of biophysical data a number of models have emerged including the formation of toroidal pores made of lipid and peptide,[34,35] the presence of a peptide 'carpet' covering the bilayer surface and ultimately leading to membrane disruption,[36] or the formation of pores through randomly arranged micelle-like aggregates within the membrane.[37] With these ideas in mind a common model has been proposed where in the presence of external stimuli such as AMPs, 'Soft Membranes Adapt and Respond, also Transiently' (SMART).[13] Within the SMART model, lipid membranes initially adapt to the disruptive properties of the peptides, but undergo macroscopic phase transitions transiently or permanently, locally or globally when the peptide concentration increases.[13]

Because the lipid physico-chemical properties play an important role in the SMART model, suggestions where peptide-induced changes in the line tension form the underlying mechanisms for antimicrobial activity involve related lines of ideas.[38,39] The membrane physico-chemical properties have also been suggested to be essential for the selectivity of cationic linear peptides for bacterial over eukaryotic cells.[40] Positively charged peptides show an orders-of-magnitude stronger interaction with membranes carrying a negative surface charge (bacteria) than with neutral ones (eukaryotes).[41,42] Furthermore, lipid-mediated mesophase-like arrangements along the membrane surface have been demonstrated to depend on anionic lipids.[43] Finally, the membrane association of multicationic antimicrobial peptides has also been shown to strongly affect a number of peripheral membrane proteins.[44]

Mechanistic investigations on the synergistic interactions of antimicrobial peptides

Magainin 2 and PGLa are part of a naturally occurring cocktail of peptides in the skin of *Xenopus laevis* frogs. They have been investigated individually but when added as a mixture they exhibit a much increased antimicrobial activity when compared to the individual peptides.[45] The synergistic activity of the magainin 2–PGLa mixture is also shown when calcein release from phospholipid liposomes is studied.[46–48] The enhanced activity of the peptide mixture was proposed to be due to a combination of fast pore formation by PGLa and increased pore stability due to magainin 2.[47]

NMR structural investigations also indicate that in the synergistic mixture both bilayer-associated PGLa and magainin are helical and adopt an alignment parallel to the surface of membranes when these carry unsaturated groups,[23,49,50] including supported bilayers made from *E. coli* lipid extracts.[51]

Cross-linking experiments testing membrane-associated PGLa and magainin 2 each carrying a GGC extension indicate the preferential formation of parallel dimers.[52] Of note, fluorescence quenching experiments reveal more densely packed mesophase arrangements of both peptides in the synergistic mixture.[14,51] A reduction in bilayer repeat distance due to the presence of the peptides has also been measured by diffraction methods.[33,53]

A number of studies are suggestive of small favorable interactions of PGLa and magainin 2 when both are membrane-associated,[47,54,55] but these experiments do not reveal if such interactions are due to direct contacts between the peptides, long-range electrostatics or driven by the lipid matrix.[28,56] Notably, synergism has been shown to strongly depend on the lipid head group composition, suggesting important involvement of the lipids.[48,51] Furthermore, mutagenesis experiments point out a role of the F16W, E19Q and carboxy-terminal sites of magainin 2.[47,57] Within PGLa, changing residues G7, G11, and L18 or the positively charged K15 and K19 sites has an effect on the synergistic enhancement.[57]

Models for the synergistic behaviour between PGLa and magainin 2 have been proposed based on activity assays and low-resolution structural methods without being conclusive.[47,51,57] However, high-resolution investigations of the structure, topology, dynamics and interactions between peptides and between peptides and lipids that could clarify how they interact in liquid crystalline membranes are missing.

Solid-state NMR spectroscopy is probably the only method to provide structural and dynamic information at atomistic resolution. To this end peptides which are uniformly labelled with stable NMR isotopes are needed.[58] These can be obtained by bacterial overexpression using a previously described fusion tag where the antimicrobial peptides are neutralized in tight inclusion bodies within the bacterial cells.[59] However, the expression in bacteria prevents carboxy-terminal amidation which normally occurs for PGLa. Moreover, in the system established in our laboratory, cleavage from the fusion protein by formic acid introduces an amino-terminal proline to the existing sequences.[59] Here, using synthesized peptides with modified termini, we investigated how such changes affect the antimicrobial activity, helix propensity and synergistic activities of the peptides. We then investigated how the activities correlate with some of the proposed models, where interactions of the termini have been suggested to be of key importance. Alternative concepts will be presented which shall be discussed during the Faraday Discussion meeting.

Results

To determine the effects of the altered termini, we produced the following seven peptides by solid-phase synthesis and tested them for antimicrobial and synergistic activities: PGLa, pPGLa, pPGLc, magainin 2a, magainin 2c, pmagainin 2a and pmagainin 2c, where p indicates an additional N-terminal proline, a indicates an amide and c indicates a free acid at the C-terminus.

Both PGLa and magainin 2 are known to adopt helical structures in membrane-mimetic and lipid bilayer environments.[60,61] The structural influence of the termini was determined by liquid-state NMR with the peptides dissolved in either aqueous Tris buffer (pH 7.4) or trifluoroethanol (TFE)/buffer (1 : 3 v/v). Assignments were performed based on two-dimensional ^{1}H–^{1}H TOCSY and ^{1}H–^{1}H NOESY spectra while natural abundance ^{1}H–^{13}C HSQC spectra were included to obtain the carbon chemical shifts for secondary structure calculations using the neighbour corrected Structural Propensity Calculator (ncSPC).[62] In good agreement with previous reports,[63,64] both PGLa and magainin 2a adopted random coil conformations in aqueous buffer (Fig. 1B and 2B). Surprisingly, slight β-sheet tendencies for extended segments of their sequences were observed

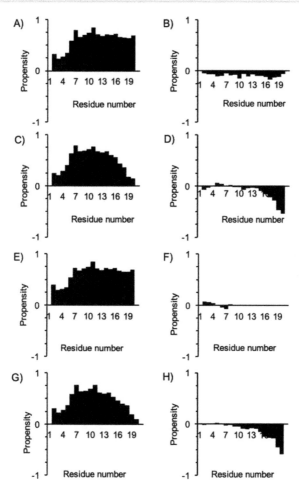

Fig. 1 Helix propensity of PGLa and effects of terminal modifications. Secondary structure propensities determined using assigned chemical shifts and the ncSPC online tool[62] range from pure sheet (−1) to pure helix (+1). Secondary structure propensity for PGLa (A) in TFE/buffer (1 : 3 v/v) and (B) in aqueous buffer. The secondary structure propensities for pPGLc, pPGLa and PGLc in TFE/buffer (1 : 3 v/v) are depicted in (C, E and G), respectively, while the differences in structural propensity when compared with PGLa are shown in (D, F and H), respectively.

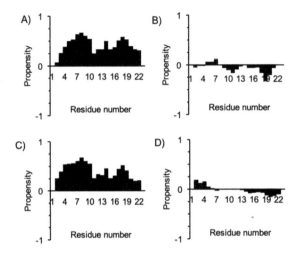

Fig. 2   Helix propensity of magainin 2 and the effect of terminal modifications. Secondary structure propensities determined using assigned chemical shifts and the ncSPC tool[62] range from pure sheet (−1) to pure helix (+1). Secondary structural propensity for magainin 2a (A) in TFE/buffer (1 : 3 v/v) and (B) in aqueous buffer. The secondary structure propensity for pmagainin 2c in TFE/buffer (1 : 3 v/v) is depicted in (C), while the difference in structural propensity when compared with magainin 2a is shown in (D).

for both peptides (Fig. 1B and 2B), suggesting that the propensity of both peptides to form β-sheet amyloids[65] is already apparent under these conditions.

The TFE/buffer mixture serves as a simple membrane mimic with helix inducing properties.[66] All peptide variants showed a dominant helix structure in TFE/buffer (1 : 3 v/v), but variations were observed (Fig. 1 and 2). The N-terminus of PGLa is mostly unstructured, whereas residues 6 to 21 adopt a helical secondary structure in TFE/buffer (1 : 3 v/v) (Fig. 1A) in agreement with a gradual increase in structuration that has been observed early on in solid-state NMR investigations of bilayer-associated PGLa.[61] When evaluating the structure of pPGLc, the N-terminus was mostly unaffected by the initial proline, whereas the carboxylic acid completely destroyed the helical structure at the C-terminus (Fig. 1C). Comparing the structural propensities of the two peptides reveals the degree of the structural changes (Fig. 1D). Evaluation of the secondary structure of pPGLa in TFE/buffer (1 : 3 v/v) confirmed that the proline induced only minor differences at the N-terminus and did not affect the structural propensity of the C-terminus (Fig. 1E and F).

Magainin 2a adopted a helical structure for most of the sequence, with a less structured segment around Lys11 and Phe12 (Fig. 2A). Introducing both the N-terminal proline and the free carboxy terminus did not induce major changes in the secondary structure (Fig. 2C and D). A slight increase in the helix propensity was observed for the N-terminus, whereas a slight decrease was observed for the C-terminus.

For all seven peptides the minimal inhibitory concentration preventing all growth (MIC_{100}) was determined as described earlier.[51] The assay was performed at least twice for each peptide, with four replicas each time. Overall, the data is very consistent as shown by the averages and standard deviations depicted in

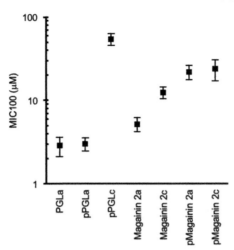

Fig. 3 Antibacterial activity of the magainin and PGLa variants investigated in this paper. The peptide concentration for 100% inhibition of *E. coli* bacterial growth (MIC_{100}) was determined for all peptide variants. The antimicrobial assays were repeated at least twice with four replicas each time. Error bars show the experimental standard deviations.

Fig. 3. For PGLa, we obtained a MIC_{100} of 2.9 ± 0.8 μM in excellent agreement with previously published values.[51] The N-terminal proline did not affect the activity significantly, as we determined a MIC_{100} of 3.0 ± 0.6 μM for pPGLa (Fig. 3). On the other hand, the carboxy-terminus of pPGLc reduced the activity by a factor of 18 when compared to PGLa, as we determined a MIC_{100} of 54 ± 9 μM. Within experimental error, pPGLc prepared by overexpression and by chemical synthesis has the same antibacterial activity (not shown).

For magainin, we found that a carboxy-terminus reduced the activity of magainin 2 by a factor of two as compared to magainin 2a, as the MIC_{100} values were 5.2 ± 1.0 μM and 12 ± 2 μM for magainin 2a and magainin 2c, respectively (Fig. 3).[51] The N-terminal proline had an even higher impact, as the activity of pmagainin 2a was reduced by a factor of four to 22 ± 4 μM. The combination of both an N-terminal proline and a carboxylic C-terminus did not result in any further reduction of the activity, as the MIC_{100} of pmagainin 2c was 24 ± 7 μM.

By combining different variants of PGLa and magainin 2, we could evaluate the effects of the peptide termini on the PGLa–magainin synergism. All combinations of the two peptides resulted in increased activities compared to those expected for purely additive systems (Fig. 4A and B). Variation of the magainin 2 termini did not influence the synergism, as all mixtures of PGLa with any of the magainin 2 variants showed MIC_{100} values of around 1 μM and synergy factors between four and five (Fig. 4A and C). Likewise, the N-terminal proline in pPGLa did not affect the synergistic activity of mixtures with variants of magainin 2 (Fig. 4, table). When combining pPGLc with magainin 2a, we obtained a similar enhancement of the activity as for PGLa–magainin 2a (Fig. 4D), but due to the significantly lower activity of pPGLc, the activity was slightly lower than that of the other mixtures with a MIC_{100} value of 2.1 ± 0.6 μM (Fig. 4B). The biggest enhancement of the antimicrobial activity, but also the lowest overall activity, was obtained when

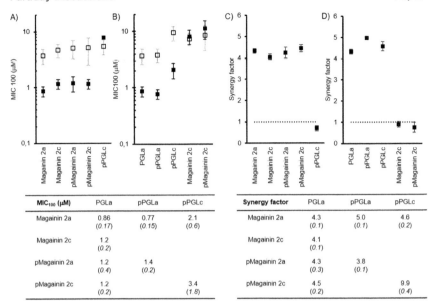

MIC₁₀₀ (µM)	PGLa	pPGLa	pPGLc
Magainin 2a	0.86 (0.17)	0.77 (0.15)	2.1 (0.6)
Magainin 2c	1.2 (0.2)		
pMagainin 2a	1.2 (0.4)	1.4 (0.2)	
pMagainin 2c	1.2 (0.2)		3.4 (1.8)

Synergy factor	PGLa	pPGLa	pPGLc
Magainin 2a	4.3 (0.1)	5.0 (0.1)	4.6 (0.2)
Magainin 2c	4.1 (0.1)		
pMagainin 2a	4.3 (0.3)	3.8 (0.1)	
pMagainin 2c	4.5 (0.2)		9.9 (0.4)

Fig. 4 Antibacterial activity of PGLa and magainin 2 mixtures. (A and B) experimental MIC₁₀₀ values for mixtures of PGLa and magainin 2a (black boxes), respectively, compared to the calculated MIC₁₀₀ values for 1 : 1 mixtures assuming the activities of the individual peptides are additive (open boxes). (C and D) calculated synergy factors for 1 : 1 mixtures containing PGLa and magainin 2a, respectively. The exact MIC₁₀₀ values (left) and calculated synergy factors (right) are listed in the tables together with the standard deviations in brackets. All experiments have been performed at least two times with four replicas each time. Error bars show either the experimental standard deviation or the calculated propagation of error.

pPGLc was mixed with pmagainin 2c. The MIC_{100} was determined to be 3.4 ± 1.8 µM and the synergy factor was calculated to be 9.9 ± 0.4 (Fig. 4, tables).

As a control, we also combined two different variants of PGLa or two different variants of magainin 2. In these cases no change in activity when compared to the expected MIC_{100} values was observed (Fig. 4A and B).

From the above-mentioned experiments, it is evident that the C-terminus plays a role in the antimicrobial activity of the individual peptides. However, the activity assays also revealed that the synergism between PGLa and magainin 2 is preserved for peptides with altered termini.

Discussion

The NMR structural characterization of peptides and proteins associated with lipid bilayers requires the introduction of isotope labels into the polypeptide sequence.[58] Bacterial overexpression remains an affordable and convenient method and offers the possibility to label the protein uniformly or selectively. To facilitate efficient bacterial expression and the purification of the resulting product, in many cases slight alterations are introduced into the native protein sequence. In the overexpression systems that we specifically designed to produce antimicrobial peptides in bacteria, tight inclusion bodies are formed due to the presence of the TAF12 fusion protein.[59] Although this system provides proven high yields of antimicrobial peptides in combination with a one-step purification, the chemical cleavage of the TAF12 fusion tag using formic acid leaves an N-terminal proline and the C-terminus is non-amidated. Because the peptide termini clearly have a role in the overall activity of the peptides, the thus obtained products were characterized here, thereby providing further insight into the mechanism of synergistic interaction of these antimicrobial peptides.

Indeed, the antimicrobial activity of pPGLc is considerably decreased when compared to PGLa (Fig. 3). This could be due to the additional negative charge at the C-terminus or the decreased propensity of this sequence for helical structures (Fig. 1). In the case of PGLa, the C-terminus appears to be more important than the non-structured N-terminus,[61] as the introduction of a proline prior to the native sequence had no effect on the activity of the peptide (Fig. 3). On the other hand, for magainin 2, the helix covers much of the sequence in membrane mimetic environments[60,67] and modifications of either terminus affect the activity of the peptide (Fig. 3). Interestingly, we did not observe an additive effect for alterations of both termini at the same time (Fig. 3).

Notably, membrane association of these peptides has been shown to be strongly dependent on electrostatic interactions and these can be modulated by changing the lipid composition[40,42,64] or the peptide charge.[68] Therefore, the reduced activity of the peptides carrying an additional negative charge at the C-terminus is probably related to a reduction of attractive electrostatic interactions at the bacterial surface.

Interestingly, the activity of the peptide mixtures pretty much follows the activity of the PGLa variant (Fig. 4B) whereas the activity of the magainin 2 variant has little influence (Fig. 4A). Indeed, it was concluded from recent calcein release experiments that PGLa preconditions the insertion of magainin into membranes with intrinsic negative curvature such as mixtures of POPE/POPG† (3/1 mole/mole).[48] In a related manner, fluorescence correlation and fluorescence quenching experiments show that fluorophore labelled PGLa helps magainin to associate with POPE/POPG† membranes by formation of loosely interacting mesophases made of the peptides.[69] Indeed, if such 'helper activities' of PGLa are reduced by alterations of its termini, overall, fewer of the peptides interact with the membrane and PGLa is expected to be the limiting factor determining the overall activity of such mixtures (Fig. 4A and B).

Several structural models explaining the synergism between PGLa and magainin 2 have been proposed. Early on, cross-linking experiments with GGC-extended PGLa and magainin sequences revealed the preferential formation of parallel heterodimers in membranes.[47] Furthermore, in these early investigations the negatively charged magainin E19 residue was found to be important. Based on coarse grain MD simulations Vacha et al. found that salt bridges between this magainin anionic charge and K12 and K15 of PGLa as well as hydrophobic interactions are important.[29] Furthermore, the peptide dimers have been found to align along the membrane surface where they further assemble into tetramers through C-terminal interactions.[33] At high concentrations, peptides of opposing

---

† Lipid abbreviations in this paper: CL: cardiolipin. DMPC: 1,2-dimyristoyl-sn-glycero-3-phosphocholine. DMPG: 1,2-dimyristoyl-sn-glycero-3-phospho-(1'-rac-glycerol). PC: phosphatidylcholine. PE: phosphatidylethanolamine. POPE: 1-palmitoyl-2-oleoyl-sn-glycero-3-phosphorylethanolamine. POPG: 1-palmitoyl-2-oleoyl-sn-glycero-3-phospho-(1'-rac-glycerol).

bilayers interact with each other, causing their adhesion. When sandwiched in between lipid bilayers in this manner, the peptides cause membrane undulations and accumulate in the resulting troughs where they form fibril-like assemblies.[33]

Such C-terminal electrostatic interactions have also been found in previous MD simulations of DMPC and DMPC/DMPG† membranes where the formation of parallel heterodimers was driven by electrostatic interactions between the anionic charges of magainin E19, its C-terminus and the lysines of PGLa at the 12, 15 and 19 positions.[70–72] This heterodimerization was required for PGLa transmembrane insertion which the authors correlated to the synergistic activities of the peptide mixture.[70,71] Thus, upon removal of the magainin negative charges, heterodimer formation is disrupted and the bilayer insertion of PGLa is inhibited.[71] In a follow-up study, Zerweck *et al.* proposed a model of a large pore made of several transmembrane PGLa homodimers, an arrangement which is held in place by C-terminal interactions with in-plane oriented magainin 2.[57]

Such an electrostatic interaction to stabilize heterodimers is unlikely when both C-termini are negatively charged. The E19 residue, the carboxy terminus and the helix dipole of magainin all add up to an overall electrical dipole on an otherwise net cationic peptide (+3). Therefore, one would expect that removal of a negative charge considerably weakens the electrostatic attraction to the even more cationic PGLa helix (+5). As the synergistic effect was retained for the pPGLc–pmagainin 2c mixture, the C-terminal interaction and its structural implications are probably not crucial for the synergism even though they cause a reduction of the antimicrobial activities of the individual peptides.

Furthermore, whereas the membrane topology of PGLa has been shown to be dependent on the fatty acyl chain saturation,[23,49] the synergistic formation of pores is strongly correlated with the negative intrinsic curvature typically observed with PE† head groups.[48,69] Therefore, when models are established, it seems wise to focus on biophysical data obtained with membranes closely matching the composition of bacterial membranes.[23,28,49,51] In membranes made of *E. coli* lipid extracts, POPE/POPG or POPE/POPG/CL,† both peptides reside on the membrane surface, therefore, realistic models of synergistic interactions should be based on this topology.[33,51,69] These models have in common that the physico-chemical properties of the lipids, such as membrane intrinsic curvature,[48] play an important role in the synergistic activity of the magainin/PGLa mixture. Notably, mesophase arrangements of the peptides oriented along the

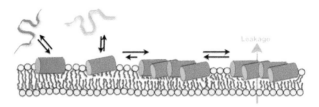

Fig. 5 The equilibria that govern the membrane association and permeabilization of membrane-associated magainin (red) and PGLa (green). The formation of mesophases, which requires PE and negatively charged lipids, depletes the pool of monomers, thus globally more peptide associates with the membrane. Furthermore, the high local peptide density of the mesophase probably enhances membrane permeabilization.

membrane surface have been detected in suitable membranes (Fig. 5) but not when the PE lipid† was replaced by PC.[51,69]

Importantly, the formation of such superstructures depletes the pool of peptide monomers and increases the total amount of membrane-associated peptides by an order of magnitude[69] which in itself can explain the increase in antimicrobial and calcein release activities (Fig. 5). Furthermore, because the synergistic action of PGLa and magainin requires the PE lipid,[48,69] peptide mixtures should not only be more active but also have an order of magnitude higher therapeutic index when compared to the individual peptides.

The mesophase formation of peptides and lipids remains a puzzling observation and currently we can only speculate about the interactions that control their assembly along the bilayer surface. Clearly, they are driven by the lipids while at the same time the interactions retain specificity with regard to the peptide sequences and the lipid composition. This is conceptually new and the critical exchange among peers during the upcoming Faraday Discussion meeting will be essential to advance such unconventional ideas. In this context, it is also important that more experimental work is performed in the future to reveal the structural details of such mesophases. Because the pPGLc and pmagainin 2c constructs keep the high level of synergism (Fig. 4), it will be possible to use them in their isotopically labelled forms for solid-state NMR investigations of the liquid crystalline lipid bilayer.

Experimental

Solid-phase peptide synthesis

Three variants of PGLa (GMASK AGAIA GKIAK VALKA L-NH$_2$) and four variants of magainin 2 (GIGKF LHSAK KFGKA FVGEI MNS-COOH) with different C-termini and/or an additional N-terminal proline were prepared by solid-phase synthesis using a Millipore 9050 automatic peptide synthesizer and Fmoc chemistry. The peptides were purified by reverse phase HPLC (Gilson, Villiers-le-bel, France) using a preparative C18 column (Luna, C-18-100 Å-5 μm, Phenomenex, Le Pecq, France) and an acetonitrile/water gradient. Their identity and purity (>90%) were checked by MALDI mass spectrometry (MALDI-TOF Autoflex, Bruker Daltonics, Bremen, Germany). The purified peptides were dissolved three times in 2 mM HCl at a 1 mg mL^{-1} concentration with subsequent lyophilization to ensure exchange of the TFA cations. Aliquots with 1 mg peptide were prepared and stored at −20 °C.

Antimicrobial assays

For all activity assays, *E. coli* bacteria (ATCC25922, ref. 0335-CRM, Thermo Fisher Scientific, Courtaboeuf, France) were grown overnight on Mueller–Hinton (MH) agar plates. A suspension of bacteria in MH medium (Millipore, Sigma Aldrich, Saint-Louis, MO, USA) was made from the plates and used to inoculate a 10 mL preculture with a starting $OD_{550} = 0.005$. The preculture was incubated overnight and then used to inoculate a culture with a starting $OD_{550} = 0.2$ (10 mL of MH). The culture was incubated until an $OD_{550} = 1.0$ was reached (after around 4 hours). From this culture, a standard bacterial suspension was prepared with $OD_{550} = 0.2$, which was used to prepare the final bacterial suspension at $OD_{550} = 0.0002$.

The antimicrobial assays were performed in 96 well microplates (F-bottom sterile non-treated polystyrene, Thermo Scientific Nunc A/S, Roskilde, Denmark). All samples were added to the first column of the plate and subsequently exposed to a 1.5-fold dilution series in 21 steps. Finally, the bacterial suspension was distributed (50 μL) to each well except the blank controls. The final peptide concentration ranged from 200 μM to 0.04 μM (after addition of bacteria).

The plates were incubated at 37 °C for 18 hours before the OD_{600} was measured. Resazurin was added to each well (0.04 mg mL^{-1} final concentration) and the plates were incubated for another two hours. The cell viability was determined based on the reduction of resazurin. The ratio of reduced resazurin was measured using the absorbance at 570 nm and 600 nm.

Data analysis

For each well, the ratio between the absorbance at 570 nm and the absorbance at 600 nm was calculated. The average value for the blank medium was then subtracted from all wells, before normalizing against the average values from wells containing bacteria but no peptide. The lowest peptide concentration where no reduction of resazurin was observed is considered the minimal inhibition concentration (MIC_{100}). Average values with standard deviations were calculated based on at least two experiments with four replicas each.

For 1 : 1 peptide mixtures, the expected MIC_{100} value was calculated under the assumption of the peptide activities being additive (no synergism or antagonism) from eqn (1):

$$MIC_{100}^{a+b} = 2 \frac{MIC_{100}^{a} \times MIC_{100}^{b}}{MIC_{100}^{a} + MIC_{100}^{b}} \qquad (1)$$

with the two peptides denoted by a and b.

The synergy factor (1/CI) was calculated based on the observed MIC_{100} of the mixtures from eqn (2):

$$CI = \frac{0.5 \times MIC_{100}^{a+b}}{MIC_{100}^{a}} + \frac{0.5 \times MIC_{100}^{a+b}}{MIC_{100}^{b}} \qquad (2)$$

with the two peptides denoted by a and b and their 1 : 1 mixture denoted by $a + b$.

Determination of helical propensity

The peptides were dissolved directly in 50 mM Tris (pH 7.4) or in a d_3-TFE/buffer (1 : 3 v/v) solution to give a final peptide concentration of 1 mg mL^{-1}. Trimethylsilylpropanoic acid (TSP) was added as an internal chemical shift reference. Two-dimensional ^1H–^1H TOCSY, ^1H–^1H NOESY and ^1H–^{13}C HSQC spectra were acquired on a 500 MHz Bruker spectrometer equipped with a BBFO cryo probe (Bruker Biospin, Rheinstetten, Germany). Assignment was based on TOCSY and NOESY cross-peaks as well as typical amino acid chemical shifts.

The secondary structural propensity was calculated based on proton (H, HA, HB) and carbon (CA, CB) chemical shifts for all residues. The neighbour corrected Structural Propensity Calculator (ncSPC)[62] online tool compensating for the neighbour effects was used for the calculation, in order to obtain the most accurate structural propensities: https://st-protein02.chem.au.dk/ncSPC/.

For the samples of PGLa and magainin 2a dissolved in aqueous buffer (50 mM Tris, pH 7.4), the assignments were based on the random coil chemical shifts calculated by the same online tool. As no correlations were observed in the NOESY spectra and only small deviations from the random coil shifts were observed, we consider the results to be reliable.

Conclusions

We tested the effect of adding a proline to the amino terminus of PGLa or magainin 2, as well as changing the chemical composition of their C-terminus. On a structural level, some of these modifications changed the helix propensity and the charge of the peptides. On a functional level, they have a strong effect on antimicrobial activity, but retain the synergistic enhancement of the antimicrobial activity of the equimolar peptide mixtures. The latter observation makes models of synergism where the two termini interact with each other unlikely.

Conflicts of interest

There are no conflicts to declare.

Acknowledgements

We gratefully acknowledge Delphine Hatey and Bruno Vincent for helping with peptide synthesis and solution NMR spectroscopy, respectively. The discussions with Martin Hof and Mariana Amaro are much appreciated. The financial contributions of the Agence Nationale de la Recherche (projects MemPepSyn 14-CE34-0001-01, Biosupramol 17-CE18-0033-3, Naturalarsenal 19-AMRB-0004-02 and the LabEx Chemistry of Complex Systems 10-LABX-0026_CSC), the University of Strasbourg, the CNRS, the Région Alsace and the RTRA International Center of Frontier Research in Chemistry are gratefully acknowledged.

Notes and references

1 H. G. Boman, *Annu. Rev. Immunol.*, 1995, **13**, 61–92.
2 M. Zasloff, *Nature*, 2002, **415**, 389–395.
3 B. Agerberth, H. Gunne, J. Odeberg, P. Kogner, H. G. Boman and G. H. Gudmundsson, *Proc. Natl. Acad. Sci. U. S. A.*, 1995, **92**, 195–199.
4 M. Pirtskhalava, A. Gabrielian, P. Cruz, H. L. Griggs, R. B. Squires, D. E. Hurt, M. Grigolava, M. Chubinidze, G. Gogoladze, B. Vishnepolsky, V. Alekseyev, A. Rosenthal and M. Tartakovsky, *Nucleic Acids Res.*, 2016, **44**, D1104–D1112.
5 G. Wang, X. Li and Z. Wang, *Nucleic Acids Res.*, 2016, **44**, D1087–D1093.
6 J. Cai, X. Li, H. Du, C. Jiang, S. Xu and Y. Cao, *Immunobiology*, 2020, 151936, DOI: 10.1016/j.imbio.2020.151936.
7 E. Yuksel and A. Karakecili, *Mater. Sci. Eng., C*, 2014, **45**, 510–518.
8 D. Yang, R. Zou, Y. Zhu, B. Liu, D. Yao, J. Jiang, J. Wu and H. Tian, *Nanoscale*, 2014, **6**, 14772–14783.
9 K. Reijmar, K. Edwards, K. Andersson and V. Agmo Hernandez, *Langmuir*, 2016, **32**, 12091–12099.

10 D. Roversi, V. Luca, S. Aureli, Y. Park, M. L. Mangoni and L. Stella, *ACS Chem. Biol.*, 2014, **9**, 2003–2007.

11 L. A. Rollins-Smith, J. K. Doersam, J. E. Longcore, S. K. Taylor, J. C. Shamblin, C. Carey and M. A. Zasloff, *Dev. Comp. Immunol.*, 2002, **26**, 63–72.

12 D. P. Tieleman, B. Hess and M. S. Sansom, *Biophys. J.*, 2002, **83**, 2393–2407.

13 B. Bechinger, *J. Pept. Sci.*, 2015, **21**, 346–355.

14 C. Aisenbrey, A. Marquette and B. Bechinger, *Adv. Exp. Med. Biol.*, 2019, **1117**, 33–64.

15 C. J. Arnusch, H. B. Albada, M. van Vaardegem, R. M. J. Liskamp, H. G. Sahl, Y. Shadkchan, N. Osherov and Y. Shai, *J. Med. Chem.*, 2012, **55**, 1296–1302.

16 C. Ghosh, N. Harmouche, B. Bechinger and J. Haldar, *ACS Omega*, 2018, **3**, 9182–9190.

17 M. Laurencin, M. Simon, Y. Fleury, M. Baudy-Floc'h, A. Bondon and B. Legrand, *Chem. – Eur. J.*, 2018, **24**, 6191–6201.

18 N. P. Chongsiriwatana, J. S. Lin, R. Kapoor, M. Wetzler, J. A. C. Rea, M. K. Didwania, C. H. Contag and A. E. Barron, *Sci. Rep.*, 2017, **7**, 16718.

19 L. A. Rank, N. M. Walsh, R. Liu, F. Y. Lim, J. W. Bok, M. Huang, N. P. Keller, S. H. Gellman and C. M. Hull, *Antimicrob. Agents Chemother.*, 2017, **61**, e00204–e00217.

20 H. G. Boman, *J. Intern. Med.*, 2003, **254**, 197–215.

21 K. Matsuzaki, O. Murase, H. Tokuda, S. Funakoshi, N. Fujii and K. Miyajima, *Biochemistry*, 1994, **33**, 3342–3349.

22 B. Bechinger, *J. Pept. Sci.*, 2011, **17**, 306–314.

23 E. Salnikov and B. Bechinger, *Biophys. J.*, 2011, **100**, 1473–1480.

24 E. Strandberg, D. Tiltak, S. Ehni, P. Wadhwani and A. S. Ulrich, *Biochim. Biophys. Acta*, 2012, **1818**, 1764–1776.

25 K. He, S. J. Ludtke, W. T. Heller and H. W. Huang, *Biophys. J.*, 1996, **71**, 2669–2679.

26 E. Salnikov, C. Aisenbrey, V. Vidovic and B. Bechinger, *Biochim. Biophys. Acta*, 2010, **1798**, 258–265.

27 C. Kim, J. Spano, E. K. Park and S. Wi, *Biochim. Biophys. Acta*, 2009, **1788**, 1482–1496.

28 N. Harmouche and B. Bechinger, *Biophys. J.*, 2018, **115**, 1033–1044.

29 M. Pachler, I. Kabelka, M. S. Appavou, K. Lohner, R. Vacha and G. Pabst, *Biophys. J.*, 2019, **117**, 1858–1869.

30 F. Y. Chen, M. T. Lee and H. W. Huang, *Biophys. J.*, 2003, **84**, 3751–3758.

31 A. Mecke, D. K. Lee, A. Ramamoorthy, B. G. Orr and M. M. Banaszak Holl, *Biophys. J.*, 2005, **89**, 4043–4050.

32 A. Farrotti, G. Bocchinfuso, A. Palleschi, N. Rosato, E. S. Salnikov, N. Voievoda, B. Bechinger and L. Stella, *Biochim. Biophys. Acta*, 2015, **1848**, 581–592.

33 I. Kabelka, M. Pachler, S. Prevost, I. Letofsky-Papst, K. Lohner, G. Pabst and R. Vacha, *Biophys. J.*, 2020, **118**, 612–623.

34 S. J. Ludtke, K. He, W. T. Heller, T. A. Harroun, L. Yang and H. W. Huang, *Biochemistry*, 1996, **35**, 13723–13728.

35 K. Matsuzaki, *Biochim. Biophys. Acta*, 1998, **1376**, 391–400.

36 Y. Shai, *Biochim. Biophys. Acta*, 1999, **1462**, 55–70.

37 H. Jenssen, P. Hamill and R. E. Hancock, *Clin. Microbiol. Rev.*, 2006, **19**, 491–511.

38 K. Hall, T. H. Lee, A. I. Mechler, M. J. Swann and M. I. Aguilar, *Sci. Rep.*, 2014, **4**, 5479.

39 J. M. Henderson, A. J. Waring, F. Separovic and K. Y. C. Lee, *Biophys. J.*, 2016, **111**, 2176–2189.

40 K. Matsuzaki, M. Harada, S. Funakoshi, N. Fujii and K. Miyajima, *Biochim. Biophys. Acta*, 1991, **1063**, 162–170.

41 T. Wieprecht, M. Beyermann and J. Seelig, *Biochemistry*, 1999, **38**, 10377–10378.

42 M. Wenk and J. Seelig, *Biochemistry*, 1998, **37**, 3909–3916.

43 C. Aisenbrey and B. Bechinger, *Langmuir*, 2014, **30**, 10374–10383.

44 M. Wenzel, A. I. Chiriac, A. Otto, D. Zweytick, C. May, C. Schumacher, R. Gust, H. B. Albada, M. Penkova, U. Kramer, R. Erdmann, N. Metzler-Nolte, S. K. Straus, E. Bremer, D. Becher, H. Brotz-Oesterhelt, H. G. Sahl and J. E. Bandow, *Proc. Natl. Acad. Sci. U. S. A.*, 2014, **111**, E1409–E1418.

45 A. Vaz Gomes, A. de Waal, J. A. Berden and H. V. Westerhoff, *Biochemistry*, 1993, **32**, 5365–5372.

46 H. V. Westerhoff, M. Zasloff, J. L. Rosner, R. W. Hendler, A. de Waal, G. Vaz, P. M. Jongsma, A. Riethorst and D. Juretic, *Eur. J. Biochem.*, 1995, **228**, 257–264.

47 K. Matsuzaki, Y. Mitani, K. Akada, O. Murase, S. Yoneyama, M. Zasloff and K. Miyajima, *Biochemistry*, 1998, **37**, 15144–15153.

48 R. Leber, M. Pachler, I. Kabelka, I. Svoboda, D. Enkoller, R. Vácha, K. Lohner and G. Pabst, *Biophys. J.*, 2018, **114**, 1945–1954.

49 E. Strandberg, J. Zerweck, P. Wadhwani and A. S. Ulrich, *Biophys. J.*, 2013, **104**, L09–L11.

50 E. S. Salnikov, C. Aisenbrey, F. Aussenac, O. Ouari, H. Sarrouj, C. Reiter, P. Tordo, F. Engelke and B. Bechinger, *Sci. Rep.*, 2016, **6**, 20895.

51 E. Glattard, E. S. Salnikov, C. Aisenbrey and B. Bechinger, *Biophys. Chem.*, 2016, **210**, 35–44.

52 T. Hara, H. Kodama, M. Kondo, K. Wakamatsu, A. Takeda, T. Tachi and K. Matsuzaki, *Biopolymers*, 2001, **58**, 437–446.

53 S. L. Grage, S. Afonin, S. Kara, G. Buth and A. S. Ulrich, *Front. Cell Dev. Biol.*, 2016, **4**, 65.

54 J. Zerweck, E. Strandberg, J. Burck, J. Reichert, P. Wadhwani, O. Kukharenko and A. S. Ulrich, *Eur. Biophys. J.*, 2016, **45**, 535–547.

55 A. Marquette, E. Salnikov, E. Glattard, C. Aisenbrey and B. Bechinger, *Curr. Top. Med. Chem.*, 2015, **16**, 65–75.

56 B. Bechinger, *J. Mol. Biol.*, 1996, **263**, 768–775.

57 J. Zerweck, E. Strandberg, O. Kukharenko, J. Reichert, J. Burck, P. Wadhwani and A. S. Ulrich, *Sci. Rep.*, 2017, **7**, 13153.

58 R. Verardi, N. J. Traaseth, L. R. Masterson, V. V. Vostrikov and G. Veglia, *Adv. Exp. Med. Biol.*, 2012, **992**, 35–62.

59 V. Vidovic, L. Prongidi-Fix, B. Bechinger and S. Werten, *J. Pept. Sci.*, 2009, **15**(4), 278–284.

60 B. Bechinger, M. Zasloff and S. J. Opella, *Protein Sci.*, 1993, **2**, 2077–2084.

61 B. Bechinger, M. Zasloff and S. J. Opella, *Biophys. J.*, 1998, **74**, 981–987.

62 K. Tamiola and F. A. Mulder, *Biochem. Soc. Trans.*, 2012, **40**, 1014–1020.

63 T. Wieprecht, O. Apostolov, M. Beyermann and J. Seelig, *J. Mol. Biol.*, 1999, **294**, 785–794.

64 T. Wieprecht, O. Apostolov, M. Beyermann and J. Seelig, *Biochemistry*, 2000, **39**, 442–452.

65 D. W. Juhl, E. Glattard, M. Lointier, P. Bampilis and B. Bechinger, *Front. Cell. Infect. Microbiol.*, 2020, **10**, 526459.

66 D. Marion, M. Zasloff and A. Bax, *FEBS Lett.*, 1988, **227**, 21–26.

67 J. Gesell, M. Zasloff and S. J. Opella, *J. Biomol. NMR*, 1997, **9**, 127–135.

68 K. Matsuzaki, A. Nakamura, O. Murase, K. Sugishita, N. Fujii and K. Miyajima, *Biochemistry*, 1997, **36**, 2104–2111.

69 C. Aisenbrey, M. Amaro, P. Pospisil, M. Hof and B. Bechinger, *Sci. Rep.*, 2020, **10**, 11652.

70 E. Strandberg, D. Horn, S. Reisser, J. Zerweck, P. Wadhwani and A. S. Ulrich, *Biophys. J.*, 2016, **111**, 2149–2161.

71 E. Han and H. Lee, *RSC Adv.*, 2015, **5**, 2047–2055.

72 A. Pino-Angeles, J. M. Leveritt III and T. Lazaridis, *PLoS Comput. Biol.*, 2016, **12**, e1004570.

Faraday Discussions

PAPER

Antimicrobial peptide activity in asymmetric bacterial membrane mimics†

Lisa Marx,[ab] Moritz P. K. Frewein, [iD][abc] Enrico F. Semeraro, [iD][ab] Gerald N. Rechberger,[ab] Karl Lohner,[ab] Lionel Porcar[c] and Georg Pabst [iD][*ab]

Received 21st June 2021, Accepted 21st July 2021

DOI: 10.1039/d1fd00039j

We report on the response of asymmetric lipid membranes composed of palmitoyl oleoyl phosphatidylethanolamine and palmitoyl oleoyl phosphatidylglycerol, to interactions with the frog peptides L18W-PGLa and magainin 2 (MG2a), as well as the lactoferricin derivative LF11-215. In particular we determined the peptide-induced lipid flip-flop, as well as membrane partitioning of L18W-PGLa and LF11-215, and vesicle dye-leakage induced by L18W-PGLa. The ability of L18W-PGLa and MG2a to translocate through the membrane appears to correlate with the observed lipid flip-flop, which occurred at the fastest rate for L18W-PGLa. The higher structural flexibility of LF11-215 in turn allows this peptide to insert into the bilayers without detectable changes of membrane asymmetry. The increased vulnerability of asymmetric membranes to L18W-PGLa in terms of permeability, appears to be a consequence of tension differences between the compositionally distinct leaflets, but not due to increased peptide partitioning.

Introduction

Antimicrobial peptides (AMPs) are widely studied compounds of the innate immune system with high potential to combat the spread of infectious diseases due to multi-resistant strains.[1,2] Compared to conventional antibiotics, AMPs translocate or impair cellular envelopes *via* unspecific molecular interactions (electrostatic, hydrophobic, entropic), although their final target might also be located in the cytosolic compartment.[3] Various models have been reported for AMP/lipid interactions, including the formation of transmembrane peptide pores, micellization, or interfacial activity, the latter of which may lead to the formation of a surface-adsorbed peptide layer (carpet) or peptide self-

[a]University of Graz, Institute of Molecular Biosciences, NAWI Graz, 8010 Graz, Austria. E-mail: georg.pabst@uni-graz.at; Tel: +43 316 380 4989

[b]Field of Excellence BioHealth, University of Graz, Graz, Austria

[c]Institut Laue-Langevin, 38043 Grenoble, France

† Electronic supplementary information (ESI) available. See DOI: 10.1039/d1fd00039j

aggregation.[4–6] Peptide pore formation has also been connected to accelerated lipid flip-flop,[7–9] and recently confirmed for asymmetric lipid membranes.[10–12]

On the other hand, it is increasingly clear that the mode of action of a given peptide depends strongly on the composition of the lipid bilayer.[13,14] That is, AMPs cannot be classified as pore-formers, or not without referring to the composition of their target membrane. Importantly, the most abundant phospholipids of bacterial membranes are phosphatidylglycerol (PG), phosphatidylethanolamine (PE) and cardiolipin.[15] For example, PGLa from the African clawed frog was reported to align differently with respect to the membrane surface, depending on lipid composition, hydration level or peptide concentration.[16,17]

It is not clear whether surface-adsorbed AMPs also lead to enhanced lipid flip-flop. Interestingly, all-atom molecular dynamics simulations have suggested that lipids may co-translocate with peptides through membranes.[18] However, experimental evidence for such a scenario is currently not available. Moreover, bacterial membranes, including cytoplasmic membranes of Gram-negative bacteria, display transbilayer compositional asymmetry,[19] whose role in the context of AMP activity is basically unknown.

This prompted us to measure lipid flip-flop in asymmetric large unilamellar vesicles (aLUVs) composed of palmitoyl oleoyl phosphatidylethanolamine (POPE) and palmitoyl oleoyl phosphatidylglycerol (POPG) using L18W-PGLa, as well as magainin 2a (MG2a) and the lactoferricin derivative LF11-215, all of which remain surface bound in PE/PG membranes.[20,21] In particular, we coupled time resolved small-angle neutron scattering (SANS) measurements of lipid flip-flop to dye-leakage and peptide partitioning using Trp emission spectroscopy. Our results reveal a peptide-concentration dependent loss of membrane asymmetry, which was most expressed for L18W-PGLa, followed by MG2a. LF11-215 in turn caused no detectable lipid flip-flop despite its high partitioning into aLUVs. This suggests that the high structural flexibility of LF11-215 enables the peptide to translocate through asymmetric membranes without noticeable effects on membrane structure. For the two linear peptides, L18W-PGLa and MG2a, lipid flip-flop instead appears to be coupled to their translocation probability. The increased permeability of aLUVS in the presence of L18W-PGLa as compared to symmetric vesicles is not due to increased peptide partitioning, but appears to be dominated by an internal lateral stress imbalance between the two leaflets.

Materials and methods

Lipids and peptides

POPE, POPG and palmitoyl-d31 oleoyl phosphatidyglycerol (POPG-d31) were purchased from Avanti Polar Lipids (Alabaster, AL) as powder and used without further purification. L18W-PGLa (GMASKAGAIAGKIAKVAWKAL-NH$_2$), MG2a (GIGKFLHSAKKFGKAFVGEIMNS-NH$_2$), and LF11-215 (FWRIRIRR-NH$_2$) were obtained in lyophilized form (purity > 95%) from PolyPeptide Laboratories (San Diego, CA). ANTS (8-aminonaphthalene-1,3,6-trisulfonic acid, disodium salt) and DPX (p-xylene-bis-pyridinium bromide) were purchased from Molecular Probes (Eugene, OR) and HEPES (purity > 99.5%) from Carl Roth (Karlsruhe, Germany). D$_2$O was obtained from Euroisotop (Saarbrücken, Germany), methyl-β-cyclodextrin (mβCD), Triton X-100 and all other chemicals (pro analysis quality) were from Sigma-Aldrich (Vienna, Austria).

Sample preparation

aLUVs with an outer leaflet enriched in POPE and an inner leaflet composed of POPG were produced using a previously reported protocol.[22] In short, outer leaflet lipids of POPG acceptor LUVs (size: $\sim$ 100 nm), suspended in HBS buffer (10 mM HEPES, 140 mM NaCl, pH 7.4), were exchanged *via* mβCD-mediated lipid transfer for POPE. For SANS experiments POPG-d31 was used instead of POPG and the HBS was replaced by buffer prepared in 100% D_2O (HBSD); for leakage experiments POPG acceptor LUVs containing ANTS/DPX were prepared as described elsewhere.[21] Outer leaflet exchange was achieved preparing first donor multi-lamellar POPE vesicles (MLVs), hydrated in HBS with 20 w/w% sucrose (0.632 M), followed by an incubation with mβCD (35 mM) at 40 °C for 2 h. Donor and acceptor vesicles were then mixed at an acceptor/donor ratio of 1 : 2 (mol/mol) and incubated at 40 °C for another 30 min. Exchange vesicles were separated from donor vesicles, mβCD and sucrose as previously detailed.[22] Vesicle size of acceptor LUVs and aLUVs was checked by dynamic light scattering (DLS) using a Zetasizer Nano ZSP (Malvern Panalytical, Malvern, UK), affirming the integrity of the produced aLUVs and absence of large donor MLVs.

The achieved lipid exchange was determined by ultra-performance liquid chromatography-tandem mass spectrometry (UPLC-MS) for protiated samples and gas chromatography (GC) for aLUVs containing POPE-d31 [see ESI† for details]. UPLC-MS results revealed an overall POPE/POPG $\sim$ 1 : 2 mol/mol ratio for aLUVs. aLUVs were converted into scrambled LUVs (ScraLUVs), *i.e.* same lipid composition, but symmetrically distributed between the two leaflets, as detailed in ref. 22. Additionally, we also prepared LUVs composed of POPE/POPG (7 : 3 mol/mol) as outer leaflet mimics (OLM) of our aLUVs. Phospholipid concentrations were determined through the Bartlett phosphate assay.[23]

Leakage assay

Measurements were performed in quartz cuvettes in 2 mL of iso-osmotic HBS buffer containing 1 mM EDTA on a Cary Eclipse Fluorescence Spectrophotometer (Varian/Agilent Technologies, Palo Alto, CA) as detailed in ref. 21. Achieved leakage after the addition of peptide was derived from

$$E_\% = \frac{I_p - I_{min}}{I_{max} - I_{min}}, \tag{1}$$

where I_{min} is the initial fluorescence without peptide, and I_{max} is the fluorescence corresponding to 100% leakage determined through the addition of a 1 vol% solution of Triton X-100.

Trp-fluorescence spectroscopy

Peptide partitioning was determined from Trp-fluorescence emission for L18W-PGLa and LF11-215 using a Cary Eclipse Fluorescence Spectrophotometer (Varian/Agilent Technologies) at an excitation wavelength of $\lambda = 280$ nm, and slit widths for incident and outgoing beams of either 5 or 10 nm, as detailed previously.[24] All samples were contained in a quartz cuvette with a magnetic stirrer and measured at 37 °C. Spectra were analyzed with a linear combination of two independent bands each fitted by a log-normal-like function.[25,26] This allowed us to extract the molar concentration of dissociated peptide, $[P]_W$, and subsequently the molar concentration of membrane-associated peptide $[P]_B = [P] - [P]_W$, where $[P]$ is the total peptide concentration in the sample.[24]

The mole fraction partitioning coefficient

$$K_x = \frac{x_B}{x_W} \simeq \frac{[P]_B}{[L]} \frac{[W]}{[P]_W}, \tag{2}$$

was then calculated for a given lipid concentration $[L]$, where x_B is the mole fraction of membrane-partitioned peptide, x_W is the mole fraction of unbound peptide and $[W]$ is the concentration of bulk water (55.3 M at 37 °C).[27,28] Additionally, we derived the ratio of bound peptide using

$$f_B = \frac{[P]_B}{[P]} = \frac{K_x}{[W]/[L] + K_x}. \tag{3}$$

Small-angle neutron scattering

SANS measurements were performed at the D22-large dynamic range small-angle diffractometer, located at the Institut Laue-Langevin in Grenoble, France, with a two-^{3}He-detector setup at a wavelength of 6 Å ($\lambda/\Delta\lambda = 10\%$), resulting in a q-range of 0.016–0.6 Å^{-1}. Flip-flop kinetics were measured with a time resolution of 2 min and sample-to-detector distances (SDD) of 1.3 and 5.6 m, and a 5.6 m collimation; low-q measurements of reference (aLUVs) and endstate measurements were conducted at SDDs of 1.3 and 17.8 m, with a 17.6 m collimation. Samples (concentration 7 mg mL^{-1} in HBSD) were measured and filled in Hellma 120-QS cuvettes of 1 mm pathway and equilibrated at 37 °C using a circulating water bath. Data, available at (DOI: 10.5291/ILL-DATA.DIR-217) were reduced using the GRASP-software, performing flat field, solid angle, dead time and transmission corrections, and were normalized by incident flux. Finally,

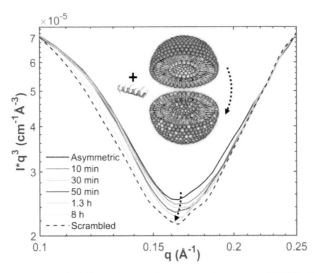

Fig. 1 Measurement principle and scattering contrast between (POPG-d31)^{in}/(POPE/POPG-d31)^{out} aLUVs and ScraLUVs in HBSD buffer, as observed by SANS at 37 °C. Scattering contrast was additionally enhanced by multiplying the scattered intensities with q^3.

contributions from the empty cell and solvent were subtracted. Data were averaged over 5 to 10 frames to achieve sufficient signal to noise ratios.

Analogously to ref. 11, lipid flip-flop was determined by measuring the peptide-induced change of scattering contrast with time (Fig. 1). Here, the contrast emerges from chain deuterated POPG-d31, which is primarily located in the inner leaflet, and fully protiated POPE, enriched in the outer leaflet. Then, the change of contrast follows[11,29]

$$\Delta\Gamma(t) = \frac{\Gamma(t) - \Gamma(\infty)}{\Gamma(0) - \Gamma(\infty)} = \exp(-k_f t), \tag{4}$$

where $\Gamma = \int Iq^3 dq$ in the shown q-range (Fig. 1), $\Gamma(0)$ corresponds to the initial aLUVs, $\Gamma(\infty)$ to ScraLUVs, and k_f is the lipid flip-flop rate.

Results

Asymmetric membranes are more vulnerable to peptide-induced dye efflux

We started our experiments by studying the kinetics of dye release induced by L18W-PGLa. Fig. 2 shows the observed permeation of aLUVs, ScraLUVs and OLM over a time of 40 min. Symmetric LUVs, mimicking the outer leaflet of our aLUVs were basically impermeable to dyes in the presence of peptides, while ScraLUVs showed initially the fastest leakage increase, but levelled off at ∼38% leakage at the end of the experiment. Instead, dye-efflux for POPE/POPG aLUVs started more gradually, but then reached final leakage values close to 100%. Changing the peptide concentration affected the leakage kinetics most (speeding-up for increasing, and slowing-down for decreasing L18W-PGLa content); a similar effect was observed upon increasing lipid concentration (Fig. S1†).

AMP partitioning depends on transbilayer lipid distribution

In order to shed some light on the increased leakage of aLUVs we first performed peptide partitioning studies making use of the Trp-residue of L18W-PGLa.

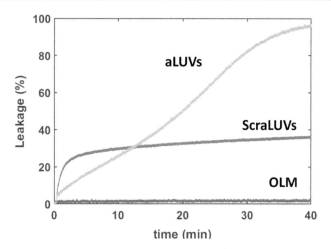

Fig. 2 Kinetics of L18W-PGLa-induced dye efflux from (POPG)^{in}/(POPE/POPG)^{out} aLUVs, ScraLUVs and OLM for [P]/[L] = 1 : 400 ([L] = 50 μM, T = 37 °C).

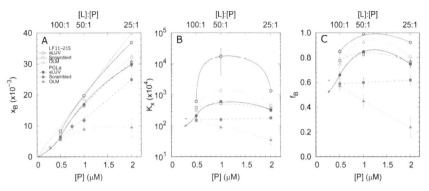

Fig. 3 Mole fraction of membrane-partitioned peptides (A), partitioning coefficient (B) and ratio of partitioned peptides (C) as a function of total LF11-215 (open symbols) and L18W-PGLa (filled symbols) concentrations and [L] = 50 μM. Data refer to aLUVs (circles), ScraLUVs (squares) and OLM (triangles). The gray line in (A) represents the limit [P]/[L] corresponding to $f_B = 1$; all other lines are guides for the eye. In the case of L18W-PGLa, the arrows indicate a realistic propagation for [P] < 0.5 μM.

Further, we included LF11-215, which also contains a Trp-residue. Fig. 3 displays the mole fraction of the partitioned peptides, partitioning coefficient, and ratio of partitioned to total number of peptides as a function of peptide concentration for [L] = 50 μM. All presented data have been taken after 60 min of incubation with the peptides. We also performed time-resolved measurements with the shortest time-interval being ~20 s after mixing, but observed no noticeable differences to the data recorded after extended incubation times.

Both peptides exhibited the highest affinity to aLUVs, followed by ScraLUVs and OLM in the studied peptide range ([P] = 0.5–2 μM). Moreover, both K_x and f_B peaked at [P] ~ 1 μM ([P]/[L] = 1 : 50), and in particular for LF11-215, where $f_B \simeq$ 1. Similar behavior, but much less pronounced was also observed for symmetric LUVs in the case of LF11-215. In turn K_x remained constant for scrambled LUVs in the presence of L18W-PGLa, mirrored also in a linear increase of x_B with [P]. In OLM, K_x and f_B decreased upon increasing L18W-PGLa concentration instead. While the non-monotonous variation of peptide partitioning might indicate a combination of cooperative (increasing K_x) and anticooperative (decreasing K_x) peptide/peptide or peptide/lipid interactions,[27] it is interesting that LF11-215 partitions more favorably into OLM than into ScraLUVs at [P] = 1 μM; a situation which is reversed for L18W-PGLa. That is, L18W-PGLa more favorably interacts with ScraLUVs than with OLM.

Our experimental set-up did not allow us to measure peptide concentrations as low as those used for leakage experiments. However, extrapolating roughly the trends observed at lower peptide concentrations ([P] → 0 ⇔ x_B → 0) suggests that L18W-PGLa partitions less into aLUVs than into both symmetric LUVs under experimental conditions used for the leakage measurements shown in Fig. 2. In order to measure peptide partitioning at [P]/[L] = 1 : 400 we increased the lipid concentration to 200 μM (Fig. S2†). Although increasing lipid concentration is known to affect peptide partitioning,[24,27] these data support the idea that L18W-PGLa does not preferentially partition into aLUVs under leakage conditions.

Lipid flip-flop in aLUVs is highly peptide specific

Finally, we determined the peptide-induced lipid flip-flop using SANS combined with a H/D contrast variation scheme that allowed us to discriminate for trans-bilayer lipid distribution. In particular, we substituted POPG by chain-perdeuterated POPG-d31 in our aLUV preparations and monitored its equilibration across both lipid leaflets by time-resolved SANS as detailed in the Materials and methods section. The scattering patterns of aLUVs and ScraLUVs were typical for single-shelled vesicles and were analysed in terms of a modified 4-slab model[30] to determine the leaflet composition (Fig. S4†). Specific care was taken to keep the peptide concentrations well below the thresholds reported for POPE/POPG (3 : 1 mol/mol) mixtures for the formation of vesicle aggregation or multilamellar vesicles.[24,31,32] This was additionally checked by inspection of the final SANS patterns after peptide addition, which did not show any signatures for changes in overall vesicle morphology or aggregation (Fig. S4†).

In the absence of peptides, no significant changes of scattering intensity were observed during the time course of the experiments (*i.e.* ~24 hours). This signifies that the produced aLUVs are sufficiently stable for all experiments presently reported. The addition of L18W-PGLa induced an equilibration of lipid distribution across both leaflets with a rate that strongly increased with peptide concentration (Fig. 4). Analysis in terms of eqn (4) yielded flip-flop half-times of $t_{1/2} \sim 500$ min at [P]/[L] = 1 : 800, which dropped to 14 min at eight times higher peptide concentration (Table 1). Interestingly, LF11-215 led at equally high [P]/[L] to no detectable lipid flip-flop (Table 1; Fig. S5†). Additionally, we studied lipid flip-flop as induced by MG2a and an equimolar mixture of L18W-PGLa and MG2a. MG2a, similar to L18W-PGLa, is supposed to remain membrane-surface aligned in the present conditions, while its equimolar mixture is well-known for its synergistic activity.[14,31,32]

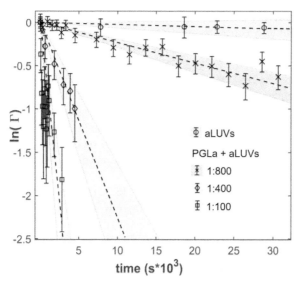

Fig. 4 Decay of scattering contrast between aLUVs and scrambled LUVs due to L18W-PGLa-mediated lipid flip-flop at [L] = 9 mM and different [L]/[P]. As a control, aLUV data in the absence of peptide are also shown.

Table 1 Flip-flop rates k_f and flip-flop half-times $t_{1/2}$ for mixtures of asymmetric vesicles with L18W-PGLa, MG2a, their equimolar mixture, and LF11-215 at different [P]/[L] ratios

Peptide	[P]/[L]	$k_f \times 10^{-5}$ (s^{-1})	$t_{1/2}$ (min)
L18W-PGLa	1 : 100	42 ± 13	14 ± 4
	1 : 400	11 ± 4	52 ± 16
	1 : 800	1.2 ± 0.4	500 ± 200
MG2a	1 : 100	1.4 ± 0.5	420 ± 140
	1 : 200	<0.6	$>10^3$
L18W-PGLa:MG2a	1 : 800	0.8 ± 0.3	700 ± 300
LF11-215	1 : 100	<0.6	$>10^3$

Our flip-flop analysis showed that MG2a is significantly less potent than L18W-PGLa in translocating lipids (Table 1; Fig. S5†). No detectable lipid flip-flop was found for [P]/[L] $= 1 : 200$ and rates at doubled MG2a concentration were comparable to L18W-PGLa at [P]/[L] $= 1 : 800$. Interestingly, the equimolar mixture of L18W-PGLa and MG2a did not exhibit a faster lipid flip-flop at [P]/[L] $= 1 : 800$ than L18W-PGLa alone. However, the equimolar peptide mixture contains only [P]/[L] $= 1 : 1600$ of either L18W-PGLa and MG2a. Considering that lipid flip-flop will drop significantly for AMP these concentrations ($t_{1/2}^{L18W-PGLa} \gtrsim 2000$ min; $t_{1/2}^{MG2a} \sim \infty$), then suggest that the measured half-time for the peptide mixture is indeed a consequence of L18W-PGLa/MG2a-synergism.

Discussion

Attempting to gain some deeper understanding of the intricate leakage behaviour of (POPG)^{in}/(POPE/POPG)^{out} aLUVs as compared to symmetric LUVs (scrambled and outer leaflet mimics) at [L] $= 50$ μM and [P] $= 125$ nM (Fig. 2) we determined peptide partitioning and peptide-induced lipid flip-flop, including LF11-215 and MG2a. All three peptides are able to inhibit bacterial growth, with reported minimum inhibitory concentrations (MICs) of 16 μM (LF11-215), $\sim$31 μM (L18W-PGLa), and $\sim$62 μM (MG2a) for *Escherichia coli* K12.[14,24] Further, while the secondary structure of L18W-MG2a and MG2a once inserted into lipid membranes can be considered as mostly α-helical,[33,34] LF11-215 due to its short amino acid sequence is expected to be structurally more flexible. In fact, acylated LF11-215, with an octanoyl chain attached to the N-terminal, was reported to form a short α-helical-like turn of five residues in micelles.[35] Delineating from this study to LF11-215, the structure of LF11-215 can be considered to be a hydrophobic wedge formed by the Phe, Trp and Ile residues, whereas the four Arg residues form a cluster of positive charge.

Thus, it might be expected that LF11-215 partitions differently into lipid membranes than L18W-PGLa; for the presently studied systems and peptide concentrations we found K_x(LF11-215) $\geq K_x$(L18W-PGLa) (Fig. 3). Intriguingly, however, LF11-215 interacted more favorably with OLM than with ScraLUVs, while the opposite partitioning behaviour was found for L18W-PGLa. Yet, differences in partitioning coefficients change significantly with peptide concentration and become negligible for low LF11-215 and L18W-PGLa content. This is a manifestation of the well-known fact that peptide partitioning is a complex non-linear interplay of intermolecular forces beyond mere electrostatic interactions between peptides and lipids. However, our findings might in part be due to the preferred interactions of L18W-PGLa with POPG,[31] which is enriched in our scrambled vesicles. In turn, insertion of the protonated N-terminal Phe-residue of LF11-215 (next to Trp) into the hydrophobic region should be easier for the less charged OLM whose composition is dominated by POPE.

Strikingly, the observed differences in leakage efficacy of L18W-PGLa are not correlated with its partitioning in membranes. Extrapolating partitioning data to low peptide concentrations, $i.e.$ matching leakage and Trp-emission experimental conditions, is most realistically done using a simple propagation of the slopes for x_B, K_x and f_B at the lowest measured peptide concentration (Fig. 3). A further constraint for the propagation is the requirement $x_B([P] = 0) \overset{!}{=} 0$. Using these simple rules suggests a lower partitioning of L18W-PGLa in aLUVs as compared to ScraLUVs or OLM. Note that even if we were to assume a constant K_x for low peptide concentrations ($i.e.$, $K_x([P] \to 0) = K_x ([P] = 0.5~\mu M)$), the actual differences between the K_x-values of aLUVs, ScraLUVs and OLM are too small to explain the significantly enhanced peptide-induced dye leakage for aLUVs (Fig. 2). Note also our additional partitioning experiments at higher lipid concentrations, which allowed us to measure at $[P]/[L] = 1 : 400$, but showed no enhanced peptide association for aLUVs (Fig. S1†). Naturally, we cannot exclude a $priori$ a further increase of K_x toward lower $[P]$ for aLUVs. This would imply a sequence of anti-cooperative $\to$ cooperative $\to$ anticooperative peptide/peptide or peptide/lipid interactions[27] with increasing peptide concentration, which appears on the basis of the available data as highly unrealistic. Also the slower onset of leakage of aLUVs as compared to ScraLUVs is unlikely to be an effect of initial anti-cooperative partitioning interactions, since we found no time-dependence in our Trp-spectroscopy data.

Instead we propose that the intriguing dye-leakage activity of L18W-PGLa in aLUVs is dominated by the elastic/structural response of the bilayer in the presence of the peptide. For example, we have previously demonstrated that insertion and translocation of linear peptides into membranes depend on the elastic energy stress stored within the lipid bilayer.[14] In particular, POPE, due to its significantly negative intrinsic lipid curvature[36] and its capability for intermolecular H-bonding[37] leads to a tightly packed polar/apolar interface and thus an increased free energy barrier for peptide insertion and translocation.

Peptide translocation has been linked to lipid flip-flop even in the absence of the peptide-induced pore formation and, importantly, also to leakage events.[18] It is therefore interesting to discuss our flip-flop measurements on aLUVs within this framework. Moreover, the high lipid concentrations used for our time-resolved SANS measurements allow us to neglect any effects originating from peptide partitioning,[27] $i.e.$ all presently studied AMPs can be assumed to be fully membrane-associated. We found significantly increased lipid flip-flop only for L18W-PGLa (Table 1). Interestingly, LF11-215 did not induce any measurable lipid flip-flop, even at $[P]/[L] = 1 : 100$, despite its high antibacterial activity (lowest MIC among all presently studied AMPs). Moreover, we previously demonstrated that LF11-215 readily translocates the bacterial envelope of $E.$ $coli.$[24] Notably, the ability of LF11-215 to induce dye-leakage from vesicles or membrane structural changes has been reported to be rather low as compared to other AMPs.[21,24] Hence, translocation of peptide through membranes, dye-leakage and lipid flip-flop are not necessarily correlated. Note also that the well-studied AMP buforin II is able to translocate through membranes without inducing lipid flip-flop.[38] In this case, the Pro-residue was reported to bestow buforin II with a higher structural flexibility to pass through the lipid bilayer. Similarly, LF11-215 is structurally much more adaptable than L18W-PGLa or MG2a, due to its short sequence where only five out of eight residues are able to form an α-helical-like turn. We thus suggest that the structural flexibility enables efficient membrane translocation of LF11-215 without measurable effects on membrane structure or lipid distribution.

On the other hand we argue that lipid flip-flop, leakage and peptide translocation are at least partially coupled in the case of L18W-PGLa and MG2a. Kabelka and Vácha reported from a computational study that the ability for linear peptides to translocate through membranes is connected to the size and distribution of their hydrophobic patches.[39] In particular, the free energy of membrane insertion was lower for peptides with an increased hydrophobic surface either along the direction of their long axis or at one of their termini. L18W-PGLa and MG2a both have an amidated, *i.e.* non-charged, C-terminus, which inserts first into the membrane upon translocation.[18] However, the hydrophobic angle of L18W-PGLa, calculated using MPEx,[40] is significantly larger than that of MG2a (Fig. S6†). Indeed, membrane surface-aligned MG2a was found to be located slightly further away from the bilayer center than L18W-PGLa in POPE/POPG bilayers.[31] Combination of these pieces of information thus suggests that MG2a is less likely to translocate POPE/POPG bilayers. Moreover, the amidated C-terminus of PGLa was reported to act like a polar brush, shuttling lipids across the bilayer.[18] A similar mechanism can be expected to apply also to MG2a. Consequently, the significantly lower lipid flip-flop rate in our aLUVs in the presence of MG2a as compared to L18W-PGLa (Table 1) most likely is a corollary of a reduced rate of peptide translocation. The apparently synergistically increased lipid flip-flop for the L18W-PGLa : MG2a equimolar mixture then suggests the facilitated peptide translocation. Although, L18W-PGLa and MG2a were shown to already form heterodimers at low peptide concentrations,[31] it appears unlikely that these dimers are sufficiently stable to translocate as one entity. Possibly, peptide translocation is assisted by enhanced spontaneous pairwise interactions of the C-termini observed for PGLa alone by MD simulations.[18]

Finally, we return to the significantly increased dye leakage from aLUVs as compared to ScraLUVs and OLM (Fig. 2). Peptides experience along their translocation path not only a free energy barrier upon entering the hydrophobic core of the membrane, but also upon exiting it in the opposing leaflet.[39] The outer leaflet of our aLUVs is enriched in POPE, while POPG exclusively populates the inner leaflet before the addition of L18W-PGLa. Based on lipid shape-packing arguments we previously reported a significantly lower free energy barrier for bilayers containing cylindrical lipids (such as POPG) as compared to conical lipids (such as POPE).[14] Hence, translocation of L18W-PGLa in $(POPG)^{in}/(POPE/POPG)^{out}$ aLUVs should be energetically easier than in OLM, which densely packs on both sides at the polar/apolar interface. Combined with the reported formation of water bridges and ion leakage during PGLa translocation,[18] this appears as a plausible scenario to explain the differences in dye-leakage between aLUVs and OLM.

Explaining the different leakage of ScraLUVs and aLUVs using the same arguments is more challenging. Here, the free energy barrier for peptide insertion from the exofacial side is lower than for aLUVs, which possibly relates to the initial more rapid increase of dye leakage. However, at the same time the barrier in ScraLUVs is higher for pushing the peptides out of the hydrocarbon regime in the inner leaflet. This might then account for the final lower leakage levels observed for ScraLUVs. Overall, non-equilibrium contributions to relaxation processes originating from peptides interfering with differential elastic stress stored in compositionally distinct membrane leaflets, plausibly play a significant role, but are difficult to quantify in the absence of a theory.

Conclusions

Membrane asymmetry adds yet another layer of complexity to antimicrobial peptide activity. Here, the used aLUVs can be seen as first order mimics of inside-out cytoplasmic membranes of Gram-negative bacteria.[19] While cytoplasmic membrane mimics with 'correct' asymmetry and composition lie ahead of some adaptions of cyclodextrin-mediated lipid exchange, the present study still entails some conclusions of physiological relevance. Firstly, and analogously to our previous finding upon including cardiolipin in (symmetric) mimics of cyto-plasmic bacterial membranes,[24] transbilayer lipid asymmetry makes bilayers more vulnerable to AMP attack due to differential tension of the membrane leaflets. However, we cannot exclude that other membrane entities (such as, *e.g.* proteins) help to counterbalance these differences. Secondly, our study corrobo-rates the idea that leakage and antimicrobial activities observed in lipid-only mimics and bacteria are not necessarily correlated. Deep understanding appears to be only within reach, when combining biophysical studies on cells and membrane mimics.[24]

Author contributions

LM performed experiments, analysed data and wrote the paper. MPKF, GNR and LP performed experiments and analysed data. EFS designed research, analysed data and wrote the paper. KL designed the research. GP designed the research and wrote the paper.

Conflicts of interest

There are no conflicts to declare.

Acknowledgements

This project was supported by the Austrian Science Funds (FWF), project no. P 30921 (to KL).

Notes and references

1 M. Zasloff, *Nature*, 2002, **415**, 389–395.
2 K. A. Brogden, *Nat. Rev. Microbiol.*, 2005, **3**, 238–250.
3 C.-F. Le, C.-M. Fang and S. D. Sekaran, *Antimicrob. Agents Chemother.*, 2017, **61**, e02340-16.

4 Z. Oren and Y. Shai, *Biopolymers*, 1998, **47**, 451–463.

5 B. Bechinger and K. Lohner, *Biochim. Biophys. Acta*, 2006, **1758**, 1529–1539.

6 W. C. Wimley, *ACS Chem. Biol.*, 2010, **5**, 905–917.

7 E. Fattal, S. Nir, R. A. Parente and F. C. Szoka, *Biochemistry*, 1994, **33**, 6721–6731.

8 K. Matsuzaki, O. Murase, N. Fujii and K. Miyajima, *Biochemistry*, 1996, **35**, 11361–11368.

9 L. Zhang, A. Rozek and R. E. Hancock, *J. Biol. Chem.*, 2001, **276**, 35714–35722.

10 T. C. Anglin, K. L. Brown and J. C. Conboy, *J. Struc. Biol.*, 2009, 37–52.

11 M. H. L. Nguyen, M. DiPasquale, B. W. Rickeard, M. Doktorova, F. A. Heberle, H. L. Scott, F. N. Barrera, G. Taylor, C. P. Collier, C. B. Stanley, J. Katsaras and D. Marquardt, *Langmuir*, 2019, **35**, 11735–11744.

12 M. H. L. Nguyen, M. DiPasquale, B. W. Rickeard, C. G. Yip, K. N. Greco, E. G. Kelley and D. Marquardt, *New J. Chem.*, 2021, **45**, 447–456.

13 E. Sevcsik, G. Pabst, A. Jilek and K. Lohner, *Biochim. Biophys. Acta*, 2007, **1768**, 2568–2595.

14 R. Leber, M. Pachler, I. Kabelka, I. Svoboda, D. Enkoller, R. Vácha, K. Lohner and G. Pabst, *Biophys. J.*, 2018, 1945–1954.

15 K. Lohner, E. Sevcsik and G. Pabst, *Advances in Planar Lipid Bilayers and Liposomes*, Elsevier, Amsterdam, 2008, Vol. 6, pp. 103–137.

16 S. Afonin, S. L. Grage, M. Ieronimo, P. Wadhwani and A. S. Ulrich, *J. Am. Chem. Soc.*, 2008, **130**, 16512–16514.

17 B. Bechinger, *J. Pept. Sci.*, 2011, **17**, 306–314.

18 J. P. Ulmschneider, *Biophys. J.*, 2017, **113**, 73–81.

19 M. Bogdanov, K. Pyrshev, S. Yesylevskyy, S. Ryabichko, V. Boiko, P. Ivanchenko, R. Kiyamova, Z. Guan, C. Ramseyer and W. Dowhan, *Sci. Adv.*, 2020, **6**, eaaz6333.

20 N. Harmouche and B. Bechinger, *Biophys. J.*, 2018, **115**, 1033–1044.

21 D. Zweytick, G. Deutsch, J. Andrä, S. E. Blondelle, E. Vollmer, R. Jerala and K. Lohner, *J. Biol. Chem.*, 2011, **286**, 21266–21276.

22 M. Doktorova, F. A. Heberle, B. Eicher, R. F. Standaert, J. Katsaras, E. London, G. Pabst and D. Marquardt, *Nat. Protoc.*, 2018, **13**, 2086–2101.

23 G. R. Bartlett, *J. Biol. Chem.*, 1959, **234**, 466–468.

24 L. Marx, E. F. Semeraro, J. Mandl, J. Kremser, M. P. Frewein, N. Malanovic, K. Lohner and G. Pabst, *Front. Med. Technol.*, 2021, **3**, 625975.

25 E. A. Burstein and V. I. Emelyanenko, *Photochem. Photobiol.*, 1996, **64**, 316–320.

26 A. S. Ladokhin, S. Jayasinghe and S. H. White, *Anal. Biochem.*, 2000, **285**, 235–245.

27 S. H. White and W. C. Wimley, *Biochim. Biophys. Acta*, 1998, **1376**, 339–352.

28 W. C. Wimley, T. P. Creamer and S. H. White, *Biochemistry*, 1996, **35**, 5109–5124.

29 D. Marquardt, F. A. Heberle, T. Miti, B. Eicher, E. London, J. Katsaras and G. Pabst, *Langmuir*, 2017, **33**, 3731–3741.

30 B. Eicher, F. A. Heberle, D. Marquardt, G. N. Rechberger, J. Katsaras and G. Pabst, *J. Appl. Crystallogr.*, 2017, **50**, 419–429.

31 M. Pachler, I. Kabelka, M.-S. Appavou, K. Lohner, R. Vácha and G. Pabst, *Biophys. J.*, 2019, **117**, 1858–1869.

32 I. Kabelka, M. Pachler, S. Prévost, I. Letofsky-Papst, K. Lohner, G. Pabst and R. Vácha, *Biophys. J.*, 2020, **118**, 612–623.

33 J. Gesell, M. Zasloff and S. J. Opella, *J. Biomol. NMR*, 1997, **9**, 127–135.

34 B. Bechinger, M. Zasloff and S. J. Opella, *Biophys. J.*, 1998, **74**, 981–987.

35 D. Zweytick, B. Japelj, E. Mileykovskaya, M. Zorko, W. Dowhan, S. E. Blondelle, S. Riedl, R. Jerala and K. Lohner, *Plos One*, 2014, **9**, e90228.

36 M. P. K. Frewein, M. Rumetshofer and G. Pabst, *J. Appl. Crystallogr.*, 2019, **52**, 403–414.

37 J. M. Boggs, *Biochim. Biophys. Acta*, 1987, **906**, 353–403.

38 S. Kobayashi, K. Takeshima, C. B. Park, S. C. Kim and K. Matsuzaki, *Biochemistry*, 2000, **39**, 8648–8654.

39 I. Kabelka and R. Vácha, *Biophys. J.*, 2018, **115**(6), 1045–1054.

40 C. Snider, S. Jayasinghe, K. Hristova and S. H. White, *Protein Sci.*, 2009, **18**, 2624–2628.

Faraday Discussions

PAPER

Virus-inspired designs of antimicrobial nanocapsules

Carlos H. B. Cruz,† Irene Marzuoli ‡ and Franca Fraternali *

Received 9th July 2021, Accepted 13th August 2021

DOI: 10.1039/d1fd00041a

Antimicrobial resistance is becoming a serious burden for drug design. The challenges are in finding novel approaches for effectively targeting a number of different bacterial strains, and in delivering these to the site of action. We propose here a novel approach that exploits the assembly of antimicrobial peptidic units in nanocapsules that can penetrate and rupture the bacterial membrane. Additionally, the chemical versatility of the designed units can be tailored to specific targets and to the delivery of genetic material in the cell. The proposed design exploits a β-annulus (sequence ITHVGGVGGSIMAPVAVSRQLVGS) triskelion unit from the Tomato Bushy Stunt Virus, able to self assemble in solution, and functionalised with antimicrobial sequences to form dodecahedral antimicrobial nanocapsules. The stability and the activity of the antimicrobial β-annulus capsule is measured by molecular dynamics simulations in water and in the presence of model membranes.

Introduction

For most of the last century, the development of new drugs has focused on the paradigm that a drug is a small inorganic compound (of mass up to 900 Da), which intervenes with a specific target (usually a protein) of a mammal or bacterial cell.

Challenges have included identifying a target unambiguously, especially when the drug binds to proteins assembled in complexes or to a number of closely related gene products.[1] An aspect of concern in the development pipeline has been the efficient delivery of the drug to the site of action: this has been puzzling pharmacologists and chemists for a while, and the aim to overcome these hurdles has inspired the work of synthetic biologists and material scientists. To compli-cate this further, the last few years have seen a dramatic increase in the

Randall Centre for Cell and Molecular Biophysics, King's College London, London, UK. E-mail: franca. fraternali@kcl.ac.uk

† Present address: Instituto de Tecnologia Química e Biológica António Xavier, Universidade Nova de Lisboa, Oeiras, Portugal.

‡ Present address: Process Chemistry and Catalysis, Biocatalysis, F. Hoffmann-La Roche Ltd, Basel, Switzerland.

development of antimicrobial resistance (AMR). The severity of the AMR threat is such that it has been raised to the status of national emergency in several countries, including the UK.[2] Indeed, strict regulations in the health, agricultural and food industry sector must be enforced to prevent the misuse of antibiotics, as we are leaving the century in which these miracle drugs were discovered, to enter a phase in which we count the number losing their efficacy.[3] AMR has become one of the major stimuli to overcome traditional approaches to drug-development.

One of the best nature-designed gene delivery systems are viruses, as they have evolved a number of efficient mechanisms to penetrate cell membranes and implement effective processes to transfer their genomic material to target cells. Recently, we have shown that engineered virus-like nanocapsules derived from synthetic multi branched peptides are able to promote bacterial membrane poration and are, at the same time, suitable for gene delivery.[4,5]

Altogether, the recent progress in understanding the mechanisms of antimicrobial resistance has helped in directing the development of new drugs. In particular, it has promoted the modification of existing compounds to overcome the resistance developed by bacteria.[6]

One crucial aspect in the efficacy of antimicrobials is their interaction and recognition of the cellular membrane and their ability to selectively perturb microbial *versus* mammalian membrane integrity.[5] Therefore, modelling such interactions is very important to select the right lipid composition mixture and to fine-tune and characterise the specific peptide–lipid interactions.[7]

We focused our attention towards membrane active peptides as host-defence peptides belonging to the class of antimicrobial peptides (AMPs). These are naturally produced by Eukarya, either as stand-alone sequences or embedded in larger proteins, as a first, weak, and broad-spectrum defence against bacteria.[8]

To exploit their potential and engineer AMP-like molecules, a careful characterisation and classification of such peptides must be done. This task has been carried out over the last few decades, but because of its complexity, at present there are still many peptides with ascertained antimicrobial activity for which the mode of action is not fully understood.[10] However, some general characteristics of these sequences and some of the mechanisms they employ have been ascertained.

Unsurprisingly, AMPs are heterogeneous in sequence, structure, targets and mode of action, to tackle the different challenges bacteria pose. Their size can vary between 6 and 59 amino acids:[11] despite being small with respect to the average size of a protein in the human body, these macromolecules are hundreds of times larger than small molecule drugs and, as such, they have a different mode of action on bacteria.

As already mentioned, the target of AMPs is the bacterial membrane. Many of them cause disruption of the microbial membrane, while others translocate into the cytoplasm to act on intra-cellular targets, and the combination of the two is not uncommon either.[12] In general, it is widely accepted that membrane interaction and the resulting membrane disruption are key factors for the antimicrobial activity of AMPs.[9]

The main determinant driving the interaction between AMPs and bacterial membranes is the positive charge that many AMPs present, as opposed to the negative charge of the membrane.[9,13]

Most AMPs have a positive charge which facilitates binding to the membrane *via* charge–charge recognition; accordingly, arginine and lysine residues are

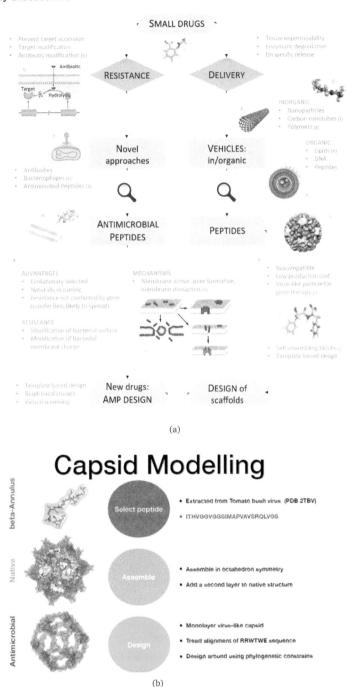

Fig. 1   (a) A Schematic of the approaches used for AM peptide design and delivery strategies. (b) Strategy adopted to build the AM-β-annulus capsule. In red is the representation of the β-annulus triskelion units within the 2TBV structure. These were extracted and modelled to form a pentagonal geometry unit. This was the constituting block for the dodecahedron supra-molecular geometry of the monolayer and bilayer capsules. These were composed of 20 units for the internal later, and 60 units in the external layer. Figures adapted from ref. 8a–g.

usually abundant in AMP sequences. However, the disruptive action takes place through the interaction of AMPs with the hydrophobic core of the membrane, therefore their sequences also contain hydrophobic aromatic residues, especially tryptophan, which favours anchoring to the lipid core.[14]

In terms of folds adopted by AMPs, these are rich in α-helix or β-sheet structures. Amphiphatic α-helices present a charged side which is tailored to face towards the phospholipid head groups, and a hydrophobic side which is favourably buried into the acyl chain core.

Several models have been proposed to describe the exact mechanism of AMPs penetration after they bind to the cytoplasmic membrane, and how their combined action leads to membrane permeabilisation.[9,11]

The picture becomes complex for oligomer-mediated insertion, *i.e.* when the action is triggered by the combined action of many copies of the peptide. At low peptide to lipid ratio, the favourable configuration is represented by peptides lying parallel to the membrane plane as described previously.[15] An increase in peptide concentration triggers the transition to an inserted state: the organisation of AMPs inside the membrane core can assume different configurations, as previously described.[9,11,13,16]

We propose an alternative to this: engineered nanocapsules that contain multiple copies of AM peptides in their structure and that display an overall positive surface. As a constituent, we focus on an AM peptide derived from bovine lactoferricin that has known antimicrobial properties and acts synergistically with other antimicrobial agents by affecting the transmembrane potential and proton-motive force, resulting in inhibition of ATP-dependent multi-drug efflux pumps.[17] A step further in engineering peptidic structures is the design of self-assembling functional blocks from first principles. Indeed, self-assembling peptides can form nanostructures ranging from nanoparticles to nanotubes, nanofibers, nanorods and hydrogels (Fig. 1).[18]

The strength of such molecular designs is in singling out the interactions that are crucial for a mechanism, clarifying whether they can be transferred to a different system or environment, and optimising their combined use in performing the desired function. Such principles can be used in the design, at the atomistic level, of novel and more effective AMPs.

We have previously described some of these engineered AM nanocapsules,[4,5,19] here we want to describe a protocol to design a self-assembling nanocapsule inspired to the Tomato Bushy Structure by extracting some essential building blocks that infer the fullerene geometry and functionalise them with peptides that carry the AM activity.

The atomistic details of the nanocapsule assembly, necessary for the antimicrobial and gene delivery activities, are not accessible by experimental techniques. Therefore, the nanocapsule stability in water and its interaction with a model membrane was studied through molecular dynamics simulations, comparing the results with the available experimental data.[4,19]

Methods

Following up from previous work,[4] we performed coarse-grained simulations of β-annulus (native-like and AM) capsules in solution and on model membranes, using the coarse-grained description provided by SIRAH.[20,21] All simulations were performed with the GROMACS software, version 5.5 and 2016.[22–24]

SIRAH simulations

The assembled capsules were solvated in an aqueous solution of sodium chloride at physiological concentration (0.15 mM) to achieve neutrality in the system. After energy minimisation, the systems were heated from 0 to 300 K over 10 ns at constant volume (10 fs time step), with protein coordinates constrained to their initial position.

Distances between the paired arms of AM peptides within the capsule were restrained with a harmonic potential of 1000 kJ $(mol\ nm^2)^{-1}$, and progressively released over 80 ns (20 fs time step), then the systems were simulated for about 1 μs. The pressure and temperature were kept constant at 1 bar and 300 K, controlled respectively using a V-rescale thermostat and a Berendsen barostat with 2 ns collision-frequency and 8 ps pressure-coupling time. Short-range interactions were treated with a 1.2 nm cut-off radius, whereas the long-range interactions used the Particle Mesh Ewald (PME) method.

We used the recent lipid parametrization in the SIRAH force field[21] united with its characteristic medium grained resolution between the atomistic and the MARTINI coarse-grained one,[20] for the simulations of antimicrobial capsules on membranes (as reported in our previous work).[5] For the simulation of the capsules on membranes, POPC and POPS (in ratio 3 : 1) were selected among the SIRAH available lipids, as POPC is zwitterionic and POPS is negatively charged, reflecting the characteristics of the DLPC and DLPG lipid used in other force-fields for representing the microbial membrane.

The SIRAH parameters and coarse-grained coordinates were assigned to the protein molecules through the cgconv script, distributed with the force field. For the membrane equilibration procedure please refer to Marzuoli et al.[5]

Virus inspired assembly: β-annulus design

We extracted the β-annulus (sequence ITHVGGVGGSIMAPVAVSRQLVGS) triskelion structure from the Tomato Bushy Stunt Virus (TBSV, PDB code: 2TBV [25]) (Fig. 2).

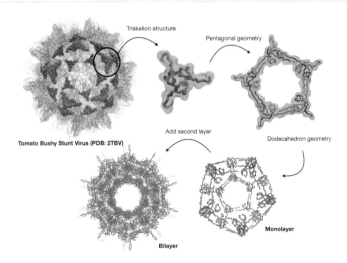

Fig. 2   Structural components of the designed capsule.

This unit has been demonstrated to assemble in a dodecahedron structure when decorated with a number of diverse nanoparticles, preserving the arrangement of the TBSV capsid structure.[26] We therefore modelled the β-annulus only assembly (native like) in a dodecahedral geometry composed of twenty triskelion units. As one can notice from the figure representing the virus structure, the β-annulus triskelion co-assembles with other viral proteins, therefore extracting it from its original environment results in many hydrophobic residues exposed on the surface of the assembled dodecahedral nanocapsule (native-like monolayer). To stabilise this assembly we therefore generated a second layer of triskelion units.

Antimicrobial β-annulus design

We aimed to design a soluble single layer antimicrobial capsule (Fig. 3). This was challenging as we saw that the native single layer β-annulus capsule was not stable *per se*. To generate an AM capsule, we inserted the antimicrobial sequence RRWTWE at the end of the 'arms' of the β-annulus sequence. This generated a hybrid sequence containing the β-annulus and the antimicrobial stretch at the branching points. This was threaded onto the single layer capsule structure *via* the suite AlignedThread from Rosetta.[27] The AM sequence RRWTWE structurally aligned with the RQLVS stretch at the C-terminal of the β-annulus sequence (Fig. 4). We joined the C-terminal fragments through anti-parallel β-sheets using the Chimera program[28] to form a stable scaffold reminiscent of the capzip capsule.[4]

Monte Carlo cycles coupled with PSSM (Position Specific Mutation Score)-based mutagenesis rounds followed by energy minimization were performed, evaluating the energy within the full capsule assembly with ROSETTA scripts.[27,29,30] Of these sampled mutations, only the ones with negative $\Delta\Delta G$ energy were accepted (Fig. 4).

The resulting structure (AM-β-annulus capsule) was simulated with the SIRAH force field in water for 1 μs (simulation conditions as before). The interaction with

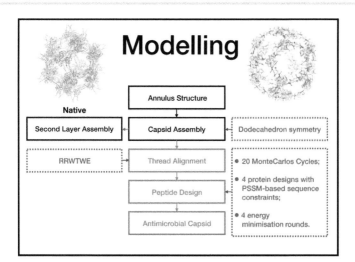

Fig. 3 Pipeline of the design procedure adopted for both *beta*-annulus capsules (native and AM).

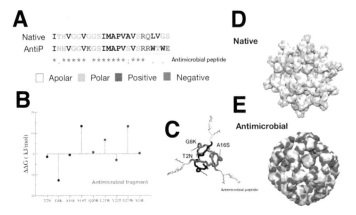

Fig. 4   (A) Sequence alignment of the native and antimicrobial peptides. The colours of the amino acids reflect their properties, with apolar, polar, positively and negatively charged amino acids coloured, respectively, white, green, blue and red. (B) single mutational energy. The mutational energy was determined by the energy difference between the native and mutant. Mutations with positive energy not belonging to the antimicrobial fragment, were accepted. (C) Triskelion structure is shown with the three composing peptides backbone coloured in black, blue and red. The antimicrobial fragment was marked in yellow with the accepted mutations indicated in the picture. (D and E) the native internal layer and the monolayer antimicrobial capsule, are shown with surface representations reflecting the residue properties (colour scheme as in A).

a microbial model membrane was tested in simulations with a POPC/POPS 3 : 1 bilayer.

The equilibrated AM-β-annulus capsule was initially placed at 2 nm of distance from the membrane surface, in a 150 mM NaCl solution (simulation conditions as above, except with semi-isotropic pressure coupling). The system was equilibrated for 0.6 μs, with the protein coordinates constraint in the initial positions. The constraints (1000 kJ $(mol\ nm^2)^{-1}$) were then released. Binding to the membrane occurred after about 1 μs, and the simulation was continued for an additional 1.5 μs to observe stability. In order to stimulate poration, a small electric field was applied across the z-axis (4 mV nm^{-1}) to accelerate the capsule penetration in the lipid bilayer (see discussion points), and the simulation was continued for 3 μs.

Results

Design of an antimicrobial capsule

The native sequence and the modified AM with the RRWTWE peptide from lactoferrin are shown aligned in Fig. 4A.

In the figure the 3D structure of the native monolayer capsule exposing the hydrophobic (apolar) residues on the surface is shown. This was used as a template for the threading procedure (Fig. 4D).

The results of the mutagenesis are displayed in the $\Delta\Delta G$ plot in the figure. The negative $\Delta\Delta G$ values point to residues that better fit in the β-annulus triskelion structure. In total, 3 mutations are necessary to better stabilise the AM-β-annulus within the capsule (T2N, G8K and A16S).

Table 1 Antibacterial activity of peptides

Peptide sequence	Start position	Score	Antibacterial activity
NHVGGVKGSIMAPVS	2	0.940	Yes
GSIMAPVSVSRRWTW	9	0.690	Yes
GVKGSIMAPVSVSRR	6	0.340	No
HVGGVKGSIMAPVSV	3	0.270	No
VKGSIMAPVSVSRRW	7	0.120	No
INHVGGVKGSIMAPV	1	0.000	No
VGGVKGSIMAPVSVS	4	0.000	No
GGVKGSIMAPVSVSR	5	0.000	No
KGSIMAPVSVSRRWT	8	0.000	No

The prediction of antimicrobial activity of the designed β-annulus sequences was performed *via* the AntiBP server,[31] the results of which are reported in Table 1. The position column refers to the starting position in the 15 aa stretch used for the prediction of antibacterial activity. The score is the sum of the antimicrobial predicted probability over each residue within the sequence fragment from the evaluated peptide sequence. The first two fragments present predicted antibacterial activity: NHVGGVKGSIMAPVS and GSIMAPVSVSRRWTW. The second one has 67% of the amino acids making up the whole external surface of the designed capsid AM-β-annulus. Therefore, we speculate that the designed nanosphere will be able to interact with a bacterial membrane and perturb its structure. There is a candidate sequence with a better score (first row in the table), but this is very different from the AM lactoferrin sequence we wanted to test, and for which we have collected experimental results.[4,19] The resulting AM capsule has dodecahedron geometry with the right balance of positive residues on the surface (Fig. 4E).

Capsule stability in water

In Fig. 5 we observe that both the native (blue continuous line) and the antimicrobial (blue dotted line) capsules' RMSD stabilises at around 1000 ns. The corresponding radii of gyration are subject to some shrinking during the first 500 ns

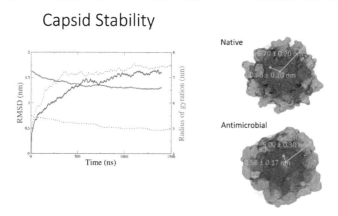

Capsid Stability

Fig. 5 Native and antimicrobial capsule stability.

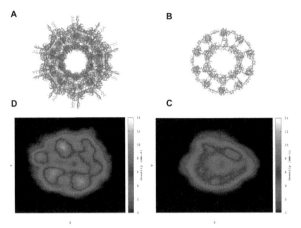

Fig. 6 Atom density projected onto the 2D plane of the designed capsules calculated along the trajectories. (A and D) Double-layer native capsid. (B and C) Single-layer designed antibacterial capsid.

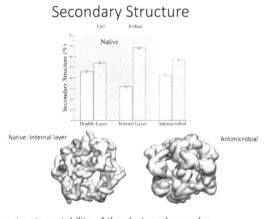

Fig. 7 Secondary structure stability of the designed capsules.

of simulation, with the AM nanocapsule radii (red dotted line) adopting smaller values. The average size of the inner pore of the native capsule measured during the trajectory is around 0.80 nm, while for the AM capsid it is around 0.6 nm. The pores and the overall geometry of the capsule appear to be stable throughout the simulated time.

Having tested the stability of the designed capsules in solution, we measured the time-averaged atom density projected onto a 2D plane for both systems *via* the GROMACS densmap tool. Colours are more intense and localised for the double layer native capsule, while the AM-β-annulus single layer capsule shows a more diffuse pattern, indicative of a relatively higher flexibility (Fig. 6).

Analysis of the secondary structure content maintained during the simulation (Fig. 7) shows that the values for the AM-β-annulus are in between the values measured for the outer and inner layer of the native β-annulus capsule. We

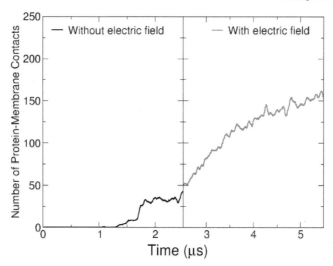

Fig. 8 Protein–lipid contact numbers across the entire simulated production time, before (black line) and after the application of the electric field (red-line).

therefore strike a balance between flexibility and overall structural stability for the designed AM capsule.

Interaction with model microbial membrane

Points for discussion: (a) the role of externally applied electric field; (b) the role of coarse-grained parametrization. The designed AM capsule displays properties of higher porosity and flexibility that may make it more amenable to

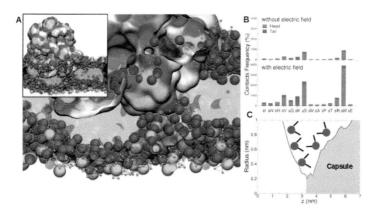

Fig. 9 (A) Poration of the model membrane. Purple spheres represent the phosphate lipid heads populating the pore. The sodium ions are represented by cyan spheres. Water molecules are represented in orange by small spheres. The capsule positioning onto the membrane at pore formation is displayed in the inset. (B) Frequency of contacts with lipid heads (red) and tails (blue) per aa (Sirah's codes) without (top) and with electric field (bottom). (C) Radius of the pore measured along the z-axis of the membrane. Purple spheres are a pictorial representation of the lipid heads.

interact with and disrupt the membrane when compared with the double layer antimicrobial capsule we studied previously.[4,5,19]

These properties were tested by placing the AM-β-annulus near a model bacterial membrane constituting POPC and POPS lipids (in ratio 3 : 1).[5]

During the course of the unrestrained simulation, the capsule moves towards the membrane and starts to interact with it at around 1 μs, as shown by the number of protein–membrane contacts as a function of time (Fig. 8). This number increases quickly to 30 after 1 μs, and remains constant till 2.5 μs, at which point the electric field is switched on. We observe the formation of a pore within 1.5 μs from the field switch, and some atoms of the capsule reach the inner core of the membrane. The number of protein–membrane contacts increases consistently during the electric field driven simulation, reaching about 150 in total, 5 times larger than that observed in the simulation in the absence of an electric field. A similar trend was observed in our previous simulations with the capzip capsule[5] in the interaction with a model membrane (DPLC/DPLG ratio 3 : 1), using a coarse-grained Martini parametrization. Clearly, the number of specific interactions is important not only for the effective attachment of the capsule on the surface but also in driving efficient penetration into the membrane (Fig. 9).

In a previous study,[5] we observed strong membrane invagination for the microbial-mimicking membrane when applying the electric field but no penetration. On the other hand, we did not observe attachment to the modelled mammal membrane, this was reassuring evidence that the capsule could discriminate between membranes increasing its effective AM activity on the microbial membrane. To evaluate the energetics of membrane penetration and the different process of penetration between these two membranes, we did resort to pulling the capsule through the two model membranes and computed the PMF profiles, highlighting a sensibly higher energy barrier for the mammalian model.

For the β-annulus simulations with and without electric field, reported here, in particular we observe lysine residues interacting favourably with the head groups of the POPS lipids, and these interactions increase in number once the capsule is in contact with the transmembrane lipidic portion.

The area per lipid (ApL), thickness of the membrane and lateral diffusion of lipid heads were calculated when the protein does not interact with the membrane (0.0–1.0 μs), after the interaction (1.5–2.0 μs) and under the effect of the electric field (4.5–5.5 μs), as shown in Table 2. The membrane shows similar ApL, thickness and lateral diffusion in the simulation without electric field before and after the capsule attachment, suggesting these interactions do not perturb

Table 2 Membrane properties under the influence of the capsule and electric field[a]

System	ApL, nm^2	Thickness, nm	LD, 10^{-6} cm^2 s^{-1}
BCI	0.633 ± 0.0005	3.90 ± 0.07	0.251 ± 0.016
ACI	0.633 ± 0.0003	3.88 ± 0.05	0.210 ± 0.001
ACI–EF	0.631 ± 0.0011	3.73 ± 0.16	0.162 ± 0.012

[a] BCI: before capsule interaction; ACI: after capsule interaction; ACI–EF: after capsule interaction with electric field.

the membrane. On the contrary, the electric field deforms the membrane and leads to a decrease in the lipid lateral diffusion in the presence of the capsule (Table 2). It is therefore speculated that the membrane–capsule interactions reduce the lateral mobility and elicit pore formation in the presence of an electric field. However, once the pore is stabilised, the order of the lipid tails shows only a slight decrease with respect to the value before the capsule attachment (Fig. 10).

Similar simulations (SIRAH force field, 4 mV nm^{-1} electric field, POPC/POPS 3 : 1 membrane) for the capsule tested in ref. 4, 5 and 19 (capzip) showed attachment to the membrane but no disruption. This is particularly relevant as capzip has experimentally proven antimicrobial activity. In this context, the simulations would suggest higher AM potency for the β-annulus capsule. This will have to be tested experimentally in future work. The capzip capsule relies on the same antimicrobial sequence, a similar triskelion structure, but a shorter linker (namely Lys–Lys–βAla). This results in a higher charge-to-hydrophobic residues ratio, which promotes attachment to the negatively charged membranes but allows for less plasticity. Additionally the capzip capsule is double layered and therefore intrinsically less bendy and malleable. The β-annulus capsule instead, displays more flexibility, and the simulation shows plastic rearrangement of its structure, contributing to pore initiation. For the capzip capsule the interaction with the membrane is especially mediated by the positively charged residues (arginine), analogously to what is observed in the present study. In both cases, the positive residues interact with the negative heads of the POPS lipids.[5]

Furthermore, it is interesting to observe how different force fields require different intensities of the electric field to achieve poration. In our previous work, where capzip molecules were simulated on a model microbial membrane, a field of 130 mV nm^{-1} was necessary to porate the membrane in an atomistic description with a value of 40 mV nm^{-1} for a MARTINI description (simulation

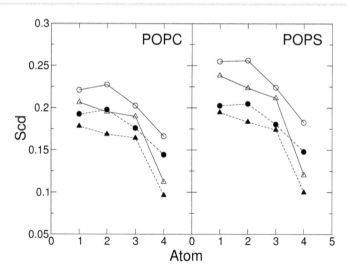

Fig. 10   Order parameter of the POPC and POPS acyl chains 1 (circles) and 2 (triangles) for the simulations without electric field and before capsule interaction (open symbols), and with electric field and a stable capsule interaction, in the last μs of simulation (filled symbols).

length of few hundreds of nanoseconds). These values were not sufficient for poration in absence of the capsule. The simulations performed with the β-annulus capsule can instead be compared with the work of Zonta and coworkers,[32] who tested different values of an electric field on a SIRAH membrane with a hemichannel structure inserted. In this work, a field of 100 mV nm^{-1} was sufficient for poration, while values in the range of 30 mV nm^{-1} were not able to perturb the simulated system. Therefore, the value used in the present work can accelerate poration in conjunction with the antimicrobial capsule, however it is unlikely to be sufficient to disrupt the membrane *per se*. The role of coarse-grained parametrization in responding to external perturbations like the electric force field explored here, should be investigated in more detail in the future.

For the data presented here, further work is needed to prove the activity of the β-annulus capsule: in line with previous work,[4,19] time-resolved imaging of the supported lipid bilayer transfected with the capsule would verify if the capsule breaks simple membranes. Such experiments are the ones most closely related to the simulations. Additionally, the viability of bacterial cells exposed to the capsule can be measured. If the compound effectively acts as an antimicrobial, high-resolution imaging would give information on how the membrane is porated.[4]

Conclusions

We have proposed a framework for the *in silico* design of an antibacterial nano-capsule that is predicted to self-assemble in solution and is able to porate bacterial membranes. The originality of the design is in the combination of already recognised self-assembling triskelion β-annulus units with antibacterial peptidic sequences that have been demonstrated to form supramolecular arrangements rich in β-sheet structures. The AM-β-annulus dodecahedral nano-capsule is stable as a monolayer assembly: this allows for increased solubility and flexibility when compared to previous designs. The designed nanocapsule is able to interact with a bacterial model membrane and to cause poration through a stable pore induced by the application of a small electric field. The pore remains stable for the last 1 μs of simulation, as measured from the protein and lipid membrane atomic contacts occurring in the simulation. Future *in silico* studies will explore the delivery of small drugs within the model membrane and, in parallel, the synthesis of the designed units and their assembly *in vitro*.

Author contributions

CHBC and IM performed the molecular design, computational mutagenesis and molecular simulations. FF conceptualized the project, supervised the methodology and data interpretation and wrote the paper with IM and CHBC.

Conflicts of interest

There are no conflicts to declare.

Acknowledgements

IM and FF acknowledge funding by the Engineering and Physical Sciences Research Council (EPSRC) through the Centre for Doctoral Training ~ Cross Disciplinary Approaches to Non-Equilibrium Systems ~ (CANES, Grant No. EP/L015854/1). CHBC acknowledges funding from CAPES – Proc. no. 88881.197985/2018-01. This project made use of the following resources: computer time on ARCHER granted *via* the UK High-End Computing Consortium for Biomolecular Simulation, HECBioSim (http://www.hecbiosim.ac.uk), supported by EPSRC (grant no. EP/R029407/1); resources provided by the Cambridge Service for Data Driven Discovery (CSD3) operated by the University of Cambridge Research Computing Service (http://www.csd3.cam.ac.uk/), provided by Dell EMC and Intel using Tier-2 funding from the Engineering and Physical Sciences Research Council (capital grant EP/P020259/1), and DiRAC funding from the Science and Technology Facilities Council (https://www.dirac.ac.uk).

Notes and references

1 R. Santos, O. Ursu, A. Gaulton, A. P. Bento, R. S. Donadi, C. G. Bologa, A. Karlsson, B. Al-Lazikani, A. Hersey, T. I. Oprea and J. P. Overington, *Nat. Rev. Drug Discovery*, 2017, **16**, 19–34.

2 J. H. Kwon and W. G. Powderly, *Science*, 2021, **373**, 471.

3 J. O'Neill, *Tackling Drug-Resistant Infections Globally: Final Report and Recommendations the Review on Antimicrobial Resistance*, https://amr-review.org/sites/default/files/160525{_g}Finalpaper{_g}withcover.pdf, 2016.

4 I. E. Kepiro, I. Marzuoli, K. Hammond, X. Ba, H. Lewis, M. Shaw, S. B. Gunnoo, E. De Santis, U. Lapinska, S. Pagliara, M. A. Holmes, C. D. Lorenz, B. W. Hoogenboom, F. Fraternali and M. G. Ryadnov, Engineering chirally blind protein pseudo-capsids into nanoprecise antibacterial persisters, *ACS Nano*, 2020, **14**(2), 1609–1622.

5 I. Marzuoli, C. H. B. Cruz, C. D. Lorenz and F. Fraternali, *Nanoscale*, 2021, **13**, 10342–10355.

6 G. Annunziato, *Int. J. Mol. Sci.*, 2019, **20**, 5844.

7 S. A. Kirsch and R. A. Böckmann, *Biochim. Biophys. Acta, Biomembr.*, 2016, **1858**, 2266–2277.

8 (*a*) J. M. A. Blair, M. A. Webber, A. J. Baylay, *et al.*, *Nat. Rev. Microbiol.*, 2015, **13**(1), 42–51; (*b*) J. Kim, *Phage as a therapeutic agent*, 2017, https://medium.com/@thryve/phage-as-a-therapeutic-agent-ed4c466302e5; (*c*) M. D. Torres, S. Sothiselvam, T. K. Lu, *et al.*, *J. Mol. Biol.*, 2019, **431**(18), 3547–3567; (*d*) The Innovation Society, 2019, http://innovationsgesellschaft.ch/en/carbon-nanotubes-can-be-toxic-to-aquatic-animals/; (*e*) Mettler Toledo, 2018, https://www.mt.com/au/en/home/applications/L1_AutoChem_Applications/L2_ReactionAnalysis/L2_Polymerization.html#publications; (*f*) Wikipedia, 2015, https://en.wikipedia.org/wiki/Liposome; (*g*) L. Schoonen and J. C. M. van Hest, *Nanoscale*, 2014, **6**(13), 7124–7141.

9 L. T. Nguyen, E. F. Haney and H. J. Vogel, *Trends Biotechnol.*, 2011, **29**, 464–472.

10 A. Ebbensgaard, H. Mordhorst, M. T. Overgaard, C. G. Nielsen, F. M. Aarestrup and E. B. Hansen, *PLoS One*, 2015, **10**, e0144611.

11 K. A. Brogden, *Nat. Rev. Microbiol.*, 2005, **3**, 238–250.

12 R. E. W. Hancock and H.-G. Sahl, *Nat. Biotechnol.*, 2006, **24**, 1551–1557.

13 M. Mahlapuu, J. Håkansson, L. Ringstad and C. Björn, *Front. Cell. Infect. Microbiol.*, 2016, **6**, 194.

14 D. I. Chan, E. J. Prenner and H. J. Vogel, *Biochim. Biophys. Acta, Biomembr.*, 2006, **1758**, 1184–1202.

15 L. Yang, T. A. Harroun, T. M. Weiss, L. Ding and H. W. Huang, *Biophys. J.*, 2001, **81**, 1475–1485.

16 T. Ebenhan, O. Gheysens, H. G. Kruger, J. R. Zeevaart and M. M. Sathekge, *BioMed Res. Int.*, 2014, **2014**, 867381.

17 J. L. Gifford, H. N. Hunter and H. J. Vogel, *Cell. Mol. Life Sci.*, 2005, **62**, 2588–2598.

18 T. Fan, X. Yu, B. Shen and L. Sun, *J. Nanomater.*, 2017, **2017**, 1–16.

19 V. Castelletto, E. de Santis, H. Alkassem, B. Lamarre, J. E. Noble, S. Ray, A. Bella, J. R. Burns, B. W. Hoogenboom, M. G. Ryadnov, J. Vandesompele and C. T. Wittwer, *Chem. Sci.*, 2016, **7**, 1707–1711.

20 M. R. Machado, E. E. B. Guisasola, F. Klein, M. Sóñora, S. Silva and S. Pantano, *J. Chem. Theory Comput.*, 2019, **15**, 2719–2733.

21 E. E. Barrera, M. R. Machado and S. Pantano, *J. Chem. Theory Comput.*, 2019, **15**, 5674–5688.

22 H. Berendsen, D. van der Spoel and R. van Drunen, *Comput. Phys. Commun.*, 1995, **91**, 43–56.

23 M. J. Abraham, T. Murtola, R. Schulz, S. Páll, J. C. Smith, B. Hess and E. Lindahl, *SoftwareX*, 2015, **1–2**, 19–25.

24 M. J. Abraham, D. van der Spoel, E. Lindahl, B. Hess and the GROMACS development team, *GROMACS User Manual*, 2021.

25 P. Hopper, S. Harrison and R. Sauer, *J. Mol. Biol.*, 1984, **177**, 701–713.

26 S. Fujita and K. Matsuura, *Nanomaterials*, 2014, **4**, 778–791.

27 S. Ovchinnikov, L. Kinch, H. Park, Y. Liao, J. Pei, D. E. Kim, H. Kamisetty, N. V. Grishin and D. Baker, *eLife*, 2015, **4**, e09248.

28 E. F. Pettersen, T. D. Goddard, C. C. Huang, G. S. Couch, D. Greenblatt, E. C. Meng and T. E. Ferrin, *J. Comput. Chem.*, 2004, **25**, 1605–1612.

29 S. J. Fleishman, A. Leaver-Fay, J. E. Corn, E.-M. Strauch, S. D. Khare, N. Koga, J. Ashworth, P. Murphy, F. Richter, G. Lemmon, J. Meiler and D. Baker, *PLoS One*, 2011, **6**, e20161.

30 A. Goldenzweig, M. Goldsmith, S. E. Hill, O. Gertman, P. Laurino, Y. Ashani, O. Dym, T. Unger, S. Albeck, J. Prilusky, R. L. Lieberman, A. Aharoni, I. Silman, J. L. Sussman, D. S. Tawfik and S. J. Fleishman, *Mol. Cell*, 2016, **63**, 337–346.

31 S. Lata, B. K. Sharma and G. P. Raghava, *BMC Bioinf.*, 2007, **8**, 1–10.

32 F. Zonta, D. Buratto, G. Crispino, A. Carrer, F. Bruno, G. Yang, F. Mammano and S. Pantano, *Front. Mol. Neurosci.*, 2018, **11**, 170.

DISCUSSIONS

Peptide–membrane interactions and biotechnology; enabling next-generation synthetic biology: general discussion

Mibel Aguilar, Patricia Bassereau, Margarida Bastos, Paul Beales, Burkhard Bechinger, Boyan Bonev, Izabella Brand, Edward Chalouhi, Ronald J. Clarke, Evelyne Deplazes, Franca Fraternali, Patrick Fuchs, Bart Hoogenboom, Reidar Lund, Najet Mahmoudi, Paula Milán Rodríguez, Paul O'Shea, Georg Pabst, Sreetama Pal, Amy Rice, John Sanderson, John Seddon, Durba Sengupta, David P. Siegel, Anand Srivastava, Johanna Utterström, Robert Vácha, Leonie van 't Hag, Aishwarya Vijayakumar and Larisa Zoranić

DOI: 10.1039/d1fd90068d

Ronald J. Clarke opened a general discussion of the paper by Mibel Aguilar: Is there anything known about the distribution of lipids across the bacterial membrane, *i.e.* is it asymmetric? When the phosphatidylglycerol content goes up in the stationary phase, is it in both leaflets or only on one side of the membrane?

Mibel Aguilar responded: There is less known about the asymmetry of bacterial membranes than mammalian membranes but evidence is emerging of a dynamic system of lipid curation. A recent study by Bogdanov *et al.*[1] showed that PE levels are different in the cytoplasmic and periplasmic layers, and that the PE levels are modulated in response to different environmental factors.

1 M. Bogdanov, K. Pyrshev, S. Yesylevskyy, S. Ryabichko, V. Boiko, P. Ivanchenko, R. Kiyamova, Z. Guan, C. Ramseyer and W. Dowhan, *Sci. Adv.*, 2020, **6**, eaaz6333.

Bart Hoogenboom asked: Have you considered looking at extracts of Gram-positive bacterial lipids, as this may avoid any confusion between lipids from the inner and outer membrane as in Gram-negative bacteria?

Mibel Aguilar responded: Yes, that is certainly what we plan to do. There is a definite need to establish a membrane lipid catalogue of the normal strains for Gram-positive and Gram negative bacteria and then use these to monitor changes in response to environmental triggers and drug interactions. There is some reported evidence of changes in lipid-synthesis enzymes in some drug-resistant strains of Gram-positive pathogens, but there is no systematic study as of yet.

Bart Hoogenboom commented: With your *E. coli* lipid extracts, do you have an idea of which lipids come from the inner membrane and which from (the inner leaflet of) the outer membrane? Do you think that could matter (or not) for the interpretation of your data?

Mibel Aguilar replied: It is definitely an important factor for future studies and we aim, for example, to study the influence of LPS. In our extraction protocol, the chemical lysis of the cells is performed using a mixture of chloroform, methanol and water, following which the LPS is pelleted by centrifugation.

Bart Hoogenboom asked: Your AFM data are recorded at room temperature, aren't they? Does the domain formation change as a function of temperature between RT and 37 °C?

Mibel Aguilar answered: Yes, membranes become more dynamic at higher temperatures, which could change the impact of AMPs through changes in fluidity, *etc.* In our current experimental set-up, the buffer volume is only 150–200 μL, and the sample can become dehydrated if imaging over 2 h at 37 °C.

Bart Hoogenboom said: We can resolve different domains (LPS-enriched and OMP enriched) by AFM on live *E. coli*, though that is of course only the outer membrane. No idea yet if/how that affects AMP action.

Mibel Aguilar replied: That is very interesting to know. I think there will be now a strong interest in monitoring and manipulating domains, especially with the most recent AFM instruments. As you infer, the next challenge is to visualise the inner membranes in live cells.

John Sanderson queried: If you extract the lipids after administration of your peptide, does the lipid profile change? Are you able to select out strains with natural resistance that have a modified lipid profile?

Mibel Aguilar replied: We have preliminary data to suggest that exposure of the bacteria to Mac1 does change the composition upon further rounds of growth. Also, there is also a certain degree of reversible lipid composition changes upon withdrawal of the peptide.

Paula Milán Rodríguez enquired: Is there a change in the bacterial size during growth that could change the degree of the curvature of the membrane and affect the AMP affinity?

Mibel Aguilar answered: That is not something that we have examined. But, given the large changes in the lipid acyl chain and head group properties, it is possible that there will be changes in the membrane curvature that could impact the susceptibility to cell lytic agents.

Paul Beales asked: I assume the lipidomics is from an extract from a large ensemble of cells. Do you have a feel for the heterogeneity in lipidomics between individual cells? Could this be greater than the average variability between different phases of the cell cycle? This could be of relevance to antimicrobial resistance mechanisms in bacterial populations, for example.

Mibel Aguilar replied: Yes, the lipid extracts are derived from cultures at OD = 1 and OD = 8 and, therefore, represent an average over the entire population. It is possible that there is more heterogeneity at OD1 (early growth phase), but OD = 8 may be more homogenous.

Paul O'Shea remarked: A major contributor to the surface charge of Gram-positive bacteria is techoic acid, so the surface electric charge may have fundamentally different origins in Gram-negative bacteria like *E. coli*. Are you suggesting that the latter originates mostly from lipids?

Mibel Aguilar responded: Techoic acid may well play a role, as Mac1 is more effective against Gram-positive bacteria. I was referring to the general observation that microorganisms, in this case bacteria, can change the overall charge of the phospholipid component of their membranes by altering the synthesis of PG and other negatively charged lipids. This can then reduce the effectiveness of agents that interact with the membrane *via* electrostatic interactions or, as we see in our current study, increase the effectiveness of Mac1.

Sreetama Pal asked: Have you looked at how different or similar these blebs are from the parent membranes in terms of the protein content, lipid composition and membrane physical properties? For example, is it possible that there are peptide subpopulations in a pseudotransmembrane or transmembrane orientation localized predominantly in the flat parts of the membrane (*i.e.*, the presence of peptides in conformations shown in both panels (b) and (h) of Fig. 8 of the paper, DOI: 10.1039/d0fd00052c)? If yes, could you repurpose your system to dissect whether distinct membrane orientations of the same peptide could lead to very different membrane remodeling (in the presence or absence of blebbing)?

Mibel Aguilar responded: We have not yet been able to extract material from each domain for further analysis. Also, the amount of lipid in each image does not provide sufficient material for analysis by mass spectrometry. Since these are lipid extracts, in this case, proteins would not contribute to the blebbing. The orientation of Mac1 within the membrane is critical and we base the transmembrane orientation on the appearance of holes in the EGP and, as we do not observe holes prior to blebbing in the SGP, we propose a surface orientation. Certainly the use of lipids to modify membrane properties is potentially a way to modulate the susceptibility to cell lytic agents.

Sreetama Pal enquired: Do the individual blebbing domains bud out completely from the parent bilayer to form separate blebs?

Mibel Aguilar responded: The SPR experiments, which only detect mass bound to the surface, are performed with a fluid cell and so any material that is displaced from the surface will be washed away. This is also a similar outcome for the SPR experiments, and this explains the drop in the baseline on the sensorgrams reflecting the loss of material during Mac1 binding. The fate of the lost material is unknown and, while some spherical vesicles may form during the blebbing, it is not possible to easily track them at this stage. It can also be noted that, while the SPR is performed with a flow system, the AFM cell is static. So the blebs may form in SPR but in AFM the static conditions allow the blebs to remain intact and visible after imaging.

Patricia Bassereau asked: What happens to the lipid composition upon division? Does the new bacteria have a different lipid composition from the mother one?

Mibel Aguilar replied: There are not many reports on the detailed lipidomic analysis of bacterial cells and the changes that occur during the growth phases. Since the lipidome differs significantly between the early and stationary phases in our study, and the cells' growth stages are not synchronised, it is likely that the membrane composition may also be different between the parent cells and the progeny.

Leonie van 't Hag said: It is interesting to see the formation of three different domains in the SPG lipid extract. Was the third domain consistently observed as a quite minor fraction at the start? Or was there any extract-to-extract variability? Do you have an idea of how many lipid species would be required in order to form more than the often observed two (liquid-ordered and -disordered) domains in three-component systems? Lipid saturation, but also alkyl chain length, probably play an important role, as you also show here.

Mibel Aguilar answered: Yes, domain 3 was consistently present in the different extracts and exhibited a consistently low percentage of the total surface.

We are intrigued that 2 or 3 domains form with such a complex mixture of phospholipids, particularly in terms of the acyl chain differences. Given that they differ in height, there may be a threshold of chain lengths that cluster to form the domains and there may be an enrichment of certain head groups. This will be important to determine in future work.

Burkhard Bechinger remarked: In a natural environment, there must be bacteria in all different growth phases, whereas at a given peptide concentration the fast growing bacteria are killed and others are exposed to the antimicrobial peptides just below the MIC of the corresponding population. Therefore, does this contribute to resistance development and how can it can be avoided?

Mibel Aguilar replied: It is becoming apparent that these membrane composition changes are metabolically regulated and potentially reversible as the bacteria respond to the environmental changes. And yes, it has been shown that the bacteria change their membrane composition in response to exposure to antimicrobial peptides by a number of mechanisms that can include reducing the overall negative charge (*e.g.* by lyslated phospholipids) and increasing the bilayer thickness. Thus, limited sequential exposure to antimicrobials that act by different mechanisms could be a method to reduce the potential for resistance to develop.

Georg Pabst opened a general discussion of the paper by Burkhard Bechinger: C-terminal amidation usually helps AMP insertion or translocation through the bilayer. Can you comment on this for your scenario? This might explain the altered activity of individual peptides, but what is its role in the synergism? Might a non amidated C-terminal additionally support the surface aligned topology and stabilize the fiber-like in-plane aggregation of PGLa and MG2a?

Burkhard Bechinger answered: Indeed, amidation of the C-terminus increases the antimicrobial activities of the peptides. Because the bacterial membranes and our corresponding models are overall negatively charged, the removal of the negative charge from the C-terminus potentially helps the peptides to associate with the membrane and to insert deeper. Interestingly, our experiments show that if one of the peptides is amidated the synergetic factor remains unchanged upon modification of the termini of the second peptide (Fig. 4C and D of the paper, DOI: 10.1039/d0fd00041h). Nevertheless, electrostatic interactions seem important because magainin without the negative charge at position 19 or PGLa with modified K15 or K19 lose synergistic activity. With regard to the formation of mesophases or fiber-like structures, we have not tested all the terminal modifications. Notably, in our fluorescence quenching experiments demonstrating mesophase arrangements, the N-terminus carries the fluorophore.[1] Such modifications have been shown to be less active in antimicrobial assays but to keep their synergistic activity.[2]

1 C. Aisenbrey, M. Amaro, P. Pospisil, M. Hof and B. Bechinger, *Sci. Rep.*, 2020, **10**, 11652.
2 A. Marquette, E. Salnikov, E. Glattard, C. Aisenbrey and B. Bechinger, *Curr. Top. Med. Chem.*, 2016, **16**, 65–75.

Sreetama Pal asked: From the binding data, is it possible to extract thermodynamic parameters reflecting the affinity of PGLa and magainin in their membrane-adsorbed forms?

Burkhard Bechinger responded: Yes, in Aisenbrey *et al.* (2020)[1] the partition isotherms have been analyzed in a quantitative manner (Table 1).

1 C. Aisenbrey, M. Amaro, P. Pospisil, M. Hof and B. Bechinger, *Sci. Rep.*, 2020, **10**, 11652.

Sreetama Pal asked: How would you expect the lipid annulus of these two peptides to evolve as each goes from an individual membrane-bound conformation to a PGLa–magainin peptide complex? In addition, are their non-annular lipids trapped between the two helices in the peptide–peptide complex?

Burkhard Bechinger responded: In general, we observe that the lipids are more disordered in the presence of in-plane oriented amphiphilic peptides. As for the mixtures, the negatively charged lipids tend to show more disorder than the neutral ones.[1] Therefore, I would expect that there are anionic lipids making a link between the cationic peptides and that there are probably not many non-annular lipids trapped in the mesophases. An observation possibly related to this is that phosphates help in fiber formation by these peptides.[2] But, with the data available at the present time, I can only speculate about such details.

1 N. Harmouche and B. Bechinger, *Biophys. J.*, 2018, **115**, 1033–1044.
2 D. W. Juhl, E. Glattard, M. Lointier, P. Bampilis and B. Bechinger, *Front. Cell. Infect. Microbiol.*, 2020, **10**, 526459.

Margarida Bastos asked : You say in the paper (DOI: 10.1039/d0fd00041h) that fluorophore labelled PGLa helps magainin to associate with POPE/POPG membranes by the formation of loosely interacting mesophases made of the peptides. Could you elaborate a bit on what the structure of these mesophases is? Are they only peptides or peptide/lipid mesophases? I believe it has to be the last one, as they depend on lipids and it is the peptide–lipid that can build up and help the second peptide's partition.

Burkhard Bechinger responded: You are correct, our models include peptides and lipids. However, we have little information as to the structural details. We speculate that the cationic charges of the peptides are screened by anionic lipids. Dipolar interactions between the peptides and salt bridges may be important (as shown in Fig. 5 in Marquette *et al.* (2016)[1]). Furthermore, it looks as if the lipid fatty acyl chains exhibit high disorder in the proximity of the peptides.

1 A. Marquette, E. Salnikov, E. Glattard, C. Aisenbrey and B. Bechinger, *Curr. Top. Med. Chem.*, 2016, **16**, 65–75.

Margarida Bastos enquired: Do you have a plan on how to elucidate this lipid/peptide combination and characterize the mesophase?

Burkhard Bechinger answered: Yes, we are performing solid-state NMR investigations in order to reveal the specific contacts between peptides and between peptides and lipids. In parallel, we are also working on revealing the size, shape and fluctuations of the supramolecular mesophases by high resolution optical microscopy techniques.

Aishwarya Vijayakumar commented: In one of the slides, the bacterial over-expression of AMP was mentioned. I have never expressed AMPs so I am curious about whether there is any affect of the AMP being expressed on the host organism. Which organism is used for overexpression of AMPs?

Burkhard Bechinger replied: This is an excellent point. We use *E. coli* laboratory strains for the overexpression of the peptides. It is indeed a difficult task to produce high amounts of antimicrobial peptides inside a bacterium. By over-expressing a fusion protein encompassing a very hydrophobic protein domain, a linker and the AMP tight inclusion bodies are formed without toxicity to the bacteria. These can be isolated and cleaved with formic acid in a one step procedure. The acid also solubilizes the inclusion bodies, thus the peptide can be purified by HPLC. The full protocol is published in ref. 1.

1 V. Vidovic, L. Prongidi-Fix, B. Bechinger and S. Werten , *J. Pept. Sci.*, 2009, **15**(4), 278–284.

Boyan Bonev asked: Did you suggest in your talk that cationic peptides act in a collective manner to pull electrostatically lipids between, say, three peptides?

Burkhard Bechinger answered: Indeed, we suggest that 'two-dimensional' supramolecular arrangements form on the surface of membranes. These are made up of peptides and lipids. At the moment we have data showing proximities on the 1 nm scale but we cannot tell yet how many peptides and lipids take part in these mesophases.

Boyan Bonev said: Do you have any evidence for transient pore formation?

Burkhard Bechinger replied: Pore formation of *e.g.* magainin was character-ized early on by electrophysiological single-channel conductance measurements and by fluorophore release from vesicles. The membrane permeation of syner-gistic mixtures has been investigated using fluorescence spectroscopy, through the monitoring of the cellular membrane potential and biological assays. These studies, including the very early ones, have recently been reviewed in ref. 1.

1 B. Bechinger, D. W. Juhl, E. Glattard and C. Aisenbrey, *Front. Med. Technol.*, 2020, **2**, 615494.

Edward Chalouhi commented: Do the terminal modifications have an effect on the structural stability of the peptides?

Burkhard Bechinger responded: Yes, the propensity of the peptides to form helixes next to the termini is affected by these modifications (Fig. 1 and 2 of the paper, DOI: 10.1039/d0fd00041h). C-terminal amidation helps to extend the helices up to the very last residues, probably because it is in concordance with the negative end of the helix dipole. The amino-terminal proline increases the helix propensity of the N-terminus of magainin but not of PGLa, probably because in lipid bilayer environments PGLa adopts a stable helix conformation only from the middle of the sequence on.[1] Notably, environments where helix formation is strong may overrule the effects of modifying the termini.

1 B. Bechinger, M. Zasloff and S. J. Opella, *Biophys. J.*, 1998, **74**, 981–987.

Larisa Zoranić asked: Is there significant peptide association prior to membrane binding, and could this be part of a synergistic mechanism?

Burkhard Bechinger answered: We have checked peptide association in solu-tion by various techniques, including NMR and DLS, and have not found indi-cations that they associate spontaneously under the conditions tested here. In aqueous buffer, the peptides are unstructured and highly charged. However, by adding bivalent anions, adjusting the pH and shaking the samples we were able to form peptide fibers.[1] Although the antimicrobial activity of the preformed fibers is reduced when compared to that of the monomers, they still exhibit synergism in antimicrobial assays. In view of the highly cationic character of the peptides, the negatively charged phospholipids are important for efficient membrane association and probably also mesophase formation within the lipid bilayer. In a related manner, in solution the phosphate ions may bridge the cationic peptides.

1 D. W. Juhl, E. Glattard, M. Lointier, P. Bampilis and B. Bechinger, *Front. Cell. Infect. Microbiol.*, 2020, **10**, 526459.

Johanna Utterström remarked: I'm also curious about the mechanism for peptide–lipid interaction in the stage leading up to the mesophase. Is it hydrophobic or electrostatic interactions or something else that is driving this? Also, I was wondering if you have tried including cholesterol and seeing how it affects this mesophase formation? Especially since cholesterol can facilitate raft formation and you mentioned that association between the peptides and negatively charged lipids has been observed.

Burkhard Bechinger responded: We know that the negative curvature of PE head groups and the negatively charged lipids are key elements for peptide association and synergism in model membranes. This suggests that electrostatics are important. These ensure a high enough P/L ratio of cationic peptides and they probably screen repulsive charges that are expected to keep the highly cationic peptides apart. The requirement for the PE lipid to observe the increased membrane partitioning of the synergistic mixtures suggests that curvature strain, tighter packing and/or the H-bonding capabilities of this lipid are also involved. Interaction contributions could also arise from changes in the (dis)order of the fatty acyl chains (lipophobic interactions). Because cholesterol significantly increases the order parameters of the lipid fatty acyl chains, it would indeed be interesting to investigate what happens in the presence of this lipid. So far, we have not performed such experiments as we focused on mimetics of bacterial membranes.

Patricia Bassereau queried: Which type of phase/organization do you expect to lead to bilayer destabilization? Is it compatible with your lipid composition?

Burkhard Bechinger answered: The ^{31}P solid-state NMR spectra of phospholipid membranes in the presence of amphipathic peptides have been analyzed in terms of distorted surfaces, 'dips', regions with high local curvature and membrane rupture (*e.g.* reviews in ref. 1 and 2 or data obtained with magainin in ref. 3). Interestingly, the synergistic membrane association of the peptides is observed in POPE/POPG but not in POPC/POPG membranes.[4] This seems counter intuitive because the inverted cone shape of POPE stabilizes the bilayer in the presence of *e.g.* magainin.[5,6] However, the higher peptide density within the mesophases and the preferential accumulation of PG lipids in the vicinity of the cationic peptides can still result in high local curvature strain and membrane disruption. This being said, at the present time we have no experimental view of the molecular details within the mesophases.

1 B. Bechinger and E. S. Salnikov, *Chem. Phys. Lipids*, 2012, **165**, 282–301.
2 C. Aisenbrey, A. Marquette and B. Bechinger, *Adv. Exp. Med. Biol.*, 2019, **1117**, 33–64.
3 B. Bechinger, *Biochim. Biophys. Acta*, 2005, **1712**, 101–108.
4 C. Aisenbrey, M. Amaro, P. Pospisil, M. Hof and B. Bechinger, *Sci. Rep.*, 2020, **10**, 11652.
5 K. Matsuzaki, Y. Mitani, K. Akada, O. Murase, S. Yoneyama, M. Zasloff and K. Miyajima, *Biochemistry*, 1998, **37**, 15144–15153.
6 C. Aisenbrey, A. Marquette and B. Bechinger, *Adv. Exp. Med. Biol.*, 2019, **1117**, 33–64.

David P. Siegel remarked: It may be helpful to do small-angle X-ray diffraction studies of your peptide–lipid mesophase.

Burkhard Bechinger replied: In the context of a binational collaboration between Austria and France such experiments were performed and are described in ref. 1 and 2. Additional features were indeed detected in the SAXS patterns.

1 M. Pachler, I. Kabelka, M. S. Appavou, K. Lohner, R. Vácha and G. Pabst, *Biophys. J.*, 2019, **117**, 1858–1869.
2 I. Kabelka, M. Pachler, S. Prevost, I. Letofsky-Papst, K. Lohner, G. Pabst and R. Vácha, *Biophys. J.*, 2020, **118**, 612–623.

Robert Vácha enquired: Have you tested the synergism with modified peptides on LUVs to test if the synergism and mechanism also remain the same?

Burkhard Bechinger responded: Synergism in model membrane systems of some peptides modified at their termini has been tested (in ref. 1–4) but not as systematically as in the paper presented here (DOI: 10.1039/d0fd00041h). We also investigated the activity of preformed fibers of magainin and/or PGLa.[5] Indeed, many more sequence variations were tested in antimicrobial assays (*e.g.* ref. 1, 3, 6 and 7).

1 K. Matsuzaki, Y. Mitani, K. Akada, O. Murase, S. Yoneyama, M. Zasloff and K. Miyajima , *Biochemistry*, 1998, **37**, 15144–15153.
2 R. Leber, M. Pachler, I. Kabelka, I. Svoboda, D. Enkoller, R. Vacha, K. Lohner and G. Pabst, *Biophys. J.*, 2018, **114**, 1945–1954.
3 J. Zerweck, E. Strandberg, O. Kukharenko, J. Reichert, J. Burck, P. Wadhwani and A. S. Ulrich, *Sci. Rep.*, 2017, **7**, 13153.
4 A. Marquette and B. Bechinger, *Biomolecules*, 2018, **8**, 18.
5 D. W. Juhl, E. Glattard, M. Lointier, P. Bampilis and B. Bechinger, *Front. Cell. Infect. Microbiol.*, 2020, **10**, 526459.
6 E. Glattard, E. S. Salnikov, C. Aisenbrey and B. Bechinger, *Biophys. Chem.*, 2016, **210**, 35–44.
7 A. Marquette, E. Salnikov, E. Glattard, C. Aisenbrey and B. Bechinger, *Curr. Top. Med. Chem.*, 2016, **16**, 65–75.

John Seddon opened a general discussion of the paper by Georg Pabst: The advantage of your lipid system seems to be that the change in SANS intensity arises purely from flip–flop and does not require lipid exchange between vesicles, making the analysis simpler and less ambiguous?

Georg Pabst responded: Indeed, SANS experiments were designed to have only one changing contrast, which is the one between the inner and outer leaflets. Although lipid transport through the aqueous phase happens, there is no contrast difference between the outer leaflets of our vesicles. Hence, there is no convolution observed between these different processes. This leaves us with only one decay constant and consequently much simpler and unambiguous data analysis. The price to pay is the fabrication of vesicles with chain deuterated lipids distributed asymmetrically across the leaflets. But there are detailed protocols available (see *e.g.* ref. 1), which should facilitate such endeavors.

1 M. Doktorova, F. A. Herbele, B. Eicher, R. F. Standaert, J. Katsaras, E. London, G. Pabst and D. Marquardt, *Nat. Protoc.*, 2018, **13**, 2086–2101.

Paul O'Shea said: The ANTS/DPX leakage assay probes a certain pore size. Have you looked at other size indicators? Does ion leakage take place at say lower concentrations or larger molecules such as dextrans at higher concentrations?

Georg Pabst replied: Unfortunately, I cannot comment on either of these questions. We applied a standard leakage assay and did not vary the size of the fluorophore and/or quencher. We also did not look into ion leakage. Your comments and interest are highly encouraging (hopefully also for others) to further pursue these lines of enquiry.

John Sanderson asked: Your dye-leakage experiments and interleaflet exchange processes have differing timescales. Marker release appears to stop after about an hour in the asymmetric LUVs, whereas exchange takes several hours more. Do you know if there are any changes in peptide orientation or partitioning occurring on a timescale comparable to that of flip–flopping?

Georg Pabst answered: The half time of lipid flip–flop in the presence of L18W-PGLa is about 1 hour at $[P]/[L] = 1/400$. However, the lipid concentration for the leakage experiments is much lower, meaning that fewer peptides will be partitioned than in flip–flop measurements. So indeed, lipid flip–flop is slower than poration of the membrane. This might suggest that pores are too short lived to the lipids to translocate to the other leaflet. Note however that lipid flip–flop will certainly continue after all of the dye has been released from the lumen of the vesicle (as transient pores will be still forming, but we will be blind to it). Hence, the time scales of leakage and lipid flip–flop are not expected to correlate. Regarding peptide orientation and partitioning: our experiments are not directly sensing peptide topology. However, based on previous reports,[1,2] peptides are expected to translocate through the membrane on this time scale and reside in the final equilibrium state (end state) with the surface aligned topologies about equally distributed in both leaflets. In order to reach this state, peptides have to transiently change their orientation with respect to the bilayer. Due to the high lipid concentration of our flip–flop measurements ($[L] = 9$ mM), peptides can be considered to be fully partitioned. We have performed time resolved Trp-emission spectroscopy measurements under conditions close to those of leakage experiments (*i.e.* where peptide partitioning can be of importance due to low lipid concentrations). We observed no time dependence in our data (as reported in the manuscript, DOI: 10.1039/d1fd00039j).

1 N. Harmouche and B. Bechinger, *Biophys. J.*, 2018, **115**, 1033–1044.
2 I. Kableka and R. Vácha, *Biophys. J.*, 2018, 115, 1045–1054.

Durba Sengupta enquired: Could you comment on differential stress in your asymmetric LUVs? And could that be related to the differential kinetics of the dye leakage (aLUVs and ScraLUVs)?

Georg Pabst responded: We have discussed our view on the contribution of differential stress in aLUVs to dye leakage in detail in the paper (DOI: 10.1039/d1fd00039j). In particular, it appears plausible that the peptides facing a more tightly packed outer leaflet due to the presence of POPE (cone-shaped, H-bonding) will have a higher free energy barrier for inserting into the hydrophobic core of the membrane. This might explain the slower leakage kinetics at the beginning. Fully translocating through the membrane, *i.e.* to a surface-aligned state in the opposing leaflet, however, also involves a second free energy barrier at the inner vesicle polar/apolar interface. In the case of aLUVs, this barrier is significantly lower than that in ScraLUVs, because of the much higher abundance of POPG (cylindrically shaped). Thus, it should be easier for peptides to 'exit' in the inner leaflet of aLUVs and ScraLUVs. This effect could explain the slowing down of leakage kinetics in ScraLUVs, finally leading to a much lower total dye-leakage. In a semi-quantitative approach, one can estimate the stored curvature free-energies per unit area in both leaflets using reported values for the intrinsic lipid curvatures[1] and bending rigidities[2] for POPE and POPG, applying molecular averaging (and assuming linear additivity). In this case, we calculate $Es \sim 4\ k_BT/nm^2$ for the outer leaflet and $\sim 3 \times 10^{-3}\ k_BT/nm^2$) for the inner leaflet of aLUVs (*i.e.* a three order of magnitude difference!). Analogously, we estimate for ScraLUVs $Es \sim 3 \times 10^{-2}\ k_BT/nm^2$). Although this certainly is too simplistic, it supports the scenario discussed above.

1 R. Leber, M. Pachler, I. Kabelka, I. Svoboda, D. Enkoller, R. Vácha, K. Lohner and G. Pabst, *Biophys. J.*, 2018, **114**, 1945–1954.
2 R. M. Venable, F. L. H. Brown and R. W. Pastor, *Chem. Phys. Lipids*, 2015, **192**, 60–74.

Edward Chalouhi asked: How do asymmetric LUVs compare to traditional LUVs in terms of stability?

Georg Pabst answered: aLUVs are out of thermal equilibrium, so lipid distribution will equilibrate over time. We have detailed this previously (see ref. 1). The stability of asymmetry certainly depends on the chosen lipid composition (and storage temperature), but is in the order of 3 days to 1 week. In the present case, we did not see any changes in the asymmetry over 24 hours in our flip–flop measurements in the absence of peptides. During this time, no fusion took place, *i.e.* we did not see any changes in the vesicle size. Regarding this aspect of the stability (say the shelf-life of the vesicle size), we would expect no significant difference from the LUVs.

1 M. Doktorova, F. A. Herbele, B. Eicher, R. F. Standaert, J. Katsaras, E. London, G. Pabst and D. Marquardt, *Nat. Protoc.*, 2018, **13**, 2086–2101.

Reidar Lund said: In the paper, you are using peptides with rather large MIC values. You are trying to describe the antimicrobial mode of action for these peptides. Still, you only report data for lower [P] : [L] ratios that are most likely well below the effective range to kill the bacteria for these peptides. The MIC values previously reported would correspond to much larger [P] : [L] values considering the amount of lipids in the bacterial membrane, see *e.g.* ref. 1 and several studies by Melo and Castanho. Can you calculate the corresponding [P] : [L] values for the MIC of these peptides? Given the experimental design, the conclusion that flip–flop is not relevant to describe the mechanism of action at [P] : [L] values of 1 : 100 therefore remains unclear. I suppose that the reason for the low [P] : [L] values is due to the stability of the vesicles upon addition of the cationic peptides. Can you stabilise them somehow? Can you see the formation of other phases at higher [P] : [L] values? How would potential interactions between vesicles affect the kinetic results? Finally, the asymmetric model bilayer contains several lipids (POPG and POPE) with different compositions in each leaflet. Yet, only one rate constant is reported. It is well-known that the flip rates depend not only on acyl chain but also on the head group, and perhaps also the local environment. Since the TR-SANS experiment reported here is only sensitive to flip–flop, different rates should perhaps be seen. Can the authors comment?

1 D. Roversi, V. Luca, S. Aureli, Y. Park, M. L. Mangoni and L. Stella, *ACS Chem. Biol.*, 2014, 9(9), 2003–2007.

Georg Pabst responded: To be clear, we do not claim to describe the antimicrobial mode of action of our peptides. All we aim to do is to detail their membrane activity. The difficulties in delineating directly from membrane interaction studies to the killing of live bacteria are manifold. Unfortunately, it is a common misconception that bare [P] : [L] ratios in live cell and liposome systems can be directly compared. For illustration, the lipid concentration in *E. coli* ATCC25922 (at a cell number density of $\sim 10^6$ cell per ml for standard MIC determination) is in the nanomolar range. The MIC values of the reported peptides are in the micromolar range, giving a [P] : [L] ratio in the order of 1000 : 1. This was summarized in Wimley and Hristova in two seminal reviews.[1,2] Note that this enormous [P] : [L] ratio is valid for most of these peptides (see *e.g.* ref. 3). In common SAXS and SANS experiments with lipid concentrations of ~10–20 mM, this would require a peptide concentration of 10–20 M and no lipid bilayer would survive that! As pointed out by Wimley, Hristova, and others, completely different experimental conditions need to be considered. Common experiments with live bacteria systems exploit an excess concentration of peptides, with a complex partitioning between the bulk, the lipid membranes and internal compartments.[4,5] That is, the scene for antimicrobial killing is not necessarily the membrane even if the peptides are membrane active. Experiments with liposomes are instead commonly performed at [P] : [L] ratios of 1 : 20 to 1 : 100, *i.e.* with a different order of magnitude, which highlights the number of peptides actually partitioned in the cell membranes. In brief, the actual ratio of membrane-partitioned-peptides per lipid in cells needs to be carefully evaluated. We have recently reported an estimate in the case of LF11-215,[5] and information about MG2a is freely available.[6] The experimental design for lipid flip–flop was indeed performed at peptide/lipid ratios that ensured vesicle stability against fusion and aggregation, eventually leading to the formation of other phases. This was observed *via* SAXS for LF11-215 [5] and for MG2a and L18W-PGLa.[7] Nonetheless, in general we do not aim to stabilize vesicles against aggregation because membrane remodeling is integral to the understanding of peptides' activity. The important take-home message from our flip–flop measurements is that peptides, although being fully partitioned in membranes, may or may not lead to lipid flip–flop. In other words, these measurements are not a unique signature of peptide activity. The experimental design of the flip–flop measurements makes the neutron data sensitive to the transleaflet diffusion of the lipids only. This is because we do have only one changing contrast difference (between the inner and outer leaflet), even if we have two lipid species! The contrast difference originates from using chain deuterated POPG and fully protiated POPE. At the beginning of the experiment, this contrast difference is the highest (because we exchange POPE into the outer leaflet of the POPG_d vesicles) and this equilibrates over time to the contrast of fully scrambled POPE/POPG membranes. Consequently, we have also only one decay constant. Observing different translocation rates for lipids would be indeed interesting, but requires a different experimental approach (we are working on it, but it is far from being trivial). Naturally, lipid transfer occurs also through the aqueous phase (*i.e.* between individual vesicles). However, there is no contrast difference between the outer leaflets of individual vesicles. Consequently, our data does not report on trans-vesicle lipid diffusion kinetics (see also ref. 8). To sum up, only one decay constant is to be expected from our experimental design and this clearly is what we observe (see Fig. 4 and S5 of the paper, DOI: 10.1039/d1fd00039j). One might expect a rapid initial decay of contrast when the peptides hit the membranes, but this would require an even faster time resolution than the current one.

1 W. C. Wimley, *ACS Chem. Biol.*, 2010, **5**(10), 905–917.
2 W. C. Wimley and K. Hristova, *J. Membr. Biol.*, 2011, **239**(1–2), 27–34.
3 M. R. Loffredo, F. Savini, S. Bobone, B. Casciaro, H. Franzyk, M. L. Mangoni and L. Stella, *Proc. Natl. Acad. Sci. U. S. A.*, 2021, **188**(21), e2014364118.
4 Y. Zhu, S. Mohapatra and J. C. Weisshaar, *Proc. Natl. Acad. Sci. U. S. A.*, 2019, **116**(3), 1017–1026.
5 L. Marx, E. F. Semeraro, J. Mandl, J. Kremser, M. P. Frewein, N. Malanovic, K. Lohner and G. Pabst, *Front. Med. Technol.*, 2021, **3**, 625975.
6 T. Kaji, Y. Yano and K. Matsuzaki, *ACS Infect. Dis.*, 2021, **7**, 2941–2945.
7 I. Kabelka, M. Pachler, S. Prévost, I. Letofsky-Papst, K. Lohner, G. Pabst and R. Vácha, *Biophys. J.*, 2020, **118**(3), 612–623.
8 D. Marquardt, F. A. Heberle, T. Miti, B. Eicher, E. London, J. Katsaras and G. Pabst, *Langmuir*, 2017, **33**, 3731–3741.

Reidar Lund commented: SAXS was used to analyse the structure of the bilayer (the ESI does not seem to be available yet online). How was the peptide included in the analysis? Did you try to compare the partition data obtained from fluorescence assays with the results from SAXS?

Georg Pabst replied: We analyzed the aLUVs using SANS not SAXS; the ESI can be found at DOI: 10.1039/d1fd00039j. SANS combined with differently contrasted leaflets allowed us to determine the distribution of deuterated/protiated lipids. Moreover, we could also show by SANS that the vesicles did not aggregate or change their morphology under the present conditions. Regarding data modelling, we recorded data at only one neutron contrast and at relatively low peptide concentrations. In this case, any inclusion of peptides in the analysis would be heavily overparameterized. The appropriate way of doing this would be using a joint analysis SAXS/SANS data using advanced optimization routines (*e.g.* a genetic algorithm or Bayesian statistics), as detailed before.[1] So, we decided to use the simplest possible model for aLUVs, which is a 4-slab model reported previously,[2] which agreed perfectly well with our scattering data (Fig. S4b in the ESI, DOI: 10.1039/d1fd00039j).

As discussed in our paper, peptide partitioning does not play a role at the high lipid concentrations used for SANS experiments ([L] = 9 mM). This can be easily seen upon application of eqn (3) in our paper (DOI: 10.1039/d1fd00039j). Use, *e.g.* $K_x = 6 \times 10^6$ (LF11-215 in aLUVs at [P] : [L] = 1 : 100) and assume that it remains constant with increasing lipid concentration from the fluorescence experiments ([L] = 50 μM) to the one used for SANS. Then, $f_B = 0.84$ for the fluorescence experiments and $f_B = 0.99 \sim 1$ for SANS. That is the amount of unbound peptide that can be neglected for SANS. In fact, this is the very reason why we discussed the absence of detectable lipid flip–flop in the case of LF11-215, despite it being fully associated with the bilayer.

1 M. Pachler, I. Kabelka, M.-S. Appavou, K. Lohner, R. Vácha and G. Pabst, *Biophys. J.*, 2021, **117**, 1858–1869.
2 B. Eicher, F. A. Heberle, D. Marquardt, G. N. Rechberger, J. Katsaras and G. Pabst, *J. Appl. Cryst.*, 2017, **50**, 419–429.

Mibel Aguilar asked: Do you see any evidence of domains forming in these asymmetric lipid bilayers? And do you think that the lipid flip–flop could occur at the edges of the domains or within the core of the domains?

Georg Pabst answered: The techniques applied in the current study are not sensitive to domain formation. However, we have performed zero-contrast SANS measurements of symmetric mixtures of POPE/POPG (3 : 1 mol/mol) in the presence of L18W-PGLa and MG2a before to exactly address this question.[1] These data unambiguously show that, despite preferential interactions between POPG and cationic peptide residues, no domains of sizes between ~1 nm to 600 nm are formed. Hence, the interactions between L18W-PGLa/MG2a and POPG are not sufficient to stabilize the domains. Recent MD simulations performed in collaboration with the Vácha group further support these findings (to be published). Although we cannot fully exclude your proposed scenario, it is highly unlikely that the (L18W-PGLa/MG2a)/POPG interactions relevant for domain formation will become long-ranged in asymmetric bilayers. However, if we theoretically assume that they are present, then lipid flip–flop may certainly also occur at the domain boundaries, given that the defect at this location is large enough.

1 I. Kabelka, M. Pachler, S. Prévost, I. Letofsky-Papst, K. Lohner, G. Pabst and R. Vácha, *Biophys. J.*, 2020, **118**, 612–623.

Margarida Bastos queried: Why did you decide to use a first order mimic of inside out cytoplasmic membranes of Gram-negative bacteria in the asymmetric model membranes?

Georg Pabst replied: This a plain case of experimental feasibility to fabricate cytoplasmic membrane mimics of Gram-negative bacteria with an enriched PE inner leaflet. The currently applied protocol needs heavy donor vesicles (sucrose loaded MLVs) and light acceptor vesicles (LUVs). Further developing protocols for aLUVs along the lines of the Heerklotz group[1] is likely to be highly promising.

1 M. Markones, C. Drechsler, M. Kaiser, L. Kalie, H. Heerklotz and S. Fiedler, *Langmuir*, 2018, **34**, 1999–2005.

Margarida Bastos commented: This is very nice work indeed. I just wanted to ask you the following – looking at amino acid sequences, one can see that the 3 peptides have the same charge, but whereas L18W-PGLa and magainin 2 (MG2a) have 4 Lys, LF11-215 has 4 Arg. Exchanging Lys and Arg leads in many cases to differences in the antimicrobial activity – we have observed and showed that Arg containing peptides were more active against *M. avium*.[1] Could you consider testing an all K LF11-215 peptide to see what structural differences in the membrane interactions that change would result in? This is a peptide that presents a significantly different behaviour as compared to those of the other two (*e.g.* large Kx, and yet no detectable changes in the membrane asymmetry, for example), and it could be enlightening to see the possible importance of K *vs.* R in these differences and structural effects. Furthermore, you see no direct relation between flip–flop and peptide translocation, and so these are not necessarily related. We also found that partition and membranolytic activity are not necessarily related.

1 T. Silva, B. Magalhães, S. Maia, P. Gomes, K. Nazmi, J. G. M. Bolscher, P. N. Rodrigues, M. Bastos and M. S. Gomes, *Antimicrob. Agents Chemother.*, 2014, **58**, 3461–3467.

Georg Pabst replied: Many thanks for your encouraging and supporting comments. Given your results, it might indeed be necessary to compare the behavior of an all K variant of LF11-215.

Burkhard Bechinger remarked: Does lipid asymmetry create an electric gradient across the membrane that could explain the enhanced dye release of these bilayers? Did you test inversion of the asymmetry (*i.e.* inside-out)?

Georg Pabst replied: Indeed, as detailed by Gurtovenko and Vattulainen,[1] the asymmetric distribution of anionic and charge neutral lipids across bilayers gives rise to a nonzero potential difference in the order of 200 mV (at least for their system). Whether and how much this contributes to the enhanced dye release needs to be tested. In our discussion (DOI: 10.1039/d1fd00039j), we mainly focused on the differential elastic stress in the membrane leaflets. However, you raise a valid additional scenario. Inverting the lipid asymmetry might be a strategy, but this is very difficult at the moment to realize experimentally. Most likely one would need to evolve the aLUV preparation variant put forward by the Heerklotz group.[2] However, even then one has to keep in mind that this is a dynamic system that has been brought out of thermodynamic equilibrium. So, the kinetics will not be straightforward to compare. Maybe MD would be more useful here, because lipid flip–flop is too slow for their time scale.

1 A. A. Gurtovenko and I. Vattulainen, *J. Phys. Chem. B*, 2008, **112**, 4629–4634.
2 M. Markones, C. Drechsler, M. Kaiser, L. Kalie, H. Heerklotz and S. Fiedler, *Langmuir*, 2018, **34**, 1999–2005.

Anand Srivastava commented: Is the peptide translocation time known? Also, how does that correlate with the flip–flop time?

Georg Pabst replied: Our experiment is not able to probe peptide translocation directly. However, as detailed in our paper (DOI: 10.1039/d1fd00039j), we think that the lipid flip–flop we observe is dominated in case of L18W-PGLa and MG2a by co-translocation with the peptides. This implies that lipid flip–flop times are correlated to peptide translocation. For example, MG2a inserts less deeply in the POPE/POPG interface than L18W-PGLa and hence has to overcome a higher free energy barrier to enter the hydrophobic interior of the bilayer for translocation. This explains the lower flip–flop rate in this case. However, lipid flip–flop needs not to be connected to peptide translocation. L11-215 readily translocates the membranes (known from previous studies[1] and also from partitioning data showing that all L11-215 will be located in the bilayer under SANS conditions), but does not induce lipid flip–flop.

1 L. Marx, E. F. Semeraro, J. Mandl, J. Kremser, M. P. Frewein, N. Malanovic, K. Lohner and G. Pabst, *Front. Med. Technol.*, 2021, **3**, 625975.

Robert Vácha asked: Is the barrier for flip–flop in the asymmetric membrane determined by both leaflets (for example by the mean of their barriers) or only by the leaflet with a higher barrier?

Georg Pabst answered: In our simplistic argument, we assumed that the leaflets would be independent/decoupled for lipid flip–flop. However, we cannot exclude transleaflet coupling, as observed before. For example, for POPE/POPC aLUVs.[1]

1 B. Eicher, D. Marquardt, F. A. Heberle, I. Letofsky-Papst, G. N. Rechberger, M.-S. Appavou, J. Katsaras and G. Pabst, *Biophys. J.*, 2019, **114**, 146–157.

Mibel Aguilar opened a general discussion of the paper by Franca Fraternali: Do these particles dis-assemble or are they degraded in microbial systems? How stable are the particles in a biological setting? Can you control their disassembly or do they get degraded?

Franca Fraternali responded: I can comment on assembling/collapsing observed evidences: It has been demonstrated via TEM and AFM experiments in solution that the nanocapsules self-assemble in minutes to few hours, have a uniform and spherical shape, and are stable up to several hours, depending on pH and other solution conditions. Nevertheless, some of these are observed as collapsed or showing a double layered structure, indicating that the supramolecular structure we observe in the simulations is essential to their stability. The efficacy of such assembly is demonstrated by the fact that upon contact with microbial membranes it would instantaneously deliver peptide concentrations that significantly exceed those necessary to rupture microbial membranes, this is rapid and irreparable damage to a microbial cell damage without the need of transitioning from unstructured monomers to membrane-active oligomers. Most importantly, the activity of these nanoscale agents is not subject to bacterial phenotypes and acquired resistance. For more details refer to Kepiro *et al.*[1]

1 I. E. Kepiro, I. Marzuoli, K. Hammond, X. Ba, H. Lewis, M. Shaw, S. B. Gunnoo, E. De Santis, U. Łapińska, S. Pagliara, M. A. Holmes, C. D. Lorenz, B. W. Hoogenboom, F. Fraternali and M. G. Ryadnov, *ACS Nano*, 2020, **4**(2), 1609–1622.

Sreetama Pal asked: Would you expect shape changes in these nanocapsules as a response to ions, pH or maybe even the presence of specific lipids or co-solutes with specific topologies? What happens if you introduce co-adducts with differential hydrophobicity or hydrophilicity?

Franca Fraternali replied: I would definitely think that adding these factors would alter the capsule shape, but probably one would have to find the right conditions to simulate these, if they are demonstrated to be experimentally viable.

Izabella Brand said: You observe membrane poration in external electric fields. Which electric field is required for poration to occur? Do the pores form on the membrane fragments with attached nanocapsules or were they independent on the membrane morphology?

Franca Fraternali answered: The electric field needed for poration varies for different force field approximations, it was in the physiological range anyway and we always tested the same electric field on the lipid membrane model without the capsule so as to confirm that in the same conditions, the lipid bilayer alone was not affected by the applied electric field.

Izabella Brand remarked: Different amino acids in the nanocapsule have different charges. Did you detect affinity of the lipid membranes for particular amino acids? Did you observe an anisotropic orientation of nanocapsules on the membrane surface? Where were attractions/repulsions during the adsorption process of the nanocapusle on the membrane surface observed?

Franca Fraternali responded: In the attachment to the membrane, the arginine residues played a key role. And yes, this implied an anisotropic orientation of the capsule during membrane penetration, but accurate measurements of this anisotropy are not feasible as one would need multiple simulations of the membrane attachment process.

Sreetama Pal said: Since these are AMP-based, I am curious to know what happens to the architecture if you try to assemble the lipids and peptide/protein units together (without any prior bilayer or capsid assembly). Do you see the formation of a nanocapsule with a lipid cover? That could be a way to protect these against microbial degradation.

Franca Fraternali answered: We have tried to self-assemble the capsule in lipid media but these simulations never led to an ordered assembly. One would have to sample a very large number of possible assemblies and collect statistics on the most visited states, which is a very computationally expensive route. Additionally, there was no experimental evidence of pre-assembled states that could help in deciding which observed semi-assembled states were effectively taking part in the capsule formation.

Najet Mahmoudi asked: Have you looked at the assembly kinetics of your antimicrobial nanocapsules? If you did, experimentally, have you seen any intermediates in the kinetics?

Franca Fraternali replied: We did not attempt this in the computational work. Experimental measures did observe capsule formation within minutes to hours. They remain quite stable for several hours and afterwards decay quickly. They were measured by dynamic light scattering and by TEM microscopy when deposited on gold nanodiscs.

Durba Sengupta asked: Is the symmetry of the capsules maintained in the simulations in water and in the membrane bound state?

Franca Fraternali replied: Yes, with some fluctuations in the secondary structure, but of course when poration occurs the capsule opens up with the penetration of the membrane. This is what one would expect as the capsule should also act as a delivery agent.

Patrick Fuchs remarked: For your CG simulations, I guess that you are using MARTINI. Is there a difference of poration if you use polarizable water or not? Also, can you come back to your all-atom results and could you get enough sampling on this huge system?

Franca Fraternali replied: We used MARTINI with polarizable water. We did not revert to atomistic simulations as these were too expensive.

We tested stability of the capsule with another newly developed force field, SIRAH, that has a different water coarse graining. This force field allowed for conformational changes. We have nevertheless not yet compared these results with MARTINI3.

Izabella Brand asked: Can you comment on the surface charge density of the nanocapsules?

Franca Fraternali replied: The surface charge is strongly positive; for example, the capzip capsule is formed by 60 molecules with 180 paired branches and with one net positive charge per branch.

Najet Mahmoudi enquired: What about the stability of the nanocapsules, the dynamics of their formation and the methods used for their characterisation?

Franca Fraternali responded: We have performed multiple simulations to check the stability of the nanocapsules in different media (water–membrane interface), as described here and in previous papers.[1,2] The methods used for their characterisation are discussed in those works and range from classical measures of antimicrobial activity to AFM, TEM and measures of the intracellular delivery of genetic material.

In terms of the dynamics of their formation, little is known, particularly because there is not a direct experimental way to look into snapshots of structural assemblies whilst the capsule is forming. That is why we preferred to focus on pre-assembled structures. We used MD to study the stability of these and their mechanism of interaction with the membrane.

1 I. E. Kepiro, I. Marzuoli, K. Hammond, X. Ba, H. Lewis, M. Shaw, S. B. Gunnoo, E. De Santis, U. Łapińska, S. Pagliara, M. A. Holmes, C. D. Lorenz, B. W. Hoogenboom, F. Fraternali and M. G. Ryadnov, *ACS Nano*, 2020, **4**(2), 1609–1622.
2 I. Marzuoli, C. H. B. Cruz, C. D. Lorenz and F. Fraternali, *Nanoscale*, 2021, **13**(23), 10342–10355.

Evelyne Deplazes said: For your all-atom system, did you only simulate part of the system? If so, did you position restrain the nanocapsules to prevent disintegration or is the system stable enough without restraints?

Franca Fraternali replied: For simulations of the nanocapsule in water, the entire system has been simulated and restraints have been imposed in the equilibration phase and then released for the actual production run simulation. This assembly was stable in multiple replica simulations of 100 ns each. For the capsule–membrane simulations, we simulated only a petameric assembly that is formed by the triskelion units and representing the bottom part of the nano-capsule in contact with the membrane. For more details, please see ref. 1.

1 I. E. Kepiro, I. Marzuoli, K. Hammond, X. Ba, H. Lewis, M. Shaw, S. B. Gunnoo, E. De Santis, U. Łapińska, S. Pagliara, M. A. Holmes, C. D. Lorenz, B. W. Hoogenboom, F. Fraternali and M. G. Ryadnov, *ACS Nano*, 2020, **4**(2), 1609–1622.

Franca Fraternali commented: Restraints were added for equilibration but then the rest of the simulations of more than 500 ns were unrestrained.

Amy Rice communicated: Is the electric field applied to the beta-annulus capsule membranes sufficient to form pores in this membrane model in the absence of the capsule?

Franca Fraternali communicated in reply: Yes, within the SIRAH force field this was sufficient.

Amy Rice communicated: What happens to the generated pore when the external electric field is removed? In other words, is the electric field required only to generate the pore or is it also necessary for maintaining the pore structure?

Franca Fraternali communicated in reply: We did not try this; we applied the electric field after equilibration and until we observed poration and a stable pore formation.

Conflicts of interest

There are no conflicts to declare.

Faraday Discussions

PAPER

Concluding remarks: peptide–membrane interactions

Patricia Bassereau [iD]

Received 25th October 2021, Accepted 29th October 2021

DOI: 10.1039/d1fd00077b

This article is based on the concluding remarks lecture given at the *Faraday Discussion* meeting on peptide–membrane interactions, held online, 8–10th September 2021.

1. Introduction

The *Faraday Discussion* meeting on peptide–membrane interactions had to be postponed by one year due to the COVID-19 pandemic and was held online since the pandemic was still not under control in September 2021. Eventually, thanks to the constant effort of the co-chairs, Paul O'Shea and John Seddon, of the Faraday Division and of the Chemistry Biology Interface Division of the Royal Society of Chemistry, the meeting took place from September 8th to 10th, 2021. The meeting was fairly animated in spite of this unusual format with many discussions as is the tradition, both after the short talks and also "in front" of the 30 posters. I would like to take the opportunity of this concluding remarks lecture to thank them all, as well as the other members of the organizing committee, Amitabha Chattopadhyay, Calum Drummond and Robert Gilbert. I feel privileged to provide the concluding remarks lecture.

As Paul O'Shea stressed in his short introduction, in spite of their apparent simplicity, peptides remain to be fully understood. They are very flexible and can explore many degrees of freedom, with many conformations and topologies, which can lead to very rich behaviour when interacting with lipid membranes (Fig. 1). In contrast, proteins with their 3D folded state exist in a more "defined structure" and change conformation to a much smaller extent. This motivated the organization of a meeting that focused for a large part on the interactions of peptides with membranes, exploring recent experimental, theoretical, computational and biotechnological aspects.

In the closing lecture and in this article, I did not follow the programme order of the 22 contributions during the meeting. I rather made the choice to group them into 4 themes: interactions with membranes of (a) anti-microbial peptides

Institut Curie, Université PSL, Sorbonne Université, CNRS UMR168, Laboratoire Physico-Chimie Curie, 75005 Paris, France. E-mail: patricia.bassereau@curie.fr

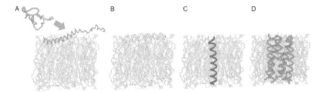

Fig. 1 Large range of conformations adopted by peptides when interacting with membranes. (A) Disordered in bulk and folded on membrane. (B) Amphiphilic helix. (C) Single transmembrane peptide. (D) Assembly into helical bundles. (From DeGrado *et al.*, DOI: 10.1039/D1FD00061F).

(AMPs), (b) simulations, (c) amphipathic peptides of trans-membrane proteins, and (d) membrane proteins.

2. Peptides in the cell-microbes battle

William DeGrado set the stage of the meeting and presented a very comprehensive introductory lecture (DOI: 10.1039/D1FD00061F) on the analysis and design of membrane-interactive peptides, interacting with the outer part of membranes or inserted in the membrane. He reminded us how it has been recognized that the membrane activity of antimicrobial peptides, in particular of melittin, depends only on their amphipathic helical structure and not on the detail of their sequence,[1] which has opened the way to rationally designed synthetic peptides with antibiotic functions,[2] as well as drugs, among them, one against COVID-19 is currently in trials. He showed the importance of back-and-forth approaches between coarse-grained simulations, electron microscopy and biochemical assays for identifying the mechanisms of action of different peptides on membranes. He next showed that the long amphipathic helices of α-synuclein, the Parkinson disease-related protein, has a peculiar feature: its hydrophobic residues create a super-helical twist when in an ideal helical conformation.[3,4] This peculiar feature could explain how α-synuclein stabilizes the fusion pore of synaptic vesicles by winding up around their neck. In the second part of his talk, he showed that the trans-membrane helix of the β3 integrin has double-sided

Fig. 2 The integrin β3 (blue) has 2 different interaction sides: with αIIb (red) and αV(green). By courtesy of W. DeGrado (background adapted from Shutterstock ID 772877308).

interactions with αIIb and αV subunits[5] (Fig. 2); others commented on the interesting possibility that this might mediate the recruitment of multiple types of integrins to focal adhesions (see Q/A). He ended his presentation by discussing the mechanism of proton transport across the membrane *via* channels using viroporins (viral ion channels) as models. He also discussed unpublished work involving *in vitro* liposome assays and the designed multimeric peptides. He concluded that a subtle balance between the number of apolar and polar residues in the core can dictate the stoichiometry of the channel, and hence selectivity for protons *versus* other ions.

Different talks evidenced the strong interplay between membrane properties and the way peptides interact with them. In addition, antimicrobial peptides and cells from the immune system exploit differences in the outer leaflets of mammalian cells (not charged, high cholesterol level) and bacterial cells (strong negative charge) to target their action.

Killer T cells form pores (based on perforins for instance), secrete cytotoxic peptides to kill cancer cells or virus infected cells (Fig. 3), and thus must develop some auto-resistance mechanism against their own peptides. Hoogenboom *et al.* (DOI: 10.1039/D0FD00043D) showed that T-cells are able to control the lipid organization of their own membrane in the immune synapse and induce the formation of a liquid ordered area resistant to perforin insertion and pore formation.[6] In addition, they also move negatively charged lipids to the external leaflet of their membrane to disrupt pore formation in that area.[6] In contrast, cancer cells have more fluid membranes and are thus more sensitive to these peptides. The mechanism used by T-cells to locally control their membrane composition while keeping the formation of the immune synapse and of membrane trafficking, in that area intact remains to be understood.

Bonev *et al.* showed (DOI: 10.1039/D1FD00036E) that the cationic peptide polymyxin B could be used in therapeutic strategies against antibiotic-resistant Gram-negative strains due its specific interaction with rLPS (lipopolysaccharide) on the outer membrane of these bacteria, which facilitates its penetration.

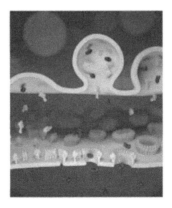

Fig. 3 Immune synapse between a killer T-cell and a target cell or virus: perforin (blue) forms pores in the target membrane. Granzymes (red) released from cytotoxic granules can penetrate in the target cell and trigger apoptosis. (From Hoogenboom *et al.*, DOI: 10.1039/D0FD00043D).

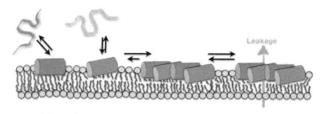

Fig. 4 Synergetic effect between magainin (red) and PGLa (green) due to a mesophase formation between the peptides, and the PE and negatively charged lipids, that favors peptide assembly and enhances membrane permeabilization. (From Bechinger *et al.*, DOI: 10.1039/D0FD00041H).

AMPs provide a first line of defence against many pathogens. It is now well established that linear cationic AMPs have a strong synergetic action on membranes, as compared to the effect of one type of AMP alone,[7] but the mechanism behind this amplification effect is not clear yet. Bechinger *et al.* showed (DOI: 10.1039/D0FD00041H), for the mixture of magainin 2 and PGLa, that synergy depends on the nature of the lipid head groups, in particular in the presence of PE lipids, but not on the C-termini of the peptides, which discards models involving interactions between termini of the peptides, in agreement with recent studies.[8] Instead, Bechinger suggested the formation of a mesophase between lipids and peptides[9] (Fig. 4), which increases the lytic efficiency by increasing the local negative curvature.

Cell membranes have asymmetric lipid distribution between the outer and inner leaflets. This is also the case for the outer membrane of Gram-negative bacteria.[10] Two groups addressed the effect of this asymmetric structure on AMP penetration using reconstituted asymmetric membranes. Brand and Khairalla stressed that melittin interacts with LPS and affects the ordering of the lipid chains (DOI: 10.1039/D0FD00039F). Pabst *et al.* showed that the asymmetry facilitates a peptides' translocation. In addition, he reported a correlation between peptide translocation capability and lipid flip-flop for rigid peptides (PGLa and magainin 2) but no flip-flop for the more flexible lactoferricin (DOI: 10.1039/D1FD00039J). He also warned us on the limitations of lipid-only mimics.

Evidently, the bacterial membrane organization does influence its interaction with AMPs. Using AFM on supported membranes of bacteria, Aguilar *et al.* explained (DOI: 10.1039/D0FD00052C) why the action of the maculatin 1.1 peptide is more efficient at the stationary growth phase of *E. coli* than in the exponential phase: the lipid composition changes dramatically between the different phases, as does the membrane organization which is much more prone to disruption in the stationary phase.

3. Recent developments in simulations

Simulations are becoming essential for revealing how peptides or proteins interact with different lipids in lipid bilayers, how these lipids are redistributed around them and how the membrane shape can change; conversely, they also show how the peptide/protein structure can be affected by lipid composition. Different presentations illustrated the key contribution of simulations at different scales.

3.1. Coarse-grained simulations

Tieleman *et al.* (DOI: 10.1039/D1FD00003A) and Vanni *et al.* (DOI: 10.1039/D0FD00058B) presented the latest development of Martini (version 3), the most popular of the MD coarse-grained models.[11] This model has already significantly contributed to the understanding of how membrane proteins or peptides and lipids interact, how they redistribute, and how the membrane shape adjusts in the lipid environments of growing complexity, developed over recent years. In the new version of the Martini model, the proteins are represented by more beads, which allows for higher flexibility, and it also includes specific interactions.[12] The new model is now being tested with more and more challenging systems. Tieleman *et al.* studied the lipid redistribution around different membrane proteins embedded into a very realistic asymmetric membrane, using carbon nanotubes as a control since they have a uniform interface with the lipids (Fig. 5). The simulations are clearly able to distinguish important differences between these proteins. Vanni *et al.* investigated peripheral proteins and peptides with transient binding to a membrane. With Martini, he satisfyingly reproduces the affinity of these molecules for membranes, and the kinetics, their sensitivity to charges or defects or their absence of binding, which altogether is very promising for the future of the model. Nevertheless, there are still some limitations with version 3

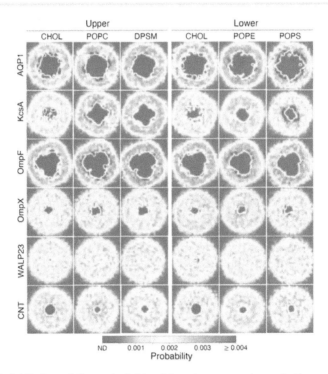

Fig. 5 Redistribution of the main lipids of the plasma membrane in the presence of different proteins in the upper leaflet (left) and the lower one (right): the membrane proteins AQP1 (aquaporin), KcsA (potassium channel), OmpF, OmpX (bacterial β-barrel outer membrane), the WALP23 peptide and carbon nanotube (CNT). (From Tieleman and Cino *et al.*, DOI: 10.1039/D1FD00003A).

that might be overcome in the future with greater computational power. For instance, only one protein at a time can be considered and the interactions between many proteins cannot be studied; it is not possible to measure the differential stresses between monolayers. But we can foresee in the coming years new insights into how peptides interact with membranes at the molecular level based on the Martini model.

3.2. All-atom MD simulations

With the COVID-19 pandemic and the development of vaccines, everybody is aware of the spike protein (S) of the SARS-CoV-2 corona virus, but much less of its other proteins such as 2 integral proteins: the matrix (M) or the envelope (E) proteins. M is essential for the assembly of new viral particles in the ERGIC. E is a viroporin and works as an ion channel for the virus. Unfortunately, no structure of these proteins is known at this stage, but 2 homology models are available for each protein.[13,14] Monje-Galvan and Voth performed long all-atom MD simulations (microseconds) with these models, using a membrane composition mimicking the endoplasmic reticulum, and compared the predictions (DOI: 10.1039/D1FD00031D). M can exist in an open and a closed conformation, and probably switches from one to the other during the virus maturation process. They show that in both models, the open conformation induces a larger deformation of the membrane (Fig. 6), which should favor the budding at the beginning of the process; moreover, the lipid distribution is different in the two conformations.

Their simulations also show that only one of the analogy models (Korkin) for the E protein corresponds to a conformation that functions as an ion channel, whereas the other one (Feig) is more open and might correspond to another function of the protein. These simulations show the urgent need for *in vitro* experiments to better understand the structure–function relations of these key viral proteins.

4.   Role of terminal peptides in trans-membrane protein function

4.1. Trans-membrane proteins

Many trans-membrane proteins contain a peptide at one terminus that can interact with membranes; however, their role is not necessarily clear, in particular for ion pumps.

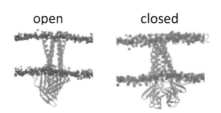

Fig. 6 Membrane deformation induced by 2 conformations of a dimer of the M protein of SARS-CoV-2, as predicted by MD simulations. (From Monje-Galvan and Voth. DOI: 10.1039/D1FD00031D).

Clarke *et al.* presented the very interesting case of two ion pumps, the Na–K-ATPase and H–K-ATPase, whose pumping rate depends on a polybasic peptide at their N-terminus (DOI: 10.1039/D0FD00040J). They proposed that their N-terminal peptides contribute to an electrostatic switch mechanism that regulates the pumping activity: the binding of the helix can be switched on and off through the phosphorylation of a serine residue by a kinase (Fig. 7). Moreover, these basic peptides can also be used by the cell as a negative lipid (PS) sensor, and thus a death sensor in order to stop pumping when the cell dies. It was also suggested during the discussion that these basic N-termini could be involved in sorting some ATP-ases as well as in the formation of dimers or tetramers.

4.2. Proteins with helical hairpin

Caveolin-1, the main component of caveolae, contains a disordered N-terminal domain, a scaffolding domain (CSD), an intramembrane domain (IMD) suggested to form a hairpin in membranes, and a C-terminus with three palmitoylations. The structure of the full protein is not known. Nevertheless, Sengupta *et al.* have used a truncated version of the protein, keeping domains interacting with the membrane – the CSD, the IMD and the C-Ter with one palmitoylation – and Martini coarse-grained simulation[15] to study the effect of caveolin-1 interacting with a complex asymmetric membrane mimicking the plasma membrane (DOI: 10.1039/D0FD00062K). They used the I-TASSER server to predict the structure of the truncated protein. Although the simulation predicts that the IMD remains parallel to the bilayer with the palmitoyl chain inserted, they found a very strong recruitment of cholesterol and sphingolipids in the opposite (extracellular) leaflet of the membrane opposed to the caveolin-1, and a clustering of PE and PS lipids around the protein in the inner leaflet. Moreover, this globally induces a nm-size curvature of the membrane. Altogether, these results are very consistent with the very strong enrichment in cholesterol and sphingomyelin,[16] as well as in PS lipids, in caveolae,[17] together with the capability of the protein to deform the membrane and form the very characteristic caveolae-omega shape. However, more work is now needed to determine caveolin-1 structure and how it inserts into a bilayer.

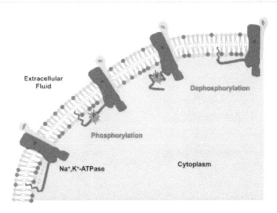

Fig. 7 Electrostatic switch mechanism involving the phosphorylation/dephosphorylation of a polybasic N-terminus of ion pumps by a kinase. Image courtesy of R. Clarke.

5.   Protein–membrane interactions

5.1.   Mitochondrial proteins (Bax)

Apoptosis is tightly regulated in cells. At the mitochondrial level, a complex interaction network determines the onset of mitochondrial permeabilization and eventually of cell death. Murad and Garcia-Saez succeeded with *in vitro* and *in cellulo* experiments to decipher how this regulation is achieved by three proteins: Bcl-xL (the regulator), tBid and Bax (that induces mitochondria lysis) (DOI: 10.1039/D0FD00045K). Bcl-xL inhibits apoptosis by opposite interactions with the other two: it enhances Bax retro-translocation in the cytosol, keeping the protein away from the membrane, while it stabilizes tBid binding to the membrane.

In addition, Barrera *et al.* showed that a single mutation on a helix of Bax (BaxE5) is sufficient to increase significantly its lytic capability, thus cell death, pointing to the importance of controlling this protein (DOI: 10.1039/D0FD00070A).

5.2.   ESCRT proteins

ESCRT-III complexes are involved in membrane scission in many processes in cells.[18] They assemble into filaments on negatively-charged membranes, or at the neck of buds where they use an ATPase for constriction and fission. The scission mechanism is still debated, but *in vitro* systems have been instrumental for progress on this issue.[19] Beales *et al.* used phase-separated giant liposomes and purified ESCRT-II and -III proteins to form buds and scission them.[20] He studied the influence of lipid domains and of the relative localization of negatively-charged lipids (PS) (in L_d or L_o phases, depending on the saturation of the PS chains) on the protein recruitment and on intraluminal vesicle (ILVs) formation (DOI: 10.1039/D0FD00042F). He showed that the binding of the ESCRTs to the L_d phase of the vesicles (by unsaturated PS) favours the formation of ILVs, but surprisingly in a L_o state. When ESCRTs are recruited to the L_o phase, very few ILVs form, and without phase selectivity. A long discussion followed this presentation (see the discussion section in the article), pointing to the need to develop a physical model for understanding these results, and to study if this lipid segregation takes place in cell membranes upon budding.

6.   Some perspectives and open questions

It was clear after these three days of discussion that, in spite being about 40 years old (see DeGrado *et al.*, DOI: 10.1039/D1FD00061F), the peptide/membrane interactions field remains very active with still many open questions on the theoretical and fundamental side, and many perspectives for novel experiments and translational applications. I will mention here some of them that came to my mind but of course, this list is far from comprehensive.

• Model membrane composition and structure

The lipid composition used in many *in vitro* experiments is rather simplistic, in order to differentiate the effect of the peptide/protein from that of the membrane, consisting often of a single lipid type (POPC, DOPC, DPPC…), sometimes a mixture with cholesterol, but generally limited to few lipids. However, many recent simulations now use lipid compositions and distribution between the two leaflets that are close to real cell membranes.[21] Different talks in this meeting (e.g., Tieleman; Vanni; Sengupta) demonstrated that with realistic membranes, it is possible to obtain definite evidence of the interplay between membrane composition and organisation on the one hand, and conformations of peptides and proteins on the other hand, and they show that redistribution occurs. One clear issue for the field now is to decide how to introduce realistic complexity to model membranes, and how far to go. Charged lipids are essential for the binding and function of many peptides and proteins, even trans-membrane proteins (see talks by Pal (DOI: 10.1039/D0FD00065E); Hoogenboom; Clarke). The asymmetry of the membranes certainly plays a role in the stresses exerted on membrane proteins, on the insertion of some proteins or on the translocation of AMPs. Different groups have developed strategies to integrate this key feature into model membranes,[22,23] as described during the talks by Brand and Pabst. For understanding synergetic effects between AMPs (see Bechinger's presentation), the role of lipids and of the structure of the peptides have been considered individually, but global aspects of membrane–peptide systems such as mechanics and stresses should be now included.

Finally, membrane proteins inserted in the bilayer can not only influence the localization of lipids but also the translocation of peptides for instance,[24] and thus should be considered in the model systems for studying how to enhance the efficiency of penetrating peptides.

• Lipids on peptides and peptides on membrane proteins

Sanderson et al. showed how peptides (here mellitin) can easily be lipidated in the presence of lysophosphatidylcholine (DOI: 10.1039/D1FD00030F), which could enhance their interaction with membranes. He suggested that this process might be involved in different diseases with amyloids related to aging, and should be considered in drug design. Also, membrane proteins can contain some cytosolic lipidated domains, as exemplified by caveolin-1 in this article (Section 4.2). These acylations are often absent in proteins purified from bacteria, but could have non-negligible effects on the interaction of these proteins with their lipid environment, and on their capability to bend membranes.

Some trans-membrane proteins contain amphipathic helices at a terminus, like some ions pumps (see Clarke et al., Section 4.1), the M2 protein from influenza virus,[25] reticulons[26] etc. The roles of these peptides are still not explored enough. They can probably induce some membrane curvature and in some conditions, fission.[27–29] They might also be important for their sorting or their localization e.g. via interactions with charged lipids (Clarke et al.). Thus, they should be investigated more systematically.

• The good old cholesterol molecule…

Cholesterol, being a small and very dynamic molecule in membranes, has been a matter of debate between biophysicists for decades. This is not the place here to summarize all questions about localization, dynamics, membrane organization, effect on protein functions etc. Nevertheless, many of them are still open. The recent developments of coarse-grain simulations show that its localization into the bilayer and its flip-flop rate are strongly influenced locally by molecular interactions with membrane proteins (Tieleman *et al.*) and other lipids, but conversely cholesterol influences its environment at a larger scale (Sengupta *et al.*). Nonetheless, these simulations predict flip-flop with characteristic times of the order of µs, whereas much longer times are observed experimentally with model membranes. It is not yet clear whether this large discrepancy originates from inadequacies of the force fields in simulations or from the too simplistic design of the model membranes. Nevertheless, this implies that we have to revisit our vision of cholesterol and continue to investigate its properties in membranes.

• Comparison between MD simulations and experiments

I have already stressed that very fast progress in MD simulations has been essential for understanding the local membrane and lipid reorganization around proteins or peptides. The main challenge now is to be able to develop these simulations with many proteins to study clustering, reorganization and membrane deformations, while keeping enough detail in the systems. This could be solved with more powerful computational systems. Another significant step would be the ability to change scale back and forth while keeping some relevant details, in order to facilitate comparison with experimental data.

• New tools

New tools are of course needed in this field to continue improving spatial and temporal resolution for a better understanding of the membrane–peptide–protein systems and the design of more efficient drugs. Based on the talks, I will mention only two examples.

Mashanov and Molloy *et al.* (DOI: 10.1039/D1FD00035G) showed how it is still difficult to interpret single particle tracking experiments and use them to decipher the cell membrane organization and dynamics. Here, development of correlative measurements coupling super-resolution microscopy and dynamics would be instrumental.

On a more applied aspect related to therapeutic developments, DeGrado *et al.*, and Barrera *et al.*, showed us that approaches coupling mutations on proteins and deep screening could be an efficient way to find drugs. For the replacement of antibiotics, Fraternali *et al.* proposed virus-inspired nanocapsules decorated with AMPs (DOI: 10.1039/D1FD00041A). The design has been obtained from simulations but the particles have to be now synthesized and tested.

Conflicts of interest

There are no conflicts to declare.

Acknowledgements

I would like to thank again the Royal Society of Chemistry and the other members of the organizing committee for this very stimulating *Faraday Discussion*. In particular, I wish to express my gratitude to Paul O'Shea and John Seddon for allowing me to have the honour of presenting the concluding remarks lecture, for the interesting discussions after each day of the meeting that contributed to drafting these remarks, and for editing this text.

References

1 W. F. DeGrado, F. Kezdy and E. Kaiser, *J. Am. Chem. Soc.*, 1981, **103**, 679–681.
2 W. F. DeGrado, in *Adv. Protein Chem.*, ed. C. B. Anfinsen, J. T. Edsall, F. M. Richards and D. S. Eisenberg, Academic Press, 1988, vol. 39, pp. 51–124.
3 R. W. Newberry, T. Arhar, J. Costello, G. C. Hartoularos, A. M. Maxwell, Z. Z. C. Naing, M. Pittman, N. R. Reddy, D. M. C. Schwarz, D. R. Wassarman, T. S. Wu, D. Barrero, C. Caggiano, A. Catching, T. B. Cavazos, L. S. Estes, B. Faust, E. A. Fink, M. A. Goldman, Y. K. Gomez, M. G. Gordon, L. M. Gunsalus, N. Hoppe, M. Jaime-Garza, M. C. Johnson, M. G. Jones, A. F. Kung, K. E. Lopez, J. Lumpe, C. Martyn, E. E. McCarthy, L. E. Miller-Vedam, E. J. Navarro, A. Palar, J. Pellegrino, W. Saylor, C. A. Stephens, J. Strickland, H. Torosyan, S. A. Wankowicz, D. R. Wong, G. Wong, S. Redding, E. D. Chow, W. F. DeGrado and M. Kampmann, *ACS Chem. Biol.*, 2020, **15**, 2137–2153.
4 R. W. Newberry, J. T. Leong, E. D. Chow, M. Kampmann and W. F. DeGrado, *Nat. Chem. Biol.*, 2020, **16**, 653–659.
5 R. I. Litvinov, M. Mravic, H. Zhu, J. W. Weisel, W. F. DeGrado and J. S. Bennett, *Proc. Natl. Acad. Sci. U. S. A.*, 2019, **116**, 12295.
6 J. A. Rudd-Schmidt, A. W. Hodel, T. Noori, J. A. Lopez, H.-J. Cho, S. Verschoor, A. Ciccone, J. A. Trapani, B. W. Hoogenboom and I. Voskoboinik, *Nat. Commun.*, 2019, **10**, 5396.
7 A. Vaz Gomes, A. De Waal, J. Berden and H. Westerhoff, *Biochemistry*, 1993, **32**, 5365–5372.
8 R. Leber, M. Pachler, I. Kabelka, I. Svoboda, D. Enkoller, R. Vácha, K. Lohner and G. Pabst, *Biophys. J.*, 2018, **114**, 1945–1954.
9 C. Aisenbrey, M. Amaro, P. Pospíšil, M. Hof and B. Bechinger, *Sci. Rep.*, 2020, **10**, 11652.
10 J. M. Boon and B. D. Smith, *Med. Res. Rev.*, 2002, **22**, 251–281.
11 S. J. Marrink and D. P. Tieleman, *Chem. Soc. Rev.*, 2013, **42**, 6801–6822.
12 P. C. T. Souza, R. Alessandri, J. Barnoud, S. Thallmair, I. Faustino, F. Grünewald, I. Patmanidis, H. Abdizadeh, B. M. H. Bruininks, T. A. Wassenaar, P. C. Kroon, J. Melcr, V. Nieto, V. Corradi, H. M. Khan, J. Domański, M. Javanainen, H. Martinez-Seara, N. Reuter, R. B. Best, I. Vattulainen, L. Monticelli, X. Periole, D. P. Tieleman, A. H. de Vries and S. J. Marrink, *Nat. Methods*, 2021, **18**, 382–388.
13 L. Heo and M. Feig, *bioRxiv*, 2020, DOI: 10.1101/2020.03.25.008904.
14 S. Srinivasan, H. Cui, Z. Gao, M. Liu, S. Lu, W. Mkandawire, O. Narykov, M. Sun and D. Korkin, *Viruses*, 2020, **12**, 360.
15 A. Krishna and D. Sengupta, *Biophys. J.*, 2019, **116**, 69–78.
16 S. Sonnino and A. Prinetti, *FEBS Lett.*, 2009, **583**, 597–606.
17 Y. Zhou, N. Ariotti, J. Rae, H. Liang, V. Tillu, S. Tee, M. Bastiani, A. T. Bademosi, B. M. Collins, F. A. Meunier, J. F. Hancock and R. G. Parton, *bioRxiv*, 2020, DOI: 10.1101/2020.01.16.909408.
18 J. Schöneberg, I.-H. Lee, J. H. Iwasa and J. H. Hurley, *Nat. Rev. Mol. Cell Biol.*, 2017, **18**, 5–17.
19 A.-K. Pfitzner, J. Moser von Filseck and A. Roux, *Trends Cell Biol.*, 2021, **31**, 856–868.

20 A. Booth, C. J. Marklew, B. Ciani and P. A. Beales, *iScience*, 2019, **15**, 173–184.

21 S. J. Marrink, V. Corradi, P. C. T. Souza, H. I. Ingólfsson, D. P. Tieleman and M. S. P. Sansom, *Chem. Rev.*, 2019, **119**, 6184–6226.

22 E. London, *Acc. Chem. Res.*, 2019, **52**, 2382–2391.

23 D. Marquardt, B. Geier and G. Pabst, *Membranes*, 2015, **5**, 180–196.

24 L. Bartoš, I. Kabelka and R. Vácha, *Biophys. J.*, 2021, **120**, 2296–2305.

25 J. S. Rossman, X. Jing, G. P. Leser and R. A. Lamb, *Cell*, 2010, **142**, 902–913.

26 N. Wang, L. D. Clark, Y. Gao, M. M. Kozlov, T. Shemesh and T. A. Rapoport, *Nat. Commun.*, 2021, **12**, 568.

27 E. Boucrot, A. Pick, G. Çamdere, N. Liska, E. Evergren, H. T. McMahon and M. M. Kozlov, *Cell*, 2012, **149**, 124–136.

28 A. Martyna, B. Bahsoun, M. D. Badham, S. Srinivasan, M. J. Howard and J. S. Rossman, *Sci. Rep.*, 2017, **7**, 44695.

29 M. A. Zhukovsky, A. Filograna, A. Luini, D. Corda and C. Valente, *Front. Cell Dev. Biol.*, 2019, **7**, 291.

Poster titles

Measuring thousands of single vesicle leakage events reveals the mode of action of antimicrobial peptides, **K. Al Nahas, M. Fletcher, K. Hammond, J. Cama, M. G. Ryadnov and U. F. Keyser**, *University of Cambridge, UK*

Association of yeast transmembrane domains in different membrane environment, **A. Alavizargar, A. Elting, R. Wedlich-Söldner and A. Heuer**, *Westfälische Wilhelms-Universität Münster, Germany*

Understanding interaction of insulin with model membranes from coarse-grained molecular dynamics, **H. Arnolds, X. Ye and J. Madine**, *University of Liverpool, UK*

Membrane permeabilization induced by antimicrobial peptides in various model membranes, **K. Beck, J. Nandy, S. Shi and M. Hoernke**, *Albert-Ludwigs-Universität Freiburg, Germany*

Structural characterization of the interaction of Homeodomains with PI(4,5)P2-containing membranes, **E. Chalouhi, G. Bouvignies, L. Carlier, P. Fuchs and O. Lequin**, *Sorbonne Université, France*

Structure of a cation-selective, peptide pore in phospholipid bilayers, **L. M. Hartmann, C. Cranfield, A. Garcia and E. Deplazes**, *University of Technology Australia, Australia*

The impact of neuronal membrane on amyloid beta: MD simulations considering in vivo conditions, **H. Fatafta, M. Khaled, M. C. Owen, A. Sayyed-Ahmad and B. Strodel**, *Institute of Biological Information Processing (IBI-7: Structural Biochemistry), Forschungszentrum Jülich, Germany*

Structural analysis of the novel fungal peptide toxin, candidalysin, **O. W. Hepworth, N. Kichik, S. Lee, R. E. Salmas, J. P. Richardson, J. R. Naglik and A. J. Borysik**, *King's College London, UK*

Understanding enhanced endosomal escape: membrane permeabilization induced by pH-sensitive polycations, **M. Hoernke, S. Schmager, and J. Brendel**, *Albert-Ludwigs-Universität Freiburg, Germany*

Lipid binding drives disorder-to-order transitions of the amyloid-β peptide, **B. Kav, H. Fatafta, B. Bundschuh, J. Loschwitz, and B. Strodel**, *Institute of Biological Information Processing (IBI-7: Structural Biochemistry), Forschungszentrum Jülich, Germany*

Investigation of lipid bilayer asymmetry effects on transmembrane domains, **F. Keller, A. Alavizargar and A. Heuer**, *Universität Münster, Germany*

Influence of transmembrane peptides on the mesoscale membrane dynamics, **E. Kelley**, *National Institute of Standards and Technology, USA*

Novel drug delivery system for antibiotic therapy using modified erythrocyte liposomes, **H. Krivic, R. Sun, S. Himbert and M. Rheinstadter**, *McMaster University, Canada*

Interaction of self-assembling diphenylalanine peptide nanotubes with bacterial membranes and biofilm, **G. Laverty, S. L. Porter, S. M. Coulter and S. Pentlavalli**, *Queen's University Belfast, UK*

Amphipathic helix folding in membranes: Markov State Models to decipher the mechanism, **P. Milán Rodríguez and P. Fuchs**, *Sorbonne Université, France*

De novo transmembrane helix-helix designs in biotechnology using evolutionary computing, **A. Nash**, *University of Oxford, UK*

Real-time visualization of phase and structural change of model lipid membranes by arginine-rich cell-penetrating peptide, **S. Park, J. Kim, S. S. Oh, and S. Q. Choi**, *Korea Institute of Science and Technology, South Korea*

Action of anti-microbial peptides on bacterial membrane-mimetic vesicles, **K. S. Nair, N. B. Raj and H. Bajaj**, *CSIR-National Institute for Interdisciplinary Science and Technology (CSIR-NIIST), India*

Bridging the activity of lactoferricin derivatives in E. coli and lipid-only membranes: partitioning and kinetics, **E. F. Semeraro, L. Marx, J. Mandl, M. K. P. Frewein, H. Bergler, K. Lohner and G. Pabst**, *University of Graz, Austria*

What we can learn from vesicle aggregation and fusion about membrane permeabilization induced by antimicrobial polycations, **S. Shi and M. Hoernke**, *Albert-Ludwigs-Universität Freiburg, Germany*

Membrane dissociation free energy of AKT1 PH domain depends on ionization states of polyphosphoinositides lipids: Insights from molecular simulations, **A. Srivastava, S. Pant, K. Jha, S. Thangamani and J. Nagesh**, *Indian Institute of Science Bangalore, India*

Modelling peptide adsorption energies on gold surfaces with an effective implicit solvent and surface model, **M. Suyetin, S. Bag, P. Anand, M. Borkowska-Panek, F. Gußmann, M. Brieg, K. Fink and W. Wenzel**, *Karlsruhe Institute of Technology, Germany*

Packing, orderliness, and defects: toward understanding the preferential protein partitioning, **M. Tripathy and A. Srivastava**, *Indian Institute of Science Bangalore, India*

Peptide-folding triggered raft formation and lipid membrane destabilization in cholesterol-rich lipid vesicles, **J. Utterström, R. Selegård, M. N. Holme, H. M. G. Barriga, M. M. Stevens and D. Aili**, *Linköping University, Sweden*

Enhanced translocation of amphiphilic peptides across membranes by transmembrane proteins, **R. Vácha, L. Bartoš and I. Kabelka**, *Masaryk University, CEITEC, Czech Republic*

The affiliation of the presenting author is given above.

The Faraday Division Poster Prize was awarded to Fabian Keller of Universität Münster, Germany for the poster on "Investigation of lipid bilayer asymmetry effects on transmembrane domains".

The RSC Chemical Biology Poster Prize was awarded to Olivia Hepworth of King's College London, UK, for the poster on "Structural analysis of the novel fungal peptide toxin, candidalysin".

List of participants

Professor Mibel Aguilar, *Monash University, Australia*
Mr Kareem Al Nahas, *University of Cambridge, United Kingdom*
Dr Azadeh Alavizargar, *Westfälische Wilhelms-Universität Münster, Germany*
Mr Siddique Amin, *Newcastle University, United Kingdom*
Dr Heike Arnolds, *University of Liverpool, United Kingdom*
Dr Helena Azevedo, *Queen Mary University of London, United Kingdom*
Professor Francisco Barrera, *University of Tennessee, USA*
Dr Patricia Bassereau, *Institut Curie, France*
Professor Margarida Bastos, *University of Porto, Portugal*
Dr Paul Beales, *University of Leeds, United Kingdom*
Professor Burkhard Bechinger, *University of Strasbourg, France*
Ms Katharina Beck, *Albert-Ludwigs-Universität Freiburg, Germany*
Mr Ian Bennett-Wright, *University of Edinburgh, United Kingdom*
Professor Renata Bilewicz, *University of Warsaw, Poland*
Mr Daniel Birtles, *University of Maryland, USA*
Dr Boyan Bonev, *University of Nottingham, United Kingdom*
Dr Claudia Bonfio, *University of Cambridge, United Kingdom*
Dr Izabella Brand, *Carl von Ossietzky Universität Oldenburg, Germany*
Mr Edward Chalouhi, *Sorbonne Université, France*
Professor Amitabha Chattopadhyay, *Centre for Cellular & Molecular Biology, India*
Dr Barbara Ciani, *University of Sheffield, United Kingdom*
Professor Ronald Clarke, *University of Sydney, Australia*
Professor Myriam Cotten, *William & Mary, USA*
Dr Ellis Crawford, *Royal Society of Chemistry, United Kingdom*
Dr Silvia Dante, *Istituto Italiano di Tecnologia, Italy*
Dr Anjali Devi Das, *Rajiv Gandhi Centre for Biotechnology, India*
Miss Kathakali De, *University of Strasbourg, France*
Professor William DeGrado, *University of California, San Francisco, USA*
Dr Evelyne Deplazes, *University of Queensland, Australia*
Dr Milka Doktorova, *University of Virginia School of Medicine, USA*
Professor Calum Drummond, *RMIT University, Australia*
Mr Omar Elfarouk Eldessouky, *Mansoura University, Egypt*
Mrs Hebah Fatafta, *Forschungszentrum Jülich, Germany*
Mr Marcus Fletcher, *University of Cambridge, United Kingdom*
Ms Feba Francis, *Indian Institute of Technology Guwahati, India*
Professor Franca Fraternali, *King's College London, United Kingdom*
Miss Giulia Frigerio, *Università degli Studi di Milano- Bicocca, Italy*
Dr Patrick Fuchs, *Sorbonne Université, France*
Professor Philip Gale, *The University of Sydney, Australia*
Professor Dr Ana J. Garcia-Saéz, *University of Cologne, Germany*
Miss Levena Gascoigne, *Eindhoven University of Technology, Netherlands*
Professor Robert Gilbert, *University of Oxford, United Kingdom*
Mr Alexander Gilchrist, *University of Sydney, Australia*

Ms Yessica Gomez, *University of California, San Francisco, USA*
Mr Stuart Govan, *Royal Society of Chemistry, United Kingdom*
Miss Olivia Hepworth, *King's College London, United Kingdom*
Professor Dr Andreas Heuer, *Universität Münster, Germany*
Dr Maria Hoernke, *Albert-Ludwigs-Universität Freiburg, Germany*
Professor Bart Hoogenboom, *University College London, United Kingdom*
Dr Fiona Iddon, *Royal Society of Chemistry, United Kingdom*
Ms Samira Jadavi, *Istituto Italiano di Tecnologia, Italy*
Mr Zachary Jarin, *National Institutes of Health, USA*
Professor Shobhna Kapoor, *Indian Institute of Technology Bombay, India*
Dr Batuhan Kav, *Forschungszentrum Jülich, Germany*
Mr Fabian Keller, *Universität Münster, Germany*
Dr Elizabeth Kelley, *National Institute of Standards and Technology, USA*
Dr Nessim Kichik, *King's College London, United Kingdom*
Mr Gabriele Kockelkoren, *University of Copenhagen, Denmark*
Ms Hannah Krivic, *McMaster University, Canada*
Dr Garry Laverty, *Queen's University Belfast, United Kingdom*
Dr Eli Lee, *Montclair State University, USA*
Dr James Lincoff, *University of California, San Francisco, USA*
Dr César López, *Los Alamos National Lab, USA*
Professor Reidar Lund, *University of Oslo, Norway*
Dr Najet Mahmoudi, *Rutherford Appleton Laboratory, United Kingdom*
Dr Gregory Mashanov, *The Francis Crick Institute, United Kingdom*
Dr James Mason, *King's College London, United Kingdom*
Mrs Paula Milán Rodríguez, *Sorbonne Université, France*
Ms Izabela Miłogrodzka, *Monash University, Australia*
Dr Justin Molloy, *The Francis Crick Institute, United Kingdom*
Dr Wendy Niu, *Royal Society of Chemistry, United Kingdom*
Professor Paul O'Shea, *Lancaster University, United Kingdom*
Professor Dr Georg Pabst, *University of Graz, Austria*
Dr Sreetama Pal, *Centre for Cellular and Molecular Biology, India*
Miss Sujin Park, *Korea Institute of Science and Technology, South Korea*
Professor Susan Perkin, *Oxford University, United Kingdom*
Mr Ivan Pires de Oliveira, *Universidade Federal de Minas Gerais, Brazil*
Ms Neethu Puthumadathil, *Rajiv Gandhi Centre for Biotechnology, India*
Dr Shuo Qian, *Oak Ridge National Laboratory, USA*
Ms Amy Rice, *National Institutes of Health, USA*
Mr Jose Alejandro Rivera Moran, *Forschungszentrum Jülich, Germany*
Ms Mary Rooney, *William & Mary, USA*
Dr Kehinde Ross, *Liverpool John Moores University, United Kingdom*
Professor Aurélien Roux, *Université de Genève, Switzerland*
Mr William Ryder, *University of Sydney, Australia*
Miss Karthika S Nair, *National Institute for Interdisciplinary Science and Technology, India*
Dr John Sanderson, *Durham University, United Kingdom*
Professor John Seddon, *Imperial College London, United Kingdom*
Dr Enrico Federico Semeraro, *University of Graz, Austria*
Dr Durba Sengupta, *National Chemical Laboratory, Pune, India*
Mr Jiale Shi, *University of Notre Dame, USA*
Ms Shuai Shi, *Albert-Ludwigs-Universität Freiburg, Germany*
Professor An-Chang Shi, *McMaster University, Canada*
Dr David P. Siegel, *Givaudan, USA*
Miss Alice Smallwood, *Royal Society of Chemistry, United Kingdom*